ECOLOGICAL ASPECTS OF PARASITOLOGY

ECOLOGICAL ASPECTS OF PARASITOLOGY

ECOLOGICAL ASPECTS OF PARASITOLOGY

Editor

C.R. KENNEDY

Department of Biological Science
University of Exeter
Exeter (Great Britain)

1976

NORTH HOLLAND PUBLISHING COMPANY—AMSTERDAM / OXFORD

Library of Congress Cataloging in Publication Data
Main entry under title:

Ecological aspects of parasitology.

Includes index.
1. Host-parasite relationships. 2. Parasitology.
I. Kennedy, Clive Russell, 1941-
QL757.E25   591.5'24   76-41889
ISBN 0-7204-0602-1

Sole distributors for the U.S.A. and Canada:

AMERICAN ELSEVIER PUBLISHING COMPANY, INC.
52 Vanderbilt Avenue
New york, N.Y. 10017

Publishers:
NORTH—HOLLAND PUBLISHING COMPANY — AMSTERDAM
NORTH—HOLLAND PUBLISHING COMPANY, Ltd. — OXFORD

PRINTED IN THE NETHERLANDS

CONTENTS

VI

PREFACE

For a long period parasitology was dominated by morphological and systematic investigations. The realisation that it could be approached from an ecological standpoint initially grew slowly, but now, largely as a result of the pioneer studies of V.A. Dogiel and his colleagues, is widely accepted, to the extent that many people consider parasitology as only a special branch of ecology. Of recent years several books have appeared that treat parasites from an ecological approach, and attention has been focussed upon the ecology of parasitic life cycles, the population biology of parasites and the behavioural aspects of parasite transmission. Despite this increasing attention, many aspects of parasite ecology have received little or no consideration and information on them has not been readily available. The aim of this book is to remedy this situation by considering in ecological terms the peculiar problems faced by parasites and the ways in which they have been overcome.

The book is divided into three sections. The first is concerned with the ecological nature of some of the problems faced by animals as a particular consequence of their adopting a parasitic mode of life, and the strategy, rather than the tactics, of their solutions to these problems. The problems, with the notable exception of host responses, are not unique to parasites; the solutions to them often are, and where possible they are compared with the solutions to the same problems adopted by free-living animals. The second section is concerned with the specific problems and difficulties posed by the major organs and tissues occupied by parasites, and the tactical solutions adopted by parasites in each of these habitats that enable them to cope with and overcome these difficulties. Within any animal, the differing physico-chemical conditions within these habitats impose different constraints upon the animals occupying them. The major organs and tissues of vertebrates are considered in turn from this point of view, and the approach is again an ecological one. The third section is concerned with population ecology, and emphasises that although individual parasites have to face and overcome particular problems, parasitism is essentially a dynamic relationship between two species populations.

The concept of the book, its ecological theme and the choice of contents are my responsibility. All the authors were advised of its aims, of the general outline and the approach required, but within this framework the treatment of each chapter was left to them. The state of knowledge is such that a uniform approach would seldom have been possible, even if it were desirable. In consideration of some organ systems and problems it is possible to make reference to a wide range of parasites, but in others only one or a few species have been studied in sufficient detail. In many respects therefore this book indicates the limits of our knowledge and the extent of what is unknown, and it is to be hoped that this may help to stimulate further research into aspects of parasite ecology and into species of which we are at present ignorant. In addition to being linked together by the general theme, each chapter also represents a major and authoritative review of its subject; in some cases this is the first time that a review of the subject has been attempted. It is hoped that this aspect as well as the ecological approach will prove of interest and value to all engaged in parasitological research.

C.R. KENNEDY

LIST OF CONTRIBUTORS

ANDERSON, R.M.
> Department of Zoology, University of London, King's College, London WC2R 2LS (Great Britain)

ARME, C.
> North Staffordshire Polytechnical Institute, Department of Chemistry, College Road, Stoke-on-Trent, Staffs. (Great Britain)

ARTHUR, D.R.
> Department of Zoology, University of London, King's College, London (Great Britain)

BAKER, J.R.
> M.R.C. Biochemical Parasitilogy Unit, Molteno Institute, Downing Street, Cambridge CB2 3EE (Great Britain)

BEFUS, A.D.
> Department of Medicine, McMaster University, Hamilton, Ontario L8S 4J9 (Canada)

CHENG, T.C.
> Institute for Pathobiology, Center for Health Sciences, Lehigh University, Bethlehem, Penn. 18015 (U.S.A.)

COX, F.E.G.
> Department of Zoology, King's College, Strand, London WC2R 2LS (Great Britain)

CROMPTON, D.W.T.
> The Molteno Institute of Cambridge, Cambridge (Great Britain)

DESPOMMIER, D.
> Division of Tropical Medicine, School of Public Health, College of Physicians and Surgeons, Columbia University, 630 West 168th Street, New York, N.Y. 10032 (U.S.A.)

FERNANDO, C.H.
> Department of Biology, University of Waterloo, Ontario (Canada)

X

HALVÖRSEN, O.

  Institute of Biology and Geology, Universoty of Tromsö, N-9001 Tromsö (Norway)

HANEK, C.

  Ministry of Agriculture and Fisheries, Nassau (Bahamas)

HOLMES, J.C.

  Department of Zoology, University of Alberta, Edmonton, Alberta (Canada)

HOWELL, M.J.

  Department of Zoology, Australian National University, P.O. Box 4, Canberra 2600, A.C.T. (Australia)

KEARN, G.C.

  School of Biological Sciences, University of East Anglia, Norwich NR4 7TJ (Great Britain)

KENNEDY, C.R.

  Department of Biological Sciences, University of Exeter, Exeter EX4 4PS (Great Britain)

LONG, P.L.

  Houghton Poultry Research Station, Houghton, Huntingdon, Cambs. (Great Britain)

MACINNES, A.J.

  Department of Biology, The University of California, Los Angeles, Calif. 90024 (U.S.A.)

O'CONNOR, G.R.

  Francis I. Proctor Foundation for Research in Ophthalmology and the Department of Ophthalmology, University of California, San Francisco, Calif. (U.S.A.)

PODESTA, R.B.

  Department of Zoology, University of Toronto, Toronto, Ontario M5S 1A1 (Canada)

ROSE, J.H.

  M.A.F.F. Central Veterinary Laboratory, Weybridge (Great Britain)

SMITHERS, S.R.

  National Institute for Medical Research, Mill Hill, London NW7 1AA (Great Britain)

WAKELIN, D.

  Wellcome Laboratories for Experimental Parasitology, University of Glasgow, Bearsden Road, Glasgow, G61 1QH (Great Britain)

WORMS, M.J.

  National Institute for Medical Research, Mill Hill, London NW7 1AA (Great Britain)

Part I

PROBLEMS FACED BY PARASITES

Ecological Aspects of Parasitology
Editor: C.R. Kennedy
© *North-Holland Publishing Company, Amsterdam, 1976*

CHAPTER 1

HOW PARASITES FIND HOSTS: SOME THOUGHTS ON THE INCEPTION OF HOST–PARASITE INTEGRATION

Austin J. MACINNIS

Department of Biology, The University of California, Los Angeles, Calif. 90024 (U.S.A.)

Contents

1.1. Introduction

By definition, parasitism and infectiousness are inseparable. In order for parasites to maintain their life cycles — to survive — they must be able consistently to reproduce themselves. In this respect they are no different to any other form of life. The unique aspect of a parasite's life cycle is its dependence on a period of close association, inhabitation or utilization of another species as its niche, whereas "free-living" organisms occupy "non-living" niches. In order for a parasite to continue this mode of life, it or its offspring must find and infect a new niche, or host. As clearly stated by Rogers (1962), "evolution of infectiousness is of prime importance in the evolution of parasitism." Of course, the reciprocal of his statement is just as true. It is also true that the processes used by parasites to find and infect hosts do not differ from the processes used by free-living organisms to find and invade their niches in the non-living environment. Since the distinction between parasite and free-living organism is thus conceptual, I will begin by defining these and related terms in a way that is useful to the understanding of how parasites find and use hosts. Following this I will present an overall view of how signals impinge on the life cycles of parasites. This overall view will establish a background for the following chapters on the ecology of parasites. If ecology is defined as the study of the life cycles of organisms within their environment, we see that this provides a logical framework for our approach in this volume.

Much ado has accrued in the literature concerning the numerous efforts to define parasite, host, and related terms. These definitions have been debated yearly at symposia and seminars, yielding most frequently only heat, not light. Therefore, I am going to ignore this literature, beginning instead with the comment by Justus Mueller (1966) who stated ". . . no satisfactory definition of parasitism exists . . .", but I acknowledge that the definitions I present are based on the significant contributions of past and present colleagues.

1.2. Definitions

The following definitions were derived and revised during the past ten years of my teaching and research on parasites. The definitions have been useful in organizing information on this subject, and have led to further experimentation. A detailed monograph on this subject will be presented elsewhere (MacInnis, 1976), a brief resume has been published (MacInnis, 1974) and these serve as the basis for the definitions presented here. Those interested in the details of this subject should consult the papers mentioned as well as the scholarly review of Starr (1975). The latter contains an excellent historical summary of such definitions and the criteria upon which they are based, but does not yield the simple, reductionist product that suffices for our purposes in this volume.

When considering any definition of parasitism it is important to remember that the associations between different species in question are the result of evolution and these associations continue to evolve.

A general theory of symbiosis (revised slightly from MacInnis, 1974)

1. Certain organisms associate in interspecific interactions, in close spatial association, subsequently evolving a union of their life cycles.
2. Some of these interacting organisms evolved life cycles in which the physiological integration became obligatory for the survival of one or both of the associates.

3. *Symbiosis* is defined as a super-set term that includes the sub-sets commensalism, parasitism and mutualism; the definition of symbiosis thus emerges from the definitions of its sub-sets, and was clearly stated by de Bary (1879): "Zusammenleben ungleichnamiger Organismen".

4. *Parasitism* is defined as the case in which one partner, the *Parasite*, of a pair of interacting species is dependent upon a minimum of one gene or its product from the other interacting species, defined as the *Host*, for survival.

5. *Mutualism* is defined as the case in which each of the interacting species functions as *both* host and parasite.

6. In *Commensalism*, the interacting species have not evolved genetic dependency, but may do so in the future, thus becoming parasitism or mutualism.

7. In order to maintain a life cycle of parasitism, new hosts must be infected by the parasite. *Infectivity* (I) may be expressed as: I = Host-finding capacity of the parasite + susceptibility of host/resistance of host.

8. *Pathogenicity* = I $\times$ number of parasites in host $\times$ damage to host / parasite.

The elegant beauty of this definition of a parasite resides in the logical, ultimate criterion: the unit of evolution. All other criteria are subservient to the information coded in the nucleic acids (the few exceptions need not concern us here). It is true that DNA may respond to some signals from the environment, but that is exactly what we must elucidate if we are to understand and eventually control host-parasite integration, including one small but essential part of this relationship: how parasites find their hosts. This definition of parasitism precludes the necessity of concern about various criteria used in past attempts to define the term, such as size, temporal aspects, harm, goodness, social, regulatory mechanisms, etc. It is obvious that all of these criteria are reducible to the common denominator of life: The Double Helix and its associates which exist in their own symbiosis. The question to ponder now is whether, by this genetic definition of parasitism, any free-living organisms exist.

Presently we must devote some of our concern to elucidation of the genetic aspects of parasitism and the consequences of parasitism. I will try to use the foregoing definition of parasitism as the cohesive thread, spiraled through the phases and mechanisms of the peregrinations parasites perform traversing the barriers to their goal, often described as "the ultimate welfare state". Perhaps we will come to realize that such a definition pervades all of parasitology, possibly explaining why some parasites reside on gills, and other parasites reside on the head. It does, in fact, yield a simple explanation of *Host Specificity*, a concept of legitimate importance to my assigned topic of host-finding as well as the following chapters. The most general case of host specificity may be defined as one taxon of parasite restricted to one taxon of host (MacInnis, 1976). Based on the genetic definition of parasitism, one may deduce that *specific hosts* must be only those that can supply the genetic information required by the parasite, and to which the parasites have access not supervened by innate resistance, ecological, geographical or similar physical barriers.

Another corollary of the genetic definition of parasitism is a logical explanation of what has been called parasitic reduction, or simplification. In essence, parasitic reduction is an adaptation, the parsimony of natural selection, which, of course, must act on DNA. If parasites become dependent by deletion of genetic material, they could have smaller genomes than related, free-living species. We have presented evidence for this (Searcy and MacInnis, 1970), and not surprisingly, against it. Clearly, parasites may accrue additional information as complex life cycles evolve, yet a single mutation in the form of a deletion or inactivation of an essential component may make obligatory dependency on an associate.

Now that we understand what is meant by parasite, host, and the essential nature of infectivity for perpetuating the relationship, we can turn our attention to the details of how parasites find hosts. The foregoing introduction reveals the reasoning for my selection of the subtitle for this chapter; the inception of host-parasite integration. Nature seldom leaves necessary aspects of life's nexuses to chance, or even to choice, but leads on, inexorably, with cues that have been selected through milennia of natural experiments, invariably resulting in sufficient parasites reaching their hosts.

Occasionally, we sense some aspect of host-finding as a weak bond, linking the union of parasite and host, and we seize upon it, studying it in ionic detail that pales the efforts of even the most respectable molecular fadists. Yet herein lies the hope that entices us to seek further revelations of Nature's secrets, from which will spring the rational design of simple ways to prevent this union of host and parasite. To paraphrase Doty, ". . . With nucleic acids, as with people, parasites and hosts, it is interesting to see how complementary pair bonds form, separate, and rejoin." How then, do parasites find hosts, or hosts find parasites?

1.3. A comprehensive scheme of host-finding, site-finding and development of parasites

There are several excellent, relatively recent reviews of host-finding and/or parasite behavior. These provide access to much of the growing, sometimes controversial, literature on this subject, and may be consulted for details and further introduction to the subject. It would be redundant to re-review this material here. The interested reader should consult the works of Ulmer (1971); Chernin (1974); Cheng (1967); Schwabe and Kilejian (1968); Canning and Wright (1972).

It is nearly impossible to review all aspects of host-finding used by the numerous and diverse groups of parasites. It would require several volumes to accomplish. Although it is exhilarating to do the impossible, it seems more useful at this time to generate some principles, formulate significant unanswered questions, and give a few specific details of advances in selected areas of host-finding during the past five years.

If one surveys the numerous and incredible ways that parasites find hosts, and how hosts acquire parasites, there appear amongst the trees a few rays of light which illuminate some general principles that are useful in describing or categorizing the life cycle of any parasite. More importantly, these principles also predict crucial stages in the life cycles where signals or cues impinge on these processes. A life cycle of a parasite is clearly nothing more than its development, differentiation, and maturation, made more complex by the requirement for a host. We do not yet know or understand many of the key problems of development, but this adds to the intrigue and challenges parasitologists or other inquisitive minds to venture into this frontier.

Once we have accepted the fact that host-finding is part of Nature's developmental processes, our task simplifies. Many of the signals, cues, instructions, or whatever name we assign to the factors which elicit responses by or in parasites, fall within the domain of inducers. I use the term inducer here in its general sense, but with a connotation of developmental and behavioral overtones, since most of the molecular mechanisms of how inducers function in the development of eukaryotes remain mysterious. The inducing function of signals in a parasite's life cycle was predicted, I believe, for the first time in 1964 by my deceased colleague Clark P. Read in a lecture at Woods Hole, Mass., entitled the "Physiological Bases of Parasitism". In this lecture he stated that parasites can be described as having a developmental plan which suffers from "interrupted coding". That is, parasites may develop on their own plan to a point, but can only proceed when supplied with the "code" available in the host. Now we can see that this is clearly a subset of the general, genetic definition of parasite described earlier. I acknowledge here my gratitude to C.P.R. for many mutualistic discussions of this subject, which induced me to continue investigations in this area.

What are the general ways that parasites find hosts or hosts acquire parasites, and how do signals impinge upon these processes? Perusal of the vast assemblage of knowledge concerning eukaryotic parasites, primarily protozoans and helminths, reveals a useful dichotomy that enables construction of a flow-chart that illustrates each crucial developmental stage and process in the life cycles of parasites, and allows demonstration of how, and which, signals impinge on each stage or process. This dichotomy separates the processes of host-finding or site-finding within or on the host, and all other processes of development of the parasite, into one of two general categories which I call *active* or *passive*. Depending on the modifiers one uses with these words, coupled with the major steps in a life cycle, one can visualize a binary scheme that encompasses the entirety of host-parasite interaction.

Such a flow chart is shown in Fig. 1. I have, in fact, chosen to present this scheme in a binary modality because this format leads readily to computer modelling. I use the term *active* in this scheme to indicate that the parasite actively responds to a signal from the environment or host. Also, the host may respond to the parasite. The term *passive* is used to indicate no observable response by parasite or host, and that encounters between parasite and host may occur randomly, perhaps by chance, but may be influenced by adaptations such as amplification of the number of parasites. Also included in this passive category are the parasite's responses to intrinsic signals generated within itself. These of course are active, but excludable by definition from the active category on the basis of their origin *.

Fig. 1 shows two major pathways of host-finding. The passive mode is concerned primarily with the flow of parasites through food-chains, predation, accidental infection, etc. and is considered in the following chapter (2) by John Holmes. Included in his chapter as well are the details of the exquisite, active modification of behaviour of hosts induced by parasites, which leads to increased predation on parasitized hosts by subsequent hosts. His chapter should be consulted for its elaboration of this field.

The active mode of Fig. 1 is our primary concern here. At each identifiable step in the life cycle where an active response or function of the parasite is observed, any or all of the signals listed as originating in the environment or host (Fig. 1) may impinge on the physiological processes of the parasite. Indeed, the scheme can be expanded by inserting additional steps as our knowledge increases, or simplified by omission of those steps not present in a particular life cycle. Depending on the stage selected in any life cycle, one can enter the flow chart at an appropriate step and proceed through a cycle, recycle portions if demanded, add reservoir hosts, accidental hosts, etc. ad infinitum. No matter how many hosts are used, or whether reproduction is sexual or asexual, the same criteria apply. Hence the term vector need not be used, nor do we need to expend ATP determining if the cycle is propagative, mechanical transmission, cyclodevelopmental, phoretic, zoonotic, etc., although some of these terms may still be useful. Note also that the sequential flow illustrated in Fig. 1 may be altered to accommodate the temporal sequence appropriate for a particular parasite. For example, development may require food-finding before mate-finding, several processes may occur concurrently, and some may follow the sequences shown.

1.4. Host-finding: response to the environment

The parasites which follow the passive pathway in Fig. 1 are those that have no free-living stage and are transmitted through food-chains, and those

---

* This circular argument is used here as a literary liberty to permit construction of the model, which, hopefully, may be useful.

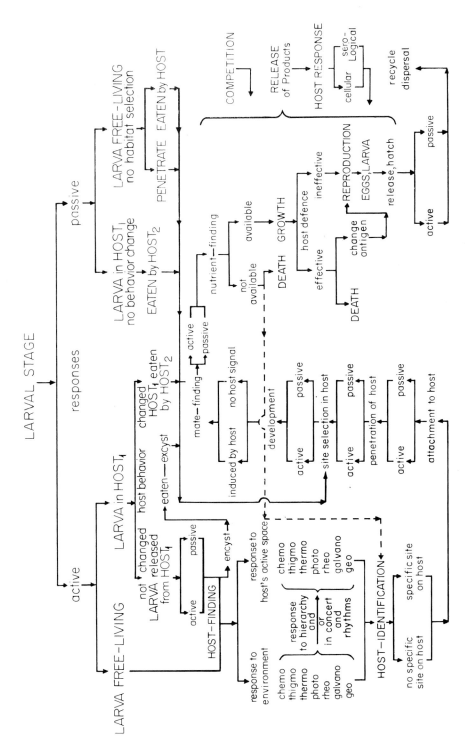

Fig. 1. A Flow-chart showing where signals may impinge on the life cycles of parasites.

that have nonmotile larval stages. However, signals from the environment may influence the host's behavior in ways that are important or essential for maintaining a parasite's life cycle. Signals such as light and temperature are obviously important in this respect in influencing host behaviour (see Holmes, for example, Chapter 2). Some larval stages (I use larval stage throughout this paper to include all non-adult forms, including eggs, cysts, etc.) can not respond to environmental signals by any active processes that might lead them to a host. An amoebic cyst has no means of selecting or moving to a habitat where it is more likely to encounter a host. Environmental factors may also influence development of a parasite existing within a host, and the process then becomes active. Temperature as an inducer or modifier of a parasite's development is a good example of such a phenomenon, but further discussion of this is beyond the realm of this chapter on host-finding.

The active branch of Fig. 1 is of primary concern to our elaboration of the details of host-finding by free-living stages of parasites. In order to become a free-living stage the parasite must be released or actively escape from a host. Thus initiation of host-finding may begin with an essential, active process induced by environmental signals that leads to the brief periods of free life for a parasite. As an example of this we can cite the active escape of schistosome cercariae from their molluscan host in response to light and, possibly, to increased temperature. Cercariae may also respond to signals of host origin that stimulate escape, but no evidence for this response is yet available. Another intriguing example of the importance of the release process to host-finding occurs in the escape of larvae from the female *Dracunculus* when the infected host enters water. Many other examples such as these exist. The two examples presented establish the importance of an active step in the production of free-living stages in some life cycles, and is discussed in the chapter (7) on dispersal by Kennedy.

Now that we have a free-living larval stage we can examine the perturbations of the environment that influence larval behavior. A teleological generality that can be stated concerning this phase of host-finding is that larval stages of parasites generally respond to environmental stimuli in ways which usually bring them closer to the habitat normally frequented by a potential host. Such behavior is clearly an adaptation easily selected in the evolution of infectiousness. Since parasites initially evolved from free-living relatives we can expect to find that parasites respond to environmental signals in much the same way as do free-living organisms. And we do find this. However, the responses may be adapted to accomplish host-finding (MacInnis, 1965). Less obvious is the possibility that the parasite may evolve entirely new responses as proposed by Davenport (1966). We should attempt to elucidate the latter since they may represent unique processes amenable to interruption in ways that will not destroy desirable aspects of the environment as we attempt to control parasitic infections.

When a larval stage of a parasite is liberated into the free world it faces many alternatives. Often it has an evolved program that limits its responses in that the choices have already been made for it: when the signal appears, it can not be ignored and hence is directed to or kept in the environment most likely to contain the next required stimulus, and eventually a host. The various signals present in the environment that have been shown to influence the behavior of larval parasites resemble a tale of love: "show me the way", they say, "I can not tell you all the ways I find thee". Under host-finding in Fig. 1, I have attempted to list all of the environmental factors which may serve as stimuli to direct or trap parasites. Not surprisingly, they are almost exactly the same as those that function within a host. Other stimuli not listed may function in host-finding, such as sound or magnetism, but I know of no examples. Parasitoid insects, with which I have no familiarity, might respond to sound produced by a potential prey (host). Examination of even a single life cycle such as that of a schistosome reveals that one larval stage such as a miracidium utilizes all of the signals listed, perhaps with the exception of galvanotaxis, which has not yet been thoroughly studied. Thus hatching of the schistosome egg may be stimulated by light, heat, and the ionic content of the water. Upon release from the egg, the miracidium may respond to gravity, light, water currents, temperature, the ionic content of the milieu, and physical boundaries.

Superimposed on the response to an individual stimulus is the possibility of a hierarchy existing among the stimuli. For example, response to light may prevail over a response to gravity. In addition, there may be synergistic or additive effects amongst stimuli, the response may change with varying intensities of stimuli, and, in the final analysis, all stimuli probably function in concert. The importance of rhythms, such as circadian rhythms, in the process of host-finding, may also be influenced or be set by environmental stimuli, especially light and temperature. A further complexity is that the level of one stimulus may dictate the response to other stimuli as shown recently by the remarkable studies of Shiff (1974). Depending on ambient temperature, *S. haematobium* miracidia may respond to light and gravity either positively or negatively, but the end result brings them to the habitat of the snail at the appropriate season of the year!

An unusual response by certain larval helminths to temperature gradients in the environment was discovered by McCue and Thorson (1964). They demonstrated that larval stages of parasitic nematodes migrated toward areas of higher temperature, as did larval stages of free-living nematodes. Unexpectedly, the larvae of the parasitic forms continued to migrate into areas of higher and higher temperature. On the other hand, the larvae of free-living forms stopped migrating when they reached the zone of temperature optimal for their well-being. On the basis of their results these authors suggested that perhaps the parasitic forms have lost feedback control mechanisms, and can only continue a positive response, leading inexorably to their death in the experimental protocol. Similar responses to other stimuli should be sought in parasites, possibly leading to the identification of stimuli suitable as practical control measures.

Further elaboration of the countless ways in which environmental stimuli influence the behavior of innumerable parasites is impractical in this chapter. However, there is one important exception to this which must be considered since it serves to bridge the gap between parasite and host, and also serves as a convenient transition between our consideration of environmental stimuli and those of host origin. This exceedingly interesting exception concerns the conditioning of the environment by the host in a way that can be perceived by the receptive parasite. The studies by Chernin (1974), Shiff (1974), Sponholtz and Short (1976), Wright (1972), MacInnis et al. (1974) and others have shown that snails elaborate into their environment substances which stimulate miracidia of schistomes, effectively increasing the active space of a potential host. This phenomenon is undoubtedly important for other trematodes, and for any larval stage of a parasite that responds to chemical or other stimuli transmitted by a potential host through the environment. Additional details of this are presented next and under the heading of recent important advances.

1.5. Host-finding: response to the host's active space

As noted above, some hosts may condition their environment in a manner that effectively increases the volume which a parasite may recognize as possibly containing a potential host. The limits of this volume are the boundaries to which stimuli emanating from a potential host extend, hence defining a host's active space. If no stimuli exude beyond the confines of the potential host's body, the active space equals the host's body, size, which can be relatively small in a big pond. Intuitively, it appears logical for a parasite to have evolved responses to the active space of potential hosts, thus effectively enhancing their probability of locating hosts. Such stimuli of host origin that condition the environment can be separated for study from those confined to the limits of the host's body, but the distinction may be the result of our reductionist approach. For our discussion it is advantageous to use this separation.

A host's active space can be increased effectively by chemicals that are released into its surrounding environment. The few chemicals that have been identified as serving this function are by-products of the host's physiological processes such as metabolism, respiration, osmoregulation, etc. CO_2 produced by a host may lure or stimulate larval stages of nematodes (reviewed in Croll, 1972). Snails condition water with several constituents that stimulate miracidia, such as amino acids (Wright and Ronald, 1972; MacInnis et al., 1974) or ions (Sponholtz and Short, 1976; Stibbs et al., 1976). More details on this are included in the section on recent advances. Phytoparasitic nematodes respond to CO_2, amino acids and sugars which condition the environment surrounding roots of the potential plant host. Cercariae may respond to chemicals elaborated from the surface of hosts (see Ulmer, 1971; Cable, 1972; MacInnis, 1969, for review), but no substantial evidence exists which shows a directed response to chemicals by cercariae, leading them over a distance to the host. The latter needs study, especially on schistosomes.

Other stimuli of host origin that may amplify its active space are heat, water or air currents, and possibly galvanotaxis. Again, evidence for the latter is meager (see Croll, 1972) but needs to be considered in future studies. Temperature differentials over short distances may be created by warm-blooded animals, and detected by some larval parasites such as schistosome cercariae (MacInnis, 1969; Hubbard, 1969). Insects and other arthropods may also function as heat-seeking missiles. Numerous species of cercariae may respond to water currents or turbulence (Ulmer; Cable; loc. cit.), and miracidia enhance their scanning capacity by orienting upstream (Webbe, 1966) or up the currents produced by the respiratory processes of molluscs. Light produced by hosts is not yet known to function in the process of host-finding, but the creation of a shadow by a potential fish host effectively increases its active space for the cercaria that responds to the shadow (Ulmer, 1971). No sound receptors have been reported in parasitic protozoa or helminths, to my knowledge, but at least one trematode is capable of producing sound incidental to the movement of its sucker (reference to this gem of trivia is available from the author by request only). I can think of no examples of how geo- or thigmotaxis might increase a host's active space. Perhaps hair or fur encountered by a parasite might extend the active space of a mammal, but this stretches the concept.

1.6. Host-finding: response to the host

Following responses to the environment, then to host-conditioned environment, the parasite must eventually contact and recognize a potential host, and be prepared for the next steps in the initiation of parasite-host integration. At this point in our analysis of host-finding it is informative to consult Fig. 1 to perceive some of the possible steps that can be identified in this overall process. Contact with the host might occur randomly, or be enhanced by signals in the host's active space. Once a potential host has been contacted, the parasite must recognize or identify the subject as host, then initiate one or more of the following steps to gain admission: select a specific site on the host, attach and penetrate. Other parasites may be eaten by the host, induce the host to eat them, or by good luck venture into an orifice in the host. The distinction between some of these processes again may be the hazy, grey line of reductionism, but, until the entire story is known, they provide a logical framework for study of host-finding, selection, and initiation of host-parasite integration. Indeed, additional steps may yet be revealed.

The foregoing paragraph illustrates the complexity of the final steps of the host-finding processes. I have pointed this out here to indicate, first of all, that in each of these active steps the parasite may respond to most or all of the signals listed in Fig. 1, and, secondly, that it is exceedingly difficult to separate responses to the host that may be concerned only with the process of host-finding (which I restrict to those responses that bring the parasite into contact with the host) and not also with subsequent steps. The processes of host-recognition, penetration, site-selection and subsequent steps are within the scope of the following chapters.

1.7. Signals as inducers

Having presented an overall view of how signals impinge on the life cycle of a parasite, we may now return to the problem of initiation of host-parasite integration and the possible role of signals as inducers of behavioural change or developmental processes in parasites. Environmental signals which modify the behaviour of parasites can be divided into those extrinsic to the host and originating in the environment, those extrinsic to the host but of host origin, and those intrinsic to the host and originating within the host. The responses to extrinsic signals, regardless of their origin, serve to keep, or bring and keep the parasite into the active space of a potential host. These responses may function in one or more of the following ways: (1) they may be directed responses to a gradient or to the source of a stimulus that decrease the distance between host and parasite; (2) they may be trapping reactions, such as a cessation of movement when the signal strikes, thus keeping the parasite within the vicinity of the potential host; (3) they may increase the random activity of the parasite which enhances probability of contact with the potential host. In this sense 2 and 3 are both kinetic responses. It is not yet known if such signals do more than induce behavioural changes, but it is possible that some may initiate developmental changes, or physiological changes, that prepare the parasite for the next step. This may be the case when the same signal functions in host-finding, host-recognition, and initiation of penetration which also may include metamorphosis of the larval stage, such as shedding of ciliary plates in miracidia, discharge of contents of penetration glands, membrane changes, loss of tail by cercariae, etc. We have yet to resolve whether separate signals are required for each of these steps. Clearly, some of these steps are inductions in the strictest developmental sense, with the host often serving as the source of inducing substances. We also know very little about receptors, sensory receptors, or binding sites for inducers in parasites. Some of these must function as receptors that can perceive stimuli at some distance from the host. Others may be short-range, or contact receptors. Surely we can learn to apply the vast and sophisticated tools of neurobioloby to these problems.

When the parasite has entered the host it may begin further response to signals of host origin, site-finding within the host, development, maturation, response to host hormones, etc. Although these processes are beyond the scope of this chapter, I take this opportunity to alert the reader to the role such signals play in the life cycles of parasites, and hope they will be aware that induction of development of a parasite follows the same pattern as a free-living organism, except for some requirements, many of which may be inducers, supplied by the host.

1.8. Recent advances on host-finding

This section will be restricted to an analysis of several recent papers concerning the release of chemicals by potential hosts and the roles these chemicals serve in the processes of host-finding by parasites. A superb resume of the earlier literature on this subject has been presented by Eli Chernin (1974).

As noted in Chernin's review, a mounting mass of evidence has accrued slowly, but inevitably leading to the inescapable conclusion that the miracidial stage of schistosomes responds to substances released into the environment by their molluscan host. The quest — seeking identification of the elusive pimpernel responsible for stimulating and attracting miracidia — has followed a pathway as tortuous as the gyrations presented by the miracidia themselves in search for their grail. Once it had been established beyond doubt that molluscan hosts were indeed elaborating one or more substances into the surrounding milieu, the mysterious substance was named *Miraxone* (see Chernin, 1974). It has been much easier to name the substance than to identify it. Some of the first efforts at identification were based on the studies of MacInnis (1965) who had shown that certain amino acids, short-chain fatty acids, and a sialic acid attracted miracidia when incorporated into agar. These chemicals also attracted miracidia to snail tissues which had been extracted with water or ethanol and subsequently conditioned in the appropriate chemical. The choice of chemicals used in these experiments was based on the likelihood of their being end-products of the snail's metabolism, utilized in osmoregulation, or found in snail mucus. Indeed, it was subsequently shown that water conditioned by snails (=SCW) contained amino acids (Wright and Ronald, 1972; MacInnis et al., 1974). We also reported the presence of a lipid fraction of SCW, which, when transesterified with methyl alcohol and analyzed .by gas-liquid chromatography, indicated the presence of at least six peaks whose identity or ability to stimulate miracidia have not yet been resolved (MacInnis et al., 1974). We also showed that the amount of amino acids recovered from SCW attracted *Schistosoma mansoni* miracidia as effectively as did fresh SCW. Wilson (1968) and Wilson and Denison (1970) have shown that mucus of *Lymnaea truncatula* contains glucose, 16 amino acids and lipid fractions which may serve to stimulate miracidia of *Fasciola hepatica*.

An interesting turn in the attempts to elucidate the nature of miraxone was taken by Glen Sponholtz, working in Professor R.B. Short's laboratory. Sponholtz reasoned that snails must be removing calcium from their environment to build their shells, and hence the levels of calcium and related ions in SCW might modify the behavior of miracidia. Subsequent experiments revealed that SCW contained less calcium and more magnesium than the starting material (Sponholtz and Short, 1976) and that when the appropriate ratio of Ca/Mg was presented to a miracidium, it responded, showing attraction as described by MacInnis (1965). From their results these authors concluded that the Ca/Mg ratio was the significant factor in attracting miracidia, since the most attraction was observed when the ratio of Ca/Mg was 0.5/3.5 mM, and increasing concentrations of calcium in the presence of constant magnesium concentration gave less attraction. They also reported that the slight increase in magnesium found in SCW (from 0.35 mM in well water to 0.43 mM in SCW) when tested as magnesium alone " . . . was not sufficient to explain the attraction".

This tale took another fascinating twist through the recent, continuing efforts of Professor Chernin's laboratory (Stibbs et al., 1976). I am greatly indebted to these authors for their generosity in sending me a pre-publication copy of their manuscript. These authors have extended the pioneering efforts of Chernin's group on the isolation and purification of miraxone. In this careful report they have demonstrated that a major component of water conditioned by *Biomphalaria glabrata* is $MgCl_2$, which also elicits an excellent response from *S. mansoni* miracidia when tested in the point inoculation assay described by Chernin. Their freshly obtained SCW contains an average of 0.24 mM magnesium and other components including calcium. They concluded from various experiments that all of the miraxonal activity in SCW could be attributed to its magnesium content. The reader of this puzzle might conclude that our years of miracidia watching have induced us to gyrate in similar circles. Here we are faced with three different reports, each purporting that their isolate from SCW accounts for 100% of its activity in attracting miracidia. In one (MacInnis et al., 1974), amino acids were obtained but ions not sought; the second (Sponholtz and Short, 1976) claims that decreasing calcium concentration with a slight increase in magnesium is important, and the third (Stibbs et al., 1976) yields good evidence for increased magnesium concentration as the miraxonal essence. Perhaps these perplexing results can be reconciled, if not to our complete satisfaction, at least to an extent that we can formulate the essential requirements of future experimentation.

In studying the procedures and results of these three papers I have noted one important aspect that may be the key to the results obtained. MacInnis et al. (1974) used commercial spring water for conditioning by snails. This water contains calcium and magnesium (and other salts) at about the same levels as those found in the well water used by Sponholtz and Short (1976) to obtain SCW. However, Stibbs et al. (1976) conditioned distilled water. The latter two labs conditioned their water for 4 and 5 h, respectively, whereas we conditioned for 18 h. Stibbs et al. used 100 snails in 100 ml; the other two groups used 100 snails in 500 ml of water. This information, coupled with the results of Thomas and Lough (1974) leads to the following explanation. Thomas and Lough have shown that the snails used in these experiments, *Biomphalaria*, regulate the flux of calcium into and out of their bodies depending upon the concentration of calcium in the external milieu, as well as on the volume in which a snail is confined. They have studied this both in closed systems and in open, or flowing, systems. Their closed system is similar to the manner in which SCW is obtained for study of miraxone. They have demonstrated in the closed system that when the initial, external calcium concentration ranges from 2.0 to 0.625 mM/l the snails *remove* calcium from the medium, but when the external calcium concentration is lower than 0.625 mM/l the snails *excrete* calcium. When no calcium was present initially in the external medium (6.25 ml/snail) the loss of calcium from the the snail to the medium was 6.0 μM/gram of snail/day over a three day experiment. Although their "conditioning" period is considerably longer than those used by the various groups to obtain SCW, their results certainly bear on the problem.

With the results of Thomas and Lough in mind, let us now reconsider the previous results. Sponholtz and Short began their conditioning regimen with well water containing 0.91 mM calcium. Based on the results of Thomas and Lough it is not surprising that Sponholtz and Short also observed a reduction in calcium in their SCW since the initial concentration was greater than 0.625 mM/l. Although Stibbs et al. did not mention the presence of calcium in their SCW obtained by conditioning distilled water, a personal communication from Stibbs indicates that calcium levels increased about the same as did magnesium. Neither Sponholtz and Short nor Stibbs et al. recovered detectable amino acids in their relatively short conditioning periods. Possibly the longer period used by MacInnis et al. can account for their demonstration of amino acids, which agrees closely with the results on another snail obtained by Wright and Ronald (1972). Thomas and Lough apparently did not study magnesium changes in their experiments. It would be most helpful if they have such data.

The take home lesson from this discussion is that we must consider carefully the nature of the initial conditioning medium used in future experiments. The ionic composition of the initial water used to obtain SCW appears crucial, as well as the volume of water and number and size of snails.

Longer conditioning periods, with water similar to that used to raise the snails in aquaria, seem essential to eliminate osmotic shock and approach the natural condition. This must be balanced with other factors such as bacterial growth. Resolution of the identity of the constituents of SCW is imminent, and at the present moment it seems likely that miraxone will consist of more than one chemical entity. If calcium and magnesium are important, we must also consider their possible effects on the ciliary beat of the miracidium. Amino acids may also chelate these ions, and this complex may be coupled to fluxes of these ions in and out of snail tissues.

Further importance of the role of ions in the life cycle of schistosomes has been demonstrated recently by Kassim and Gilbertson (1976). I am indebted to these authors for their generosity in providing me with a prepublication copy of their results. They have shown a slight reduction in the percentage of *S. mansoni* eggs that hatch in artificial pond water containing no calcium (38.9%), or no calcium or sodium (32.3%) as compared to the well water control of 43.3%. In addition, they also noted that hatched miracidia soon lost their motility in the absence of sodium, but continued to swim in the absence of calcium or magnesium.

All of these studies focus increased importance on the role of ions in the host-parasite puzzle of snail and miracidium. Much more is known about the role of these ions in the lives and distribution of *Biomphalaria* (reviewed in Thomas et al., 1974) than is known about their effects on the parasite. Their role in schistosomiasis is just beginning to be unveiled. Is it possible that control of a single ion can lead to control of a disease with world-wide impact? The carrot appears closer each day!

1.9. Acknowledgements

This paper is dedicated to my colleagues Eli Chernin, Frank Etges, the late Clark P. Read, Clive Shiff, and Robert B. Short, who provide encouragement and stimulating criticism that are priceless. Membership in this World Union of Miracidial Watchers is a cherished honor, and it's free to those seeking membership. I am indebted to the National Science Foundation, and the NIH for support of my research described herein, and to the Clark Foundation for current support which enabled preparation of this manuscript.

1.10. REFERENCES

Cable, R.M. (1972). In: "Behavioral Aspects of Parasite Transmission." (E.U. Canning and C.A. Wright, eds.) pp. 1—18. Academic Press, New York.
Canning, E.U. and C.A. Wright, Eds. (1972). "Behavioral Aspects of Parasite Transmission." Academic Press, New York.

Cheng, T.C. (1967). In: "Advances in Marine Biology." Vol. 5. (F.S. Russell, ed.) pp. 16—134. Academic Press, New York.

Chernin, E. (1974). "Some Host-finding Attributes of *Schistosoma mansoni* Miracidia." Am. J. Trop. Med. and Hyg. 23, 320—327.

Croll, N.A. (1972). In: "Behavioral Aspects of Parasite Transmission." (E.U. Canning and C.A. Wright, eds.) pp. 31—52. Academic Press, New York.

Davenport, D. (1966). In: "Symbiosis." (S.M. Henry, ed.) pp. 381—430. Academic Press, New York.

De Bary, A. (1879). "Die Erscheinung der Symbiose." Verlag von Karl J. Trübner, Strassburg.

Hubbard, J.R. (1969). "On the Behavior of Cercariae of *Schistosoma mansoni*." Masters Thesis. University of California.

Kassim, K. and D.E. Gilbertson. (1976). "Hatching of *Schistosoma mansoni* Eggs and Observations on Motility of Miracidia." J. Parasit. (submitted, cited by permission of authors).

MacInnis, A.J. (1965). "Responses of *Schistosoma mansoni* Miracidia to Chemical Attractants." J. Parasit. 51, 731—746.

MacInnis, A.J. (1969). "Identification of Chemicals Triggering Cercarial Penetration Responses of *Schistosoma mansoni*." Nature 224, 1221—1222.

MacInnis, A.J. (1974). "A General Theory of Parasitism." Proc. 3rd Int. Cong. of Parasit. Vol. 3, pp. 1511—1512. Facta Publications, Munich.

MacInnis, A.J. (1976). "Symbiosis Defined." In preparation.

MacInnis, A.J., W.M. Bethel and E.M. Cornford. (1974). "Identification of Chemicals of Snail Origin that Attract *Schistosoma mansoni* Miracidia." Nature 248, 361—363.

McCue, J.F. and R.E. Thorson. (1964). "Behavior of Parasitic Stages of Helminths in a Thermal Gradient." J. Parasit. 50, 67—71.

Mueller, J.F. (1966). In: "Host-Parasite Relationships." (J.E. McCauley, ed.) pp. 11—14. Oregon State University Press, Corvalis.

Rogers, W.P. (1962). "The Nature of Parasitism." pp. 287. Academic Press, New York.

Schwabe, C.W. and A.K. Kilejian. (1968). "Chemical Aspects of the Ecology of Platyhelminths." Chem. Zool. 2, 467—549.

Searcy, D.G. and A.J. MacInnis. (1970). "Measurements by DNA Renaturation of the Genetic Basis of Parasitic Reduction." Evolution 24, 796—806.

Shiff, C.J. (1974). "Seasonal Factors Influencing the Location of *Bulinus (Physopsis) globosus* by Miracidia of *Schistosoma haematobium* in Nature." J. Parasit. 60, 578—583.

Sponholtz, G.M. and R.B. Short. (1976). "*Schistosoma mansoni* Miracidia: Stimulation by Calcium and Magnesium." J. Parasit. 62, 155—157.

Starr, M.P. (1975). In: "Symbiosis." (D.H. Jennings and D.L. Lee, eds.). pp. 1—20, Symposia of the Soc. Exp. Biol. Vol. XXIX. Cambridge University Press, London.

Stibbs, H.H., E. Chernin, S. Ward and M.L. Karnovsky. (1976). "Magnesium Emitted by Snails Alters Swimming Behavior of *Schistosoma mansoni* Miracidia." Nature 260, 702—703.

Thomas, J.D., M. Benjamin, A. Lough and R.H. Aram. (1974). "The Effects of Calcium in the External Environment on the Growth and Natality Rates of *Biomphalaria glabrata* (Say)." J. Anim. Ecol. 43, 839—860.

Thomas, J.D. and A. Lough. (1974). "The Effects of External Calcium Concentration on the Rate of Uptake of this Ion by *Biomphalaria glabrata* (Say)." J. Anim. Ecol. 43, 861—871.

Ulmer, M.J. (1971). In: "Ecology and Physiology of Parasites." (A.M. Fallis ed.). pp. 123—160. University of Toronto Press, Toronto.

Webbe, G. (1966). "The Effect of Water Velocities on the Infection of *Biomphalaria sudanica tanganyicensis* Exposed to Different Numbers of *Schistosoma mansoni* Miracidia." Ann. Trop. Med. Parasit. 60, 85—89.

Wilson, R.A. (1968). "An Investigation into Mucus Produced by *Lymnaea truncatula*, the Snail Host of *Fasciola hepatica*." Comp. Biochem. Physiol. 24, 629—633.

Wilson, R.A. and J. Denison. (1970). "Short Chain fatty acids as stimulants of Turning acitivity by the Miracidium of *Fasciola hepatica*." Comp. Biochem. Physiol. 32, 511—517.

Wright, D.G.S. and K. Ronald. (1972). "Effects of Amino Acids and Light on the Behavior of Miracidia of *Schistosomatium douthitti*." Can. J. Zool. 50, 855—860.

Ecological Aspects of Parasitology
Editor: C.R. Kennedy
© *North-Holland Publishing Company, Amsterdam, 1976*

CHAPTER 2

HOST SELECTION AND ITS CONSEQUENCES

John C. HOLMES

Department of Zoology, University of Alberta, Edmonton, Alberta (Canada)

Contents

2.1. Introduction

It is a basic tenet of evolutionary ecology that different combinations of characteristics (morphological, physiological, or behavioural) confer different fitness for survival or reproduction on the organisms possessing them, and that natural selection tends to drive populations toward higher fitness levels (Emlen, 1973). The combination of characteristics and the ecological context in which they function have been referred to as the "adaptive strategy" of the population or species (Levins, 1968).

One of the more important parts of the adaptive strategy of any organism is the selection of a suitable place (habitat or host) in which to live. (Unfortunately, the term "selection" is widely used in two quite different contexts, both of which are important in this chapter. I use "select", "host selection" or "habitat selection" to denote the "choice" or use of one host or habitat instead of another, and "natural selection" to denote the genetic, evolutionary process.) Most of the ideas on habitat selection by animals deal with the proportion of time spent in different habitats or the conditions under which an animal should move out of a particular habitat and select a different one. With the exception of many of the ectoparasites, parasites, once established, cannot move out and search again. In this respect, they are similar to plants or to sessile animals; therefore, strategies of host selection may be expected to be similar to the strategies of habitat selection of plants or sessile animals.

There are obviously two types of strategies for host selection by parasites — those appropriate to infective stages which can actively move to and attach to or invade the host, and those appropriate to infective stages which must reach their host by passive means. The former are essentially similar to strategies of site selection by planktonic larvae, on which there is a large body of literature (see Meadows and Campbell, 1972; Doyle, 1975; Moore, 1975); applications to parasites have been covered by MacInnis in the preceding chapter. In this chapter, I shall consider strategies of infective stages which reach the host by passive means. I shall focus primarily on helminths, especially those with complex life cycles, because that is the group I know best.

2.2. Suitable vs. unsuitable hosts

From a parasite's point of view, a "suitable host" must be one in which the parasite can continue its life cycle: for an intermediate host one in which it can develop to an infective stage; for a definitive host, one in which it can mature and reproduce. The mere presence of a parasite in a host, even in considerable numbers, does not denote that the host is a suitable one. For example, in Cold Lake, Alberta, *Metechinorhynchus salmonis* makes up about 70% of the parasites of *Catastomus catastomus*, but no mature females have been found (Table 1, from Leong, 1975), *C. catastomus* are not suitable hosts.

It has long been recognized that suitable hosts are not equally suitable. As Dogiel et al. (1964) put it;

"When a parasite occurs in more than one host, it is almost always possible to observe that in one of them it is most frequent, grows to the largest size, reaches maturity most rapidly, produces the greatest number of eggs and, generally, appears to be best adapted to the conditions in it . . . Although present also in other hosts, the parasite is, in them, less common and less abundant; its growth is retarded and it is exposed to considerable resistance by the host . . . In other hosts, the parasite occurs very rarely and as a rule develops in them only with difficulty" (p. 437).

Dogiel et al. refer to these three groups as the main, secondary, and accidental hosts, respectively, and give similar terms used by others. The variables Dogiel et al. use are not always correlated, however, as the data in Table 1 show. It is clear that the salmonids are more suitable hosts for *M. salmonis* than are *Lota* or *Esox*. However, it is not clear whether *Coregonus clupea-*

TABLE 1

Hosts of *Metechinorhynchus salmonis* in Cold Lake, Alberta, Canada (Data from Leong, 1975).

	Number examined	Percent infected	Mean no. worms	Percent gravid female worms
Salmonidae				
Coregonus clupeaformis	863	99	170	14
Coregonus artedii	757	32	9	50
Salvelinus namaycush	35	100	413	24
Oncorhynchus kisutch	288	100	187	30
Esocidae				
Esox lucius	62	84	31	3
Gadidae				
Lota lota	29	100	174	2
Catastomidae				
Catastomus commersonii	36	8	4	0
Catastomus catastomus	12	83	31	0
Percidae				
Stizostedion vitreum	12	67	3	0
Gasterosteidae				
Pungitius pungitius	1,083	19	2	0

formis or *Salvelinus namaycush*, in which the prevalence and intensity of infection were high but the maturation rate was low, are more suitable than *Coregonus artedii*, in which the prevalence and intensity were low but the maturation rate was high. Suitability apparently is a complex function, responding to several independent variables.

Many authors have pointed out that there is also variability in susceptibility within an individual species of host. This variability has been studied most extensively in mosquito vectors of malaria and in snail intermediate hosts of schistosomes, it occurs both between (mosquitos: Ward, 1963; snails: Files and Cram, 1949; Paraense and Corrêa, 1963) and within (mosquitos: Huff, 1927; Ward, 1963; snails: Brooks, 1953; Pan, 1963) host populations. Part of this variability undoubtedly is due to a multi-factor "genetic liability" of the individual host animal (Newton, 1953; Richards, 1973), similar to the genetic liability of humans to various diseases (Falconer, 1965). Part is apparently due to a similar variation in infectivity of the individual parasite (Saoud, 1965). Basch (1975) has suggested that "suitability" is based upon a concordance of compatible phenotypes between an individual host and an individual parasite, an hypothesis applied previously to plants and their parasites (see review in Day, 1974). Certainly, the number of places where host signals impinge on the life cycle of a parasite (see chapter by MacInnis) and the complexity of the interactions between host and parasite in some of those places (see chapters by Arme, Wakelin, and Befus and Podesta) provide plenty of scope for genetic interaction.

Sprent (1963) has provided an instructive example of host specificity in relation to the life history of the ascaridoid, *Amplicaecum robertsi*, a parasite of Australian pythons. The eggs hatched and larvae persisted, but did not grow, in a variety of invertebrates and cold-blooded vertebrates. In birds, the eggs hatched, and the larvae grew slightly. In indigenous Australian mammals, the eggs hatched and the larvae grew rapidly, attaining a length at which they would mature in carpet snakes (*Morelia spilotes*) within two months. In introduced mammals, the larvae grew more slowly, and fewer reached the critical size for subsequent maturation in pythons. Applying the criteria outlined above, the indigenous mammals are suitable intermediate hosts, introduced mammals less suitable, and birds and the poikilotherms unsuitable intermediate hosts. However, members of the last group are useful as transport hosts, since the larvae transferred readily when fed to mammals, in which the required development can take place.

Sprent also found that infective third stage larvae would moult and develop to the adult stage in three species of predatory lizards, but not in a herbivorous lizard. Reproducing adults, however, were found only in the carpet snake. Carpet snakes occasionally feed on large lizards (Worrell, 1963), and adults from lizards can mature when fed to carpet snakes. Therefore, lizards harbouring immature adults may sometimes act as transport hosts. The system makes an interesting one in which to examine the behaviour of a parasite in an unsuitable host.

2.3. Parasites in unsuitable hosts

Fig. 1 is a generalized scheme of the stages through which an ingested larva or juvenile (referred to as "larva" hereafter) must pass to develop into a mature reproducing adult, or, with a few obvious changes, an infective stage. It is similar in concept to part of MacInnis' diagram in the previous chapter, but is less detailed; I shall use it to focus on the ecological reasons for evolving appropriate mechanisms at points of divergence rather than the mechanisms themselves—the strategy rather than the tactics. I shall also focus, not on the "normal" development in suitable hosts, shown by the double arrows, but on that in unsuitable hosts.

Eggs of *A. robertsi* hatch in a wide variety of hosts, including arthropods, in which the larvae apparently could not penetrate the gut wall. Larvae of *A. robertsi* also appear to be activated in a wide range of reptiles, birds, and mammals. The conditions necessary for activation have not been studied, but

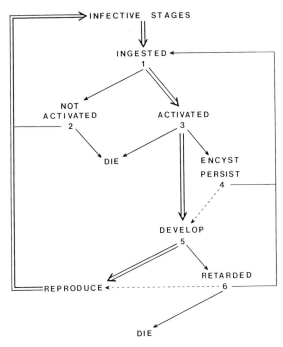

Fig. 1. Fate of an ingested parasite in suitable (double arrows) and unsuitable hosts.

they appear to be rather non-specific, as in other ascaridoids (Rogers and Sommerville, 1968).

Conditions required for activation (hatching, excystation or exsheathment) of other helminths appear to be correlated with the outcome at other divergence points, especially 2 and 3. Such a correlation should be expected, since the various adaptations of an organism are not independent, but part of a co-ordinated adaptive strategy developed over many generations of selection.

The outcome at point 1 apparently depends primarily on the probability of surviving passage through an unsuitable host and remaining in a condition infective to a suitable host. Eggs which have not been activated can sometimes survive such a passage and later infect a suitable host (Crompton, 1970; Pawlowski and Schultz, 1972). The probability that a larva released from an ingested intermediate host could do so must be very small; a cyst might pass through the gut, but the advantages of the intermediate host (see below) would be lost. For larvae ingested with their intermediate hosts, errors of not being activated in a potentially suitable host would be more costly than errors of being activated in an unsuitable host, so natural selection would be expected to favour relatively non-specific conditions for activation. For eggs, natural selection would be expected to favour relatively specific conditions for activation. In general, conditions for hatching of eggs do appear to be more specific (see Asanji and Williams, 1974; Laws, 1968, and reviews by Crompton, 1970, and Smyth, 1966, 1969). However, where at least some of the unsuitable hosts can act as transport hosts, as in *A. robertsi*, hatching conditions appear to be less specific.

Two major groups of factors, physiological and ecological, appear to affect the outcome at point 3. Larvae may die in hosts with special defenses (such as the chitinized gut of insects for *A. robertsi*), begin developing in hosts similar to suitable ones (birds and mammals for larval *A. robertsi*, predatory reptiles for adults), and migrate out of the gut to persist as resting stages in the tissues of the others, which can then function as transport hosts.

The major ecological factor affecting the probability of persisting as a resting stage appears to be the probability of success of a transport host. Parasites of omnivores, carnivores, or piscivores are likely to persist in a transport host. Parasites of herbivores or insectivores (or generally, those which feed on invertebrates) are less likely to persist.

Certain groups of parasites have become well known for their use of transport hosts. Alariine trematodes (Pearson, 1956); protocephalid, pseudophyllid, and trypanorhynch cestodes (Freeman, 1973); many acanthocephalans (Crompton, 1970); and ascaridoid, spirurid and dioctophymatid nematodes (Chitwood, 1969) are good examples.

In some of these groups, both mature adults (in the intestine) and immature stages (in the tissues or the body cavity) can sometimes be found in an individual, suitable definitive host. For those species which do not have an obligate tissue migration, the usual explanation is that larvae which have not yet reached the infective stage migrate to the tissues as in a transport host, and that these larvae are at a dead end, unless the host is cannibalistic (DeGiusti, 1949). However, Fischer and Freeman (1969) have shown that in one such species, *Proteocephalus ambloplitis*, plerocercoids in the viscera, stimulated by a rise in temperature in spring, can penetrate into the gut and mature (shown by the broken arrow at point 4 in Fig. 1). They also showed that plerocercoids ingested by the definitive host during the summer did not remain in the gut, but must have invaded the viscera or died. Their results suggest that seasonal factors (temperature?) can affect the outcome at point 3 in "suitable" hosts.

The primary factors determining the outcome at point 5 appear to be physiological and/or immunological (Sprent, 1966, 1969). It is here that the differences in suitability are expressed, both between and within host species. At this point, ecological factors appear to play their strongest role in normally suitable hosts, in which seasonal factors and/or a large population of parasites in the host may result in retarded development, which may be released later, allowing maturation and reproduction (see review by Michel, 1974).

The general picture, then, is that painted by Dogiel: a parasite will have one (or a few) host(s) in which its growth and reproduction (or its development into an infective stage) are maximal; others in which its growth and reproduction are reduced; and a larger number in which it can establish, but not mature. Thus far, the picture is identical to one for plants or sessile animals. Some parasites, however, have an additional feature without parallel in plants or other animals: the hosts in which their development is inhibited may function as transport hosts.

The system appears to be one in which the host selects its parasites, not one in which the parasite selects its hosts, as the title of this chapter would imply. The host certainly plays a strong role. However, since reaching a suitable host is of primary importance in determining the reproductive output of a parasite, there should be strong natural selection for any adaptation in the parasite which increases its chances of reaching a suitable host, and these adaptations may be thought of as ways in which the parasite does "select" its hosts. Some of these adaptations, especially those which pertain to the use of intermediate (or transport) hosts, will be covered in the next section.

2.4. Intermediate hosts and selection of suitable definitive hosts

Intermediate hosts fulfill a number of functions in the life of a parasite. They provide a suitable milieu for part of the parasite's development. That development may include an amplification of numbers, reproducing a genome successful in reaching the intermediate host. This amplification is particularly well developed in digeneans, where variable numbers of cercariae, highly adapted for transmission, are produced. In many intermediate hosts, the parasite is in a resting stage with an extended life span, increasing the time available for infecting the definitive host (Chapter 7), and acting as a hedge against extinction of local populations (Anderson, 1974) or as a method of overwintering or passing some other period of unfavorable environmental conditions. These are all important functions, but none contributes directly to host selection.

The remaining function, which can contribute to host selection, is to facilitate infection of the definitive host by using food web relationships. In this regard, transport hosts (in which no development takes place) are identical to intermediate hosts. "Intermediate hosts" will be used in the remainder of this chapter to refer to either or both.

Parasites use two types of food web relationships—the feeding patterns of vectors, usually blood-feeding invertebrates, or the relationships between predatory definitive hosts and their prey, the intermediate hosts. In each case, the evolution of the parasite is dependent upon the characteristics of the intermediate host, in much the same fashion as a plant which depends upon an animal for dispersal of its seeds. Levin and Kerster (1974) have pointed out two general adaptations in such plants: they synchronize their fruiting period with the time of the greatest activity of the vector, and they provide a nutritional reward for the vector. Van der Pijl (1969) has emphasized two more: they locate their fruit in the region of greatest activity of the vector (or devise adaptations so that it will fall there), and they provide the fruit with attractants adapted to the sensory mechanisms used by the vector. Parasites also use these adaptations, as evidenced below.

2.4.1. Vectors and their behavior

The most obvious adaptation of vector-transmitted parasites is the synchrony shown between the presence of the parasites and the activity of the vectors. Two types of synchrony are apparent, one involving a daily cycle, the other an annual one.

A daily (or diel) cycle in the numbers of microfilariae of *Wuchereria bancrofti* in the peripheral blood of infected humans was first noted by Manson in 1878; he recognized that it was related to the biting time of the vector, which led to his discovery of mosquito transmission (Hawking, 1975). Since that time, diel cycles have been noted in a wide variety of filarids and blood protozoans (see review by Hawking, 1975).

Where such systems have been investigated, they show clear relationships to the biting habits of the vectors. The clearest relationship is shown by *Wuchereria bancrofti*. Over most of its range, its vectors are night-biting mosquitos, and the microfilariae show a very marked rhythm, with peak numbers in the peripheral blood at night, and virtually none there during the day. In some areas, however, particularly in the South Pacific, its vectors are day-biting mosquitos, and the microfilariae show a less marked, but still obvious, diurnal peak.

Such differences can also arise in the same geographical area, but in different definitive-intermediate host pairs. In the Cameroons rain forest, the microfilariae of *Loa loa* in man show a diurnal peak, and are transmitted by day-biting *Chrysops silacea* and *C. dimidiata*. In three species of monkeys, the microfilariae have a nocturnal peak, and are transmitted by crepuscular or night-biting, canopy-dwelling *C. langi* and *C. centurionis*. The two strains of filarid appear to be speciating; they could still be hybridized in the laboratory, but hybridization in the field appeared unlikely, and self-eliminating (Duke, 1972).

What may be the end result of such a process was reported by Seed and co-workers. Near New Orleans, *Rana clamitans* harbour at least two species of *Trypanosoma*. A large species (*T. rotatorium*) is present in large numbers in the peripheral blood during the day, when the frogs are in the water, but rare at night, when they forage on land (Southworth et al., 1968). A smaller species is present in large numbers during the night, but is scarce during the day. Seed (1970) suggested that they were transmitted by a leech and by a night-biting insect, respectively.

These rhythms, at least in the microfilariae, are known to be due to the activity of the parasites themselves (Hawking, 1975). The migration into the peripheral blood is clearly an adaptation for transmission; the migration out of the peripheral blood is less clear. Hawking has suggested that conditions for survival are poorer in the peripheral circulation than in the viscera, and has presented arguments that the higher oxygen tensions in the peripheral circulation may be the reason, at least for microfilariae of *W. bancrofti*. The argument doesn't hold for other groups of microfilariae, however (Hawking, 1975). An alternative hypothesis is suggested by the analysis of plant dispersal methods by van der Pijl (1969); he points out that strategies for both pollination and seed dispersal by animals are "partly positive (attracting legitimate visitors) and partly negative (excluding others)" (p. 41). Perhaps the migration out of the peripheral blood is an adaptation to exclude other vectors.

Annual rhythms of infective stages in the blood are also well known. Some, such as *Leucocytozoon simondi* of mallards (and other avian haemosporidians), are due to a resumption in schizogany and the production of gametocytes on stimulation by hormonal changes in the host in the spring (Chernin, 1952). Others, such as the increase in numbers of microfilariae of *Dirofilaria immitis* in the peripheral blood of dogs in late summer, are correlated with high ambient temperatures (Katamine et al., 1970), and are considered more likely to be due to a redistribution of microfilariae (Hawking, 1975). Seasonal differences in the depth of microfilariae of *Onchocerca* spp. in the skin of their hosts (Eichler, 1973; Mellor, 1973) also appear to be due to activity of the microfilariae. All are correlated with the availability of the vectors.

Tissue-inhabiting microfilariae may also show preferential locations correlated with the preferred biting habits of their vectors. Eichler (1971) found that 90% of the microfilariae of *Onchocerca gutturosa* in cattle migrated to the umbilical region, which was bitten by 80% of the *Simulium ornatum* vectors. Mellor (1974) found that over 95% of the microfilariae of *O. cervicalis* in horses migrated to the abdominal midline, where 85% of the *Culicoides nubeculosus* vectors bit. Mellor (1974) has also summarized the information on location of *O. volvulus* in human hosts in areas with different simuliid vectors. In each case, the adults and the microfilariae are found predominantly in the part of the body preferred by the vector.

Microfilariae have also been shown to be attracted to the specific site of feeding vectors (see review by Nelson, 1970: p. 200). De Leon and Duke (1966) showed that Guatemalan *Simulium ochraceum* ingested 20—25 times as many microfilariae of a Guatemalan strain of *O. volvulus* as of an African strain; African *S. damnosum* ingested 2—4 times as many African as Guatemalan strain microfilariae (Duke et al., 1967); the results suggest that the attraction can be strain-specific.

Apparently, two of the four general adaptations of plants to seed dispersal by animals elucidated by Levin and Kerster or van der Pijl (synchrony, both annual and diel, and site co-ordination) are also well developed in vector-borne parasites. The host already provides a nutritional reward and attractants for the vector, so those adaptations might be considered superfluous in the parasite. (However, it might be interesting to see if vectors are more strongly attracted to infected than non-infected hosts.) Attractants do play a role, but they attract the parasite to the vector, not vice versa. The basic adaptive strategies of vector-borne parasites and animal-dispersed plants seem remarkably similar.

These strategies apply to reaching the vector, with subsequent dispersal of seeds or transmission to another host dependent upon the normal behavior of the vectors. For the latter, the major strategy of most vector-borne parasites seems to be to avoid causing damage to the vector that would interfere with its normal behaviour patterns (De Leon and Duke, 1966). Vectors that transmit the parasites by being swallowed by the host (such as trypanosomes of amphibians in mosquitos or other biting arthropods—Bardsley and Harmsen, 1973) might be an exception. The strategies outlined below might be appropriate, but this possibility does not appear to have been investigated.

2.4.2. Predator—prey relationships

Adaptations for the use of predator—prey relationships in increasing the possibility of transmission of a parasite seem to fall into two groups: adaptations to reach the prey, and adaptations to increase the chances that the prey will be taken by the right predator.

The strategies used by parasites to reach the prey seem to be essentially the same as those used by animal dispersed plants or vector-borne parasites. Synchrony of the reproductive period of the parasite with the active period of the intermediate host is not as obvious as in the filarids or blood protozoans, but does exist, both on a diel and an annual basis (see Hawking, 1975 for a review). The latter is most obvious in parasites with short-lived intermediate hosts, such as copepods.

It should be pointed out that some of these synchronies involve not just two, but three or more species. An instructive example is that of the cestode *Triaenophorus crassus* in lakes in central Canada. All proglottids mature synchronously, and entire gravid worms pass out when the pike (*Esox lucius*) move into shallow water to spawn in early spring. Development through the infective procercoid in the littoral *Cyclops bicuspidatus* is rapid, so that the second intermediate hosts, *Coregonus clupeaformis* and *C. artedii*, are infected before they leave the shallow water in early summer (Miller, 1952). In the southern part of the region, all host species are present, but *T. crassus* is not; the lakes warm up faster in the spring, and the coregonids presumably migrate out of the shallow water before procercoids could become infective.

Adaptations to ensure that parasites will reach the places where the intermediate hosts are active include the well known motility of gravid proglottids of species of *Taenia*, *Raillietina*, and several other tapeworms (see review by Freeman, 1973), which migrate away from the fecal deposits, which are avoided by their intermediate hosts. The rich variety in shape, size, and density of eggs produced by cestodes of aquatic birds are also such adaptations (Jarecka, 1961). For example, the filiform eggs of various species of *Diorchis* allows them to settle on the needle-shaped leaves of *Ceratophyllum*, where the ostracod intermediate hosts live and feed. Cestodes with benthic-feeding amphipod or annelid intermediate hosts have large, heavy eggs or have their eggs enclosed in large, heavy egg packets, which fall to the bottom and can be eaten only by large invertebrates.

The outer membranes of the same eggs have high concentrations of various nutrients, including glycogen, fats, and various proteins (reviewed by Rybicka, 1966), which may provide a nutritional reward for ingesting the eggs. More obvious nutritional rewards must be provided by the slime balls of *Dicrocoelium dendriticum* (Krull and Mapes, 1952), the motile proglottids of *Raillietina* spp. (Bartel, 1965), and the "egg capsule" — the dried body of the female — of *Skrjabinoptera phrynosoma* (Lee, 1957), all of which are carried by foraging ants to their nests, where they are fed to the larvae.

A related adaptation is to mimic the normal food of the intermediate host, such as some hymenolepidid eggs which mimic the diatoms normally eaten by the ostracod host (Jarecka, 1961), or the egg packets of other hymenolepidids, which mimic the filamentous algae eaten by the amphipod host (Podesta and Holmes, 1970).

Once in an intermediate host, the normal operation of the food web will greatly enhance the probabilities of reaching some potential definitive hosts and greatly decrease the probabilities of reaching others. An excellent example is provided by Combes (1968). *Gorgodera euzeti* and *Gorgoderina vitelliloba* both develop in fingernail clams (*Pisidium*) which live in shallow burrows. The cercariae of *G. euzeti* emerge, but remain attached in small groups to the shell of the clam, where they wave rhythmically at the bottom of the burrow. The cercariae of *G. vitelliloba* migrate to the surface of the mud, where they also wave. The former are eaten by the larvae of the insect *Sialis*, the latter by tadpoles of *Rana temporaria*. Male *R. temporaria* remain near the lake during the summer and feed heavily on emerging *Sialis*, acquiring *G. euzeti*. The females move away from the lake until fall, when they return and feed heavily on metamorphosing tadpoles, acquiring *G. vitelliloba*. In another area, where *Rana ridibunda* is present, *R. temporaria* is less aquatic; there, *R. ridibunda* is the major host for *G. euzeti* (Combes and Gerbeau, 1970).

In most cases, predators of a particular species eat only a small proportion of each generation of a given prey species. Thus, there should be strong natural selection in a parasite for any adaptation which increases the chances of the infected prey animal being among those eaten by a suitable definitive host. Holmes and Bethel (1972) have shown that such adaptations are intimately related to the behavior of the predators; the details to what Holling (1968) called the predator's "tactics", the broad outlines to what various authors have termed "feeding strategies". (See Emlen, 1966; Schoener, 1971 for further discussion of feeding strategies.)

Most authors investigating the theory of predation assume that a basic aspect of an optimum feeding strategy is for the predator to maximize the food value (energy content) obtained per unit time (or energy) expended in finding, catching, and handling the prey. A corollary is that inefficiency in catching the prey can be tolerated only if the prey is relatively large. Holmes and Bethel (1972) have suggested that when the predator is efficient, the strategy of the parasite should be to make the prey more conspicuous, and that when the predator is inefficient, the parasite should make the prey not only more conspicuous, but also easier to catch, by reducing its stamina or its defensive responses to the predator.

Holmes and Bethel discussed several examples of greater conspicuousness of infected intermediate hosts, the most obvious of which was the pulsatile, brightly coloured sporocysts of *Leucochloridium* and related genera in the tentacles of their snail hosts. More recently, Hindsbo (1972) reported a depigmentation in amphipods infected with *Polymorphus minutus*, and Seidenberg (1973) and Muzzall and Rabalais (1975a) described similar depigmentations in isopods infected with *Acanthocephalus dirus* and *A. jacksoni*. In all three cases, the depigmented arthropods were conspicuous against the dark background of their habitat, and were assumed to be more vulnerable to the duck (*P. minutus*) or the fish hosts. Plateaux (1972) described a modified pigmentation in ants, *Leptothorax nylanderi*, infected with dilepidid cysticercoids. The yellow infected ants differed sufficiently in morphological and behavioural features to be first considered a separate species of social parasite. The infected ants did not leave the nest, so would be vulnerable only to a predator which broke open the nest.

A wide variety of parasites have been reported to debilitate or to reduce the stamina of their intermediate hosts (Holmes and Bethel, 1972). Among the best examples are the plerocercoids of *Ligula intestinalis* in various forage fishes and *Schistocephalus solidus* in sticklebacks. Dence (1958) showed that *Ligula*-infected *Notropis cornutus* were sluggish and less gregarious than uninfected shiners; other authors have observed similar effects in other fishes; all have suggested that the behavioural changes would make the infected fish more vulnerable to fish-eating birds. Van Dobben (1952) substantiated the suggestions, finding that 30% of the *Rutilus rutilus* taken by *Phalacrocorax carbo*, but only 6.5% of the general population, were infected.

Lester (1971) found that *Gasterosteus aculeatus* infected with plerocercoids of *S. solidus* could not swim continuously at speeds over one body length per second, used more oxygen at rest than uninfected fish, and had a much more rapid rise in oxygen consumption with increased speed than did uninfected fish, all of which suggest reduced stamina. Many authors have commented that infected sticklebacks could easily be caught by hand, or, presumably, by birds.

A number of parasites alter the responses of their host to predators (Holmes and Bethel, 1972), ranging from the sluggishness of fish infected with *Ligula* or *Schistocephalus* to the complete disorientation of fish with metacercariae of *Austrodiplostomum mordax* in the cerebellum and optic centers of the brain (Szidat and Nani, 1951). The latter are blind, uncoordinated, and very easy prey for surface-feeding birds, the definitive hosts. The best example is that of *Baylisascaris procyonis* larvae in mice and squirrels. Larvae migrating in the central nervous system produced loss of coordination, a tendency to circle erratically, and loss of fear toward larger animals making the infected mice easy prey to the raccoon definitive hosts (Tiner, 1953). Tiner (1954) calculated that 5% of the mice in his study area would have died from the direct effects of the larvae; these mice were undoubtedly taken preferentially by raccoons, the most abundant predators in the study area.

The adaptations described thus far would appear to make the infected intermediate hosts more susceptible to predators in general, not specifically to a suitable definitive host. However, Herting (1966) showed that the differences in foods of two species of predatory fish were due to a combination of the behaviour of the predators and that of their potential prey, and concluded that organisms highly susceptible to one predator may be nearly invulnerable to another. Both theoretical and field investigations suggest that predators faced with competition from other predators specialize in the microhabitats used (in the short run) and their method of feeding (over evolutionary time) (Schoener, 1971). For a parasite, the obvious corollary is that any adaptation which modifies the behavior of its intermediate host so as to bring it into the "feeding niche" of the definitive host should be selected for.

Ants infected with metacercariae of *Dicrocoelium dendriticum* respond to a drop in temperature by migrating to the tips of blades of grass or other vegetation, locking their mandibles to the plant, and remaining torpid until the temperature rises, when they resume normal behaviour (Hohorst and Graefe, 1961; Anokhin, 1966; Grus, 1966). This pattern keeps the infected ants near the top of the vegetation during the early morning and late evening feeding periods of the definitive hosts (sheep), but allows them to move to less exposed areas during the hot, dry midday.

Ants infected with another dicrocoelid, *Brachylecithum mosquensis*, do not show their normal photophobic responses, are active later in the fall, and are found much more frequently in open, rocky areas than are uninfected ants (Carney, 1969). They also remain motionless, or slowly circling, on tops of rocks in open areas. These modifications obviously increase their vulnerability to insectivorous birds, such as robins (the definitive host), which feed in such places.

The bivalve *Macoma balthica* normally remains buried in the sediment and well concealed. However, Swennen (1969) has found that some individuals heavily infected with gymnophallid metacercariae crawled along just under the surface of tidal flats in the high intertidal zone, leaving conspicuous tracks which indicated their position. Swennen and Ching (1974) identified the causal organism as *Parvatrema affinis*, which matures in oystercatchers, gulls, and waterfowl. Swennen (1969) suggested that the infected bivalves were more vulnerable to visually-hunting predators (such as their bird definitive hosts), although Hulscher (1973) has indicated that oystercatchers may reject heavily parasitized *Macoma*.

Holmes and Bethel (1972) outline several other cases in which altered responses of infected animal hosts seem to increase their vulnerability to predation by the definitive hosts. Since then, Hindsbo (1972) has shown that *Gammarus lacustris* harboring *Polymorphus minutus* were found more frequently in the upper, lighted area of his experimental tank, and were 2.5 times more vulnerable than uninfected *G. lacustris* to predation by domestic ducklings. Similarly, isopods harboring *Acanthocephalus jacksoni* spend more time wandering on top of leaves, increasing their vulnerability to a variety of fishes (Muzzall and Rabalais, 1975b).

Bethel and Holmes (1973) have shown that each of three species of acanthocephalans modify the behaviour of their infected intermediate hosts, but each in a different way. Uninfected amphipods avoid lighted areas; when disturbed, they dive away from light and (if possible) burrow into the mud bottom. *Gammarus lacustris* infected with *Polymorphus marilis* select lighted areas, but show a (normal) negative phototaxis to strong sudden light, and normal evasive behavior (although they return to the lighted area within a much shorter time than controls).

Gammarus lacustris infected with *Polymorphus paradoxus* select areas of most intense illumination when undisturbed; they are hypersensitive to disturbance and their evasive behaviour is markedly modified. After disturbance, they are positively phototactic, skimming along the surface of the water, making a conspicuous surface disturbance. When they contact any solid material at the surface, they cling to it with their gnathopods.

Uninfected *Hyalella azteca* show a marked affinity for vegetation, but otherwise have essentially the same behaviour as uninfected *G. lacustris*. Those infected with cystacanths of the polymorphid *Corynosoma constrictum* selected areas of highest illumination and showed a variable phototaxis. About 60% showed a strong positive response to sudden, bright light; the rest showed a strong negative response. After disturbance, these proportions were reversed. Those which did show a negative phototaxis soon returned to the lighted area.

The behavioural alterations, especially in the evasive behaviour, produced by the three acanthocephalans appear to be well adapted to the feeding habits of their respective definitive hosts. The diving of *P. marilis*-infected amphipods brings them into the feeding area of the definitive host, lesser scaup (*Aythya affinis*), a diving duck. The surface activity of those infected with *P. paradoxus* makes them vulnerable to feeding mallards (*Anas platyrhynchos*), and by clinging to the vegetation they become vulnerable to accidental ingestion by muskrats (*Ondatra zibethica*) and beaver (*Castor canadensis*). All three are suitable definitive hosts. The varied behavior of *Hyalella* infected with *C. constrictum* makes part of the population vulnerable to surface-feeding ducks, part to divers. This species has a wide host range, including mallards, muskrats, and lesser scaup.

It should be apparent that the strategies adopted by many parasites are complex, and include more than one of the features discussed above. For example, *Polymorphus paradoxus* and *Parvatrema affinis* both produce behavioural changes that alter the microhabitat occupied by the intermediate host, make it more conspicuous, reduce its stamina, and (at least in the case of *P. paradoxus*) modify its response to potential predators.

Acknowledgements

This chapter was written while on sabbatical leave at the Department of Parasitology, University of Queensland. I thank H.M.D. Hoyte for providing facilities, and C. Dobson, G.C. Kearn and J.F.A. Sprent for constructive criticisms.

2.5. REFERENCES

Anderson, R.M. (1974). "Mathematical models of host-helminth parasite interactions." In: "Ecological stability." (M.B. Usher and M.H. Williamson, eds.) pp. 43—69. Chapman and Hall, London.

Anokhin, I.A. (1966). "Daily rhythm in ants infected with metacercariae of *Dicrocoelium lanceatum.*" Dokl. Akad. Nauk SSSR 166, 757—759.

Asanji, M.F. and Williams, M.O. (1974). "Studies on the excystment of trematode metacercariae in vivo." J. Helminth. 48, 85—91.

Bardsley, J.E. and Harmsen, R. (1973). "The trypanosomes of Anura." Adv. Parasit. 11, 1—73.

Bartel, M.H. (1965). "The life cycle of *Raillietina (R.) loeweni* Bartel and Hansen, 1964 (Cestoda) from the black-tailed jackrabbit, *Lepus californicus melanotis* Mearns." J. Parasit. 51, 800—806.

Basch, P.F. (1975). "An interpretation of snail-trematode infection rates: specificity based on concordance of compatible phenotypes." Int. J. Parasit. 5, 449—452.

Bethel, W.M. and Holmes, J.C. (1973). "Altered evasive behavior and responses to light in amphipods harboring acanthocephalan cystacanths." J. Parasit. 59, 945—956.

Brooks, C.P. (1953). "A comparative study of *Schistosoma mansoni* in *Tropicorbis havanensis* and *Australorbis glabratus.*" J. Parasit. 39: 159—165.

Carney, W.P. (1969). "Behavioral and morphological changes in carpenter ants harboring dicrocoeliid metacercariae." Am. Midl. Nat. 82, 605—611.

Chernin, E. (1952). "The relapse phenomenon in the *Leucocytozoon simondi* infection of the domestic duck." Am. J. Hyg. 56, 101—118.

Chitwood, M.B. (1969). "The systematics and biology of some parasitic nematodes." In: "Chemical Zoology, Volume III, Echinodermata, Nematoda, and Acanthocephala." (M. Florkin and B.T. Scheer, eds.) pp. 223—244. Academic Press, New York.

Combes, C. (1968). "Biologie, Ecologie des cycles et Biogéographie de Digènes et Monogènes d'Amphibiens dans l'Est des Pyrénées." Mém. Mus. natn. Hist. nat., Paris. Ser. A, Zool. 51, 1—195.

Combes, C. and Gerbeaux, M.T. (1970). "Recherches écoparasitologiques sur l'helminthofaune de *Rana ridibunda perezi* (Amphibien Anoure) dans l'Est des Pyrénées." Vie Milieu 21(C), 121—158.

Crompton, D.W.T. (1970). "An ecological approach to acanthocephalan physiology." pp. 1—125. Cambridge University Press.

Day, P.R. (1974). "Genetics of host-parasite interaction." pp. 92—110. Freeman, San Francisco.

DeGiusti, D.L. (1949). "The life cycle of *Leptorhynchoides thecatus* (Linton), an acanthocephalan of fish." J. Parasit. 35, 437—460.

DeLeon, J.R. and Duke, B.O.L. (1966). "Experimental studies on the transmission of Guatemalan and West African strains of *Onchocerca volvulus* by *Simulium ochraceum*, *S. metallicum* and *S. callidum.*" Trans. R. Soc. trop. Med. Hyg. 60, 735—752.

Dence, W.A. (1958). "Studies on *Ligula*-infected common shiners (*Notropis cornutus* Agassiz) in the Adirondacks." J. Parasit. 44, 334—338.

Dobben, W.H. van (1952). "The food of the cormorant in the Netherlands." Ardea 40, 1—63.

Dogiel, V.A., Polyanski, Yu. I. and Kheisin, E.M. (1964). "General parasitology." Z. Kabata (transl.) pp. 1—516. Oliver and Boyd, Edinburgh.

Doyle, R.W. (1975). "Settlement of planktonic larvae: a theory of habitat selection in varying environments." Am. Nat. 109, 113—126.

Duke, B.O.L. (1972). "Behavioural aspects of the life cycle of *Loa.*" In: "Behavioural aspects of parasite transmission." (E.U. Canning and C.A. Wright, eds.) pp. 97—105. Academic Press, London.

Duke, B.O.L., Moore, P.J. and DeLeon, J.R. (1967). "*Onchocerca—Simulium* complexes. V. The intake and subsequent fate of microfilariae of a Guatemalan strain of *Onchocerca volvulus* in forest and Sudan-savannah forms of West African *Simulium damnosum.*" Ann. trop. Med. Parasit. 61, 332—337.

Eichler, D.A. (1971). "Studies on *Onchocerca gutturosa* (Neumann, 1910) and its development in *Simulium ornatum* (Meigen, 1818). II. Behaviour of *S. ornatum* in relation to the transmission of *O. gutturosa.*" J. Helminth. 45, 259—270.

Eichler, D.A. (1973). "Studies on *Onchocerca gutturosa* (Neumann, 1910) and its development in *Simulium ornatum* (Meigen, 1818). III. Factors affecting the development of the parasite in its vector." J. Helminth. 47, 73—88.

Emlen, J.M. (1966). "The role of time and energy in food preference." Am. Natur. 100, 611—617.

Emlen, J.M. (1973). "Ecology: an evolutionary approach. " pp. 1—493. Addison—Wesley, Reading, Massachusetts.

Falconer, D.S. (1965). "The inheritance of liability to certain diseases, estimated from the incidence among relatives." Ann. Hum. Genet., Lond. 29, 51—71.

Files, V.S. and Cram, E.B. (1949). "A study on the comparative susceptibility of snail vectors to strains of *Schistosoma mansoni.*" J. Parasit. 35, 555—560.

Fischer, H. and Freeman, R.S. (1969). "Penetration of parenteral plerocercoids of *Protocephalus ambloplitis* (Leidy) into the gut of smallmouth bass." J. Parasit. 55, 766—774.

Freeman, R.S. (1973). "Ontogeny of cestodes and its bearing on their phylogeny and systematics." Adv. Parasit. 11, 481—557.

Grus, I. (1966). "Prilog poznavanju epizootiologije dikrocelioze i drugog prelaznog domacina na terenima srbije." Acta vet., Beogr. 16, 249—255.

Hawking, F. (1975). "Circadian and other rhythms of parasites." Adv. Parasit. 13, 123—182.

Herting, G.E. (1966). "Effects of behavior on vulnerability of forage organisms to predation by bowfin and spotted gar." M. Sc. Thesis, University of Missouri, Columbia.

Hindsbo, O. (1972). "Effects of *Polymorphus* (Acanthocephala) on color and behaviour of *Gammarus lacustris.*" Nature, Lond. 238, 333.

Hohorst, W. and Graefe, G. (1961). "Ameisen-obligatorische zwischenwirte des lanzettegels (*Dicrocoelium dendriticum*)." Naturwissenschaften 48, 229—230.

Holling, C.S. (1968). "The tactics of a predator." Symp. Roy. Entomol. Soc. London 4, 47—58.

Holmes, J.C. and Bethel, W.M. (1972). "Modification of intermediate host behaviour by parasites." In: "Behavioural aspects of parasite transmission." (E.U. Canning and C.A. Wright, eds.) pp. 123—149. Academic Press, London.

Huff, C.G. (1927). "Studies on the infectivity of plasmodia of birds for mosquitoes with special reference to the problem of immunity in the mosquito." Am. J. Hyg. 7, 706—734.

Hulscher, J.B. (1973). "Burying-depth and trematode infection in *Macoma balthica.*" Neth. J. Sea Res. 6, 141—156.

Jarecka, L. (1961). "Morphological adaptations of tapeworm eggs and their importance in the life cycles." Acta Parasit. Polon. 9, 409—426.

Katamine, D., Aoki, Y. and Iwamoto, I. (1970). "Analysis of microfilarial rhythm." J. Parasit. 56 (4, Section 2, Part 1), 181.

Krull, W.H. and Mapes, C.R. (1952). "Studies on the biology of *Dicrocoelium dendriticum* (Rudolphi, 1819) Looss, 1899 (Trematoda: Dicrocoeliidae), including its relation to the intermediate host, *Cionella lubrica* (Müller). VII. The second intermediate host of *Dicrocoelium dendriticum.*" Cornell Vet. 42, 603—604.

Laws, G.F. (1968). "The hatching of taeniid eggs." Expl. Parasit. 23, 1—10.

Lee, S.H. (1957). "The life cycle of *Skrjabinoptera phrynosoma* (Ortlepp) Schultz, 1927 (Nematoda: Spiruroidea), a gastric nematode of Texas horned toads, *Phrynosoma cornutum.*" J. Parasit. 43, 66—75.

Leong, T.S. (1975). "Metazoan parasites of fishes of Cold Lake, Alberta: A community analysis." Ph.D. Thesis, University of Alberta, Edmonton.

Lester, R.J.G. (1971). "The influence of *Schistocephalus* plerocercoids on the respiration of *Gasterosteus* and a possible resulting effect on the behaviour of the fish." Can. J. Zool. 49, 361—366.

Levin, D.A. and Kerster, H.W. (1974). "Gene flow in seed plants." Evolutionary Biology 7, 139—220.

Levins, R. (1968). "Evolution in changing environments: some theoretical explorations." Monogr. Pop. Biol. 2, 1—120.

Meadows, P.S. and Campbell, J.I. (1972). "Habitat selection by aquatic invertebrates." Adv. Mar. Biol. 10, 271—382.

Mellor, P.S. (1973). "Studies on *Onchocerca cervicalis* Railliet and Henry, 1910: I. *Onchocerca cervicalis* in British horses." J. Helminth. 47, 97—110.

Mellor, P.S. (1974). "Studies on *Onchocerca cervicalis* Railliet and Henry, 1910: IV. Behaviour of the vector *Culicoides nubeculosus* in relation to the transmission of *Onchocerca cervicalis.*" J. Helminth. 48, 283—288.

Michel, J.F. (1974). "Arrested development of nematodes and some related phenomena." Adv. Parasit. 12, 279—366.

Miller, R.B. (1952). "A review of the *Triaenophorus* problem in Canadian lakes." J. Fish. Res. Board Canada, Bull. no. 95.

Moore, P.G. (1975). "The role of habitat selection in determining the local distribution of animals in the sea." Mar. Behav. Physiol. 3, 97—100.

Muzzall, P.M. and Rabalais, F.C. (1975a). "Studies on *Acanthocephalus jacksoni* Bullock 1962 (Acanthocephala: Echinorhynchidae). II. An analysis of the host-parasite relationship of larval *Acanthocephalus jacksoni* in *Lirceus lineatus* (Say)." Proc. Helminth. Soc. Wash. 42, 35—38.

Muzzall, P.M. and Rabalais, F.C. (1975b). "Studies on *Acanthocephalus jacksoni* Bullock 1962 (Acanthocephala: Echinorhynchidae). III. The altered behaviour of *Lirceus lineatus* (Say) infected with cystacanths of *Acanthocephalus jacksoni.*" Proc. Helminth. Soc. Wash. 42, 116—118.

Nelson, G.S. (1970). "Onchocerciasis." Adv. Parasit. 8, 173—224.

Newton, W.L. (1953). "The inheritance of susceptibility to infection with *Schistosoma mansoni* in *Australorbis glabratus.*" Expl. Parasit. 2, 242—257.

Pan, C.T. (1963). "Generalized and focal tissue responses in the snail, *Australorbis glabratus*, infected with *Schistosoma mansoni.*" Ann. N.Y. Acad. Sci. 113, 475—485.

Paraense, W.L. and Corrêa, L.R. (1963). "Variation in susceptibility of populations of *Australorbis glabratus* to a strain of *Schistosoma mansoni.*" Rev. Inst. Med. trop. S. Paulo 5, 15—22.

Pawlowski, Z. and Schultz, M.G. (1972). "Taeniasis and cysticercosis (*Taenia saginata*)." Adv. Parasit. 10, 269—343.

Pearson, J.C. (1956). "Studies on the life cycles and morphology of the larval stages of *Alaria arisaemoides* Augustine and Uribe, 1927, and *Alaria canis* LaRue and Fallis, 1936 (Trematoda: Diplostomidae)." Can. J. Zool. 34, 295—387.

Pijl, L. van der (1969). "Principles of dispersal in higher plants." pp. 1—154. Springer-Verlag, New York.

Plateaux, L. (1972). "Sur les modifications produites chez une Fourmi par la présence d'un parasite Cestode." Annls. Sci. nat. (Zool), Ser 12. 14, 203—220.

Podesta, R.B. and Holmes, J.C. (1970). "Hymenolepidid cysticercoids in *Hyalella azteca* of Cooking Lake, Alberta: life cycles and descriptions of four new species." J. Parasit. 56, 1124—1134.

Richards, C.S. (1973). "Susceptibility of adult *Biomphalaria glabrata* to *Schistosoma mansoni* infection." Am. J. Trop. Med. Hyg. 22, 748—756.

Rogers, W.P. and Sommerville, R.I. (1968). "The infectious process, and its relation to the development of early parasitic stages of nematodes." Adv. Parasit. 6, 327—348.

Rybicka, K. (1966). "Embryogenesis in cestodes." Adv. Parasit. 4, 107—186.

Saoud, M.F.A. (1965). "Susceptibilities of various snail intermediate hosts of *Schistosoma mansoni* to different strains of the parasite." J. Helminth. 39, 363—376.

Schoener, T.W. (1971). "Theory of feeding strategies." Ann. Rev. Ecol. Syst. 2, 369—404.

Seed, J.R. (1970). "Diurnal and seasonal rhythms in parasitaemia levels of some trypanosomes infecting *Rana clamitans* from Louisiana." J. Parasit. 46 (4, Section 2, Part 1): 311—312.

Seidenberg, A.J. (1973). "Ecology of the acanthocephalan, *Acanthocephalus dirus* (Van Cleave, 1931), in its intermediate host, *Asellus intermedius* Forbes (Crustacea: Isopoda)." J. Parasit. 59, 957—962.

Smyth, J.D. (1966). "The physiology of trematodes." pp. 1—256. Oliver and Boyd, Edinburgh.

Smyth, J.D. (1969). "The physiology of cestodes." pp. 1—279. Oliver and Boyd, Edinburgh.

Southworth, G.C., Mason, G. and Seed, J.R. (1968). "Studies on frog trypanosomiasis. I.

A 24-hour cycle in the parasitaemia level of *Trypanosoma rotatorium* in *Rana clamitans* from Louisiana." J. Parasit. 54, 255—258.

Sprent, J.F.A. (1963). "The life history and development of *Amplicaecum robertsi,* an ascaridoid nematode of the carpet python (*Morelia spilotes variegatus*). II. Growth and host specificity of larval stages in relation to the food chain." Parasitology 53, 321—337.

Sprent, J.F.A. (1966). "The components of host specificity in infections with ascaridoid nematodes." Proc. First Int. Congr. Parasit. (Rome, 21—26 Sept. 1964). 1, 15—18.

Sprent, J.F.A. (1969). "Evolutionary aspects of immunity in zooparasitic infections." In: "Immunity to parasitic animals, Vol. I." (G.J. Jackson, R. Herman and I. Singer, eds.) pp. 3—62. Appleton-Century-Crofts, New York.

Swennen, C. (1969). "Crawling-tracks of trematode-infected *Macoma balthica* (L.)." Neth. J. Sea Res. 4, 376—379.

Swennen, C. and Ching, H.L. (1974). "Observations on the trematode *Parvatrema affinis,* causative agent of crawling tracks of *Macoma balthica.*" Neth. J. Sea Res. 8, 108—115.

Szidat, L. and Nani, A. (1951). "Diplostomiasis cerebralis del pejerrey." Rev. Inst. nac. Invest. Cienc. nat. 1, 323—384.

Tiner, J.D. (1953). "Fatalities in rodents caused by larval *Ascaris* in the central nervous system." J. Mammal. 34, 153—167.

Tiner, J.D. (1954). "The fraction of *Peromyscus leucopus* fatalities caused by raccoon ascarid larvae." J. Mammal. 35, 589—592.

Ward, R.A. (1963). "Genetic aspects of the susceptibility of mosquitos to malarial infection." Expl. Parasit. 13, 328—341.

Worrell, E. (1963). "Reptiles of Australia." p. 100. Angus and Robertson, Sydney.

Ecological Aspects of Parasitology
Editor: C.R. Kennedy
© North-Holland Publishing Company, Amsterdam, 1976

CHAPTER 3

ENTRY INTO THE HOST AND SITE SELECTION

D.W.T. CROMPTON *

Division of Nutritional Sciences, Cornell University, Ithaca, N.Y. (U.S.A.)

Contents

3.1. Introduction

Recent appraisals of parasitological literature have concluded that many species of endoparasite occupy relatively precise sites within their hosts (Table 1; Ulmer, 1971; Crompton, 1973; Holmes, 1973; Mettrick and Podesta, 1974; Crompton and Nesheim, 1975). This generalization often seems to apply irrespective of the developmental stage of the parasite or the status of its hosts. For example, the filarial nematode *Dirofilaria aethiops* (Table 1) occupies at least four sites or microhabitats (Holmes, 1973) during its development. The adult worms are usually found in connective tissue surrounding the muscle bundles in the thighs of monkeys (Hawking and Webber, 1955). The unsheathed microfilariae live in the monkey's circulatory system until they are ingested by mosquitoes in which they develop to the third stage in

---

* Permanent address: The Molteno Institute, University of Cambridge, Cambridge (Great Britain).

connective tissue before transferring to the salivary glands, whence they are transmitted to monkeys (Webber, 1955).

The purpose of this chapter is to discuss how parasites enter their hosts and reach and change sites within and between the habitats therein. Ideally, such an assignment requires a consideration of the processes of entry, identification of the routes taken within hosts when moving to sites, knowledge of the adaptations of the parasites for travel and navigation within the host and information about the influence of the habitat on all these activities. When an ecological approach is adopted in studying endoparasitism, each organ system of the body may be assigned the status of habitat on the basis of an idea developed by Andrewartha and Birch (1954). They considered a habitat to be an area which seems to possess a certain uniformity with respect to some quality which the investigator decides is important in his study. Thus the parasitologist may argue that the physiological function of an organ system of the host provides the necessary uniformity. It is conceivable, therefore, that when a parasite moves from a site in the gastric mucosa to one in the peribronchial connective tissue of the same host, as occurs during the development of *Filaroides martis* (Table 1), it undergoes a change of habitat not unlike that experienced by a developing amphibian.

Most parasitologists have referred to movements of parasites within their hosts as migrations. In this discussion, the term "emigration" is used for journeys from the place of entry to a site in a host or when a change of site occurs within a host. "Migration" is reserved for movements involving phases of coming and going between or within sites (Heape, 1931; Allee et al., 1949). The reason for distinguishing emigrations from migrations is to emphasize that different biological functions appear to be associated with the movement of parasites. Thus, it is not intended to suggest that the use of the term migration should be restricted or that this arbitrary scheme should be adopted. Too few controlled investigations have been carried out and insufficient information is available to support a detailed classification of the movements of parasites. Although many observations have been made on the sites occupied by endoparasites, most studies of emigrations and migrations have involved helminths and the protozoan genus *Eimeria* (see Chapter 20). This bias is reflected in the very restricted lists of parasites (Table 1, 2 and 3) chosen for consideration here.

3.2. Entry into the host

Nutritional and behavioural relationships between hosts, chance contact and contamination, the behaviour of parasites and the behavioural changes they induce in hosts are some of the factors involved in the successful transmission of parasites. In many cases, transmission and the process of entry into the host often involve the same developmental stage and, although host

TABLE 1

Observations on the sites occupied by some endoparasitic helminths in their definitive hosts. A = Acanthocephalan; C = Cestode; N = Nematode; T = Digenetic Trematode; a = amphibian; b = bird; f = fish; m = mammal; r = reptile.

Parasite		Host	Observations on sites	References
LUNGS				
N. *Filaroides martis*	m.	*Mustela vison* (Mink)	Peribronchial connective tissue; also in tissue surrounding arteries	Stockdale (1970); Stockdale and Anderson (1970)
N. *Protostrongylus stilesi*	m.	*Ovis canadensis* (Bighorn sheep)	Usually in parenchymal tissue of lungs	Forrester (1971)
T. *Haematoloechus medioplexus*	a.	*Rana pipiens* (adult frog)	Attached to wall of lungs	Cort (1915); Krull (1931)
T. *Paragonimus uestermani*	m.	Domestic cat	Usually within cysts in pulmonary tissue	Yokogawa (1965)
BODY CAVITIES				
N. *Litomosoides carinii*	m.	*Sigmodon hispidus litoralis* (Cotton rat)	Massed together in the pleural cavity; occasionally inhabits peritoneal space	Williams (1948)
MUSCULATURE				
N. *Dirofilaria aethiops* (= *D. corynodes*, Hawking, 1975)	m.	*Cercopithecus aethiops* (monkey)	Majority in the thighs; in the deep fasciae, on or between muscle bundles	Hawking and Webber (1955)
LIVER AND DIGESTIVE ORGANS				
C. *Hymenolepis microstoma*	m.	Laboratory mouse	Biliary duct and duodenum	Dvorak et al. (1961); De Rycke (1966); Caley (1974)
T. *Eurytrema vulpis*	m.	*Vulpes vulpes* (fox)	Pancreatic ducts	Stunkard (1947)
T. *Fasciola hepatica*	m.	Domestic sheep	Biliary system	Dawes and Hughes (1964)
T. *Plagioporus sinitsini*	f.	*Nocomis biguttatus* (chub)	Gall-bladder	Dobrovolny (1939)

TABLE 1 (continued)

ALIMENTARY CANAL

Parasite	Host	Observations on sites	References
A. *Echinorhynchus truttae*	f. *Salmo trutta* (trout)	Posterior part of intestine	Awachie (1966)
A. *Moniliformis dubius*	m. Laboratory rat	Anterior small intestine	Burlingame and Chandler (1941); Holmes (1961)
A. *Polymorphus minutus*	b. Domestic duck	Posterior part of small intestine	Crompton and Whitfield (1968)
C. *Echinococcus granulosus*	m. Domestic dog	Duodenum; between villi	Smyth (1969); Smyth et al. (1970)
C. *Raillietina cesticillus*	b. Domestic fowl	Anterior part of small intestine	Foster and Daugherty (1959); Gray (1972a,b)
C. *Spirometra mansonoides*	m. Domestic cat	Scolex usually attached in middle third of small intestine	Mueller (1965)
N. *Ascaris lumbricoides*	m. Domestic pigs	Usually anterior half of small intestine	Roberts (1934); (see Makidono, 1956)
N. *Dochmoides stenocephala*	m. Domestic dog	Posterior part of small intestine	Gibbs (1961)
N. *Haemonchus contortus*	m. Domestic sheep	Abomasum	Veglia (1915); Stoll (1929)
N. *Heterakis gallinarum*	b. Domestic fowl	Caeca	Clapham (1933); Roberts (1937); Nath and Pande (1963)
N. *Kalicephalus inermis coronellae*	r. *Coluber constrictor* (Black racer-snake)	Anterior part of oesophagus	Schad (1962)
N. *Nippostrongylus brasiliensis*	m. Laboratory rat	Anterior part of small intestine	Brambell (1965); Alphey (1970); Jenkins (1974); Connan (1974)
N. *Spirocerca lupi*	m. Domestic dog	Submucosa of oesophagus	Bailey (1972)

TABLE 1 (continued)

Parasite		Host	Observations on sites	References
N. Trichuris muris	m.	Laboratory mouse	Caecum	Fahmy (1954)
T. Crepidostomum cooperi	f.	Perca flavescens (Yellow perch)	Pyloric caecal region of small intestine	Cannon (1972)
T. Diplodiscus temperatus	a.	Rana pipiens	Rectum (in young tadpoles and frogs) and stomach (in metamorphosing tadpoles)	Herber (1939)
T. Psilostomum ondatrae	b.	Domestic fowl	Mucosa of proventriculus	Newsom and Stout (1933); Beaver (1939)
T. Renifer aniarum	r.	Natrix sipedon (water snake)	Mouth cavity	Byrd (1935)
T. Sphaeridiotrema globulus (= S. spinoacetabulum)	b.	Domestic duck	Caeca	Macy et al. (1968)
URINOGENITAL SYSTEM				
N. Dioctophyma renale	m.	Mustela vison	Kidneys; usually right organ	Fyvie (1971); (see Hallberg, 1953)
T. Gorgodera amplicava	a.	Rana palustris (frog)	Bladder and urogenital ducts	Goodchild (1948)
T. Prosthogonimus macrorchis	b.	Domestic fowl (adult female)	Oviducts	Macy (1934)
SENSORY AND NERVOUS SYSTEM				
N. Oxyspirura mansoni	b.	Domestic fowl	Eyes; beneath nictitating membrane, in conjunctival sacs and naso-lacrimal ducts	Schwabe (1951)
T. Halipegus eccentricus	a.	Rana pipiens	Eustachian tubes (ears)	Thomas (1939)
VASCULAR AND LYMPHATIC SYSTEMS				
N. Angiostrongylus cantonensis	m.	Laboratory rat	Pulmonary arteries	Mackerras and Sandars (1955); Alicata and Jindrak (1970)

TABLE 1 (continued)

Parasite		Host	Observations on sites	References
N.	*Dirofilaria immitis*	m. Domestic dog	Usually in right ventricle and pulmonary arteries	Lindsay (1962)
N.	*Wuchereria bancrofti*	m. Man	Lymphatic vessels and glands	Manson-Bahr (1966)
T.	*Aporocotyle macfarlani*	f. *Sebastes caurinus* (rockfish)	Afferent branchial arteries, ventral aorta and rarely in chambers of heart	Holmes (1971)
T.	*Sanguinicola kalmathensis*	f. *Salmo clarkii henshawi* (cutthroat trout)	Usually in the efferent renal vein	Wales (1958)
T.	*Schistosoma mansoni*	m. Laboratory mouse (and man)	Liver, and portal and mesenteric veins	Standen (1949)

TABLE 2

Some infective stages of endoparasites and their routes of entry into their hosts. L_n represents the appropriate larval stage in the developmental cycle of a nematode; MF represents microfilaria.

	Parasite	Infective stage	Susceptible host	Resulting stages	References
ORAL ROUTE					
Protozoa	*Eimeria* spp.	Sporulated oocyst (Fig. 5)	Domestic fowl	Entire cycle	Levine (1942); Doran and Farr (1962); Crompton and Nesheim (1975)
	Trichomonas gallinae	Trichomonad in "pigeons's milk"	Domestic pigeon	Entire cycle	Stabler (1954)
Acantho-cephala	*Moniliformis dubius*	Shelled acanthor [= egg] (Fig. 7)	*Periplaneta americana*	Cystacanths	Edmonds (1966); Lackie (1972)
	Polymorphus minutus	Cystacanth in *Gammarus* spp. (Fig. 19)	Domestic duck	Adults	Lingard and Crompton (1972); Lackie (1974)
Cestoda	*Echinococcus granulosus*	Protoscolex from hydatid cyst in sheep (Fig. 16)	Domestic dog	Adults	Smyth et al. (1970)
	Hymenolepis diminuta	Shelled oncosphere [= egg] (Fig. 13)	Beetles (*Tribolium* spp. and *Tenebrio molitor*)	Cysticercoids	Lethbridge (1971); Lethbridge and Gijsbers (1974)
	H. microstoma	Cysticercoid in *Tribolium* spp. (Fig. 15)	Laboratory mouse	Adults	Caley (1974)
	Spirometra mansonoides	Procercoid in *Cyclops vernalis*	Laboratory rodents	Plerocercoids [= spargana]	Mueller (1965)
	Taenia saginata	Cysticercus in ox	Man	Adults	Lapage (1968)

TABLE 2 (continued)

	Parasite	Infective stage	Susceptible host	Resulting stages	References
Nematoda	*Ascaris lumbricoides* var. *suis*	Shelled L_2 [= egg] (Fig. 20)	Domestic pigs	Adults	Fairbairn (1961); Rogers (1966)
	Dirofilaria aethiops	L_1 [= unsheathed MF] (Fig. 22)	*Aedes aegypti*	L_3	Webber (1955)
	Dioctophyma renale	Larva in *Ameiurus melas* (fish)	*Mustela vison*	Adults	Hallberg (1953)
	Haemonchus contortus	Ensheathed L_3 (Fig. 24)	Domestic sheep	Adults	Rogers (1966)
	Oxyspirura mansoni	L_3 in *Pycnoscelus surinamensis*	Domestic fowl	Adults	Schwabe (1951)
	Uncinaria lucasi	L_3 in maternal milk	*Callorhinus ursinus* (pups)	Adults	Olsen (1967)
Trematoda	*Haematoloechus medioplexus*	Shelled miracidium [= egg] (see Fig. 8)	*Planorbula armigera*	Cercariae	Krull (1931)
	H. medioplexus	Metacercaria in *Sympetrum* spp. (see Fig. 10)	*Rana pipiens*	Adults	Krull (1931)
	Fasciola hepatica	Metacercaria on vegetation (Fig. 12)	Domestic sheep	Adults	Dawes (1963); Dixon (1965, 1966)

CUTANEOUS ROUTE

	Parasite	Infective stage	Susceptible host	Resulting stages	References
Protozoa	*Trypanosoma brucei* group	Metacyclic tryp. in feeding apparatus	Man, cattle and laboratory rodents	Bloodstream forms	Gordon et al. (1956); Vickerman (1971)
	Trypanosoma (*Schizotrypanum*) *cruzi*	Metacyclic tryp. in rectum of reduviid bugs	Man and reservoir hosts	Extra- and intracellular forms	Weinman (1968)
	Plasmodium spp.	Sporozoite in salivary gland of ♀ mosquitoes	Man	Gametes	Levine (1973)

TABLE 2 (continued)

Parasite	Infective stage	Susceptible host	Resulting stages	References
Nematoda				
Angiostrongylus cantonensis	L_1	Numerous slugs and snails	L_3	Mackerras and Sandars (1955); Alicata and Jindrak (1970)
Nippostrongylus brasiliensis	Sheathed L_3 (see Fig. 23)	Laboratory rats	Adults	Gharib (1955); Twohy (1956); Stirewalt (1966)
Trematoda				
Fasciola hepatica	Miracidium (see Fig. 8)	*Lymnaea truncatula*	Cercariae	Wilson et al. (1971)
Schistosoma mansoni	Cercaria (Fig. 9)	Laboratory rodents and man	Adults	Stirewalt (1966)
DIAPLACENTAL ROUTE				
Nematoda				
Protostrongylus stilesi	? L_1 via mother	*Ovis canadensis*	? Adults	Forrester and Senger (1964)
Toxocara canis	? L_2/L_3 via mother	Domestic dogs	Adults	Webster (1958)
VENERAL ROUTE				
Protozoa				
Trichomonas vaginalis	Trichomonad	Man	Entire cycle	Jirovec and Petru (1968)
Trypanosoma equiperdum	Flagellate	Horse	Entire cycle	Hoare (1972)

TABLE 3

Some examples of emigrations and migrations undertaken by endoparasitic helminths during development in vertebrate hosts (refer to Tables 1 and 2). p.i. represents post infection; M represents migration; L_n represents the appropriate larval stage in the development of a nematode.

Parasite	Host	Observations	References
ACANTHOCEPHALA			
Echinorhynchus truttae	Salmo trutta	Immature acanthocephalans become established in anterior part of small intestine; maturity is reached as population moves into the posterior part of the intestine [several weeks].	Awachie (1966)
Moniliformis dubius [M]	Laboratory rat	Immature worms become established in posterior part of the small intestine; maturity is reached about 3 weeks later by which time the population is in the anterior half of the small intestine. Gradually the mature worms spread throughout most of the small intestine.	Burlingame and Chandler (1941); Holmes (1961); Crompton and Nesheim (unpubl. obsns)
CESTODA			
Hymenolepis diminuta [M]	Laboratory rat	Between 7 to 10 days p.i., the worms have moved anteriorly from the middle of the small intestine to the duodenum. A circadian response correlated with the host's feeding regime is also occurring during the course of the infection and the worms tend to spread posteriorly after maturity has been reached.	Mettrick and Podesta (1974)
H. microstoma	Laboratory mouse	Cysticercoids excyst in duodenum (Caley, 1974) and young worms are found in anterior small intestine. Three days later some individuals are attached to lining of bile duct and by next day p.i. most worms are in the bile duct.	De Rycke (1966)

TABLE 3 (continued)

Parasite	Host	Observations	References
Echinococcus granulosus	Rabbit and sheep	Oncospheres penetrate villi and can lyse tissues. In non-ruminants they enter the venules and tend to reach the liver via the hepatic portal system. In ruminants, the central lacteal of each villus is entered and the oncospheres tend to reach the lungs via the lymphatic system.	Heath (1971)
Raillietina cesticillus	Domestic fowl	Initially, young worms become established in duodenum; by 24 h p.i., many worms have moved about 10 cm posteriorly, to usual site of the adult cestodes.	Foster and Daugherty (1959); Gray (1972b)
Spirometra mansonoides	Laboratory mouse	Procercoids bore through intestinal wall, usually on the side of greater curvature opposite to the mesentery. After a period in the abdominal cavity, they enter the body wall and occupy the musculature or subcutaneous fascia of the pelvic area and thigh muscles.	Mueller (1965)
Taenia pisiformis	Laboratory and wild rabbits	Oncospheres penetrate tissue of the small intestine and travel via the venous system to the site where the cysticercus develops. Young rabbits are more susceptible than older rabbits and the anterior small intestine is penetrated more readily than the posterior.	Heath (1971)
NEMATODA *Angiostrongylus cantonensis*	Laboratory rat	Upon ingestion, L_3 penetrate gut wall and within 1 h majority enter venules and reach liver via hepatic portal system: travel to right artrium of heart via post. vena cava and thence to lungs via pulmonary artery and back to main arterial circulation from heart. Larvae now dispersed in body but especially in CNS where L_3 moult to L_4, 5 days p.i.; final moult occurs 5 to 6 days later. Young adults pass into cerebral veins, return to heart and lungs and become established in pulmonary arterial system [26—33 days p.i.].	Mackerras and Sandars (1955); Alicata and Jindrak (1970)

TABLE 3 (continued)

Parasite	Host	Observations	References
Dioctophyma renale	*Mustela vison* (natural) *M. putorius*—ferret—(exptl.)	Larvae believed to penetrate duodenal wall and cross abdominal cavity at point where duodenum is situated very near to the curvature of the hilus of the right kidney.	Hallberg (1953); Fyvie (1971)
Dochmoides stenocephala	Domestic dog	Larvae invade mucosa of stomach and duodenum which they abandon after about 48 h. By about 10 days p.i., most worms are established in posterior half of small intestine.	Gibbs (1961)
Dracunculus medinensis	Domestic dog	After oral ingestion in crustacean intermediate host, L_3 penetrate intestinal wall by about 13 h, cross the body cavity and reach muscles of abdominal and thoracic wall by about 15 days. They then move into the sub-cutaneous connective tissue.	Muller (1968, 1971)
Filaroides martis	*Mustela vison*	L_3 enter gastric glands, moult to L_4 3 to 4 days later. By 11 days p.i., L_4 have moulted to young adults which pass via adventitia of gastric arteries and pancreato-duodenal artery to junction of coeliac and dorsal aorta. Worms pass through diaphragm in adventitia of dorsal aorta to base of heart. They transfer to adventitia of pulmonary arteries from which they invade parenchyma of lungs and peribronchial connective tissue [15 days p.i.].	Stockdale and Anderson (1970)
Heterakis gallinarum	Domestic fowl	Eggs hatch in anterior small intestine and within 24 to 48 h p.i. the larvae can be recovered from the caecal contents. Further development may occur in caecal tissues before the adults become established in the caecal lumen.	Uribe (1922); Clapham (1933)
Kalicephalus inermis coronellae	*Coluber constrictor*	After infection of host, L_3 and L_4 emigrate from duodenum and stomach to the anterior part of the oesophagus.	Schad (1956; 1962)

TABLE 3 (continued)

Parasite	Host	Observations	References
Nippostrongylus brasiliensis	Laboratory rat	L_3 reach the lungs about 15 h after penetrating the skin. They moult and by about 59 h p.i. most L_4 have reached the small intestine via the respiratory system, oesophagus and stomach. The final moult is completed by about 120 h p.i.	Haley (1962)
Oxyspirura mansoni	Domestic fowl	L_3 released from intermediate hosts in the crop, crawl up the oesophagus to the roof of the mouth and invade the eyes by moving up the naso-lacrimal ducts [some minutes after release from int. host].	Schwabe (1951)
Spirocerca lupi	Domestic dog	L_3 penetrate mucosa of stomach, bore into walls of gastric arteries and move on into wall of aorta in ca. 3 weeks. After $2\frac{1}{2}$ to 3 months in wall of aorta worms have become young adults, which emerge from aorta, penetrate through the oesophageal wall into the lumen and then move back into the wall [102 to 124 days p.i.].	Bailey (1972)
Toxocara canis	Domestic dog (pups)	L_2 penetrate intestinal wall and majority reach liver having entered the portal system via the lymphatic and venous systems of intestinal wall. Some larvae reach the lungs via the heart and pulmonary artery. Some of the larvae reach the trachea and then the oesophagus, the second moult having occurred. Third moult occurs in the stomach and final moult in the duodenum [c. 13 days p.i.].	Webster (1958)
DIGENETIC TREMATODA *Clonorchis sinensis*	Rabbit	Within 6 h of ingestion of metacercariae, young flukes can be recovered from bile duct, presumably moving to the intrahepatic biliary vessels. Some evidence indicates that the intrahepatic ducts can be colonized even after ligation of the bile duct before the metacercariae are swallowed.	Wykoff and Lepes (1957); (see Komiya, 1966)

TABLE 3 (continued)

Parasite	Host	Observations	References
Dicrocoelium dendriticum	Hamster	Immature flukes recovered from biliary system within 1 h of ingestion of metacercariae; immature flukes enter bile duct of in vitro preparations.	Krull (1958)
Diplodiscus temperatus [M]	*Rana pipiens*	After excystation, flukes become established in the rectum of young tadpoles and adult frogs. When the host intestine shortens during metamorphosis, flukes moved from the rectum to the stomach. On completion of metamorphosis, the flukes return to the rectum of the adult frog and become mature.	Herber (1939)
Fasciola hepatica	Laboratory mouse	After excystation in the small intestine, young flukes tunnel through the intestinal wall and pass through the body cavity; they penetrate into the liver tissue where they reside for some time before becoming mature in the biliary vessels.	Dawes (1963); (see Dawes and Hughes, (1964)
Gorgodera amplicava	*Rana palustris*	Metacercariae excyst in the small intestine and the young flukes crawl along the mucosa of the small and large intestine and enter the reproductive and excretory ducts [8 h p.i.].	Goodchild (1948)
Haematolechus medio-plexus	*Rana pipiens*	Young flukes travel from the stomach, up the oeso-phagus to the larynx and so to the lungs. They feed on blood during the journey [24 h p.i.].	Krull (1931)
Halipegus eccentricus	*Rana pipiens*	Metacercariae are ingested by tadpoles and young flukes remain, with little development, in cardiac end of stomach. As metamorphosis occurs, flukes move into and up the oesophagus and thence into the Eustachian tubes.	Thomas (1939)
Paragonimus westermani	Laboratory rat	Many young flukes in peritoneal cavity 30 min p.i., having passed through wall of small intestine assisted by collagenase and proteolytic secretions. Many reach lungs after entering the pleural cavity by penetrating the diaphragm.	Yokogawa (1965)

TABLE 3 (continued)

Parasite	Host	Observations	References
Renifer aniarum	*Natrix sipedon*	Metacercariae excyst in duodenum where young flukes remain for several days before passing through the stomach and up the oesophagus to the mouth cavity [28 days p.i.].	Byrd (1935)
Schistosoma mansoni	Laboratory mouse	Cercarial passage through the skin is usually completed within 24 h. Schistosomulae reach the right heart and lungs via the lymphatic and venous vessels. The liver is invaded, probably via the blood system, and the mature flukes move from the liver into the vessels of the portal system.	Jordan and Webbe (1969)

56

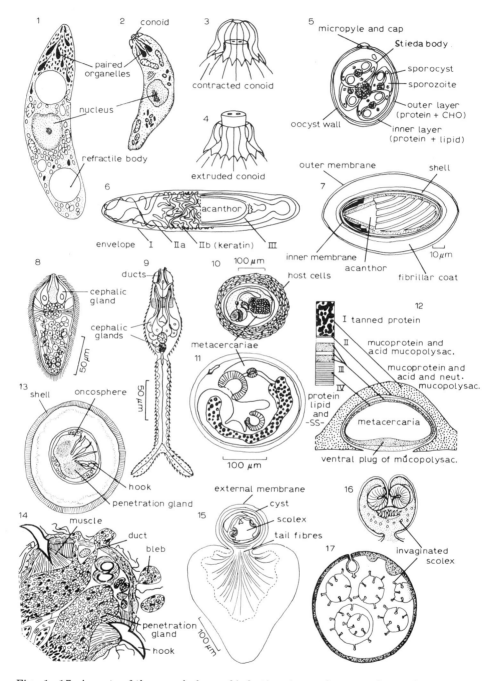

Figs. 1—17. Aspects of the morphology of infective stages of some endoparasites.
(1) Representation of the structure of a sporozoite of *Eimeria* spp. (after Scholtyseck, 1973, Fig. 9). (2) Representation of the structure of a merozoite of *Eimeria tenella*. (3) Contracted conoid apparatus of *E. tenella*. (4) Extruded conoid apparatus of *E. tenella* (Figs. 2, 3 and 4 after McLaren and Paget, 1968, Fig. 1a, c and d). (5) Sporulated oocyst

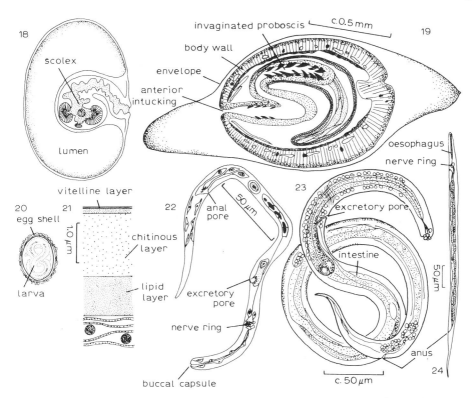

Figs. 18—24. Aspects of the morphology of infective stages of some endoparasites.
(18) Cysticercus of *Taenia* spp. (after Manson and Shipley, 1907, Fig. 117). (19) Cystacanth of *Polymorphus minutus* (after Lackie, 1975, Fig. 6). (20) Infective (embryonated egg) of *Ascaris lumbricoides* (after Christenson et al., 1940, Fig. 137 H). (21) Details of the structure of the egg shell of *Ascaris* (after Bird, 1971, Fig. 76 C). (22) Representation of the morphology of a microfilarial nematode (after McLaren, 1972, Fig. 1). (23) Infective third-stage larva of *Ancylostoma* spp. (after Nichols, 1956, plate VII). (24) Infective third-stage larve of *Haemonchus contortus* (after Dickmans and Andrews, 1933, plate 1, Fig. 4).

---

of *Eimeria* spp. (after Levine, 1973, Fig. 8.2). (6) Representation of the shelled acanthor of *Polymorphus minutus* (after Whitfield, 1973, Fig. 1). (7) Representation of the shelled acanthor of *Moniliformis dubius* (after Whitfield, 1971, Figs. 1 and 2). (8) Miracidium of *Schistosoma* spp. (9) Cercaria of *Schistosoma* spp. (Figs. 8 and 9 after Brown and Belding, 1964). (10) Encapsulated metacercaria from the haemocoele of an insect (after Salt, 1963, Fig. 2c). (11) Metacercaria originating from *Cercaria helvetica* xxxiii (after Pike, 1967, Fig. 5). (12) Details of the structure of the metacercarial cyst of *Fasciola hepatica* (after Dixon, 1965, Fig. 1A and B). (13) Shelled oncosphere of *Hymenolepis* spp. (after Lackie, 1975, Fig. 3). (14) Representation of the events occurring during secretion by the penetration glands of oncospheres of *Hymenolepis diminuta* (after Lethbridge and Gijsbers, 1974, Fig. 3). (15) Cysticercoid of *Hymenolepis microstoma* (after Caley, 1974, Fig. 3). (16) Invaginated scolex from a hydatid cyst of *Echinococcus granulosus* (after Faust et al., 1970, Fig. 31—32A). (17) Representation of a hydatid cyst of *E. granulosus* (after Wardle and McLeod, 1952, Fig. 49E).

finding is not within the scope of this chapter, it should be remembered that the events occurring during host finding may prime or stimulate the parasite for host entry in an irreversible manner. A developmental stage, which is adapted for host finding and entry, may also be adapted to withstand adverse environmental conditions until contact is made with a susceptible host. The morphology of a selection of infective stages is shown in Figs. 1 to 24 and an attempt has been made to illustrate something of the protective and infective features which have evolved.

The infective agents of most protozoan and helminth endoparasites enter their hosts by one of four routes (Table 2). There exists a multitude of interesting observations on routes of entry involving many more species of parasite than the representative types cited in the table. For example, there is evidence that man may become infected with *Enterobius vermicularis* either by inhaling the eggs (Faust et al., 1970) or by a process of retrofection in which eggs hatch on the perianal skin to release larvae capable of crawling back into the host's large intestine (Schüffner and Swellengrebel, 1949). In the case of the protozoon *Toxoplasma gondii*, the host may swallow the parasite after faecal contamination of food, through carnivorism and during suckling (Frenkel, 1973). The possibility also exists that the entry of *T. gondii* may be achieved by venereal and diaplacental transmission, by means of helminth eggs and through blood transfusions or even organ transplants; the evidence for some of these modes of entry is indirect and equivocal (Frenkel, 1973). Other aspects of infection processes involving protozoa have been discussed recently by Newton (1972).

3.2.1. Oral infection

A general conclusion from the results of many studies on infective stages of the type which are usually swallowed by the appropriate host (Figs. 5—7, 10—22,24) is that a combination of environmental factors and parasite responses are involved in the activation and release of the parasite from its enclosing tissues (see Lackie, 1975). For example, infective eggs of *Moniliformis dubius* (Fig. 7), from the body cavity of mature female worms, are stimulated to hatch in vitro provided that the molarity and pH of the incubation medium and the carbon dioxide concentration of the gas phase are set within limits similar to the conditions believed to exist in the gut of *Periplaneta americana* (Edmonds, 1966). The acanthors respond to this stimulation with muscular activity and the secretion of chitinase. In birds and mammals, the excystation of sporulated oocysts of *Eimeria* spp. (Fig. 5) consists of the release of the four sporocysts from each oocyst and the two sporozoites from each sporocyst. Mechanical grinding (Goodrich, 1944) or the presence of carbon dioxide (Hibbert and Hammond, 1968) or a combination of both factors exposes the sporocysts to the effects of bile and pancreatic secretions (Doran and Farr, 1962; Doran, 1966) and results in the emergence of the sporozoites from the surrounding debris (see Ryley, 1972, 1973; Marquardt, 1973; Crompton and Nesheim, 1976).

The eggs of ascaridid nematodes (Fig. 20) and the ensheathed third-stage larvae of the nematode *Haemonchus contortus* (Fig. 24) are stimulated to exsheath in vitro, and probably in the tracts of their hosts also, by exposure to certain concentrations of carbon dioxide and undissociated carbonic acid under reducing conditions at a particular temperature and pH (Fairbairn, 1961; Rogers, 1960, 1966; Rogers and Sommerville, 1963). The activation of the eggs and the metacercariae and larval stages of platyhelminths which become established in vertebrates (Table 2) also requires environmental stimuli in the form of digestive secretions (Smyth, 1969), gastric juice and bile often being important (Read, 1970). Bile salts appeared to enhance the activation and evagination of protoscoleces of *Echinococcus granulosus* (Fig. 16); bile collected from unfavourable hosts of *E. granulosus* damaged the cestode's surface and led Smyth to suggest that bile may be one of the factors determining host specificity.

Several features associated with the oral route of entry into the host are illustrated by studies on the infection of ducks with cystacanths of *Polymorphus minutus* (Fig. 19) contained in the haemocoeles of *Gammarus pulex*. Cystacanths were observed to be released from the tissues of *G. pulex* in the portion of the tract of the duck anterior to the pylorus (Lingard and Crompton, 1972). Occasionally damaged cystacanths were found in the ventriculus, but there was no sign of activation or proboscis eversion until the parasite had passed the pylorus. Cystacanths were found with everted proboscides after introduction directly into the small intestine by surgical techniques (Lingard and Crompton, 1972).

Lackie (1974) demonstrated that activation occurred in vitro at a temperature of 42—44°C, in a balanced salt solution of suitable osmotic pressure provided that the pH was not lowered below 5. Activation was observed to be markedly enhanced when cystacanths were incubated for a few minutes in diluted duck bile, bile salts or chromatographically pure sodium taurocholate (Lackie, 1974). The bile salt was not essential for activation, but without its enhancing effect the parasites would probably have been propelled by the host's intestinal motility beyond their normal site in the small intestine (Crompton and Whitfield, 1968) before their proboscides had become functional. Similarly, if activation occurred below pH 5.0, in the normal hydrogen concentration found in the host's proventriculus and ventriculus (Farner, 1942), evagination might have taken place and many proboscides might have been damaged by ventricular grinding.

3.2.2. Cutaneous infection

Details of the morphology and penetrative apparatus of the larval stages of helminths which enter their hosts by active passage through the integument are shown in Figs. 8, 9 and 23.

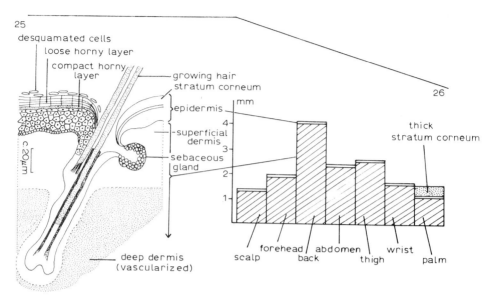

Figs. 25 and 26. The structure of mammalian skin.

(25) Representation of mammalian epidermis (after Spearman, 1973, p. 115). (26) Variation in the thickness of the layers of the skin from different regions of the human body (after Rushmer et al., 1966, Fig. 1C).

The penetration of skin by cercariae of schistosomes and third-stage larvae of hookworms has been investigated in detail using the skin of man and various rodents. It is difficult, however, to make generalizations about the details of penetration because the structure (Fig. 25), properties and physiology of the skin of mammals vary between species, with the age and physiological status of the host and with its location on the body (Fig. 26) (Rushmer et al., 1966). In man, whose evolution has resulted in hair loss, the stratum corneum has increased and that of the palms and soles is about 600 μm thick (Rushmer et al., 1966). Presumably the soles of the feet represent a region of vulnerability to penetration by hookworm larvae.

Cercariae of S. mansoni seem to explore the skin surface for entry sites which vary according to the topography of the skin in question (Stirewalt and Hackey, 1956; Stirewalt, 1966, 1971). Wrinkle crevices and hair projections appear to be the most favoured sites for entry and skin lipids are considered to provide an important stimulus for the parasite (Stirewalt, 1971). Next, the cercariae secrete adhesive material from their post-acetabular glands and become anchored perpendicular to the surface of the skin. By the combined effects of muscular activity and further secretions, the stratum corneum is pierced (Fig. 25). The cercariae now lose their tails and become known as schistosomulae which release further histolytic material after a resting period. These secretions are assumed to soften the deeper skin tissues

(Fig. 25) and the schistosomulae usually move parallel to the basement membrane until they enter a sebaceous gland which forms a natural pathway into the dermis. The electron micrographs of Rifkin (1971) illustrate the intimacy of the schistosomula-host skin interaction. On the basis of histochemical observations, Lewert and Lee (1954) concluded that major but localized alterations occurred in skin glycoproteins and that extra-cellular fluid increased during penetration of rodent skin by schistosomes. The schistosomulae appear to be able to move efficiently in the dermis of susceptible hosts and speeds varying from about 20 to 50 μm per min have been estimated for *S. mansoni* in the skin of 5-day-old rats (Stirewalt and Hackey, 1956). The evidence shows that cercariae can secrete substances which not only alter tissues in general but also hyaluronic acid, streptococcal capsules, heparin, mucopolysaccharide, collagen, elastin, polypeptides, casein and haemoglobin (Lewert and Lee, 1954; Stirewalt, 1963, 1966). The secretions vary between closely related species and may be considered as adaptations to the structural diversity found in the cutaneous barriers of different hosts (see Stirewalt, 1963).

The penetration of mouse skin by third-stage larvae of *Nippostrongylus brasiliensis* can occur within 5 min and the larvae appear to select their point of entry (Gharib, 1955). Twohy (1956) noticed that larvae of *N. brasiliensis* often did not burrow into the skin of rats for at least an hour and that fewer worms were retrieved from rats which had been entered naturally than from those which received a sub-cutaneous injection of infective larvae. Lewert and Lee (1954) found that skin penetration by *N. brasiliensis* was not accompanied by the degradation of host glycoproteins although there was an accumulation of extracellular fluid. In contrast to the cercariae of *S. mansoni*, the larvae of *N. brasiliensis* did not appear to secrete collagenase. Lee (1972) acquired some evidence for the breakdown of collagen in rodent skin during the entry of *N. brasiliensis*, but he concluded that collagenase of parasitic origin was not necessarily involved. Both Lee (1972) and Matthews (1972), who studied the penetration of cat skin by larvae of *Ancylostoma tubaeforme*, emphasized the role of surface water in skin penetration and their observations indicate that mechanical rather than enzymatic processes are important in skin penetration by hookworm larvae.

The integument of those regions of a host's body which are exposed in the environments of the infective stages or their vectors are likely to be the common points of entry of skin-penetrating parasites. For example, the miracidia of the blood fluke *Spirorchis parvus* attacked the skin around the tentacles of gastropod snails of the genus *Helisoma*, while the cercariae tended to penetrate the thin membranes at the margins of the eyes, nose and mouth of the turtle, *Chrysemys picta* (Wall, 1941). Information about places of entry is of interest because these regions mark the beginning of many of the complex emigrations undertaken by parasites (Table 3).

Emigrations do not always start beneath the skin. Many of the larval parasites which are released in the alimentary tract during oral entry (Table 2), including acanthors (Figs. 6 and 7), oncospheres (Figs. 13 and 14; Heath, 1971; Lethbridge and Gijsbers, 1974), procercoids (Mueller, 1965) and various nematode (see Otto, 1966) and trematode larvae (see Dawes and Hughes, 1964), penetrate the intestinal walls of invertebrate and vertebrate hosts. Dawes (1963) concluded that the passage of young flukes of *Fasciola hepatica*, which had been freed from metacercariae (Fig. 12) in the small intestine of mice, involved dissolution of the epithelium, connective tissue and muscle layers of the intestinal wall. The parasites fed on the breakdown products as they burrowed into the abdominal cavity. It is not known, however, how the young *F. hepatica* are directed to leave the intestinal tissues for the liver (Dawes and Hughes, 1964); intestinal tissue ought to be as nutritious as hepatic or biliary material.

3.2.3. Diaplacental infection

The infection of mammals in utero during gestation is of interest because it is usually achieved by larval stages which must have emigrated in the body of the mother. Some idea of the emigratory route and the time involved for *Toxocara canis* to become established in the small intestine of pups under 3 months of age is given in Table 3. When the hosts are older, the moulting sequence of the worms is disrupted, they become dispersed in other somatic tissues and do not reach maturity in their normal site. The larvae remain dormant in the tissues but, if the host is a pregnant bitch, some aspect of pregnancy activates them and they pass through the placenta into the foetal dog. The foetus may also acquire *T. canis* if infective eggs are ingested by the bitch (Webster, 1958).

3.3. Emigration during development and site selection

Observations on several emigratory routes travelled over varying periods of time by a selection of endoparasitic helminths in vertebrate hosts are summarized in Table 3 and the references cited should be consulted for full details and qualifications. Emigrations appear to occur more commonly before maturity has been attained. It is difficult to consider emigrations without discussion of site finding or site selection or to explain how parasites complete and survive their journeys during exposure to the various forms of host resistance.

Studies on parasitic emigrations in the mammalian gastrointestinal canal have indicated that the helminth's changing metabolic requirements may activate responses to the prevailing physico-chemical or nutritive gradients so that a change of location occurs (Mettrick and Podesta, 1974). Most of this work concerns *Hymenolepis diminuta* and it may be argued that this parasite migrates in the small intestine of the rat (Table 3; Cannon and Mettrick, 1970) in addition to the circadian migrations which appear to be a response to the host's feeding routine (Read and Kilejian, 1969; Mettrick and Podesta, 1974). The idea, however, that the emigrations of other intestinal parasites may be associated with environmental gradients in the lumen is attractive. Similar hypotheses could be advanced to account for the emigratory activity of parasites in and between the other habitats of the body. The surgical transplantation of larval nematodes (Table 3) into abnormal sites within suitable hosts might indicate whether tissues must be encountered in a certain sequence and whether a particular tissue provides optimal conditions for a vital process or satisfies nutritional requirements. Dawes (1963) has already linked the feeding proclivities and emigrations of immature *F. hepatica* (Table 3). By analogy with ecological studies of free-living animals (Andrewartha and Birch, 1954), observations on the diets of endoparasites and how they obtain an adequate nutrient supply appropriate to their stage of development would be expected to contribute to our understanding of their distribution. Schad (1963) has demonstrated how observations on the diet and feeding activities of helminths can help to explain their ecology.

By further analogy with observations on free-living animals and on man in particular, it appears that the nature of the terrain usually affects the route and logistics of a journey. This obvious conclusion may prove helpful when Table 3 is examined or attempts are made to investigate emigrations. Holmes (1973) has discussed observations which may be interpreted to show how a host's internal morphology affects emigrations and site selection. In a sample of several hundred wild mink, *Mustela vison*, infected with *Dioctophyma renale* (Table 3), 86% of the mink had parasites in the right kidney only (Fyvie, 1971), presumably because the portion of the duodenum through which the worms pass on leaving the alimentary tract is closer to the right kidney than to the left in mink (Hallberg, 1953). Miller (1952) considered that more plerocercoids of *Triaenophorus crassus* were located in the muscles on the right side of ciscoes, *Leucichthys* spp., because the stomach lies against the right body wall of the fish. Heath (1971) suggested that the frequent occurrence of cysts of *Echinococcus granulosus* (Table 3) in the lungs of ruminants and the livers of non-ruminants is connected with the diameter of the lacteals of the villi of the hosts. The relatively large oncosphere of *E. granulosus* can enter the relatively large lacteals of the ruminant and so reach the lungs via the lymphatic system whereas its entry into the smaller lacteals in a non-ruminant is less likely and the portal route to the liver is to be expected. With regard to site selection, it is conceivable that a host's mucosal morphology, which Williams (1960) has suggested may have a role in determining host specificity, may also be involved in the site specificity of intestinal parasites.

In addition to internal morphology, the species, age, sex and physiological status of the host are factors which appear to influence site selection, emigrations and migrations by endoparasites. The cestode *Schistocephalus solidus* occupies different sites in the small intestines of domestic ducks, fowls and pigeons (McCaig and Hopkins, 1963). The digenetic trematode *Nudacotyle novicia* lives in the bile duct of meadow mice, *Microtus pennsylvanicus*, and the duodenum of musk rats, *Ondatra zibethica* (Ameel, 1944). Metacercariae of *Paragonimus westermani* (Table 1) excyst in the small intestine of dogs, cats and rats, but the routes and times taken for the emigrations of the young flukes to the lungs (Table 3) differ in the different hosts (Yokogawa, 1965). The observations of Webster (1958) on *T. canis* (Tables 2 and 3) in dogs show how emigrations differ according to the age, sex and physiological status of the host. The larvae of *T. canis* followed a tracheal route back to the small intestine in young dogs and a somatic route to the muscles in older dogs; about 30% more larvae were recovered from the muscles of female than from male dogs (Webster, 1958). The diaplacental infection of foetal dogs by larvae of *T. canis* in pregnant bitches illustrates how an emigration may be influenced by the physiological status of the host. In young tadpoles and adults of *Rana pipiens*, *Diplodiscus temperatus* (Table 1) is found in the rectum. During the host's metamorphosis, the flukes move to the stomach where they remain until metamorphosis has been completed and the frog has adopted a carnivorous diet (Herber, 1939). These observations suggest that the changing physiological state of *R. pipiens* may initiate a true migration by *D. temperatus*.

Compelling circumstantial and indirect evidence for site selection by endoparasites is suggested by their distribution within their hosts (Table 1), by the fact that many emigrations terminate when a site has been reached (Table 3) and by their range of sensory structures and varied behavioural responses (see Ulmer, 1971; Croll, 1970, 1975). If they are to carry conviction, explanations for the arrival of *Eurytrema vulpis*, *Prosthogonimus macrorchis* and *Halipegus eccentricus* in their sites in the pancreatic ducts, oviducts and Eustachian tubes of their respective hosts (Table 1) need to rely on more than random processes or chance encounters.

Results from experimental studies suggest that the directed behaviour of some intestinal parasites is probably involved in site selection (Tables 1, 2 and 3). Seven-day-old adults of *Nippostrongylus brasiliensis* were retrieved from their normal site in the jejunum of uninfected rats after surgical transplantation into either the anterior or posterior part of the small intestine (Alphey, 1970). Adults of *Ancylostoma caninum* appeared to be able to find their normal site in the jejunum of dogs irrespective of their point of surgical introduction into the small intestine (Roche, 1966). *Hymenolepis diminuta* can also be recovered from a predicted region of the small intestine of uninfected rats after experimental manipulation and transplantation (Goodchild, 1958; Bråten and Hopkins, 1969) and *H. microstoma* moves into the bile ducts of recipient mice after having been removed from this site in donor mice (see Mettrick and Podesta, 1974). After transplantation, adult *Spiro-*

cerca lupi emigrated from the thoracic cavity of a dog to their normal site in the oesophageal wall, indicating in this case that site selection or recognition is not necessarily a property of developing parasites (Bailey, 1972). The remarkable site specificity of protozoan parasites of the genus *Eimeria* in the post-pyloric part of the alimentary tract of domestic birds is described in detail in Chapter 20.

3.4. Emigrations and migrations by mature parasites

One manifestation of certain intraspecific and interspecific reactions between mature parasites in the vertebrate alimentary tract is the extension of a parasite's site (see Crompton, 1973; Marquardt, 1973) and equivalent extensions of site may occur in other habitats in the host. Emigrations within sites or possibly resulting in changes of site by mature parasites may be associated with the release of reproductive products, the presence of other species of parasite and the immune responses of the host. Migrations with changing metabolic requirements of the parasites concerned or with major physiological changes within the host (Table 3). Examples of these migrations have already been mentioned and the reader is referred to the detailed review of Mettrick and Podesta (1974).
ferred to the detailed review of Mettrick and Podesta (1974).

The release of larval stages into water by the guinea worm, *Dracunculus medinensis*, which inhabits the subcutaneous connective tissue of man and experimental animals, usually occurs after the female worm has emerged through the skin of a lower limb of its host (Muller, 1971). If *D. medinensis* can live in any part of the host's subcutaneous connective tissue, this type of movement may be considered as an emigration within a site; how the worms find the lower limbs has not been resolved (Muller, 1971). A similar but shorter emigration may accompany the deposition of larval stages of *F. martis* (Table 1) on the bronchial epithelium of mink (Stockdale and Anderson, 1970). The egg-laying behaviour of female *Enterobius vermicularis* may also involve an emigration (Faust et al., 1970).

The trematode *Haematoloechus medioplexus* lives firmly attached to the walls of the lungs of frogs (Table 1). Nematodes of the genus *Rhabdonema* are also common in the lungs of frogs but some form of incompatibility appears to exist between *Haematoloechus* and *Rhabdonema* so that the trematodes may be confined to one lung and nematodes to the other (Krull, 1931). The locations of populations of the blood flukes *Aporocotyle macfarlani* and *Psettarium sebastodorum* changed during concurrent infections in the vascular system of the rockfish (Holmes, 1971, 1973) and populations of the cestode *Proteocephalus filicollis* and of immature forms of the acanthocephalan *Neoechinorhynchus rutili* appeared to move apart in the small intestine of the three-spined stickleback, *Gasterosteus aculeatus* (Chappell,

1969). These examples illustrate emigrations or movements associated with the presence of other parasites. Observations on populations of *Tricho-strongylus colubriformis* in guinea-pigs (Connan, 1966) and *N. brasiliensis* (Table 1) in rats of varying developmental and physiological state (Connan, 1974; Jenkins, 1974) indicate that the onset of the hosts' immune responses may initiate emigrations by the nematodes.

3.5. Conclusions

The evidence indicates that in the process of infection, the specially adapted infective stages receive and respond to activating stimuli and enhancing conditions associated with susceptible hosts. Other observations on the emigrations of developing stages of parasites within their hosts suggest that these movements should not be attributed to random processes alone. The observed distribution of parasites within their hosts is difficult to explain unless some form of directed site selection by the parasites is assumed. More experimental work now needs to be done to support or refute these opinions. Until the results of these studies are available, it is considered to be unwise and premature to compare the emigrations, migrations and site-selection activities of endoparasites with those of free-living organisms.

Acknowledgements

I am grateful to the authors who kindly allowed me to use their figures in this chapter. I also thank the following copyright holders: Academic Press (Fig. 21); American Association for the Advancement of Science (Fig. 26); American Microscopical Society (Fig. 24); American Society of Parasitologists (Fig. 23); Cambridge University Press (Figs. 2, 3, 4, 6, 7, 10, 11, 12, 13, 14, 15, 19, 22, 25); Lea and Febiger Publishers (Fig. 16); Meredith Publishing Division (Figs. 8, 9); Society of Protozoologists (Fig. 1); University of Minnesota (Fig. 17); University Park Press, Baltimore (Fig. 20).

3.6. REFERENCES

Alicata, J.E. and Jindrack, K. (1970). "Angiostrongylosis in the Pacific and Southeast Asia". C.C. Thomas, Springfield, Ill.

Allee, W.C., Emerson, A.E., Park, O., Park, T. and Schmidt, K.P. (1949). "Principles of Animal Ecology". Saunders, Philadelphia.

Alphey, T.W. (1970). "Studies on the distribution and site location of *Nippostrongylus brasiliensis* within the small intestine of laboratory rats". Parasitology 61, 449—460.

Ameel, D.J. (1944). "The life history of *Nudacotyle novicia* Barker, 1916 (Trematoda: Notocotylidae)". J. Parasit. 30, 257—263.

Andrewartha, H.G. and Birch, L.C. (1954). "The Distribution and Abundance of Animals". University of Chicago Press.

Awachie, J.B.E. (1966). "The development and life-history of *Echinorhynchus truttae* Schrank 1788, (Acanthocephala)". J. Helminth. 40, 11—32.

Bailey, W.S. (1972). "*Spirocerca lupi:* a continuing enquiry". J. Parasit. 58, 3—22.

Beaver, P.C. (1939). "The morphology and life history of *Psilostomum ondatrae* Price, 1931 (Trematoda: Psilostomidae)". J. Parasit. 25, 383—393.

Bird, A.F. (1971). "The Structure of Nematodes". Academic Press, London.

Brambell, M.R. (1965). "The distribution of a primary infestation of *Nippostrongylus brasiliensis* in the small intestine of laboratory rats". Parasitology 55, 313—324.

Bråten, T. and Hopkins, C.A. (1969). "The migration of *Hymenolepis diminuta* in the rat's intestine during normal development and following surgical transplantation". Parasitology 59, 891—905.

Brown, H.W. and Belding, D.L. (1964). "Basic Clinical Parasitology", (2nd edition). Appleton-Century-Crofts, New York.

Burlingame, P.L. and Chandler, A.C. (1941). "Host-parasite relations of *Moniliformis dubius* (Acanthocephala) in albino rats, and the environmental nature of resistance to single and superimposed infections with this parasite". Am. J. Hyg. 33, 1—21.

Byrd, E.E. (1935). "Life history studies of Reniferinae (Trematoda, Digenea) parasitic in Reptilia of the New Orleans area". Trans Am. microsc. Soc. 54, 196—225.

Caley, J. (1974). "The functional significance of scolex retraction and subsequent cyst formation in the cysticercoid larva of *Hymenolepis microstoma*". Parasitology 68, 207—227.

Cannon, C.E. and Mettrick, D.F. (1970). "Changes in the distribution of *Hymenolepis diminuta* (Cestoda: Cyclophyllidea) within the rat intestine during prepatent development". Can. J. Zool. 48, 761—769.

Cannon, L.R.G. (1972). "Studies on the ecology of the papillose allocreadiid trematodes of the yellow perch in Algonquin Park, Ontario". Can. J. Zool. 50, 1231—1239.

Chappell, L.H. (1969). "Competitive exclusion between two intestinal parasites of the three-spined stickleback, *Gasterosteus aculeatus* L". J. Parasit. 55, 775—778.

Christenson, R.O., Jacobs, L. and Wallace, F.G. (1940). In: "An Introduction to Nematology." (B.G. and M.B. Chitwood, eds.) pp. 174—189. M.B. Chitwood, Publisher, Babylon, New York.

Clapham, P.A. (1933). "On the life-history of *Heterakis gallinae*". J. Helminth. 11, 67—86.

Connan, R.M. (1966). "Experiments with *Trichostrongylus colubriformis* (Giles, 1892) in the guinea-pig. 1. The effect of the host response on the distribution of the parasites in the gut". Parasitology 56, 521—530.

Connan, R.M. (1974). "The distribution of *Nippostrongylus brasiliensis* in prolonged infections in lactating and neonatally infected rats". Parasitology 68, 347—354.

Cort, W.W. (1915). "North American lung flukes". Trans. Am. microsc. Soc. 34, 203—240.

Croll, N.A. (1970). "The Behaviour of Nematodes". Edward Arnold (Publishers), London.

Croll, N.A. (1975). In: "Advances in Parasitology, 13". (B. Dawes, ed.) pp. 71—122. Academic Press, London.

Crompton, D.W.T. (1973). "The sites occupied by some parasitic helminths in the alimentary tract of vertebrates". Biol. Rev. 48, 27—83.

Crompton, D.W.T. and Nesheim, M.C. (1976). In: "Advances in Parasitology, 14". (B. Dawes, ed.) pp. 95—194. Academic Press, London.

68

Crompton, D.W.T. and Whitfield, P.J. (1968). "The course of infection and egg production of *Polymorphus minutus* (Acanthocephala) in domestic ducks". Parasitology 58, 231—246.

Dawes, B. (1963). "The migration of juvenile forms of *Fasciola hepatica* L. through the wall of the intestines in the mouse, with observations on food and feeding". Parasitology 53, 109—122.

Dawes, B. and Hughes, D.L. (1964). In: "Advances in Parasitology, 2". (B. Dawes, ed.) pp. 97—168. Academic Press, London.

De Rycke, P.H. (1966). "Development of the cestode *Hymenolepis microstoma* in *Mus musculus*". Z. Parasitkde. 27, 350—354.

Dickmans, G. and Andrews, J.S. (1933). "A comparative morphological study of the infective larvae of the common nematodes parasitic in the alimentary tract of sheep". Trans Am. microsc. Soc. 52, 1—25.

Dixon, K.E. (1965). "The structure and histochemistry of the cyst wall of the metacercaria of *Fasciola hepatica* L". Parasitology 55, 215—226.

Dixon, K.E. (1966). "The physiology of excystment of the metacercaria of *Fasciola hepatica* L". Parasitology 56, 431—456.

Dobrovolny, C.G. (1939). "Life history of *Plagioporus sinitsini* Mueller and embryology of new cotylocercous cercariae (Trematoda)". Trans. Am. microsc. Sci. 58, 121—155.

Doran, D.J. (1966). "Location and time of penetration of duodenal epithelial cells by *Eimeria acervulina* sporozoites". Proc. helminth. Soc. Wash. 33, 43—46.

Doran, D.J. and Farr, M.M. (1962). "Excystation of the poultry coccidium, *Eimeria acervulina*". J. Protozool. 9, 154—161.

Dvorak, J.A., Jones, A.W. and Kuhlman, H.H. (1961). "Studies on the biology of *Hymenolepis microstoma* (Dujardin, 1845)". J. Parasit. 47, 833—838.

Edmonds, S.J. (1966). "Hatching of eggs of *Moniliformis dubius*". Expl. Parasit. 19, 216—226.

Fahmy, M.A.M. (1954). "An investigation of the life cycle of *Trichuris muris*". Parasitology 44, 50—57.

Fairbairn, D. (1961). "The in vitro hatching of *Ascaris lumbricoides* eggs". Can. J. Zool. 39, 153—162.

Farner, D.S. (1942). "The hydrogen ion concentration in avian digestive tracts". Poult. Sci. 21, 445—450.

Faust, E.C., Russell, P.F. and Jung, R.C. (1970). "Clinical Parasitology", (8th edition). Lea and Febiger, Philadelphia.

Forrester, D.J. (1971). In: "Parasitic Diseases of Wild Animals". (J.W. Davis and R.C. Anderson, eds.) pp. 158—173. The Iowa State University Press, Ames, Iowa.

Forrester, D.J. and Senger, C.M. (1964). "Prenatal infection of Bighorn Sheep with protostrongylid lungworms". Nature, Lond. 201, 1051.

Foster, W.B. and Daugherty, J.W. (1959). "Establishment and distribution of *Raillietina cesticillus* in the fowl and comparative studies on amino acid metabolism of *R. cesticillus* and *Hymenolepis diminuta*". Expl. Parasit. 8, 413—426.

Frenkel, J.K. (1973). In: "The Coccidia". (D.M. Hammond with P.L. Long, eds.) pp. 343—410. University Park Press, Baltimore.

Fyvie, A. (1971). In: "Parasitic Diseases of Wild Animals". (J.W. Davis and R.C. Anderson, eds.) pp. 258—262. The Iowa State University Press, Ames, Iowa.

Gharib, H.M. (1955). "Observations on skin penetration by the infective larvae of *Nippostrongylus brasiliensis*". J. Helminth. 29, 33—36.

Gibbs, H.C. (1961). "Studies on the life cycle and developmental morphology of *Dochmoides stenocephala* (Railliet 1884) (Ancylostomidae: Nematoda)". Can. J. Zool. 39, 325—348.

Goodchild, C.G. (1948). "Additional observations on the bionomics and life history of

Gorgodera amplicava Looss, 1899 (Trematoda: Gorgoderidae)". J. Parasit. 34, 407–427.

Goodchild, C.G. (1958). "Transfaunation and repair of damage in the rat tapeworm, *Hymenolepis diminuta*". J. Parasit. 44, 345–351.

Goodrich, H.P. (1944). "Coccidian oocysts". Parasitology 36, 72–79.

Gordon, R.M., Crewe, W. and Willett, K.C. (1956). "Studies on the deposition, migration and development to the blood forms of trypanosomes belonging to the *T. brucei* group. Ann. trop. Med. Parasit. 50, 426–437.

Gray, J.S. (1972a). "The effect of host age on the course of infection of *Raillietina cesticillus* (Molin, 1858) in the fowl". Parasitology 65, 235–241.

Gray, J.S. (1972b). "Studies on the course of infection of the poultry cestode *Raillietina cesticillus* (Molin, 1858) in the definitive host". Parasitology 65, 243–250.

Haley, A.J. (1962). "Biology of the rat nematode, *Nippostrongylus brasiliensis* (Travassos, 1914). II. Preparasitic stages and development in the laboratory rat". J. Parasit. 48, 13–22.

Hallberg, C.W. (1953). "*Dioctophyma renale* (Goeze, 1782). A study of the migration routes to the kidneys of mammals and resultant pathology". Trans Am. microsc. Soc. 72, 351–363.

Hawking, F. (1975). In: "Advances in Parasitology, 13". (B. Dawes, ed.) pp. 123–182. Academic Press, London.

Hawking, F. and Webber, W.A.F. (1955). "*Dirofilaria aethiops* Webber, 1955, a filarial parasite of monkeys: 2. Maintenance in the laboratory". Parasitology 45, 378–387.

Heape, W. (1931). "Emigration, Migration and Nomadism". W. Heffer and Sons Ltd., Cambridge.

Heath, D.D. (1971). "The migration of oncospheres of *Taenia pisiformis*, *T. serialis* and *Echinococcus granulosus* within the intermediate host". Int. J. Parasit. 1, 145–152.

Herber, E.C. (1939). "Studies on the biology of the frog amphistome, *Diplodiscus temperatus* Stafford". J. Parasit. 25, 189–195.

Hibbert, L.E. and Hammond, D.M. (1968). "Effects of temperature on in vitro excystation of various *Eimeria* species". Expl. Parasit. 23, 161–170.

Hoare, C.A. (1972). "The Trypanosomes of Mammals". Blackwell, Oxford.

Holmes, J.C. (1961). "Effects of concurrent infections on *Hymenolepis diminuta* (Cestoda) and *Moniliformis dubius* (Acanthocephala). I. General effects and comparison with crowding". J. Parasit. 47, 209–216.

Holmes, J.C. (1971). "Habitat segregation in sanguinicolid blood flukes (Digenea) of scorpaenid rockfishes (Perciformes) on the Pacific Coast of North America." J. Fish. Res. Board Can. 28, 903–909.

Holmes, J.C. (1973). "Site selection by parasitic helminths: interspecific interactions, site segregation, and their importance to the development of helminth communities". Can. J. Zool. 51, 333–347.

Jenkins, D.C. (1974). "*Nippostrongylus brasiliensis:* the distribution of primary worm populations within the small intestine of neonatal rats". Parasitology 68, 339–345.

Jirovec, O. and Petru, M. (1968). In: "Advances in Parasitology, 6". (B. Dawes, ed.) pp. 117–188. Academic Press, London.

Jordan, P. and Webbe, G. (1968). "Human Schistosomiasis". Heinemann, London.

Komiya, Y. (1966). In: "Advances in Parasitology, 4". (B. Dawes, ed.) pp. 53–106. Academic Press, London.

Krull, W.H. (1931). "Life-history studies on two frog lung flukes, *Pneumonoeces medioplexus* and *Pneumobites parviplexus.*" Trans. Am. microsc. Soc. 50, 215–277.

Krull, W.H. (1958). "The migratory route of the metacercaria of *Dicrocoelium dendriticum* (Rudolphi, 1819) Looss, 1899 in the definitive host: Dicrocoeliidae". Cornell Vet. 48, 17–24.

Lackie, A.M. (1974). "The activation of cystacanths of *Polymorphus minutus* (Acanthocephala) in vitro". Parasitology 68, 135—146.

Lackie, A.M. (1975). "The activation of infective stages of endoparasites of vertebrates". Biol. Rev. 50, 285—323.

Lackie, J.M. (1972). "The course of infection and growth of *Moniliformis dubius* (Acanthocephala) in the intermediate host *Periplaneta americana*". Parasitology 64, 95—106.

Lapage, G. (1968). "Veterinary Parasitology", (2nd Edn.). Oliver and Boyd, Edinburgh.

Lee, D.L. (1972). "Penetration of mammalian skin by the infective larva of *Nippostrongylus brasiliensis*". Parasitology 65, 499—505.

Lethbridge, R.C. (1971). "The hatching of *Hymenolepis diminuta* eggs and penetration of the hexacanths in *Tenebrio molitor* beetles". Parasitology 62, 445—456.

Lethbridge, R.C. and Gijsbers, M.F. (1974). "Penetration gland secretion by hexacanths of *Hymenolepis diminuta*". Parasitology 68, 303—311.

Levine, N.D. (1973). "Protozoan Parasites of Domestic Animals and of Man", (2nd Edn.). Burgess, Minneapolis.

Levine, P.P. (1942). "Excystation of coccidial oocysts of the chicken". J. Parasit. 28, 426—428.

Lewert, R.M. and Lee, C.-C. (1954). "Studies on the passage of helminth larvae through host tissues. I. Histochemical studies on the extracellular changes caused by penetrating larvae. II. Enzymatic activity of larvae in vitro and in vivo". J. infect. Dis. 95, 13—51.

Lindsey, J.R. (1962). "Diagnosis of filarial infections in dogs. II. Confirmation of microfilarial indentifications". J. Parasit. 48, 321—326.

Lingard, A.M. and Crompton, D.W.T. (1972). "Observations on the establishment of *Polymorphus minutus* (Acanthocephala) in the intestines of domestic ducks". Parasitology 65, 159—165.

Mackerras, M.J. and Sandars, D.F. (1955). "The life history of the rat lung-worm, *Angiostrongylus cantonensis* (Chen) (Nematoda: Metastrongylidae)". Aust. J. Zool. 3, 1—21.

Macy, R.W. (1934). "Studies on the taxonomy, morphology, and biology of *Prosthogonimus macrorchis* Macy, a common oviduct fluke of domestic fowls in North America". University of Minnesota Agricultural Experiment Station. Technical Bulletin 98, 1—71.

Macy, R.W., Berntzen, A.W. and Benz, M. (1968). "In vitro excystation of *Sphaeridiotrema globulus* metacercariae, structure of cyst and the relationship to host specificity". J. Parasit. 54, 28—38.

Makidono, J. (1956). "Observations on *Ascaris* during fluoroscopy". Am. J. trop. Med. Hyg. 5, 699—702.

Manson, P. and Shipley, A.E. (1907). In: "System of Medicine, volume 2, Part II." (T.C. Allbutt and H.D. Rolleston, eds.) pp. 829—863. Macmillan, London.

Manson-Bahr, P.H. (1966). "Manson's Tropical Diseases", (16th Edn.). Baillière, Tindall and Cassell, London.

Marquardt, W.C. (1973). In: "The Coccidia". (D.M. Hammond with P.L. Long, eds.) pp. 23—43. University Park Press, Baltimore.

Matthews, B.E. (1972). "Invasion of skin by larvae of the cat hookworm, *Ancylostoma tubaeforme*". Parasitology 65, 457—467.

McCaig, M.L.O. and Hopkins, C.A. (1963). "Studies on *Schistocephalus solidus*. II. Establishment and longevity in the definitive host". Expl Parasit. 13, 273—283.

McLaren, D.J. (1972). "Ultrastructural studies on microfilariae (Nematoda: Filariodea)". Parasitology 65, 317—332.

McLaren, D.J. and Paget, G.E. (1968). "A fine structural study on the merozoite of *Eimeria tenella* with special reference to the conoid apparatus". Parasitology 58, 561—571.

Mettrick, D.F. and Podesta, R.B. (1974). In: "Advances in Parasitology, 12". (B. Dawes, ed.) pp. 183—278. Academic Press, London.

Miller, R.B. (1952). "A review of the *Triaenophorus* problem in Canadian lakes". The Fisheries Research Board of Canada. Bulletin 95, 1—42.

Mueller, J.F. (1965). In: "Host-Parasite Relationships". (J.E. McCauley, ed.) pp. 11—58. 26th Annual Biology Colloquium, Oregon State University.

Muller, R. (1968). "Studies on *Dracunculus medinensis* (Linnaeus). I. The early migration route in experimentally infected dogs". J. Helminth. 42, 331—338.

Muller, R. (1971). In: "Advances in Parasitology, 9". (B. Dawes, ed.) pp. 73—151. Academic Press, London.

Nath, D. and Pande, B.P. (1963). "Lesions associated with some of the nematodes parasitic in the alimentary canal of Indian domestic fowls". Res. vet. Sci. 4, 390—396.

Newsom, I.E. and Stout, E.N. (1933). "Proventriculitis in chickens due to flukes". Vet. Med. 28, 462—463.

Newton, B.A. (1972). In: "Microbial Pathogenicity in Man and Animals". (H. Smith and J.H. Pearce, eds.). Symposium 22, Society for General Microbiology pp. 269—301. Cambridge University Press.

Nichols, R.L. (1956). "The etiology of visceral larva migrans II. Comparative larval morphology of *Ascaris lumbricoides*, *Necator americanus*, *Strongyloides stercoralis* and *Ancylostoma caninum*. J. Parasit. 42, 363—399.

Olsen, O.W. (1967). "Animal Parasites", (2nd Edn.). Burgess, Minneapolis.

Otto, G.F. (1966). In: "Biology of Parasites". (E.J.L. Soulsby, ed.) pp. 85—99. Academic Press, London.

Pike, A.W. (1967). "Some stylet cercariae and a microphallid type in British freshwater molluscs". Parasitology 57, 729—754.

Read, C.P. (1970). "Parasitism and Symbiology". Ronald Press, New York.

Read, C.P. and Kilejian, A.Z. (1969). "Circadian migratory behaviour of a cestode symbiote in the rat host". J. Parasit. 55, 574—578.

Rifkin, E. (1971). In: "The Biology of Symbiosis". (T.C. Cheng, ed.) pp. 25—43. Butterworths, London.

Roberts, F.H.S. (1934). "The Large Roundworm of Pigs, *Ascaris lumbricoides*". David White, Government Printer, Brisbane.

Roberts, F.H.S. (1937). "Studies on the life history and economic importance of *Heterakis gallinae* (Gmelin, 1790; Freeborn, 1923), the caecum worm of fowls. Aust. J. exp. Biol. Sci. 15, 429—439.

Roche, M. (1966). "Influence of male and female *Ancylostoma caninum* on each other's distribution in the intestine of the dog". Expl. Parasit. 19, 327—331.

Rogers, W.P. (1960). "The physiology of infective processes of nematode parasites; the stimulus from the animal host". Proc. R. Soc. B 152, 367—386.

Rogers, W.P. (1966). In: "Biology of Parasites". (E.J.L. Soulsby, ed.) pp. 33—40. Academic Press, London.

Rogers, W.P. and Sommerville, R.I. (1963). In: "Advances in Parasitology, 1". (B. Dawes, ed.) pp. 109—177. Academic Press, London.

Rushmer, R.F., Buettner, K.J.K., Short, J.M. and Odland, G.F. (1966). "The skin". Sci. 154, 343—348.

Ryley, J.F. (1972). In: "Comparative Biochemistry of Parasites". (H. Van den Bossche, ed.) pp. 359—381. Academic Press, London.

Ryley, J.F. (1973). In: "The Coccidia". (D.M. Hammond with P.L. Long, eds.) pp. 145—181. University Park Press, Baltimore.

Salt, G. (1963). "The defence reactions of insects to metazoan parasites". Parasitology 53, 527—642.

Schad, G.A. (1956). "Studies on the genus *Kalicephalus* (Nematoda: Diaphanocephalidae) I. On the life histories of the North American species *K. parvus*, *K. agkistrodontis*, and *K. rectiphilus*". Can. J. Zool. 34, 425—452.

Schad, G.A. (1962). "Studies on the genus *Kalicephalus* (Nematoda: Diaphanocephali-dae) II. A taxonomic revision of the genus *Kalicephalus* Molin, 1861". Can. J. Zool. 40, 1035—1165.

Schad, G.A. (1963). "Niche diversification in a parasitic species flock". Nature, Lond. 198, 404—406.

Scholtyseck, E. (1973). In: "The Coccidia". (D.M. Hammond with P.L. Long, eds.) pp. 81—141. University Park Press, Baltimore.

Schüffner, W. and Swellengrebel, N.H. (1949). "Retrofection in oxyuriasis. A newly dis-covered mode of infection with *Enterobius vermicularis*". J. Parasit. 35, 138—146.

Schwabe, C.W. (1951). "Studies on *Oxyspirura mansoni*, the tropical eyeworm of poul-try. II. Life history. Pacific Sci. 5, 18—35.

Smyth, J.D. (1969). "The Physiology of Cestodes". Oliver and Boyd, Edinburgh.

Smyth, J.D., Gemmell, M. and Smyth, M.M. (1970). "Establishment of *Echinococcus granulosus* in the intestine of normal and vaccinated dogs". H.D. Srivastava Comme-morative Volume, 167—178.

Spearman, R.I.C. (1973). "The Integument". University Press, Cambridge.

Stabler, R.M. (1954). "*Trichomonas gallinae:* a review". Expl. Parasit. 3, 368—402.

Standen, O.D. (1949). "Experimental schistosomiasis II. Maintenance of *Schistosoma mansoni* in the laboratory, with some notes on experimental infection with *S. haema-tobium*". Ann. trop. Med. Parasit. 43, 268—283.

Stirewalt, M.A. (1963). "Chemical biology of secretions of larval helminths". Ann. N.Y. Acad. Sci. 113, 36—53.

Stirewalt, M.A. (1966). In: "Biology of Parasites". (E.J.L. Soulsby, ed.) pp. 41—59. Aca-demic Press, London.

Stirewalt, M.A. (1971). In: "The Biology of Symbiosis". (T.C. Cheng, ed.) pp. 1—23. But-terworths, London.

Stirewalt, M.A. and Hackey, J.R. (1956). "Penetration of host skin by cercariae of *Schis-tosoma mansoni.* I. Observed entry into skin of mouse, hamster, rat, monkey and man. J. Parasit. 42, 565—580.

Stockdale, P.H.G. (1970). "Pulmonary lesions in mink with a mixed infection of *Fila-roides martis* and *Perostrongylus pridhami*". Can. J. Zool. 48, 757—759.

Stockdale, P.H.G. and Anderson, R.C. (1970). "The development, route of migration, and pathogenesis of *Filaroides martis* in mink". J. Parasit. 56, 550—558.

Stoll, N.R. (1929). "Studies with the strongylid nematode, *Haemonchus contortus.* I. Acquired resistance of hosts under natural reinfection conditions out-of-doors". Am. J. Hyg. 10, 384—418.

Stunkard, H.W. (1947). "A dicrocoeliid trematode, *Eurytrema vulpis* n.sp. provis., from the pancreatic ducts of the red fox". J. Parasit. 33, 459—466.

Thomas, L.J. (1939). "Life cycle of a fluke, *Halipegus eccentricus* n.sp., found in the ears of frogs". J. Parasit. 25, 207—221.

Twohy, D.W. (1956). "The early migration and growth of *Nippostrongylus muris* in the rat". Am. J. Hyg. 63, 165—185.

Ulmer, M.J. (1971). In: "Ecology and Physiology of Parasites". (A.M. Fallis, ed.) pp. 123—160. University of Toronto Press.

Uribe, C. (1922). "Observations on the development of *Heterakis papillosa* Bloch in the chicken". J. Parasit. 8, 167—176.

Veglia, F. (1915). "The anatomy and life-history of the *Haemonchus contortus* (Rud)". 3rd and 4th Rep. vet. Res. Un. S. Afr. 293—345.

Vickerman, K. (1971). In: "Ecology and Physiology of Parasites". (A.M. Fallis, ed.) pp. 58—91. University of Toronto Press.

Wales, J.H. (1958). "Two new blood parasites of trout". California Fish and Game 44, 125—136.

Wall, L.D. (1941). "*Spirorchis parvus* (Stunkard) its life history and the development of its excretory system (Trematoda: Spirorchiidae)". Trans Am. microsc. Soc. 60, 221—260.

Wardle, R.A. and McLeod, J.A. (1952). "The Zoology of Tapeworms". University of Minnesota Press.

Webber, W.A.F. (1955). *"Dirofilaria aethiops* Webber, 1955, a filarial parasite of monkeys III. The larval development in mosquitoes". Parasitology 45, 388—399.

Webster, G.A. (1958). "On prenatal infection and the migration of *Toxocara canis* Werner, 1782 in dogs". Can. J. Zool. 36, 435—440.

Weinman, D. (1968). In: "Infectious Blood Diseases of Man and Animals, Vol. 2". (D. Weinman and M. Ristic, eds.) pp. 144—171. Academic Press, London.

Whitfield, P.J. (1971). "The locomotion of the acanthor of *Moniliformis dubius* (Acanthocephala)". Parasitology 62, 35—47.

Whitfield, P.J. (1973). "The egg envelopes of *Polymorphus minutus* (Acanthocephala)". Parasitology 66, 387—403.

Williams, H.H. (1960). "The intestine in members of the genus *Raja* and host specificity in the Tetraphyllidea". Nature, Lond. 188, 514—516.

Williams, R.W. (1948). "Studies on the life cycle of *Litomosoides carinii*, filariid parasite of the cotton rat, *Sigmodon hispidus litoralis".* J. Parasit. 34, 24—43.

Wilson, R.A., Pullin, R. and Denison, J. (1971). "An investigation of the mechanism of infection by digenetic trematodes: the penetration of the miracidium of *Fasciola hepatica* into its snail host *Lymnaea truncatula".* Parasitology 63, 491—506.

Wykoff, D.E. and Lepes, T.J. (1957). "Studies on *Clonorchis sinensis.* I. Observation on the route of migration in the definitive host". Am. J. trop. Med. Hyg. 6, 1061—1065.

Yokogawa, M. (1965). In: "Advances in Parasitology, 3". (B. Dawes, ed.) pp. 99—158. Academic Press, London.

Ecological Aspects of Parasitology
Editor: C.R. Kennedy
© *North-Holland Publishing Company, Amsterdam, 1976*

CHAPTER 4

FEEDING

C. ARME *

Department of Zoology, Queen's University, Belfast (N. Ireland)

Contents

* Present address: North Staffordshire Polytechnical Institute, Department of Chemistry, College Road, Stoke-on-Trent, Staffs. (Great Britain).

4.1. Introduction

Parasitism may be examined within the broad framework of ecology, although a thorough understanding of the association is difficult since the environment of the parasite is another living organism, the host. Because of the complex nature of the host-parasite relationship, its analysis frequently involves observations on isolated host-organs and parasites, in vitro. Invaluable though such studies are in providing basic data, it is usually difficult to predict the applicability of the results so obtained to an in vivo situation. Thus, when one attempts to extrapolate from information gained from isolated hosts and parasites in order to obtain a perspective of the intact symbiosis, it is clear that the characteristics of the whole may differ from a mere sum of the properties of the parts. This is because in vivo the parasite and host may interact in a variety of ways, the extent of which is only now becoming apparent in those organisms which have been extensively studied. Thus, following infection with the cestode *Hymenolepis diminuta*, the gastro-intestinal physiology of the laboratory rat undergoes marked changes, with the result that the infected host differs, in several important respects, from the non-parasitised rat. In the case of the parasite, recently discovered properties of the glycocalyx (particularly its ability to adsorb host protein), together with the migratory behaviour of the worm which exposes it to regions of the intestine with differing physico-chemical properties, may well result in nutritional patterns which differ from those observed in isolated worms in vitro (see Arme, 1975).

The essence of the parasitic mode of life is the nutritional relationship within the symbiosis. It is important to appreciate, however, that most of the data described in this chapter have been derived from in vitro studies, and must therefore be subject to the qualifications mentioned above. In a review of this length it is not possible to attempt a comprehensive survey of feeding in parasitic animals. Rather, specific examples have been chosen to illustrate the principle modes of feeding, and references are given in the text to the relevant literature for related organisms.

4.2. Protozoa

The host-parasite interface in protozoa is the plasma-membrane, and it is the properties of this membrane which largely control the movement of substances in and out of the organism. A detailed consideration of the functional morphology of the plasma-membrane and its associated structures, together with its role in homeostatic regulation, is outside the scope of this chapter. A useful introduction to the subject, however, can be found in articles by Curtis (1972), Vickerman (1972) and Smyth (1973).

Two principal modes of nutrition are found in the parasitic protozoa viz.

osmotrophy in which nutrients in solution are absorbed across membranes, and *phagotrophy*, which involves the engulfing of parts of the external environment via invaginations in the parasite plasma-membrane. Some authors restrict the term phagocytosis to processes which involve intake of dense materials, and use a second term, pinocytosis to describe the formation of food vacuoles which contain fluids. Such a distinction may be important to cell biologists but, in terms of nutrition, phagocytosis and pinocytosis are synonomous since, in the latter case, dissolved solutes are often of more significance than the solvents themselves. It is now recognised that many parasitic protozoa have patterns of feeding which utilise both the above mechanisms to varying degrees.

4.2.1. Osmotrophy

The term osmotrophy is perhaps inappropriate, since it implies that movement of nutrients across membranes is a simple process, dependent on differences in osmotic pressure between intra- and extra-cellular environments, or upon diffusion. For many parasitic protozoa this is demonstrably not the case, and a variety of mechanisms for nutrient uptake across membranes have been described in several species. The kinetics of transport phenomena can be analysed with the Michaelis-Menten equations commonly used in the study of enzyme-catalysed reactions. It must be emphasised, however, that the similarities between enzyme and transport kinetics do not necessarily mean that enzymes are involved in transport. The principal uptake mechanisms for soluble nutrients are active transport, facilitated diffusion and diffusion. The kinetic characteristics of these processes have been described by Pappas (1975).

Organisms may utilise all three uptake mechanisms for the absorption of different nutrient classes, and often a single nutrient may be absorbed by a combination of uptake processes. For example, absorption at low substrate concentrations may occur predominantly by a mediated mechanism, but this may be masked at higher concentrations by a large diffusion component of uptake.

4.2.1.1. Amoebae

Studies on free-living amoebae have shown that specific transport systems for glucose and amino acids are absent, and that their absorption occurs by pinocytosis (Holter, 1959; Chapman-Andreson, 1962; Bowers and Olszewski, 1972; Wander and Hilgard, 1975). In contrast, glucose uptake in three strains of the parasite *Entamoeba histolytica* (strains DKB; 200-NIH and Laredo) has been shown to occur by two processes, pinocytosis and, more important, by a mediated transport system (Serrano and Reeves, 1974, 1975). Why parasitic amoebae should possess well developed glucose transport systems when they are absent in free living forms is not known. The apparent lack of mitochondria in *E. histolytica* (El Hashimi and Pittman, 1970; Proctor and Gregory, 1973) may result in an increased dependence upon glucose in this species; references to recent studies on the metabolism of *E. histolytica* are given in Weinbach and Diamond (1974).

4.2.1.2. Flagellates

The nutrition of parasitic flagellates has recently been reviewed (Trager, 1974), and additional information is provided by Bowman (1974) and Trager (1975). The culture media used for in vitro maintenance are complex and it will clearly be some time before the nutritional requirements of parasitic flagellates are known. The cultivation of salivarian trypanosomes produces forms which are unable to reinfect the vertebrate host and which resemble in many ways, both morphologically and physiologically, the trypanosomes found in insect vectors. For example, during culture the mitochondrion is activated and changes in oxidative metabolism occur.

Although glucose and amino acids are utilised by certain trypanosomes in their metabolism, relatively few studies on the uptake of these compounds have been undertaken. A criticism of those studies which have been made is that no account has been taken of possible uptake via the cytosome and flagellar pocket.

Most of the information on transport processes in flagellates has been derived from blood-stream forms of the genus *Trypanosoma* and culture forms of *T. cruzi*, although limited data are available for other genera. For example exchange of metabolites between *Leishmania donovani* and host-cells has been shown to occur (Bhattacharya and Janovy, 1975), although the mechanisms whereby these exchanges are effected are not known. Preliminary studies on the uptake of carbohydrates in trypanosomes by von Brand et al. (1967), Seed et al. (1965) and Min (1965, 1966) have been extended by Southworth and Read (1969, 1970), Sanchez and Read (1969), Sanchez (1974), and Ruff and Read (1974).

Southworth and Read (1969) confirmed the suggestions of previous authors of a mediated component of carbohydrate uptake in *Trypanosoma gambiense*. Using kinetic analyses, based on short-interval incubations in a variety of substrates, they demonstrated that glucose, mannose, glycerol, fructose and 2-deoxyglucose were readily absorbed by bloodstream forms obtained from rats via two uptake systems. The rate of uptake was non-linear with respect to substrate concentration and exhibited saturation kinetics. These preliminary observations were extended by Southworth and Read (1970), who examined the specificity of carbohydrate transport systems in *T. gambiense* in more detail. The requirements for binding of glucose to the transport system were shown to be highly specific for carbon atoms 1, 3, 4 and 6.

Sanchez and Read (1969) demonstrated that glucose absorption in the bloodstream forms of *Trypanosoma lewisi* occurred by a saturable transport process, which could be inhibited by a variety of low molecular-weight metabolites, and Sanchez (1974) studied the rates of glucose and glucosamine accumulation by *T. lewisi* at different ages post-innoculation. After 4 days, glucose uptake was twice that of glucosamine, but uptake rates of both fell until, by day 13, they were equal. These results, indicating age variations in the quantitative aspects of transport, may be viewed together with the data of Sanchez and Dusanic (1968) and Entner and Gonzales (1966), who described metabolic and antigenic changes in *T. lewisi* during the course of the infection in rats.

Amino acids are readily absorbed by the bloodstream forms of *T. lewisi*, and culture forms of *T. cruzi*. Mediated transport mechanisms are apparently involved, although diffusion plays an important role in uptake, particularly at high substrate concentrations. For example, lysine uptake in cultured *T. cruzi* was found by Hampton (1970a,b) to be complex, involving at least three systems. Since, in other animal cells, several basic amino acids have been shown to share the same uptake system, it was unusual to find that lysine uptake in *T. cruzi* was not susceptible to inhibition by other basic amino acids. In marked contrast, significant partially competitive inhibition of lysine transport by neutral amino acids was noted. A further unique feature of lysine absorption by *T. cruzi* was its inhibition by several amines (1-aminopentane, glucosamine, methylamine, 5-hydroxytryptamine and methenamine), suggesting a lack of requirement of carboxyl groups in the transport system.

Arginine transport in cultured *T. cruzi* is equally complex (Hampton, 1971). A large diffusion component was accompanied by at least two mediated systems, one of which was so specific for arginine that homoarginine did not interact with it.

Although the bloodstream form of *T. lewisi* appears not to absorb aspartic acid, the mediated uptake of a variety of other amino acids was observed by Manjra and Dusanic (1972). Valine and phenylalanine were taken up by a saturable system with negligible diffusion, whereas above a concentration of approximately 0.5 mM, arginine was absorbed primarily by diffusion. An interesting and unusual observation was that although strong evidence was presented that L-valine and L-phenylalanine shared a transport site, only L-phenylalanine was inhibited by its D-isomer, D-valine having no effect on the uptake of L-valine.

Mediated uptake of amino acids by *T. gambiense* has been described by Southworth and Read (1972), and Manjra and Dusanic (1973) studied nucleoside uptake by bloodstream forms of *T. lewisi*.

Prior to information on transport system becoming available it was recognised that there were fundamental differences between stercorarian trypanosomes such as *T. lewisi* and salivarian forms like *T. brucei*. These differences seem to be also reflected in the scanty information on their transport systems; for example *T. cruzi* and *T. lewisi* (Stercoraria) apparently resemble one another more than they do *T. gambiense* (Salivaria). Hampton (1971) has suggested that whereas the evolution of the bloodstream forms of the African trypanosomes has involved modifications in their metabolic pathways, in the Stercoraria, the development of membrane transport systems may have taken precedence. Nevertheless, all the trypanosomes examined inhabit nutrient-rich environments, in which nutrient uptake by diffusion might have been considered sufficient for maintenance of essential metabolic functions. The fact that these parasites possess, and presumably utilise, specific membrane mechanisms in their nutrition might be related to their need to maintain an intracellular environment of different composition, in terms of carbohydrates, amino acids, etc., to that of the external environment. The development of transport systems possessing various degrees of affinity for different compounds may also have evolved in order to compensate for metabolic deficiencies or specialisations in the parasite. For example, it would clearly be of interest to evaluate possible changes in the relative importance of amino acid and glucose uptake systems in bloodstream and culture forms of salivarian trypanosomes.

4.2.1.3. Plasmodia

A major difficulty encountered in attempts to assess the extensive literature on malaria parasites is the diversity of species studied by different authors. However, in nutritional terms, the problems faced by erythrocytic stages of all malaria parasites are similar. They may feed upon the contents of red blood cells by phagotrophy, and this may be supplemented to varying degrees by absorption of soluble nutrients supplied by the plasma. The study of the latter process is complicated by the fact that compounds in the plasma need to cross both erythrocytic and parasite membranes before they become available to the plasmodia. In vitro experiments have indicated that, although most amino acids required by the malaria parasites are supplied by digestion of host erythrocyte contents, these by themselves may not be sufficient to maintain growth. Despite these interesting observations there have been surprisingly few studies of nutrient absorption by plasmodia.

The preliminary observations of Sherman et al. (1967) and McCormick (1970) were supplemented by a more detailed study of the transport and incorporation of amino acids into normal and malaria-infected duck erythrocytes and erythrocyte-free *P. lophurae* by Sherman and Tanigoshi (1972). Using short term (5 min) incubations, little metabolism of glutamic acid, lysine and arginine was detected. In erythrocyte-free preparations of *P. lophurae*, the uptake of glycine, alanine, isoleucine and methionine was a linear function of concentration. This, together with low distribution ratios previously established for these amino acids in isolated parasites (Sherman et al., 1967), strongly suggests that they are absorbed by diffusion. Arginine, lysine, glutamic acid and aspartic acid, on the other hand, entered by a saturable process.

With normal duck erythrocytes, all six amino acids studied were found to be absorbed by saturable transport systems. When the cells were infected with *P. lophurae*, however, a remarkable change in the mode of amino acid uptake was observed. With the exception of glycine, all amino acids investigated entered the red cell by diffusion. These striking changes in erythrocyte permeability are clearly associated with the presence of the parasite, but the precise way whereby they are effected is not known. Sherman and Tanigoshi (1972) suggest that one component that might be involved is ATP. The importance of ATP for growth of malarial parasites has been demonstrated for several species (Brewer and Powell, 1965; Brewer and Coan, 1969; Trager, 1967, 1972) and this need is reflected in reduced ATP levels in infected erythrocytes. Since ATP is implicated in the operation of the erythrocyte sodium-pump, reduced intraerythrocytic ATP may be associated with the increased sodium content described by Dunn (1969) and Sherman and Tanigoshi (1971) in red cells of infected animals. It has been demonstrated that uptake of amino acids by erythrocytes is dependent upon differences in the concentration of intra- and extra-erythrocytic sodium so that in plasmodium-infected red-cells, the normal sodium-coupled mediated transport systems for amino acids may not be able to function. As a consequence, in order to obtain plasma nutrients, the permeability of the plasma-membrane of the host erythrocyte becomes changed, possibly being induced by material elaborated by the parasite. A substance which affects the osmotic fragility of duck erythrocytes has been isolated from *P. lophurae*-infected cells (Herman, 1969). Moulder (1962) based his permeability-defect hypothesis, in part, on the ability of erythrocyte-free plasmodia to take up highly polar substances like ATP and coenzyme A, to which most cells are quite impermeable. The results of Sherman and co-workers support this view and enables the hypothesis to be extended to the parasitised erythrocyte. However, evaluation of Moulder's views must include an assessment of the relative roles of phagotrophy and purely membrane mechanisms in the nutrition of plasmodia.

4.2.2. Phagotrophy

4.2.2.1. Amoebae

Although phagocytosis in amoebae is well known, an interesting membrane structure, possibly involved in feeding, has been recently described in *Entamoeba histolytica* by Eaton et al. (1969). They described a cup-shaped, surface lysosome, in the centre of which was a trigger mechanism. The trigger has been postulated to contain a lipase which may initiate damage to host cells (Eaton et al., 1970; Proctor and Gregory, 1972). Eaton et al. (1970) suggested that the operation of the trigger involved a depolarisation of the membrane. However, Lushbaugh and Miller (1974) found that all regions of the surface lysosomes and triggers of *E. histolytica* bound similar amounts of cationic iron. They concluded, therefore, that no special charge relationships existed within the organelle.

4.2.2.2. Flagellates

All members of the Kinetoplastida possess a deep invagination of the plasma-membrane at the origin of the flagellum. This flagellar pocket or reservoir has been shown to be the site of protein ingestion by endocytosis in several trypanosome species. Many free living kinetoplastid flagellates, parasites of poikilotherms and *Trypanosoma cruzi* also possess a cytosomal-cytopharynx complex, which arises in the region of the flagellar pocket, and which is also active in protein uptake (see for example Brooker, 1971, Preston, 1969). In contrast, cytostomal complexes have not been detected in salivarian trypanosomes (Vickerman, 1969, 1974; Langreth and Balber, 1975) and the only route for ingestion of macromolecules appears to be via coated vesicles arising from the flagellar pocket. Although coated vesicles were described by Langreth and Balber for bloodstream forms of *T. brucei* no such vesicles were noted in culture forms. The reason for the lack of a pinocytotic system involving coated vesicles in cultured *T. brucei* is not known. It may reflect environmental features which are not condusive to the induction of pinocytosis, or imply a reduced nutritional reliance on the uptake of macromolecules in insect forms. If reduced protein uptake is shown to be a general feature in all the midgut and culture forms of haemoflagellates, then it is possible that the membrane transport of amino acids, and other low molecular weight compounds, will be of considerable importance in the nutrition of these parasites. Could this be the explanation for the extreme specificity shown by certain of the amino acid transport systems in culture *T. cruzi*, described above?

4.2.2.3. Plasmodia

The concept of intracellular phagotrophy in the erythrocytic stages of *Plasmodium lophurae* was introduced by Rudzinska and Trager (1957). Since then phagotrophy has been demonstrated in many species of plasmodia, and the literature in this field has been reviewed by Aikawa (1971), Aikawa and Sterling (1974) and Rudzinska and Vickerman (1968). Although the occurrence of phagotrophic feeding in malaria parasites is established beyond doubt, there is some question as to the mechanism of food-vacuole formation. Rudzinska and Trager (1957) observed random invaginations of host cytoplasm over the surface of *P. lophurae*, and they assumed that the double membrane-bound vacuoles which were present in the cytoplasm were food vacuoles produced from these invaginations. This view has been disputed by Aikawa and co-workers who claim that the double membrane-bound vesicles are artifacts produced by the sectioning of protrusions of host-cell cytoplasm into the parasite and that feeding occurs via a specialised organelle, the cytostome. Digestion of ingested host-cell cytoplasm occurs in the food vacuoles, resulting in the production of malarial pigment granules of haemozoin. References to details of the digestive process and the effects of chemotherapeutic agents on digestion and autophagy in plasmodia can be found in Aikawa (1971), Warhurst (1973), Warhurst et al. (1974) and Davies et al. (1975). Although the cytosome is the main route of ingestion in plasmodia, there is some evidence that random pinocytosis occurs around the plasma-membrane of *Plasmodium vinckei*, since small peripheral vacuoles were found to contain haemozoin (Cox and Vickerman, 1966).

4.3. Helminths

Of the major helminth groups, two, the trematodes and nematodes, possess differentiated organ systems for digestion, although, in some species, uptake of soluble nutrients across the outer body wall has been demonstrated. In contrast, cestodes and acanthocephalans lack a digestive tract, and therefore nutrient absorption takes place solely through the body surface. The excellent review of Lumsden (1975) should be consulted for further information on the surface structure of helminths.

4.3.1. Role of tegument in cestode nutrition

With few exceptions, adult cestodes inhabit the alimentary canal of vertebrates, deriving their food supply from its contents. An understanding of the physico-chemical properties of the host gastrointestinal tract is essential in any consideration of cestode physiology. Although brief mention of this aspect of host physiology will be made below, interested readers should consult the reviews of Read (1950, 1971), Mettrick and Podesta (1974), Arme (1975) and Chapter 15 for further information.

The rat tapeworm, *Hymenolepis diminuta*, is the doyen of parasitologists working in the field of cestode nutrition. Techniques for the in vitro and in vivo maintenance of this parasite were described by Read et al. (1963) and, as a result of their work, a standardised experimental procedure, yielding highly reproducible results, has been developed. During the preparation of this chapter, Pappas and Read (1975) published a review of membrane transport in helminths and this should be consulted for detailed information.

4.3.1.1. Nutrient uptake

Adult *Hymenolepis diminuta* is apparently impermeable to peptides, proteins and other macromolecules (Lumsden et al., 1970), although a variety of soluble, low molecular-weight organic nutrients are readily absorbed, often by mediated transport systems. The original studies by Read et al. (1963) demonstrated that methionine was accumulated against a concentration gradient by a saturable uptake system, exhibiting specificity, and affected by alterations in pH and the presence of certain metabolic poisons. More recently, the active uptake of proline, histidine, cycloleucine and α-aminoisobutyric acid has been demonstrated (Kilejian, 1966; Woodward and Read, 1969; Harris and Read, 1969).

Glucose uptake in *H. diminuta* also occurs by a temperature-dependent active transport process, exhibiting saturation kinetics (Phifer, 1960; Pappas et al., 1974; Read et al., 1974). Galactose is also transported via the glucose system (Read, 1961), but the worm is virtually impermeable to fructose (Arme and Read, 1970). The uptake of glycerol occurs by a combination of a mediated system and diffusion (Pittman and Fisher, 1972). 1 : 2 propanediol, at a high I : S ratio of 100 : 1, competitively inhibited glycerol uptake, but a variety of sugars, sugar alcohols, fatty acids and amino acids had no effect, thus demonstrating the specificity of the glycerol site. Uglem et al. (1974) extended the observations of Pittman and Fisher (1972), and demonstrated that glycerol uptake in *H. diminuta* occurred by two distinct systems. One was sodium-dependent and interacted with 1 : 2 propanediol, the other operated in sodium-free media and had no affinity for the glycerol analogue. Arme and Read (1968) demonstrated that acetate was absorbed by *H. diminuta* by a combination of diffusion and a mediated transport system. The acetate system was specific for short-chain fatty acids and was distinct from a mechanism for the uptake of long chain fatty acids, which was subsequently characterised by Chappell et al. (1969). *Hymenolepis diminuta* has also been shown to have a complex system for the absorption of purines and pyrimidines. Limitations of space prevent a detailed consideration of this uptake system, but a recent publication by Page and MacInnis (1975) contains a useful summary of previous work. Although *H. diminuta* has a specific growth requirement for pyridoxine (Roberts and Mong, 1973), pyridoxine and nicotinamide were absorbed only by diffusion (Pappas, 1972; Pappas and Read, 1972a). In contrast separate, specific mediated mechanisms exist for the uptake of thiamine (Pappas and Read, 1972b) and riboflavin (Pappas and Read, 1972a).

The above studies have utilised only one stage in a complex life cycle — the adult worm. In addition there is an egg, which is essentially free living, and a cysticercoid larva which develops in the haemocoele of insects. The cyst wall of the larval tapeworm has been shown to be similar in structure to the tegument of the adult worm (Ubelaker et al., 1970). Studies on the absorption of amino acids, glucose and sodium acetate by the cysticercoid have revealed the presence of transport systems which bear a striking qualitative and quantitative similarity to those of the adult worm (Arme and Coates, 1971, 1973; Arme, Middleton and Scott, 1973).

The above observations have all been made using in vitro systems and there have been relatively few studies on nutrient uptake by *H. diminuta* in vivo. Glucose uptake by *H. diminuta* in vivo has been studied by Overturf (1966) and Podesta and Mettrick (1974) using perfusion techniques, in closed intestinal loops. Glucose absorption was demonstrated in both investigations, but the kinetics of uptake were difficult to analyse. Podesta and Mettrick demonstrated that the relative importance of solvent drag-coupled glucose uptake and mediated glucose transport, varied with changes in pH. At pH 7, approximately 2—3 times more glucose was absorbed by a mediated process whereas at pH 6.0, both uptake mechanisms contributed equally to glucose absorption. Recent studies on the in vivo absorption of amino acids by *H. diminuta* have been reviewed by Arme (1975). The available data suggests that amino acid fluxes which have been demonstrated in vitro occur in a similar manner in vivo.

The uptake of low molecular-weight organic nutrients has also been demonstrated for several other species of tapeworm. Data for *Hymenolepis citelli*, *H. microstoma*, *Taenia crassiceps*, *T. taeniaeformis* and *Calliobothrium verticillatum* have been reviewed by Pappas and Read (1975).

An important feature in the biology of intestinal parasites of elasmobranch fishes is the presence, in their environment, of urea concentrations of up to 0.5 M. Read et al. (1959) demonstrated that urea was important for osmoregulation in *Calliobothrium* and Simmons (1961) described high urease activity in two out of three trypanorhynchs examined, but none in seven species of tetraphyllids. Despite the importance of urea in the biology of trypanorhynchs and tetraphyllids, urea uptake by elasmobranch cestodes appears to occur by simple diffusion (Simmons et al., 1960).

4.3.1.2. Digestion

Arme and Read (1970) emphasised that the tegument of *H. diminuta* could be regarded as a digestive-absorptive surface having many features in common with the intestinal mucosa of vertebrates. The capacity to absorb nutrients has been described above but, in addition, there are enzymes present in the tegument. These are membrane-bound hydrolases, which function as intrinsic digestive enzymes, and also extrinsic enzymes of host origin, which are adsorbed onto the tegument, and which participate in membrane or contact digestion as originally postulated by Ugolev (1965). The roles of intrinsic and extrinsic digestive enzymes have been recently reviewed by Arme (1975).

The brush border of mammalian mucosal cells contains intrinsic membrane-bound phosphohydrolases and disaccharidases (Granger and Baker, 1950; Miller and Crane, 1961b). No intrinsic carbohydrases have been demonstrated for *H. diminuta*; indeed, in vitro, the worm is unable to metabolise carbohydrates other than galactose and glucose (Read, 1967). In contrast, intrinsic phosphohydrolase activity has been demonstrated by Lumsden et al. (1968), Arme and Read (1970), Dike and Read (1971a,b) and Pappas and Read (1974). These intrinsic enzymes apparently function in a digestive capacity by hydrolysing a variety of phosphate esters, to which cells are normally impermeable, releasing free phosphate and an organic component, both of which can be absorbed. Miller and Crane (1961a) demonstrated that the products resulting from the activity of intrinsic disaccharidase and phosphohydrolase activity possessed a kinetic advantage, in their subsequent absorption, over identical substances free in the environment. Dike and Read (1971b) and Pappas and Read (1974) suggested that in *H. diminuta*, the hydrolases which utilised phosphorylated sugars and nucleosides as substrates, and the transport systems for sugars and nucleosides, were located close together on the tegumental surface. Thus, as was demonstrated for the intestinal mucosa, the products of hydrolysis can be transported before they diffuse into the environment, and they thus gain a kinetic advantage in uptake over similar, non-esterified substrates, in the medium. The role of phosphatases in transport processes has been the subject of some debate (see Lumsden, 1975; Arme, 1975). However, there is no doubt, from the results described above, that the transport systems and sites of phosphohydrolase activity are separate.

Like the vertebrate intestine, *H. diminuta* also interacts with extrinsic pancreatic enzymes (see Arme, 1975). Thus pancreatic amylase is adsorbed onto the surface of the parasite and by presumed changes in its molecular configuration, its catalytic activity is enhanced (Taylor and Thomas, 1968; Read, 1973). In contrast the activities of trypsin, α- and β-chymotrypsin and lipase are reduced. For trypsin and chymotrypsin the inhibition of activity is apparently irreversible (Pappas and Read, 1972c,d) but lipase inhibition is readily reversible (Ruff and Read, 1973). The ability to avoid digestion by host proteases is clearly of advantage to the tapeworm and it would be of interest to investigate this phenomenon in other intestinal parasites.

Pappas et al. (1973) described ribonuclease activity associated with the surface of intact *H. diminuta* but failed to demonstrate conclusively whether the enzyme was of intrinsic or extrinsic origin.

It is apparent therefore that the tegument of *H. diminuta* and the intestinal mucosa have much in common in terms of their morphology and physiology and functionally they are both absorptive-digestive-protective surfaces. It is necessary, however, to inject a note of caution concerning the interpretation of the results relating to the nutrition of *H. diminuta*. The majority of these data have originated from in vitro studies which, although contributing greatly to our understanding of basic mechanisms, may not be applicable in their entirety to worm nutrition in the intact symbiosis. For example the peripatetic nature of the worm and its consequent exposure to regions of the intestine with differing physiological properties must be considered. In this connection the findings of Mettrick (1970, 1971), in which linear gradients in the intestine of pH and chemical composition were demonstrated, may be of importance. Recently discovered properties of the glycocalyx, particularly its ability to adsorb host protein, may also influence nutrition in vivo. Unpublished data obtained by undergraduates in my laboratory in which preincubation with polyions was found to affect subsequent uptake of amino acids, suggests that this is an aspect of cestode nutrition which deserves further consideration.

4.3.2. Acanthocephala

Like cestodes, acanthocephalans lack a gut, and low molecular-weight organic nutrients are absorbed across the tegument. Mediated transport systems for a variety of amino acids have been demonstrated by Rothman and Fisher (1964), Edmonds (1965), Branch (1970a) and Uglem and Read (1973). Sugar uptake was demonstrated by Crompton and Lockwood (1968) and the uptake of inorganic cations was described by Branch (1970b). In contrast to findings with *Hymenolepis diminuta*, described above, the presence of *Moniliformis dubius* does not alter the catalytic activities of pancreatic trypsin, α- and β-chymotrypsin, lipase or amylase (Ruff et al., 1973). Nonetheless amylase of host origin was loosely adsorbed onto the surface of the parasite and its presence would presumably ensure the production of maltose close to the worm surface. In this connection it may be relevant to note that Laurie (1957) showed that *M. dubius* was able to metabolise maltose, but not a variety of other disaccharides. Byram and Fisher (1974) reported interesting observations concerning the uptake of macromolecules by *Moniliformis dubius*. They demonstrated that horseradish peroxidase was taken into the parasite via the surface crypts, and then transferred to lysosomes in the adjacent cytoplasm, presumably in vesicles formed by a pinocytotic mechanism. It is of interest that pinocytosis has not been demonstrated in adult *H. diminuta*, and the relative importance of mediated uptake and pinocytosis in acanthocephalan nutrition remains to be ascertained.

4.3.3. Trematoda

In adult Trematoda, the processes of absorption and digestion are more complex than in cestodes because of the presence of a differentiated gut, a tegument across which absorption may occur and, in some species, the occurrence of extra-corporeal digestion. Limitations of space permit only a consideration of nutrition in the adults of three trematode species, *Schistosoma mansoni*, *Fasciola hepatica*, and *Diclidophora merlangi*. References to nutrition in other species and in larval stages may be found in Erasmus (1972), Nollen et al. (1974) and McManus and James (1975). Electron microscopic observations have revealed that the morphology of the gut and tegument of many trematodes is consistent with their having possible digestive, absorptive or secretory roles. However, surprisingly little work has been undertaken to elucidate the relative importance of tegument and gut in the nutrition of these parasites.

4.3.3.1. Role of tegument in digenean nutrition
The uptake of methionine by schistosomula, 21-day old and adult *Schis-*

tosoma mansoni was demonstrated by Chappell (1974). In the adult worm, no significant differences were detected between uptake rates in normal flukes, and those which had been ligatured in the pharyngeal region (between the oral and ventral suckers). Chappell therefore concluded that uptake occurred primarily across the tegument. The results of kinetic studies enabled two components of the methionine uptake system to be distinguished. Diffusion operated predominantly at high substrate concentrations, whilst uptake at low concentrations was mainly via a mediated system, specific for amino acids. Methionine was not absorbed against a concentration gradient although previously Isseroff and Levy (1972) had claimed that several amino acids, including methionine, were absorbed by an active transport system in adult *S. mansoni*. Asch and Read (1975) confirmed the transtegumental uptake of amino acids by adult male *S. mansoni*, using glycine and proline as substrates. Preliminary observations on carbohydrate uptake across the tegument of *S. mansoni* by Isseroff, Bonta and Levy (1972) suggested that glucose, galactose, fructose, 3-O-methylglucose and ribose were absorbed by facilitated diffusion.

Mansour (1959) demonstrated the transtegumental uptake of glucose in ligatured adult *Fasciola hepatica*. Kinetic analysis of sugar uptake in *Fasciola* (Isseroff and Read, 1974) showed that the uptake of glucose, galactose, fructose, mannose, glucosamine and ribose occurred by facilitated diffusion, possibly involving two transport sites. Xylose appeared to be absorbed by diffusion.

In contrast to the majority of sugars studied, all the amino acids examined by Isseroff and Read (1969) were taken up by diffusion. Nevertheless, adult *Fasciola* may obtain significant amounts of amino acids from bile, since the free amino acids in bile from infected hosts may be up to ten times higher than in non-infected animals (Isseroff, Tunis and Read, 1972).

The data of Wright and Isseroff (1973) show that acetate uptake in *Fasciola* was inhibited by certain other short chain fatty acids at high I : S ratios, suggesting the presence of a mediated system for fatty acid absorption.

4.3.3.2. Role of gut in digenean nutrition

There is a paucity of information regarding the role of the gut in schistosome nutrition. The adult worm undoubtedly ingests blood and degrades it, but experimental evidence for the absorption of the products of haemoglobin breakdown across the gut is not available. Thus a gut protease isolated from *S. mansoni* was apparently unable to hydrolyse globin to its constituent amino acids (Grant and Senft, 1971; Sauer and Senft, 1972) and studies on oesophageal morphology of *S. mansoni* by Dike (1971) provided no indication that pinocytosis of intact protein occurred.

Whether adult *Fasciola hepatica* is a blood or tissue feeder has been a matter of some controversy (see Dawes and Hughes, 1964, 1970). Supporters of either view may have attributed a degree of selectivity in the feeding of the parasite that is not warranted and the worm may have a mixed nutrition. As for schistosomes, the experimental data concerning the role of the gut in digestion in *Fasciola* is circumstantial. Proteases have been demonstrated histochemically and biochemically (Halton, 1967, 1963) and Howell (1973) demonstrated that a major source of proteolytic activity in the fluke resided in the caecal region. Using the electron microscope, Robinson and Threadgold (1975) described secretory activity in the gastrodermis of *F. hepatica*. Additionally the presence of lamellae-like projections bearing microvilli and the possible presence of a lysosomal system suggested a digestive-absorptive role for the gastrodermal cells.

4.3.3.3. Monogenea

Detailed studies on the possible role of the tegument in the nutrition of monogenetic trematodes are lacking and the ectoparasitic habit of many Monogenea suggests that the gut may be the main route for nutrient uptake. Feeding is effected by the action of a muscular pharynx (Halton and Jennings, 1965) and histochemical and light microscopical investigations indicated that the diet of monoposthocotyleans was mucus and epidermal tissue, whereas in the polyopisthocotylea, the diet consisted primarily of blood. The nutrition of *Diclidophora merlangi*, a polyopisthocotylean parasite found on the gills of whiting (*Gadus merlangus*), has been extensively studied by Halton and co-workers. Whereas in the Digenea, the predominantly extra-cellular digestion of blood renders the task of residual haematin elimination relatively easy, in the Monogenea, blood breakdown occurs intracellularly, resulting in an accumulation of haematin residues in the gastrodermis. Halton and Jennings (1965), using the light microscope, suggested that haematin was eliminated by a process of cell sloughing. However, recent studies with the electron microscope have indicated that this is not the case (Halton et al., 1968; Halton, 1974, 1975). Digestion of blood is initiated by enzymes secreted into the lumen of the prepharynx, so that the blood meal becomes a homogeneous and non-cellular mixture on reaching the intestinal caeca. Blood proteins and haemoglobin are taken up into haematin cells by endocytosis involving coated vesicles, which release their contents into a channel system into which Golgi-derived digestive enzymes are also discharged. Following digestion in this lysosomal system, haematin residues are extruded into the gut lumen by exocytosis. In contrast to the above findings for *Diclidophora*, Rohde (1973) has shown that haematin elimination in two species of *Polystomoides* involves cell loss.

4.3.4. Nematoda

The wide variety in habitats and types of food utilised by nematodes precludes a detailed review of the feeding methods in this group. However, mention may be made of certain aspects of nematode biology which relate to matters discussed previously for other helminths.

The cuticle of adult worms appears to be impermeable to all molecules with the exception of water and certain organic anthelminthics. Thus Castro and Fairbairn (1969) and Weatherby et al. (1963) failed to demonstrate transcuticular uptake of organic nutrients by adult *Ascaris* and *Ascaridia galli.* A possible mediated uptake system for glucose absorption was described in larval *Mermis negrescens* by Rutherford and Webster (1974). The gut of nematodes is permeable to a variety of low molecular-weight organic nutrients and references to gut absorption and the effects of anthelminthics may be found in Beames et al. (1974), Gentner et al. (1972), van den Bossche and Borgers (1973), van den Bossche and de Nollen (1973) and Zam et al. (1963). The production and function of antienzymes has been discussed by Rhoades and Romanowski (1974). Additional relevant information may be found in Lee (1965, 1972) and Bird (1971).

4.4. REFERENCES

Aikawa, M. (1971). *"Plasmodium:* the fine structure of malarial parasites." Expl. Parasit. 30, 284—320.

Aikawa, M. and Sterling, C.S. (1974). "Intracellular parasitic protozoa." Academic Press, New York.

Arme, C. (1975). In "Symp. Soc. exp. Biol. No. 29 — Symbiosis." (Jennings, D. and Lee, D.L. eds.). pp. 505—532. Cambridge University Press, Cambridge.

Arme, C. and Coates, A. (1971). "Active transport of amino acids by cysticercoid larvae of *Hymenolepis diminuta."* J. Parasit. 57, 1369—1370.

Arme, C. and Coates, A. (1973). *"Hymenolepis diminuta:* active transport of α-amino-isobutyric acid by cysticercoid larvae." Int. J. Parasit. 3, 553—560.

Arme, C., Middleton, A. and Scott, J.P. (1973). "Absorption of glucose and sodium acetate by the cysticercoid larvae of *Hymenolepis diminuta."* J. Parasit. 59, 214.

Arme, C. and Read, C.P. (1968). "Studies on membrane transport: The absorption of acetate and butyrate by *Hymenolepis diminuta* (Cestoda)." Biol. Bull. 135, 80—91.

Arme, C. and Read, C.P. (1970). "A surface enzyme in *Hymenolepis diminuta* (Cestoda)." J. Parasit. 56, 514—516.

Asch, H.L. and Read, C.P. (1975). "Transtegumental absorption of amino acids by male *Schistosoma mansoni."* J. Parasit. 61, 378—379.

Battacharya, A. and Janovy, Jn., J. (1975). *"Leishmania donovani:* autoradiographic evidence for molecular exchanges between parasites and host cells." Expl. Parasit. 37, 353—360.

Beames, G.C., Jnr., Bailey, H.H., Schanbacher, L.M. and Rock, C.O. (1974). "Movement of triglycerides and monoglycerides across the intestine of *Ascaris suum."* Comp. Biochem. Physiol. 47A, 889—896.

Bird, A.F. (1971). "The structure of Nematodes." Academic Press, New York.

Bowers, B. and Olszewski, T.E. (1972). "Pinocytosis in *Acanthamoeba castellanii."* J. Cell Biol. 53, 681—694.

Bowman, I.B.R. (1974). In "Trypanosomiasis and Leishmaniasis with special reference to Chaga's disease." pp. 255—271 Ciba Fdn. Symp. No. 20. Elsevier, Amsterdam.

Branch, S.I. (1970a). "Accumulation of amino acids by *Moniliformis dubius* (Acanthocephala)." Expl. Parasit. 27, 95—99.

Branch, S.I. (1970b). "*Moniliformis dubius* and *Macracanthorhynchus hirudinaceus:* Na, K, Ca and Mg content and Na and K active transport." Expl. Parasit. 27, 33—43.

Von Brand, T., Tobie, E.J. and Higgins, H. (1967). "Hexose and glycerol absorption by some trypanosomatidae." J. Protozool. 14, 8—14.

Brewer, G.J. and Coan, C.C. (1969). "Interaction of red cell ATP levels and malaria, and the treatment of malaria with hypoxia." Milit. Med. 134, 1056—1067.

Brewer, G.J. and Powell, R.D. (1965). "A study of the relationship between the content of ATP in human red cells and the course of falciparum malaria: a new system that may confer protection against malaria." Proc. Nat. Acad. Sci. U.S. 54, 741—745.

Brooker, B.E. (1971). "The fine structure of *Crithidia fasciculata* with special reference to the organelles involved in the ingestion and digestion of protein." Z. Zellforsch. 116, 532—563.

Byram, J.E. and Fisher, F.M. Jnr. (1974). "The absorptive surface of *Moniliformis dubius* (Acanthocephala). II. Functional aspects." Tissue and Cell, 6, 21—42.

Castro, G.A. and Fairbairn, D. (1969). "Comparison of cuticular and intestinal absorption of glucose by adult *Ascaris lumbricoides.*" J. Parasit. 55, 13—16.

Chapman-Andresen, C. (1962). "Studies on pinocytosis in amoebae." C.R. Trav. Lab. Carlsberg ser. Chim. 33, 73—264.

Chappell, L.H. (1974). "Methionine uptake by larval and adult *Schistosoma mansoni.*" Int. J. Parasit. 4, 361—369.

Chappell, L.H., Arme, C. and Read, C.P. (1969). "Studies on membrane transport. V. Transport of long chain fatty acids in *Hymenolepis diminuta* (Cestoda)." Biol. Bull. 136, 313—326.

Cox, F.E.G. and Vickerman, K. (1966). "Pinocytosis in *Plasmodium vinckei.*" Ann. trop. Med. Parasit. 60, 293—296.

Crompton, D.W.T. and Lockwood, A.P.M. (1968). "Studies on the absorption and metabolism of D-(U-^{14}C) glucose by *Polymorphus minutus* (Acanthocephala) in vitro." J. exp. Biol. 48, 411—425.

Curtis, A.S.G. (1972). In "Functional aspects of parasite surfaces." (A.E.R. Taylor and R. Muller, eds.), pp. 1—18. Blackwell, Oxford.

Davies, E.E., Warhurst, D.C. and Peters, W. (1975). "The chemotherapy of rodent malaria, XXI. Action of quinine and WR 122,455 (a 9-phenanthrenemethanol) on the fine structure of *Plasmodium berghei* in mouse blood." Ann. trop. Med. Hyg. 69, 147—153.

Dawes, B. and Hughes, D.L. (1964). In "Advances in Parasitology" (B. Dawes, ed.) 2, pp. 97—168. Academic Press, London and New York.

Dawes, B. and Hughes, D.L. (1970). In "Advances in Parasitology" (B. Dawes, ed.), 8, pp. 259—274. Academic Press, London and New York.

Dike, S.C. (1971). "Ultrastructure of the esophageal region in *Schistosoma mansoni.*" Am. J. trop. Med. Hyg. 20, 552—568.

Dike, S.C. and Read, C.P. (1971a). "Tegumentary phosphohydrolases of *Hymenolepis diminuta.*" J. Parasit. 57, 81—87.

Dike, S.C. and Read, C.P. (1971b). "Relation of tegumentary phosphohydrolase and sugar transport in *Hymenolepis diminuta.* J. Parasit. 57, 1251—1255.

Dunn, M.J. (1969). "Alterations of red cell sodium transport during malarial infection." J. Clin. Inv. 48, 674—684.

Eaton, R.D., Meerovitch, E. and Costerton, J.W. (1969). "A surface-active lysosome in *Entamoeba histolytica.*" Trans. R. Soc. trop. Med. Hyg. 63, 678—680.

Eaton, R.D., Meerovitch, E. and Costerton, J.W. (1970). "The functional morphology of pathogenicity in *Entamoeba histolytica.*" Ann. trop. Med. Parasit. 64, 299—304.

Edmonds, S.J. (1965). "Some experiments on the nutrition of *Moniliformis dubius* Meyer (Acanthocephala)." Parasitology, 55, 337—344.

El-Hashimi, W. and Pittman, F. (1970). "Ultrastructure of *Entamoeba histolytica* trophozoites obtained from the colon and from in vitro cultures." Am. J. Trop. Med. Hyg. 19, 215—226.

Entner, E. and Gonzalez, C. (1966). "Changes in antigenicity of *Trypanosoma lewisi* during the course of infection in rats." J. Protozool. 13, 642—645.

Erasmus, D.A. (1972). "The Biology of Trematodes." Edward Arnold, London.

Evans, D.A. and Brown, R.C. (1972). "The utilisation of glucose and proline by culture forms of *Trypanosoma brucei.*" J. Protozool. 19, 686—690.

Gentner, H., Savage, W.R. and Castro, G.A. (1972). "Disaccharidase activity in isolated brush border of *Ascaris suum.*" J. Parasit. 58, 247—251.

Granger, B. and Baker, R.F. (1950). "Electron microscope investigation of the striated border of intestinal epithelium." Anat. Rec. 107, 423—442.

Grant, C.T. and Senft, A.W. (1971). "Schistosome proteolytic enzyme." Comp. Biochem. Physiol. 38, 663—678.

Halton, D.W. (1963). "Some hydrolytic enzymes in two digenetic trematodes." Proc. 16th Int. Cong. Zool. Washington D.C. 1, 29.

Halton, D.W. (1967). "Observations on the nutrition of digenetic trematodes." Parasitology, 57, 639—660.

Halton, D.W. (1974). "Hemoglobin absorption in the gut of a monogenetic trematode." J. Parasit. 60, 59—66.

Halton, D.W. (1975). "Intracellular digestion and cellular defecation in a monogenean, *Diclidophora merlangi.*" Parasitology, 70, 331—340.

Halton, D.W., Dermott, E. and Morris, G.P. (1968). "Electron microscope studies of *Diclidophora merlangi* (Monogenea: Polyopisthocotylea). I. Ultrastructure of the cecal epithelium." J. Parasit. 54, 909—916.

Halton, D.W. and Jennings, J.B. (1965). "Observations on the nutrition of monogenetic trematodes." Biol. Bull. 129, 257—272.

Hampton, J.R. (1970a). "Lysine transport in the culture form of *Trypanosoma cruzi:* kinetics and inhibition of uptake by structural analogues." Int. J. Biochem. 6, 706—714.

Hampton, J.R. (1970b). "Lysine uptake in cultured *Trypanosoma cruzi:* interactions of competitive inhibitors." J. Protozool. 17, 597—600.

Hampton, J.R. (1971). "Arginine transport in the culture form of *Trypanosoma cruzi.*" J. Protozool. 18, 701—703.

Harris, B.G. and Read, C.P. (1968). "Studies on membrane transport. III. Further characterisation of amino acid systems in *Hymenolepis diminuta* (Cestoda)." Comp. Biochem. Physiol. 26, 545—552.

Herman, R. (1969). "Osmotic fragility of normal duck erythrocytes as influenced by extracts of *Plasmodium lophurae*, *P. lophurae*-infected cells and plasma." J. Parasitol. 55, 626—632.

Holter, H. (1959). "Problems of pinocytosis with special reference to amoebae." Ann. N.Y. Acad. Sci. 78, 524—537.

Howell, M.J. (1973). "Localisation of proteolytic activity in *Fasciola hepatica.*" J. Parasit. 59, 454—456.

Isseroff, H., Bontá, C.Y. and Levy, M.G. (1972a). "Monosaccharide absorption by *Schistosoma mansoni*. I. Kinetic characteristics." Comp. Biochem. Physiol. 43A, 849—858.

Isseroff, H. and Levy, M. (1972). "Amino acid transport in *Schistosoma mansoni.*" Am. Zool. 12, 681.

Isseroff, H. and Read, C.P. (1969). "Studies on membrane transport. VI. Absorption of amino acids by fascioliid trematodes." Comp. Biochem. Physiol. 30, 1153—1159.

Isseroff, H. and Read, C.P. (1974). "Studies on membrane transport. VIII. Absorption of monosaccharides by *Fasciola hepatica.*" Comp. Biochem. Physiol. 47A, 141—152.

Isseroff, H., Tunis, M. and Read, C.P. (1972b). "Changes in amino acids of bile in *Fasciola hepatica* infections." Comp. Biochem. Physiol. 41B, 157—163.

Kilejian, A. (1966). "Permeation of L-proline in the cestode, *Hymenolepis diminuta.*" J. Parasit. 52, 1108—1115.

Langreth, S.G. and Balber, A.E. (1975). "Protein uptake and digestion in bloodstream and culture forms of *Trypanosoma brucei.* J. Protozool. 22, 40—53.

Laurie, J.S. (1957). "The in vitro fermentation of carbohydrates by two species of cestodes and one species of Acanthocephala." Expl. Parasit. 6, 245—260.

Lee, D.L. (1965). "The Physiology of Nematodes." Oliver and Boyd, Edinburgh and London.

Lee, D.L. (1972). "The structure of the helminth cuticle." Adv. Parasit. 10, 347—379.

Lumsden, R.D. (1975). "Surface ultrastructure and cytochemistry of parasitic helminths." Expl. Parasit. 37, 267—339.

Lumsden, R.D., Gonzalez, G., Mills, R. and Viles, J. (1968). "Cytological studies on the absorptive surfaces of cestodes. III. Hydrolysis of phosphate esters. J. Parasit. 54, 524—535.

Lumsden, R.D., Threadgold, L.T., Oaks, J. and Arme, C. (1970). "On the permeability of cestodes to colloids: an evaluation of the transmembranosis hypothesis." Parasitology 60, 185—193.

Lushbaugh, W.B. and Miller, J.H. (1974). "Fine structural topochemistry of *Entamoeba histolytica* Schaudinn, 1903." J. Parasit. 60, 421—433.

Manjra, A.A. and Dusanic, D.G. (1972). Mechanisms of amino acid transport in *Trypanosoma lewisi.*" Comp. Biochem. Physiol. 41A, 897—903.

Manjra, R. and Dusanic, D.G. (1973). "Transport of nucleosides in *Trypanosoma lewisi.*" Comp. Biochem. Physiol. 44B, 587—593.

Mansour, T. (1959). "Studies on the carbohydrate metabolism of the liver fluke *Fasciola hepatica.*" Biochim. Biophys. Acta 34, 456—464.

McCormick, G.J. (1970). "Amino acid transport and incorporation in red blood cells of normal and *Plasmodium knowlesi*-infected Rhesus monkeys." Expl. Parasit. 27, 143—149.

McManus, D.P. and James, B.L. (1975). "The absorption of sugars and organic acids by the daughter sporocysts of *Microphallus similis* (Jäg)." Int. J. Parasit. 5, 33—38.

Mettrick, D.F. (1970). "Protein nitrogen, amino acid and carbohydrate gradients in the rat intestine." Comp. Biochem. Physiol. 37, 517—541.

Mettrick, D.F. (1971). "The microbial fauna, nutritional gradients, and physicochemical characteristics of the small intestine of uninfected and parasitised rats." Can. J. Physiol. Pharmac. 49, 972—984.

Mettrick, D.F. and Podesta, R.B. (1974). In "Advances in Parasitology" (B. Dawes, ed.) 12, pp. 183—278. Academic Press, New York and London.

Miller, D. and Crane, R.K. (1961a). "The digestive function of the epithelium of the small intestine. I. An intracellular locus of disaccharide and sugar phosphate ester hydrolysis." Biochem. Biophys. Acta, 52, 281—293.

Miller, D. and Crane, R.K. (1961b). "The digestive function of the epithelium of the small intestine. II. Localisation of disaccharide hydrolysis in the isolated brush border portion of intestinal epithelial cells. Biochem. Biophys. Acta. 52, 293—298.

Min, H.S. (1965). "Studies on the transport of carbohydrate in *Crithidia luciliae.*" J. cell comp. Physiol. 65, 243—248.

Min, H.S. (1966). "Effects of inhibitor, competitors and temperature on transport of carbohydrates in *Crithidia luciliae.*" J. cell. Physiol. 68, 237—240.

Moulder, J.W. (1962). "The Biochemistry of Intracellular Parasitism." pp. 13—42. Univ. Chicago Press, Chicago.

Nollen, P.M., Pyne, J.L. and Bajt, J.E. (1974). "*Megalodiscus temperatus:* absorption and incorporation of tritiated tyrosine, thymidine and adenosine." Expl. Parasit. 35, 132—140.

Overturf, M. (1966). "In vivo and in vitro uptake and distribution of ^{14}C-labelled glucose by *Hymenolepis diminuta,"* Comp. Biochem. Physiol. 17, 705—713.

Page, C.R. III and MacInnis, A.J. (1975). "Characterisation of nucleoside transport in hymenolepidid cestodes." J. Parasit. 61, 281—290.

Pappas, P.W. (1972). *"Hymenolepis diminuta:* absorption of nicotinamide." Expl. Parasit. 32, 403—406.

Pappas, P.W. and Read, C.P. (1972a). "The absorption of pyridoxine and riboflavin by *Hymenolepis diminuta."* J. Parasit. 58, 417—421.

Pappas, P.W. and Read, C.P. (1972b). "Thiamine uptake by *Hymenolepis diminuta."* J. Parasit. 58, 235—239.

Pappas, P.W. and Read, C.P. (1972c). "Trypsin inactivation by *Hymenolepis diminuta."* J. Parasit. 58, 864—871.

Pappas, P.W. and Read, C.P. (1972d). "Inactivation of α- and β-chymotrypsin by intact *Hymenolepis diminuta* (Cestoda)." Biol. Bull. 173, 605—616.

Pappas, P.W. and Read, C.P. (1974). "Relationship of nucleoside transport and surface phosphohydrolase activity in the tapeworm, *Hymenolepis diminuta."* J. Parasit. 60, 447—452.

Pappas, P.W. and Read, C.P. (1975). "Membrane transport in helminth parasites: a review." Expl. Parasit. 37, 469—530.

Pappas, P.W., Uglem, G.L. and Read, C.P. (1973). "Ribonuclease activity associated with intact *Hymenolepis diminuta."* J. Parasit. 59, 824—828.

Pappas, P.W., Uglem, G.L. and Read, C.P. (1974). "Anion and cation requirements for glucose and methionine accumulation in *Hymenolepis diminuta* (Cestoda)." Biol. Bull. 146, 56—66.

Phifer, K.O. (1960). "Permeation and membrane transport in animal parasites: the absorption of glucose by *Hymenolepis diminuta."* J. Parasit. 46, 51—62.

Pittman, R.G. and Fisher, F.M. Jnr. (1972). "The membrane transport of glycerol by *Hymenolepis diminuta."* J. Parasit. 58, 742—749.

Podesta, R.B. and Mettrick, D.F. (1974). "Components of glucose transport in the host-parasite system, *Hymenolepis diminuta* (Cestoda) and the rat intestine." Can. J. Phys. Pharmac. 52, 183—197.

Preston, T.M. (1969). "The form and function of the cytosome-cytopharynx of the culture forms of the elasmobranch haemoflagellate *Trypanosoma raiae* Laveran and Mesnil." J. Protozool. 16, 320—333.

Proctor, E.M. and Gregory, M.A. (1972). "The observation of a surface active lysosome in the trophozoites of *Entamoeba histolytica* from the human colon." Ann. trop. Med. Parasit. 66, 339—342.

Proctor, E.M. and Gregory, M.A. (1973). "Ultrastructure of *E. histolytica* — strain NIH 200." Int. J. Parasit. 3, 457—460.

Read, C.P. (1950). "The vertebrate small intestine as a habitat for parasitic helminths." Rice Inst. Pam. 37, 1—94.

Read, C.P. (1967). "Carbohydrate metabolism in *Hymenolepis* (Cestoda)." J. Parasit. 53, 1023—1029.

Read, C.P. (1971). In "Ecology and physiology of parasites." (A.M. Fallis ed.) pp. 188—200. University Toronto Press, Toronto and Buffalo.

Read, C.P. (1973). "Contact digestion in tapeworms." J. Parasit. 59, 672—677.

Read, C.P., Douglas, L.T. and Simmons, J.E. Jr. (1959). "Urea and osmotic properties of tapeworms from elasmobranchs." Expl. Parasit. 8, 58—75.

Read, C.P., Rothman, A.H. and Simmons, J.E. Jr. (1963). "Studies on membrane transport with special reference to host-parasite integration." Ann. N.Y. Acad. Sci. 113, 154—205.

Read, C.P., Stewart, G.L. and Pappas, P.W. (1974). "Glucose and sodium fluxes across the brush border of *Hymenolepis diminuta* (Cestoda)." Biol. Bull., 147, 146—162.

Rhoades, M.L. and Romanowski, R.D. (1974). "The secretory nature of the excretion gland cells of *Stehanurus denatatus*. III. Proteinase inhibitors." Expl. Parasit. 35, 363–368.

Riding, I.L. (1970). "Microvilli on the outside of a nematode." Nature (Lond.) 226, 179–180.

Roberts, L.S. and Mong, F.N. (1973). "Developmental physiology of cestodes XIII. Vitamin B_6 requirement of *Hymenolepis diminuta* during in vitro cultivation." J. Parasit. 59, 101–104.

Robinson, G. and Threadgold, L.T. (1975). "Electron microscope studies of *Fasciola hepatica*. XII. The fine structure of the gastrodermis." Expl. Parasit. 37, 20–36.

Rohde, K. (1973). "Ultrastructure of the caecum of *Polystomoides malayi* Rohde and *P. renschii* Rohde (Monogenea: Polystomatidae)". Int. J. Parasit. 3, 461–466.

Rothman, A.H. and Fisher, F.M. Jnr. (1964). "Permeation of amino acids in *Moniliformis* and *Macracanthorhynchus* (Acanthocephala)." J. Parasit. 50, 410–414.

Rudzinska, M.A. and Trager, W. (1957). "Intracellular phagotrophy by malaria parasites." J. Protozool. 4, 190–199.

Rudzinska, M.A. and Vickerman, K. (1968). In "Infectious blood diseases of man and animals." (D. Weisman and M. Ristic, eds.), pp. 217–306. Academic Press, New York and London.

Ruff, M.D. and Read, C.P. (1973). "Inhibition of pancreatic lipase by *Hymenolepis diminuta.*" J. Parasit. 59, 105–111.

Ruff, M.D. and Read, C.P. (1974). "Specificity of carbohydrate transport in *Trypanosoma equiperdum.*" Parasitology 68, 103–115.

Ruff, M.D., Uglem, G.L. and Read, C.P. (1973). "Interactions of *Moniliformis dubius* with pancreatic enzymes." J. Parasit. 59, 839–843.

Rutherford, T.A. and Webster, J.M. (1974). "Trans-cuticular uptake of glucose by the entomophilic nematode, *Mermis nigrescens.*" J. Parasit. 60, 804–808.

Sanchez, G. (1974). "The effect of some amino acids on carbohydrate uptake by *Trypanosoma lewisi.*" Comp. Biochem. Physiol. 47A, 553–558.

Sanchez, G. and Dusanic, D.G. (1968). "Respiratory activity of *Trypanosoma lewisi* during several phases of infection in the rat." Expl. Parasit. 23, 361–370.

Sanchez, G. and Read. C.P. (1969). "Carbohydrate transport in *Trypanosoma lewisi.*" Comp. Biochem. Physiol. 28, 931–937.

Sauer, M.C.V. and Senft, A.W. (1972). "Properties of a proteolytic enzyme from *Schistosoma mansoni.*" Comp. Biochem. Physiol. 42, 205–220.

Seed, J.R., Baquero, M.A. and Duda, J.F. (1965). "Inhibition of hexose and glycerol utilisation by 2-deoxy-D-glucose in *Trypanosoma gambiense* and *Trypanosoma rhodesiense.*" Expl. Parasit. 16, 363–368.

Serrano, R. and Reeves, R.E. (1974). "Glucose transport in *Entamoeba histolytica.*" Biochem. J. 144, 43–48.

Serrano, R. and Reeves, R.E. (1975). "Physiological significance of glucose transport in *Entamoeba histolytica.*" Expl. Parasit. 37, 411–416.

Sherman, I.W., Raghunath, A.V. and Ruble, J.A. (1967). "The accumulation of amino acids by *Plasmodium lophurae* (avian malaria)." Comp. Biochem. Physiol. 23, 43–57.

Sherman, I.W. and Tanigoshi, L. (1971). "Alterations in sodium and potassium in red blood cells and plasma during the malaria infection (*Plasmodium lophurae*)". Comp. Biochem. Physiol. 40A, 543–546.

Sherman, I.W. and Tanigoshi, L. (1972). "Incorporation of ^{14}C-amino acids by malaria (*Plasmodium lophurae*). V. Influence of antimalarials on the transport and incorporation of amino acids." Proc. Helminth. Soc. Wash. 39, 250–260.

Simmons, J.E. Jnr. (1961). "Urease activity in trypanorhynch cestodes." Biol. Bull. 121, 535–546.

Simmons, J.E. Jnr., Read, C.P. and Rothman, A.H. (1960). "Permeation and membrane transport in animal parasites: permeation of urea into cestodes from elasmobranchs." J. Parasit. 46, 43—50.

Smyth, J.D. (1973). "Some interface phenomena in parasitic protozoa and platyhelminths." Can. J. Zool. 51, 367—377.

Southworth, G.C. and Read, C.P. (1969). "Carbohydrate transport in *Trypanosoma gambiense.*" J. Protozool. 16, 720—723.

Southworth, G.C. and Read, C.P. (1970). "Specificity of sugar transport in *Trypanosoma gambiense.*" J. Protozool. 17, 396—399.

Southworth, G.C. and Read, C.P. (1972). "Absorption of some amino acids by the haemoflagellate, *Trypanosoma gambiense.*" Comp. Biochem. Physiol. 41A, 905—911.

Taylor, E.W. and Thomas, J.N. (1968). "Membrane (contact) digestion in the three species of tapeworm, *Hymenolepis diminuta, Hymenolepis microstoma* and *Moniezia expansa.*" Parasitology 58, 535—546.

Trager, W. (1967). "Adenosine triphosphate and the pyruvic and phosphoglyceric acid kinase of the malaria parasite *Plasmodium lophurae.*" J. Protozool. 14, 110—113.

Trager, W. (1972). In "Comparative Biochemistry of Parasites." (H. van den Bossche, ed.), pp. 343—350. Academic Press, New York and London.

Trager, W. (1974). In "Trypanosomiasis and Leishmaniasis, with special reference to Chaga's disease." pp. 225—245. Ciba Fdn. Symp. No. 20. Elsevier, Amsterdam.

Trager, W. (1975). "On the cultivation of *Trypanosoma vivax:* a tale of two visits in Nigeria." J. Parasit. 61, 3—11.

Ubelaker, J.E., Cooper, N.B. and Allison, V.F. (1970). "The fine structure of the cysticercoid of *Hymenolepis diminuta.* I. The outer wall of the capsule." Z. f. Parasitenk. 34, 258—270.

Uglem, G.L., Pappas, P.W. and Read, C.P. (1974). "Na⁺-dependent and Na⁺-independent glycerol fluxes in *Hymenolepis diminuta* (Cestoda)." J. Comp. Physiol. 93, 157—171.

Uglem, G.L. and Read, C.P. (1973). "*Moniliformis dubius:* uptake of leucine and alanine by adults." Expl. Parasit. 34, 148—153.

Ugolev, A.M. (1965). "Membrane (contact) digestion." Physiol. Rev. 45, 555—595.

van den Bossche, H. and Borgers, M. (1973). "Subcellular distribution of digestive enzymes in *Ascaris suum* intestine." Int. J. Parasit. 3, 59—65.

van den Bossche, H. and De Nollen, S. (1973). "Effects of mebendazole on the absorption of low molecular weight nutrients by *Ascaris suum.*" Int. J. Parasit. 3, 401—407.

Vickerman, K. (1969). "The fine structure of *Trypanosoma congolense* in its bloodstream phase." J. Protozool. 16, 54—69.

Vickerman, K. (1972). In "Functional aspects of parasite surfaces." (A.E.R. Taylor and R. Muller, eds.). pp. 71—92. Blackwell, Oxford.

Vickerman, K. (1974). In "Trypanosomiasis and Leishmaniasis with special reference to Chaga's disease." Ciba Fed. Symp. No. 20, pp. 171—190.

Wander, R.H. and Hilgard, H.R. (1975). "The specificity of amino acid uptake by *Amoeba proteus.*" Comp. Biochem. Physiol. 50C, 67—70.

Warhurst, D.C. (1973). In "Chemotherapeutic agents in the study of parasites." (A.E.R. Taylor and R. Muller, eds.), pp. 1—28. Blackwell, Oxford.

Warhurst, D.C., Homewood, C.A. and Baggaley, V.C. (1974). "The chemotherapy of rodent malaria. XX. Autophagic vacuole formation in *Plasmodium berghei* in vitro." Ann. trop. Med. Parasit. 68, 265—281.

Weatherly, N.F., Hanson, M.F. and Moser, H.C. (1963). "In vitro uptake of ¹⁴C-labelled alanine and glucose by *Ascaridia galli* (Nematoda) of chickens." Expl. Parasit. 14, 37—48.

Weinbach, E.C. and Diamond, L.S. (1974). "*Entamoeba histolytica.* I. Aerobic metabolism." Expl. Parasit. 35, 232—243.

Woodward, C.K. and Read, C.P. (1969). "Studies on membrane transport. VIII. Transport of histidine through two distinct systems in the tapeworm *Hymenolepis diminuta.*" Comp. Biochem. Physiol. 30, 1161—1177.

Wright, R.W. and Isseroff, H. (1973). "Further studies on the absorption of acetate by *Fasciola hepatica.*" Comp. Biochem. Physiol. 45B, 95—99.

Zam, S.G., Martin, W.E. and Thomas, L.J. Jnr. (1963). "In vitro uptake of Co^{60}-vitamin B_{12} by *Ascaris suum.*" J. Parasit. 49, 190—196.

Ecological Aspects of Parasitology
Editor: C.R. Kennedy
© *North-Holland Publishing Company, Amsterdam, 1976*

CHAPTER 5

NEGATIVE INTERACTION AMONGST PARASITES

O. HALVÖRSEN

Institute of Biology and Geology, University of Tromsö, N-9001 Tromsö (Norway)

Contents

5.1. Concepts of negative interaction

Organisms living in the same community may interact with each other so that one organism or population is negatively affected by the other. The mechanisms involved in these interactions and their ecological results are often complex, but the different relationships are usually categorized as amensalism, predation, parasitism or competition (Odum, 1971). It has proved difficult to define parasitism and to draw the line between parasitism and predation. Competition has also proved difficult to define strictly as is illustrated by Birch (1957), Milne (1961), DeBach (1966), Miller (1967), Odum (1971), and Krebs (1972) who all define competition differently.

Historical reviews of the meaning of competition in biological literature and discussion of its definition have been given by Birch (1957), Milne (1961), DeBach (1966) and Miller (1967). Competition is sometimes used in a very broad sense and may include predation. Most authors, however, attempt to restrict the meaning of competition to interaction among organisms for obtaining a common limited resource (Birch 1957, Milne 1961, Miller 1967, Ayala 1972). According to Milne (1961) most of the things for which animals compete can, in addition to mates, be classed as either food or space.

5.2. Evidence for, and examples of, negative interaction amongst parasites

5.2.1. Interspecific interaction

5.2.1.1. Exclusion

Examples of exclusion or reduction in numbers of one or more parasite species in the presence of another are given by many authors. Some of these examples are referred in the following discussion. Cross-immunity will not be included in detail, as it will be considered elsewhere (Chapter 6).

Negative interaction amongst ectoparasites of fish has been studied by Wilson (1916) and Paperna (1964). Wilson (1916) exposed crappies, *Pomoxis annularis*, of which 25 out of 100 were infected with the copepod *Ergasilus coeruleus*, to infection with glochidia of *Lampsilis recta*. The presence of even a small number of copepods (10 or less) upon the gills reduced the number of glochidia which established to about one-fourth of that found on the controls free of copepods. When there were 200 copepods or more on a gill very few, or no, glochidia became established. Wilson claimed that the antagonism may also work in the opposite direction, as no copepods were found on fish naturally infected with 300 or more glochidia. Paperna (1964) investigated the occurence of the monogeneans *Dactylogyrus extensus* and *D. vastator* on the gills of reared carp. In cases of heavy infestation with *D. vastator* (50 or more per fish), infestation with *D. extensus* was slight (one or two per fish) and the latter species was found only on the basal part of the gill arches close to the insertion on the pharynx. The number of fish that were infected with a single species only was much greater than expected. Laboratory tests showed that when fish initially infected with *D. extensus* were subsequently exposed to *D. vastator* there was a gradual disappearance of the former species with increasing numbers of the latter. When the fish became free of *D. vastator*, they were again prone to infection by *D. extensus*. A similar interaction between *D. vastator* and *D. anchoratus* was reported by Paperna and Kohn (1964).

Indications of negative interaction leading to exclusion or reduction of one species have also been observed among intestinal parasites of fish. Cross (1934) investigated ciscoes infected with the cestode *Proteocephalus exiguys* and acanthocephala of the genus *Neoechinorhynchus*. He found that fish having 15 or more acanthocephala had a very low infection with tapeworms, and those harbouring more than 25 tapeworms contained few or no acanthocephala. Thomas (1964), who studied the populations of helminth parasites in brown trout, found negative correlation in the occurrence of several pairs of parasites; *Crepidostomum metoecus* and *Dacnitis truttae*, *Crepidostomum farionis* and *Neoechinorhynchus rutili*, and *C. farionis* and *D. truttae*. The most clearly defined cases of negative correlations occur only between species occupying the same microhabitat. Reichenbach-Klinke (1966) examined a number of fish species from the Donau for parasites, and recorded that infections with one parasite species predominated. When the acanthocephalan *Pomphorhynchus laevis* occurred in intense infections of 80—200, it was nearly always the only species.

With mammals a number of experimental studies have been undertaken to elucidate the effects of concurrent infections with species of parasites. Larsh and Donaldson (1944) infected mice subcutanously with 2000 larvae of the nematode *Nippostrongylus muris* and within an hour thereafter each mouse was given 4200 or 2150 eggs of the tapeworm *Hymenolepis nana* var. *fraterna*. Compared with controls the results indicated that infection with *N. muris* reduced the number of cysticercoids of *H. nana* that were able to develop.

Cox (1952) showed in experiments with mice that a previous infection with the nematode *Ancylostoma caninum* inhibited the development of *Trichinella spiralis*. Goulson (1958) extended these experiments and found that there was a significant reduction in the number of adult *T. spiralis* establishing only when the *A. caninum* infection was administered 24 to 48 hours before the infection with *T. spiralis*. The adult *T. spiralis* population was not reduced when the interval between the infections was shorter than 24 hours or longer than 48 hours.

Keeling (1961) studied experimentally the relationship between the nematodes *Trichuris muris* and *Aspiculuris tetraptera* in the albino mouse. The results showed that the number of hosts in which *T. muris* established was greatly reduced when *A. tetraptera* was established in them first. Some exclusion was also noted when simultaneous infection was carried out, but when an infection with *A. tetraptera* was superimposed on an infection with *T. muris*, the reduction in the number of the latter was statistically insignificant.

Species of gasterointestinal nematodes of sheep have been reported to interact negatively with each other. Stewart (1955) reported that intake of larvae of the abomasal parasites *Haemonchus contortus, Ostertagia circumcincta* and *Trichostrongylus axei* could cause elimination of each other, and of *T. colubriformis* in the duodenum. Turner et al. (1962) infected lambs experimentally with larvae of *H. contortus, O. circumcincta* and *T. axei.* Three groups of lambs were employed; those given a single species infection, those simultaneously infected with any combination of two species, and those simultaneously infected with all three species. The results showed that *H. contortus* was adversely affected, especially by *T. axei* infections but also by infections with *O. circumcincta. O. circumcincta* tended to be moderately adversely affected by infections of one or both of the other two species, while *T. axei* appeared to be slightly enhanced by infections with the other two.

Kisilewska's (1970a) results from her studies of parasites of the bank vole, *Clethrionomys glareolus* provide indications of negative interactions amongst parasites of a natural mammal population. The analysis of the occurrence of 8 species of cestodes and 10 species of nematodes in the rodent intestine suggests that negative interactions leading to exclusion or reduction take place between several pairs of the parasite species.

The examples given above comprise mainly ectoparasites and parasites of the alimentary canal and represent open systems (Kennedy, 1975). There are also several examples of negative interaction leading to exclusion or reduction in closed systems (Kennedy, 1975) where the parasite is normally not released from the host except for the continuation of its life cycle.

Among investigators of parasites in the circulatory system of vertebrates, Schilling (1936) observed that *Trypanosoma brucei* and *T. congolense* could supersede each other in mixed infections. In most cases *T. congolense* appeared to be the "stronger" of the two species. Von Brand and Tobie (1960) showed experimentally that when an unaltered strain of trypanosomes was mixed with a strain that either had lost its kinetoplast or that had been rendered arsenic-resistant, the strain with changed characteristics invariably, and more or less rapidly, disappeared from a population maintained by serial transfer in rats or mice. The observed rate of disappearance closely followed the rate of disappearance calculated from the multiplication rates of the parent strains. Hamsters previously inoculated with virulent strains of *Leishmania infantum* were found by Adler (1954) to be strikingly less susceptible to *Plasmodium berghei* than normal hamsters. The infection with *L. infantum* inhibited the multiplication of the malarian parasite but the reciprocal situation was not confirmed.

Investigators of snail infections with trematode larvae have noted the rarity of concurrent infections in the same host. Sewell (1922) and Dubois (1929) appear to be among the first to have published observations. Cort et al. (1937) found in their study of cercariae in *Stagnicola emarginata* that the most common plagiorchid species occurred in double infections with the most common strigeids at a frequency less than expected, as also did combinations of plagiorchids. The echinostome *Echinostomum revolutum* was never found in combination with other species. Wesenberg-Lund (1934) observed that daughter rediae of echinostomes within the mother redia often devoured other daughter rediae or young cercariae of the same species, and he also published a photograph of an echinostome redia with its oral sucker having a firm grip on another redia. Nasir (1962) found mutilated rediae of *Echinostoma nudicaudatum* in the intestine of other rediae of the same species and suggested that the rediae were feeding on each other.

Lie et al. (1965) reported antagonism between two species of larval trematodes when double infections were experimentally produced in the same snail. In a series of different combinations of trematodes of which an echinostome was one of the species, larvae of the other species diminished in number and eventually disappeared completely. Lie et al. (1965) suggested that the missing sporocysts might have been eaten by the large active echinostome rediae. The phenomenon was investigated further in a series of publications, and the work was reviewed by Lie et al. (1968) and by Lim and Heyneman (1972). It was confirmed that echinostome rediae would prey upon other trematode larvae in the same host, and Lie et al. (1968) suggested that this interaction, which they termed direct antagonism, may be a feature of rediae in general. In addition to direct antagonism, interactions which Lie et al. (1968) termed indirect antagonism were observed in multispecies trematode infections in snails. In most combinations of trematode species used in the experiments, indirect antagonism was found to work both ways in contrast to direct antagonism where one species would always be dominant. Indirect antagonism leads to retardation of development, regression, degeneration or suppression of an affected species. Indirect antagonism is effected by sporocysts and rediae. Results in accordance with those of Lie and associates have been obtained by Anteson (1970) and Dönges (1972) with other species combinations of trematodes and snails.

5.2.1.2. Segregation

In addition to results and observations which show that infection with one species of parasite may exclude or reduce infection with another species, it has been recorded that concurrent infection may influence the localisation of the parasites within the host organism. Holmes (1973) examined some of the basic features of microhabitat specificity of parasites and explored its relationships with interspecific interactions and the development of helminth communities. He gave several examples, of which some are included below, of how concurrent infection with two parasite species may influence the localisation of one or both of them.

Holmes (1961, 1962) investigated experimentally concurrent infection of rats with the acanthocephalan *Moniliformis dubius* and the cestode *Hymenolepis diminuta*. When rats already infected with mature acanthocephalans were given cystericercoids, the anteriad ontogenetic migration of the tapeworm did not occur. When acanthocephalans were given to rats with mature tapeworms, the tapeworms moved posteriorly when the growing acanthocephalans migrated into the anterior part of the intestine.

Chappell (1969) observed that when the cestode *Proteocephalus filicollis* occurred concurrently with the acanthocephalan *Neoechinorhynchus rutili* in the three-spined stickleback, the distribution of each species in the gut was significantly different from when these species occurred alone. MacKenzie and Gibson (1970) found that the nematode *Contracaecum heterochrous* was fairly evenly distributed throughout the gut of flounder except when *C. minutus* was also present. In the latter case nearly two-thirds of the *C. heterochrous* were found in the rectum of the fish. In the intestine of plaice the localization of the digenean *Podocotyle* sp. was different when the closely related *Plagioporus varius* was present, compared with its position when *Podocotyle* sp. occurred alone. Halvorsen and MacDonald (1972) examined the distribution of helminths in the intestine of brown trout. They recorded that the distribution of the digenean *Crepidostomum metoecus* was changed in the presence of the cestode *Cyathocephalus truncatus*, while the localization of the latter was unaltered.

5.2.2. Intraspecific interaction

In many tapeworm infections, the size of the worms in the intestine of the final host shows an inverse relationship to the number of worms present (Woodland, 1924; Shorb, 1933; Hunninen, 1935; Wardle and Green, 1941; Read, 1959; Roberts and Mong, 1968; Andersen, 1972 among others). In *Diphyllobothrium dendriticum* crowding will also delay strobilisation, and socalled primary strobila will develop (Halvorsen and Andersen, 1974).

Amongst pseudophyllidean cestodes, growth and development of the procercoid in the first intermediate host have been found to be retarded or completely hampered as the number of larvae increases, and the mature procercoid will be smaller (Halvorsen, 1966; Guttowa, 1967).

The effects of crowding described above in tapeworm infections have in most cases been noted in single species infections, but Read and Phifer (1959) and Guttowa (1967) observed similar effects in concurrent infections with two tapeworm species.

Effects of crowding have also been observed in digenetic trematodes. Tandon (1973) reported that the development of the reproductive organs and the size of the amphistomes *Gastrolhylax crumenifer* and *Fischoederius elongatus* were reduced in crowded situations in the rumen of buffalo. Basch (1970) described effects of crowding on the formation of tetracotyles of *Cotylurus lutzi* in the snail *Biomphalaria glabrata*. When the number of larvae in the ovotestis was less than about 50, metamorphosis from postcercariae to tetracotyles was rapid; when they numbered several hundred or more, development was greatly retarded; and when larger numbers of cercariae penetrated the snail their development was arrested and tetrocotyles never formed.

In nematodes, size and rate of development appear not to be influenced by crowding, but like many other parasites, survival may be affected. Haley and Parker (1961) studied the effect of population density on adult worm survival in primary *Nippostrongylus brasiliensis* infections in the rat, and found adult worms survived longest in the rats that had the smallest initial worm burden.

Helminth infections with possible intrinsic homeostatic mechanisms for the avoidance of crowding have been described by Halvorsen and Williams (1968) and Williams and Halvorsen (1971), who examined the infections of the cestodarian *Gyrocotyle* in *Chimaera monstrosa* and the tapeworm *Abothrium gadi* in cod. With both parasites several worms were characteristically present in larval infections whereas generally only two mature *Gyrocotyle* and one mature *A. gadi* were found.

5.3. Mechanisms of negative interaction amongst parasites: Comparison with free-living animals

Parasites in the same host organism (idiohostal units; Kisielewska, 1970b) may, as demonstrated above, interact negatively with each other. These investigations cover a broad taxonomic range of both parasites and hosts, but are only a few among all the records which are available. This indicates that negative interactions are a significant phenomenon in the ecology of parasites.

5.3.1. Classification of interactions

Negative interaction takes the form of predation when echinostome rediae are involved in multispecies trematode infections of molluscs. The other examples of negative interaction may be categorized as competition where the host and the resources it supplies or the site in the host is the thing (Milne, 1961) for which the parasites are competing.

Basch (1970) described interaction between larval strigeids and echinostomes which involved direct antagonism (predation), indirect antagonism (competition), and hyperparasitism.

As with free-living animals, one has in most cases to accept certain observed numerical, developmental, and distributional patterns of individuals and species as indications of competition having taken place, since knowledge about the mechanisms and processes which have led to the observed situation is lacking. It is important to realise, however, that similar results, i.e. negative correlation in the occurrence of parasite species, may be caused also by other factors such as food preference and habitat selection by the host. The patterns of parasites in idiohostal units where competition is believed to have taken place are very similar to those situations which are normally interpreted as resulting from competition by students of free-living animals. Most of the examples given above are examples of exclusion or reduction in numbers of one species by another, which are the most often quoted results of competition. The examples given by Holmes (1961, 1962, 1973), Chappell (1969), MacKenzie and Gibson (1970) and Halvorsen and MacDonald (1972) of changes in site of parasites in multispecies, as compared to single species, infections, fall within the interactive segregation of Brian (1956) and Nilsson (1967). Interactive segregation is believed to be replaced in the course of evolution by genetically regulated segregation, called selective segregation by Brian (1956). Segregation of related species of parasites into different organs in the host is relatively common and may be interpreted as evidence for selective site segregation (Holmes, 1973). Segregated distribution of species along the gut is also common; the pattern of related nematodes in the small intestine of ruminants is a well known example of this. In addition, there may be radial segregation and differences in the associations with host tissue among related parasites in the intestine. Segregation in function through morphological adaptation has been discussed by Schad (1963) for nematodes and by Cannon (1972) for digeneans.

The consequences of crowding of parasites are in many respects particularly similar to those observed in aquatic animals confined in limited habitats (Rose and Rose, 1961), and also in larvae of *Drosophila* (Miller, 1967), and many other animal groups. Wynne-Edwards (1964) argued that intraspecific competition may act as a homeostatic mechanism, restricting population size and relating it to the food supply. The infections of *Gyrocotyle* in *Chimaera monstrosa* and *A. gadi* in cod described by Halvorsen and Williams (1968) and Williams and Halvorsen (1971) show similarities to the results predicted by Wynne-Edwards if competitive homeostasis operated. Stromberg and Crites (1975) speculated that a feedback system operating directly through the worms or indirectly through the host may reduce numbers of the nematode *Camallanus oxycephalus* in white bass.

5.3.2. Mechanisms of competition

As already pointed out, the mechanisms and processes involved in competition are rarely known. Those who have observed results indicative of competition among parasites have often suggested that the competition has involved mechanical interference, acquired immunity, and/or the production of materials by one species which are deleterious to the other. In most cases there is no evidence that these are the mechanisms behind the observed situations. In some cases, however, the investigations have penetrated deeper into the causal mechanisms. A more specific suggestion as to the mechanisms involved was made by Basch (1970) who studied the relationships of larval strigeids and echinostomes. He hypothesized that the crowding phenomenon of postcercarial forms, and the inhibition of development in the vicinity of active sporocysts, were manifestations of the same basic mechanism which also controlled metamorphosis and encystment in digenetic trematodes.

5.3.2.1. Food uptake

Utilisation of dietary resources has been shown to be a mechanism involved in competition amongst parasites. Read and Phifer (1959) demonstrated that host dietary carbohydrate quantity and quality might be the limiting factor in interaction between the tapeworms *Hymenolepis citelli* and *H. diminuta*, and among individuals of *H. diminuta*, leading to stunted specimens in crowded situations. Holmes (1959) showed that the interaction of *Moniliformis dubius* and *H. diminuta* was mediated through consumption of the available carbohydrates, at least when these were limited.

5.3.2.2. Reproduction

The outnumbering of one species by another as a result of differences in reproductive potential is regarded as a mechanism of competition by Hardin (1960), Beament (1961) and DeBach (1966) among others. Von Brand and Tobie (1960) observed that the rate of disappearance of the strain of trypanosomes with changed characteristics in mixed infections with unaltered strains closely followed the rate of disappearance calculated from the multiplication rates of the parent strains.

5.3.2.3. Host-mediated interaction

The indirect antagonism observed in multispecies trematode infections in snails includes an inhibitory effect and degenerative changes. Lim and Heyneman (1972) discussed the possible mechanisms involved in this, and suggested that the effects may be caused by direct toxic or competitive inhibition or by activation of tissue response by the snail.

The exclusion of other monogeneans from the gills of carp by *Dactylogyrus vastator* was shown by Paperna (1964) to be caused by histological changes, mainly hyperplasia of the epithelial lining of the mucus goblet cells, provoked by *D. vastator*.

Cox (1952) concluded that the effect of *Ancylostoma caninum* on concurrent infection with *Trichinella spiralis* was not the product of cross immunity between the two species, and suggested that inflammation and tissue sensitivity might be involved. This was supported by Goulson (1958). Stewart (1953) demonstrated that "self-cure" in nematode infections in sheep is essentially a reaction of the host associated with allergic sensitization and accompanied by the release of histamine. The reaction is accompanied by a local lesion in the mucous membrane. The administration of *Haemonchus contortus* larvae produced the change either in the abomasum or in the small intestine according to whether the sheep was infected with *H. contortus* or

Trichostrongylus spp. Allergic inflammation as a mechanism for the expulsion of worms from tissues was reviewed by Larsh and Race (1975). Their hypothesis for the mechanism of expulsion of adult *Trichinella spiralis* is that an immunologically specific delayed hypersensitivity reaction between antigensensitive T-cells and antigen results in tissue injury. This is followed by an immunologically nonspecific inflammatory reaction that results in tissue changes unfavourable to maintenance of the worms.

Damian (1964) suggested that antigen sharing by two parasite species may represent convergent evolution by the parasites toward greater antigenic compatibility with a common host, while Schad (1966) alternatively proposed that when co-occurring parasites are likely competitors, nonreciprocal cross immunity may be a device evolved to limit the abundance of a competing species. A detailed discussion of cross immunity is given in Chapter 6.

Even if the mechanisms involved in host-mediated competition among parasites are necessarily different from those operating in "habitat-mediated" competition among free-living animals, of which Rose and Rose (1961), Milne (1961) and Nilsson (1967) give examples, the principal similarity between the two situations appears strong. It is, however, necessary to consider that host-mediated reactions are also directed toward the species provoking them. The statement by Wilson and Bossert (1964) that there is no reason to doubt that the essential characteristics of intraspecific chemical communication are also true of interspecific chemical communication is relevant in this context. Miller (1967) expressed doubt as to whether mechanisms involved in interspecific territoriality would have been evolved unless they had first arisen among members of the same species. This is also most likely to be the case for the mechanisms involved in host-mediated competition among parasites.

5.3.3. Interference and exploitation

The process of competition may, according to Miller (1967) and others, involve interference and/or exploitation. Interference refers to any activity which either directly or indirectly limits a competitor's access to a necessary resource or requirement. It usually operates in a spatial context. Exploitation is the utilization of a resource once access to it has been achieved. Miller (1967) pointed out that exploitation is the dominant and characteristic form of competition in simple metazoans where interference mechanisms are poorly developed. In competition among animal parasites, however, interference appears to be the most important element. Only in those instances where intestinal parasites compete for limited food resources (Holmes 1959, Read and Phifer 1959) or where one species outnumbers the other through reproduction (von Brand and Tobie 1960) does exploitation appear to be the most important element.

5.3.4. Competition and phylogenetic relationship

Competition is believed to be of special importance among closely related species (Williams, 1951). It is therefore interesting that so many examples of competition among parasites involve species that are not closely related to each other. They therefore support Levin (1970), who used the hypervolume niche concept of Hutchinson (1957) to create a model which showed that certain dimensions of the hypervolumes were of paramount importance in the competition between two species, and therefore species which appear to fill different niches may be serious competitors.

5.4. The ecological implications of negative interaction amongst parasites

Investigations of negative interactions amongst parasites have been restricted to idiohostal units, but there have been speculations that the results may have applications at other levels of ecological complexity. Wilson (1916) suggested that fish in hatcheries might be protected against copepod infections if they were deliberately infected with glochidia. Reichenbach-Klinke (1966) and Kisielewska (1970a) also indicated this kind of possible use of parasite competition in fish management and animal farming respectively.

Heyneman and Umathevy (1967) and Lie et al. (1974) have carried out experimental field trials to elucidate possible ecological effects of negative interaction among trematode larvae. Heyneman and Umathevy (1967) found that the infection with *Schistosoma spindale* in *Indoplanorbis exustus* was eliminated presumably by *Echinostoma malayanum*, which had been introduced into the same habitat, and that the bulk of snails harbouring both infections died. Similar results were reported by Lie et al. (1974). If it is practically possible to apply negative interaction among larval trematodes in this way it may have a role to play in the control of human schistosomes.

Interspecific competition, and other forms of interspecific negative interaction in idiohostal units leading to reduced survival or reproduction of one or more parasite species in these units, will theoretically lead to fewer parasites of these species within the host population (synhostal units, Kisielewska 1970b). The distribution pattern and the occurrence of the parasites are, together with the separate and combined effects of the parasite species on the growth rate of the host population, important factors in determining the ecological effect of the negative interaction among the parasites. If the combined influence of the parasites is strong and the probability of concurrent infections high, negative interaction among the parasites may be important in reducing the stress on the host population.

The reduction in individual size with increasing density in idiohostal units in tapeworms, the so-called crowding phenomenon, may not automatically lead to reduced growth rate of the parasite population, which consists of all stages of the parasite species within the community. Up to a certain density of the idiohostal unit, rate of parasite maturation appears to be unaffected and the production of cestode progeny per host may not be reduced. In experimental infections of golden hamsters and common gulls the survival rate of *Diphyllobothrium dendriticum* has been found to increase with increasing size of the idiohostal units. This is related to the overdispersion of plerocercoids in the intermediate host population (Halvorsen and Andersen, 1974). The size of the idiohostal unit which gives the maximum growth rate of the parasite population may therefore be above the size where reduction in individual somatic growth and reproduction is seen. Above the optimal size of the idiohostal unit for the growth of the parasite population the situation turns into overcrowding.

Overcrowding, as well as intrinsic regulation of the size of idiohostal units, can function as a feedback control upon parasite population growth rate. The possible effect of this on the growth rate of the host population has not been fully analysed, but will depend on whether the point of overcrowding or density regulated idiohostal unit size is above or below the point where the parasite negatively affects the growth rate of the host population. This will in turn vary with variations in the total stress on the host population resulting from the other factors such as the nutritional state of the host animals.

Negative interactions among parasites leading to reduction of idiohostal unit size, and thus in reduction of the demands of the parasites on the hosts, may have population dynamics effects similar to those of reduction of idiohostal unit size caused by host immune reactions. Crofton (1971) discussed the difficulties in assessing the role of immunological reactions on populations. For the purpose of model-making, Crofton (1971) assumed that a minimal stimulation, expressed in numbers of parasites, was needed to evoke an immunological response which would either remove the parasites or render them sterile, and that the immunological response could be swamped by high infection pressures and host death ensue before an immunological reaction was manifested. By making these assumptions, Crofton found that the introduction of an immune factor into a mathematical model of host-parasite relationships produced an oscillating system, and not invariably higher population levels of host.

5.5. REFERENCES

Adler, S. (1954). "The Behaviour of *Plasmodium berghei* in the Golden Hamster *Mesocricetus auratus* Infected with Visceral Leishmaniasis". Trans. R. Soc. Trop. Med. Hyg. 48, 431—440.

Andersen, K. (1972). "Studies of the Helminth Fauna of Norway. XXI. The Influence of Population Size (Intensity of Infection) on Morphological Characters in *Diphyllobothrium dendriticum* Nitzsch in the Golden Hamster (*Mesocricetus auratus* Waterhouse)". Norw. J. Zool. 20, 1—7.

Anteson, R.K. (1970). "On the Resistance of the Snail, *Lymnaea catascopium pallida* (Adams) to Concurrent Infections with Sporocysts of the Strigeid Trematodes, *Cotylurus flabelliformis* (Faust) and *Diplostomum flexicaudum* (Cort and Brooks)". Ann. Trop. Med. Parasit. 64, 101—107.

Ayala, F.J. (1972). "Competition between Species". Am. Sci. 60, 348—357.

Basch, P.F. (1970). "Relationships of Some Larval Strigeids and Echinostomes (Trematoda): Hyperparasitism, Antagonism, and "Immunity" in the Snail Host". Exp. Parasit. 27, 193—216.

Beament, J.W.L. (1961). In "Mechanisms in Biological Competition". (F.L. Milthorpe, ed.). pp. 62—71. The University Press, Cambridge.

Birch, L.C. (1957). "The Meanings of Competition". Am. nat. 91, 5—18.

Brand, T. von and Tobie, E.J. (1960). "The Mechanism of Elimination of Certain Strains or Species of Trypanosomes when Mixed in Experimental Infections". J. Parasit. 46, 129—136.

Brian, M.V. (1956). "Segregation of Species of the Ant Genus *Myrmica*". J. Anim. Ecol. 25, 319—337.

Cannon, L.R.G. (1972). "Studies on the Ecology of the Papillose Allocreadiid Trematodes of the Yellow Perch in Algonquin Park, Ontario". Can. J. Zool. 50, 1231—1239.

Chappell, L.H. (1969). "Competitive Exclusion between two Intestinal Parasites of the Three-Spined Stickleback, *Gasterosteus aculeatus* L". J. Parasit. 55, 775—778.

Cort, W.W., McMullen, D.B. and Brackett, S. (1937). "Ecological Studies on the Cercariae in *Stagnicola emarginata angulata* (Sowerby) in the Douglas Lake Region, Michigan". J. Parasit. 23, 504—532.

Cox, H.W. (1952). "The Effect of Concurrent Infection with the Dog Hookworm, *Ancylostoma caninum*, on the Natural and Acquired Resistance of Mice to *Trichinella spiralis*". J. Elisha Mitchell. Sci. Soc. 68, 222—235.

Crofton, H.D. (1971). "A Model of Host-Parasite Relationships". Parasitology 63, 343—364.

Cross, S.X. (1934). "A Probable Case of Non-Specific Immunity between two Parasites of Ciscoes of the Trout Lake Region of Northern Wisconsin". J. Parasit. 20, 244—245.

Damian, R.T. (1964). "Molecular Mimicry: Antigen Sharing by Parasite and Host and its Consequences". Am. Nat. 98, 129—149.

DeBach, P. (1966). "The Competitive Displacement and Coexistence Principle". Annu. Rev. Entomol. 2, 183—212.

Dubois, G. (1929). "Les Cercaires de la Région de Neuchatel." Bull. Soc. Neuchâtel Sci. Nat. 53, 1—177.

Dönges, J. (1972). "Double Infection Experiments with Echinostomatids (Trematoda) in *Lymnaea stagnalis* by Implantation of Rediae and Exposure to Miracidia". Int. J. Parasit. 2, 409—423.

Goulson, H.T. (1958). "Studies on the Influence of a Prior Infection with *Ancylostoma caninum* on the Establishment and Maintenance of *Trichinella spiralis* in Mice". J. Elisha Mitchell. Sci. Soc. 74, 14—23.

Guttowa, A. (1967). "Experimental Coinfection of Copepoda Naturally Infected with *Proteocephalus* sp. with the Larvae of *Diphyllobothrium latum* (L)". Acta Parasit. Pol. 14, 399—404.

Hardin, G. (1960). "The Competitive Exclusion Principle". Science (Wash. D.C.) 131, 1292—1297.

Haley, A.J. and Parker, J.C. (1961). "Effect of Population Density on Adult Worm Sur-

112

vival in Primary *Nippostrongylus brasiliensis* Infections in the Rat". Proc. Helminth. Soc. Wash. 28, 176—180.

Halvorsen, O. (1966). "Studies of the Helminth Fauna of Norway. VIII. An Experimental Investigation of Copepods as First Intermediate Hosts for *Diphyllobothrium norvegicum* Vik (Cestoda)". Nytt Mag. Zool. 13, 83—117.

Halvorsen, O. and Andersen, K. (1974). "Some Effects of Population Density in Infections of *Diphyllobothrium dendriticum* (Nitzsch) in Golden Hamster (*Mesocricetus auratus* Waterhouse) and Common Gull (*Larus canus* L.)". Parasitology, 69, 149—160.

Halvorsen, O. and Macdonald, S. (1972). "Studies of the Helminth Fauna of Norway XXVI: The Distribution of *Cyathocephalus truncatus* (Pallas) in the Intestine of Brown Trout (*Salmo trutta* L.)". Norw. J. Zool. 20, 265—272.

Halvorsen, O. and Williams, H.H. (1968). "Studies of the Helminth Fauna of Norway. IX. *Gyrocotyle* (Plathyhelminthes) in *Chimaera monstrosa* from Oslo Fjord, with Emphasis on its Mode of Attachment and a Regulation in the Degree of Infection". Nytt. Mag. Zool. 15, 130—142.

Heyneman, D. and Umathevy, T. (1967). "A Field Experiment to Test the Possibility of Using Double Infections of Host Snails as a Possible Biological Control of Schistosomiasis". Med. J. Malaya. 21, 373.

Holmes, J.C. (1959). "Competition for Carbohydrates between the Rat Tapeworm, *Hymenolepis diminuta,* and the Acanthocephalan, *Moniliformis dubius. J. Parasit.* 45, (Suppl.), 31.

Holmes, J.C. (1961). "Effects of Concurrent Infections on *Hymenolepis diminuta* (Cestoda) and *Moniliformis dubius* (Acanthocephala). I. General Effects and Comparison with Crowding". J. Parasit. 47, 209—216.

Holmes, J.C. (1962). "Effects of Concurrent Infections on *Hymenolepis diminuta* (Cestoda) and *Moniliformis dubius* (Acanthocephala). II. Growth". J. Parasit. 48, 87—96.

Holmes, J.C. (1973). "Site Selection by Parasitic Helminths: Interspecific Interactions, Site Segregation, and their Importance to the Development of Helminth Communities". Can. J. Zool. 51, 333—347.

Hunninen, A.V. (1935). "Studies on the Life History and Host-Parasite Relations of *Hymenolepis fraterna* (*H. nana* var. *fraterna* Stiles) in White Mice". Am. J. Hyg. 22, 414—443.

Hutchinson, G.E. (1957). "Concluding Remarks". Cold Spring Harbor Symp. Quant. Biol. 22, 415—427.

Keeling, J.E.D. (1961). "Experimental Trichuriasis. I. Antagonism between *Trichuris muris* and *Aspiculuris tetraptera* in the Albino Mouse". J. Parasit. 47, 641—646.

Kennedy, C.R. (1975). "Ecological Animal Parasitology". Blackwell, Oxford.

Kisielewska, K. (1970a). "Ecological Organization of Intestinal Helminth Groupings in *Clethrionomys glareolus* (Schreb.) (Rodentia). V. Some Questions concerning Helminth Groupings in the Host Individuals". Acta. Parasit. Pol. 18, 197—208.

Kisielewska, K. (1970b). "On the Theoretical Foundations of Parasitosynecology". Bull. Acad. Pol. Sci. Ser. Sci. Biol. 18, 103—106.

Krebs, C.J. (1972). "Ecology. The Experimental Analysis of Distribution and Abundance". Harper and Row, New York.

Larsh, J.E. Jr. and Donaldson, A.W. (1944). "The Effect of Concurrent Infection with *Nippostrongylus* on the Development of *Hymenolepis* in Mice". J. Parasit. 30, 18—20.

Larsh, J.E. and Race, G.J. (1975). "Allergic Inflammation as a Hypothesis for the Expulsion of Worms from Tissues: A Review". Exp. Parasit. 37, 251—266.

Levin, S.A. (1970). "Community Equilibria and Stability, and an Extension of the Competitive Exclusion Principle". Am. Nat. 104, 413—423.

Lie, K.J., Basch, P.F., Heyneman, D., Beck, A.J. and Audy, J.R. (1968). "Implications for Trematode Control of Interspecific Larval Antagonism within Snail Hosts". Trans. R. Soc. Trop. Med. Hyg. 62, 299—319.

Lie, K.J., Basch, P.F. and Umathevy, T. (1965). "Antagonism between Two Species of Larval Trematodes in the Same Snail". Nature (Lond.) 206, 422—423.

Lie, K.J., Schneider, C.R., Sormani, S., Lanza, G.R. and Impand, P. (1974). "Biological Control by Trematode Antagonism. I. A Succesful Field Trial to Control *Schistosoma spindale* in Northeast Thailand". Southeast Asian J. Trop. Med. Public Health. 5, 45—59.

Lim, H.K. and Heyneman, D. (1972). In "Advances in Parasitology" (B. Dawes, ed.) 10, 191—268. Academic Press. London.

MacKenzie, K. and Gibson, D.I. (1970). In "Aspects of Fish Parasitology". (A.E.R. Taylor and R. Muller, eds.) pp. 1—42. Blackwell Scientific Publications. Oxford.

Miller, R.S. (1967). In: "Advances in Ecological Research" (J.B. Cragg, ed.) 4, pp. 1—74. Academic Press, London.

Milne, A. (1961). In: "Mechanisms in Biological Competition". (F.L. Milthorpe, ed.). pp. 40—61. The University Press, Cambridge.

Nasir, P. (1962). "Further Observations on the Life Cycle of *Echinostoma nudicandatum* Nasir, 1960 (Echinostomatidae: Trematoda)". Proc. Helminth. Soc. Wash. 29, 115—127.

Nilsson, N.A. (1967). In "The Biological Basis of Freshwater Fish Production" (S.D. Gerkin, ed.). pp. 295—313. Blackwell Scientific Publications. Oxford.

Odum, E.P. (1971). "Fundamentals of Ecology". W.B. Saunders Company, Philadelphia.

Paperna, I. (1964). "Competitive Exclusion of *Dactylogyrus extensus* by *Dactylogyrus vastator* (Trematoda, Monogenea) on the Gills of Reared Carp". J. Parasit. 50, 94—98.

Paperna, I. and Kohn, A. (1964). "Studies on the Host-Parasite Relations between Carps and Populations of Protozoa and Monogenetic Trematodes in Mixed Infestations". Rev. Bras. Biol. 24, 269—276.

Read, C.P. (1959). "The Role of Carbohydrates in the Biology of Cestodes. VIII. Some Conclusions and Hypotheses". Exp. Parasit. 8, 365—382.

Read, C.P. and Phifer, K. (1959). "The Role of Carbohydrates in the Biology of Cestodes VII. Interaction Between Individual Tapeworms of the Same and Different Species". Exp. Parasit. 8, 46—50.

Reichenbach-Klinke, H-H. (1966). "Die gegenseitige Beeinflussung Verschiedener Parasitenarten am Beispiel der Fischhelminthen". Z. Parasitenkd. 28, 95—98.

Roberts, L.S. and Mong, F.N. (1968). "Developmental Physiology of Cestodes. III. Development of *Hymenolepis diminuta* in superinfections". J. Parasit. 54, 55—62.

Rose, S.M. and Rose, F.C. (1961). In "Mechanisms in Biological Competition". (F.L. Milthorpe, ed.). pp. 207—218. The University Press, Cambridge.

Schad, G.A. (1963). "Niche Diversification in a Parasitic Species Flock". Nature (Lond.) 198, 404—406.

Schad, G.A. (1966). "Immunity, Competition and Natural Regulation of Helminth Populations". Am. Nat. 100, 359—364.

Schilling, C. (1936). "Versuche zur Schutzimpfung gegen Tsetsekrankheit". Z. Immunitaetsforsch. Exp. Ther. 87, 482—518.

Sewell, R.S. (1922). "Cercariae Indicae". Indian. J. Med. Res. 10, Suppl., 1—370.

Shorb, D.A. (1933). "Host-Parasite Relations of *Hymenolepis fraterna* in the Rat and the Mouse". Am. J. Hyg. 18, 74—113.

Stewart, D.F. (1953). "Studies on Resistance of Sheep to Infestation with *Haemonchus contortus* and *Trichostrongylus* spp. and on the Immunological Reactions of Sheep Exposed to Infestation. V. The Nature of the "Self-Cure" Phenomenon". Aust. J. Agric. Res. 4, 100—117.

Stewart, D.F. (1955). " "Self-Cure" in Nematode Infestations of Sheep". Nature (Lond.) 176, 1273—1274.

Stromberg, P.C. and Crites, J.L. (1975). "Population Biology of *Camallanus oxycephalus*

Ward and Magath, 1916 (Nematoda: Camallanidae) in White Bass in Western Lake Erie". J. Parasit. 61, 123—132.

Tandon, R.S. (1973). "Studies on "Crowding Effect" on *Gastrothylax crumenifer* and *Fischoederius elongatus*, the Common Amphistome Parasites of Ruminants, Observed Under Natural Conditions". Res. Bull. Meguro Parasit. Mus. 7, 12—14.

Thomas, J.D. (1964). "Studies on Populations of Helminth Parasites in Brown Trout (*Salmo trutta* L.)". J. Anim. Ecol. 33, 83—95.

Turner, J.H., Kates, K.C. and Wilson, G.I. (1962). "The Interaction of Concurrent Infections of the Abomasal Nematodes, *Haemonchus contortus*, *Ostertagia circumcincta*, and *Trichostrongylus axei* (Trichostrongylidae), in Lambs". Proc. Helminth. Soc. Wash. 29, 210—216.

Wardle, R.A. and Green, N.K. (1941). "The Rate of Growth of the Tapeworm *Diphyllobothrium latum* (L)". Can. J. Res. (D). 19, 245—251.

Wesenberg-Lund, C. (1934). "Contribution to the Development of the Trematoda Digenea. II. The Biology of the Freshwater Cercariae in Danish Freshwaters". Kgl. Danske Vidensk. Selsk. Skr. Nat. Mat. Afd. Ser. 9. 3, 1—223, Plate I—XXXIX.

Williams, C.B. (1951). "Intrageneric Competition as illustrated by Moreau's Records of East African Bird Communities". J. Anim. Ecol. 20, 246—253.

Williams, H.H. and Halvorsen, O. (1971). "The Incidence and Degree of Infection of *Gadus morhua* L. 1758 with *Abothrium gadi* Beneden, 1871 (Cestoda: Pseudophyllidea)". Norw. J. Zool. 19, 193—199.

Wilson, C.B. (1916). "Copepod Parasites of Fresh-water Fishes and their Economic Relation to Mussel Glochidia". Bull. U.S. Bur. Fish. 34, (1914), 131—374.

Wilson, E.O. and Bossert, W.H. (1964). "Chemical Communication among Animals". Recent Prog. Horm. Res. 19, 673—716.

Woodland, W.N.F. (1924). "On the life cycle of *Hymenolepis fraterna* (*H. nana* var. *fraterna* Stiles) in the White Mouse". Parasitology 16, 69—83.

Wynne-Edwards, V.C. (1964). "Population Control in Animals". Sci. Am. 211, 124—130.

Ecological Aspects of Parasitology
Editor: C.R. Kennedy
© *North-Holland Publishing Company, Amsterdam, 1976*

CHAPTER 6

HOST RESPONSES

D. WAKELIN

Wellcome Laboratories for Experimental Parasitology, University of Glasgow, Bearsden Road, Glasgow, G61 1QH (Great Britain)

Contents

6.1. Introduction

Many aspects of the lives of parasitic animals have exact parallels in the ecology of free-living organisms, as other contributions to this volume will emphasize. One aspect, however, has no parallel and that is the capacity of the environment of a parasite to respond to that parasite in a manner which can be detrimental to its survival. Although ecologists have coined the term "environmental resistance" (Chapman, 1928) for those factors in the environment that limit the survival and reproductive potential of an organism, there is, it seems to me, a qualitative difference between such factors and the responses which living organisms make to invading parasites, even though the end result for the parasite may be the same. Nowhere in the environment of a free-living animal is there a parallel for the exquisite mechanisms by which hosts can recognize "self" and "not-self" and which provide the basis for many of the responses to be discussed in this chapter. The existence of this capacity in host organisms has been a dominant influence in the evolution of parasites and it plays a major role in determining host specificity, i.e. in determining the range of environments in which a parasite can survive, and in regulating the size of parasite populations.

6.2. Scope of the chapter

This chapter will be limited largely to a consideration of the responses made by molluscs, arthropods and homoiothermic vertebrates to invasion by protozoan and metazoan endoparasites and the term "host-responses" will be interpreted as alterations in the environment provided by the host which are parasite-induced and host-mediated. This interpretation will therefore include examples of both natural and acquired immune responses, as these are usually defined (Humphrey and White, 1970) but will exclude consideration of certain other factors which contribute to natural immunity, e.g. naturally occurring antibodies, antiparasitic factors and behavioural responses. It is not the intention to provide a detailed account of particular examples of host responses, but rather to introduce the general nature of host responses, to discuss their effectiveness against parasites and to outline their significance for both host and parasite.

6.3. Basis of host responses

Almost every phylum of animals contains species which have become adapted for life as parasites; equally animals of almost every phylum are known to act as hosts for parasites. The variety of host-parasite combinations is therefore enormous and it might be expected that a similar variety would occur in the responses that host animals make to parasite infection. However, the majority of responses that will be considered here contain a fundamental similarity of mechanism that rests on the ability of cells to distinguish "self" from "not-self". The ability, which enables the organism to differentiate between its own cells (= "self") and foreign cells (= "not-self") and thus to preserve the integrity of its body in the face of invasion by parasitic organisms (in the widest sense) and proliferation of mutant cells, was an essential evolutionary requirement for even the earliest multicellular organisms and is the characteristic property of specialized cell populations in all extant phyla (Boyden, 1963). Common to all such populations is the ability to recognize and phagocytose "not-self" particulate material, a phenomenon whose occurrence and significance in bodily defence was first recognized by Metchnikov (1884).

In invertebrate animals phagocytic cell populations remain the major means by which responses (as defined earlier) can be made to invading parasites, although the responses are not limited to phagocytosis *per se*. Vertebrates have retained this ancestral system as a component of their much more elaborate defence mechanisms, the inflammatory response and the immune response. An essential difference between the responses of invertebrates and vertebrates lies in the fact that, while the former can differentiate between "self" and "not-self", vertebrates have "the ability to distinguish among different kinds of not-self" (Salt, 1970a). This more subtle ability, characteristic of the immune response, is the property of a cell system, the lymphocyte system, which is not found in invertebrates; it rests on the production by lymphocytes of recognition units (immunoglobulin molecules) capable of interacting, at the cell surface or at a distance, with an enormous range of "not-self" molecules. The effectiveness of the immune response is amplified in a variety of ways and its survival value is enhanced by the unique property of memory, that is, retention in the body, after one experience of a "not-self" molecule, of an enlarged pool of lymphocytes bearing recognition units for that molecule. The presence of these memory cells enables the animal to respond to subsequent experience of the molecule at an accelerated rate and to an increased extent. A further advantage is seen in the mammals, where maternal transfer of immunoglobulins enables newborn animals to be given "ready-made" defences against a number of the infections with which they come into contact during the neonatal period. The strategy of the vertebrate response is, as Salt (1970a) and Tripp (1974) have pointed out, to protect the individual and ensure survival for reproduction. Invertebrates, on the other hand, show poorer protection for the individual and a greater reliance on high reproductive potential to replace diseased individuals.

6.4. Elicitation of responses

Parasites may elicit host responses in a variety of ways, but it is possible to systematize this variety into two major categories: (i) damage to host tissues as a result of mechanical interference or by the release of toxic substances, (ii) presentation of soluble factors or surfaces which are recognized as "not-self" by the host. Damage to host tissues is a common stimulus of host responses in vertebrates, where the inflammatory response is a well developed component of defence against infection, but is less important in invertebrates, in which most responses appear to be made to the presence of "not-self" material. Recognition of "not-self", through the phagocytic and immune response systems, is of major importance in vertebrates.

Despite a considerable volume of research the nature, and very often the origin, of the "not-self" stimuli associated with parasites remains obscure in all but a few cases. The stimulus that elicits haemocytic responses to metazoan parasites in arthropods is known to derive from the surface characteristics of the parasite and Vinson (1974) has suggested that surface charges may be one of the characteristics recognized as "not-self". In the context of vertebrate host responses, the "not-self" components of parasites probably include both surface characteristics and soluble factors released during metabolism. Since such components frequently elicit an adaptive immune response they may correctly be termed antigens. It is likely that all parasites present a wide variety of antigens to which the host responds, but it will be obvious that not all antigens will have equal significance in eliciting responses detrimental to the continued growth, development, survival or reproduction of the parasite concerned. Thus it is necessary to distinguish between antigens which elicit *irrelevant* responses and antigens which elicit *protective* responses, i.e. responses which provide immunity to infection or reinfection. Whereas antigens of the first category have been extensively studied and characterised (Capron, 1970; Kent, 1963; Soulsby, 1963) antigens that elicit protective responses, *functional* antigens in the sense of Soulsby (1963), have been identified much less commonly and their origin in the parasite is known in only a few cases, e.g. the surface coat of trypanosomes (Vickerman and Luckins, 1969), the tegumental membranes of *Schistosoma mansoni* (Sher et al., 1974) and the stichocytes of *Trichinella spiralis* (Despommier and Muller, 1970) and *Trichuris muris* (Wakelin, 1974).

6.5. Responses in invertebrates

Although a number of workers have applied the term immunity to the defensive responses of invertebrates, it can be used only in the sense of natural immunity — freedom from infection (Humphrey and White, 1970) or in the wider sense of "the sum of all factors which prevent an animal from being infected" (Sprent, 1969). There is no parallel in the invertebrate phyla for the adaptive immunity, defined in terms of antigen specificity, antibody production, hypersensitivity, graft rejection and memory (Good and Papermaster, 1964) that is characteristic of vertebrates, although graft rejection and memory have been described in earthworms (Hostetter and Cooper, 1974). Defence responses, in the sense used earlier (p. 116), are almost entirely limited to the recognition of "not-self" by phagocytic cells (amoebocytes, coelomocytes, haemocytes) which have the capacity to phagocytose and digest foreign material and to encapsulate large particles and organisms. It is known that the body fluids of a number of invertebrates precipitate, agglutinate and lyse introduced artificial antigens and bacteria (see Maramarosch and Shope, 1975) and may also show antiparasite activity, although there is relatively little information on that point (e.g. Weathersby, 1960; Beckett and MacDonald, 1971; Feng, 1974). There is only a little firm evidence that the factors responsible for such phenomena are induced by the material against which they act (e.g. Michelson, 1963; Gingrich, 1964) and none for any memory or enhanced production on subsequent experience of the material, even though such claims have, in the past, been made (Cantacuzene, 1916). In short, there is no evidence in invertebrates for the existence of components comparable to vertebrate antibody.

Much of what is now known of invertebrate responses to parasites has come from studies made in molluscan and arthropodan (particularly insect) hosts. For the purposes of this review they make an interesting comparison in mechanisms of defence against parasites, although in both cases the effectors are similar. Molluscs are soft-bodied animals with greatly restricted coelomic and other internal body cavities, invasion of the body can occur at any exposed point and defence responses are largely localized in the tissue through which parasites move. Arthropods, in contrast, have a rigid exoskeleton, which provides considerable protection against the penetration of parasites; entry of parasites is commonest via the intestine, and defence responses are largely confined to the extensive haemocoelic body cavity.

6.5.1. Molluscs

The predominant response to invasion is one of phagocytosis or encapsulation by amoebocytes with subsequent fibroblast activity, leading to the formation of capsules or granulomata and the immobilization and death of the parasite (Tripp, 1963; Brooks, 1969; Cheng, 1967). Such responses characterize invasion of abnormal hosts by larval Digenea and nematodes, but do not occur in the normal host (Newton, 1952; Chernin, 1962). There is as yet no information about the factors which determine whether invading parasites are recognized as "not-self" but, by analogy with responses in arthropods, surface characteristics may well be involved. Not all stages of a parasite necessarily elicit responses; Kinoti (1971) found that whereas *Bulinus truncatus* rapidly destroyed invading miracidia of *Schistosoma mattheei*, *B. africanus* showed no response to miracidia or sporocysts but did respond to cercariae.

The question of acquired immunity, to homologous or heterologous infection, has been raised on a number of occasions. Winfield (1932) reported that *Lymnaea stagnalis* infected with sporocysts of the strigeid digenean *Cotylurus flabelliformis* were resistant to subsequent penetration by cercariae of the same species. In addition, it is relatively common to find that molluscs harbour only one species of digenean even though they may be suitable hosts for a number of species (Kendall, 1964). Basch (1970) was unable to substantiate Winfield's findings and it is becoming evident that a variety of factors other than those associated with host responses, e.g. interspecies competition, regulate parasitization with digeneans in molluscs (Lim and Heyneman, 1972). There remains, however, the intriguing situation of snails such as *Biomphalaria glabratus* which, when infected with *Schistosoma mansoni*, produces a factor, "miracidial immobilizing substance", capable of immobilizing homologous miracidia (Michelson, 1963). Although the factor was present to a low degree in uninfected snails, was elicited to a considerable extent by infection with the nematode *Daubaylia potomaca* and immobilized miracidia of *Fasciola hepatica*, there was nevertheless a clear element of specificity in the response.

Mechanisms for the evasion of effective host responses have been little investigated in molluscs, although it is possible that the secretion of cysts by invading digenean metacercariae may play a role in masking "not-self" surface characteristics. Cort et al. (1941) found that when cercariae of *Cotylurus flabelliformis* invaded uninfected physid or planorbid snails the cercariae were killed, but when the snail was infected with unrelated digenean larvae the metacercariae penetrated into these larvae and completed their development into the tetracotyle stage. *C. flabelliformis* therefore evaded the defensive responses of the host by hiding beneath the surface of a parasite which did not elicit a host response.

6.5.2. Arthropods

The responses of arthropods to endoparasitic infection are mediated almost exclusively by phagocytic cells (haemocytes) in the haemocoel. The haemocytes of insects and crustaceans have been extensively studied (Rabin, 1970; Arnold, 1974) and it is known that they comprise a variety of cell types, each of which probably plays a specific role in internal homeostasis and defence. Much of our detailed knowledge of arthropodan responses has come from studies of insects experimentally infected with insect parasitoids, a field that owes much to a stimulating series of papers by Salt of Cambridge. Valuable reviews of Salt's work and that of more recent workers on parasitoids and other parasites have been given by Salt (1963, 1970a), Briggs (1964), Poinar (1969), Shapiro (1969), Whitcombe et al. (1974). There has been surprisingly little published on the defence responses of arthropods to protozoan parasites (Garnham, 1964; Salt, 1969).

The characteristic reactions of insect haemocytes to the presence of "not-self" material are phagocytosis, when the material is small in relation to the cells, a response implicated in the defence of the reduviid *Triatoma infestans* against *Trypanosoma rangeli* (Zeledon and Monge, 1966), or encapsulation when the material is relatively large. In the latter process haemocytes apply themselves closely to the "not-self" surface and accumulate in concentric layers to form an investing capsule. Encapsulation has been demonstrated against material of very diverse origins, both animate and inanimate, and represents a response to any surface which differs from that of the connective tissue lining the haemocoel (Salt, 1970b). In the case of "not-self" material of animate origin, encapsulation is often accompanied by melanisation, the innermost layer of haemocytes forming, by their breakdown, a layer of melanin over the surface of the foreign object. Melanization has been considered as a process incidental to encapsulation and only secondarily concerned with any harmful effects exerted against invading parasites (Salt, 1970a). However, a number of workers now consider that the enzymatic processes involved in melanization are themselves directly harmful (Nappi, 1973; Götz and Vey, 1974) and indeed Taylor (1969) has suggested that the polyphenol-phenolxidase system may be a primitive and widespread method by which organisms defend themselves against invasion. In certain groups of insects (e.g. chironomids and culicids), where the number of haemocytes is low, melanization of parasites is known to take place without participation of cells, a process termed "humoral encapsulation" by Götz (1969), and has been shown to be an effective defensive response against nematodes (Bronskill, 1962; Götz, 1969). In unsuitable mosquito hosts, malarial parasites and filariid nematodes are destroyed by processes which suggest, or which are known to involve, humoral encapsulation mechanisms (Weathersby, 1960; Esslinger, 1962; Poinar, 1969).

Cellular encapsulation is the universal fate of parasites entering abnormal hosts and is normally followed by melanization and death, perhaps by asphyxiation (Salt, 1970a) although more direct antiparasitic effects may also be involved. Encapsulation of parasites in normal hosts is also common, but in this situation parasites usually survive (Poinar, 1969). In a number of cases, parasites are not encapsulated by their normal hosts and there are a variety of ways in which this can be brought about. One of the most intriguing of these is the masking of "not-self" to avoid recognition. The wasp *Nemeritis canescens*, which parasitizes the lepidopteran *Ephestia kuehniella*, coats her eggs before oviposition with a particulate material that is synthesized in the calyx region of the reproductive tract. This material inhibits any recognition of the egg and first instar larva by the host's haemocytes (Rotheram, 1967). Vinson (1972), using the wasp *Cardiochiles nigriceps*, found that calyx fluid conferred the property of non-recognition upon DEAE—Sephadex, a weakly basic anionic ion-exchanger, rapidly encapsulated by the normal host *Heliothis virescens*. However, non-recognition was apparent only in the normal host; DEAE—Sephadex coated with the calyx fluid of *C. nigriceps* was still encapsulated by the abnormal host *H. zea*. Non-recognition of parasite surfaces has been described in other parasites; Poinar (1974) has reported that larval *Galleria* encapsulate adult individuals of the nematode *Mesodiplogaster heritieri* but do not respond to the dauer larva of this parasite, which is enclosed in a lipophilic second-stage cuticle.

Parasites that are recognized as foreign may escape encapsulation by preventing full development of the cellular changes involved in the process (Nappi, 1975) or by penetrating into tissues where haemocytes are inoperative. For example, larval *Filipjevimermis leipsandra* avoid encapsulation in first or second instar larvae of the beetle *Diabrotica* by penetrating rapidly

122

into a ganglion of the nervous system. After a period of development, the nematode re-emerges into the body cavity of the beetle (now in its third instar) and is not encapsulated (Poinar et al., 1968). If the nematode penetrates directly into a third instar beetle, however, it is rapidly encapsulated, melanized and killed (Fig. 1). Survival in the face of initial encapsulation has been described for the acanthocephalan *Moniliformis dubius* in the cockroach

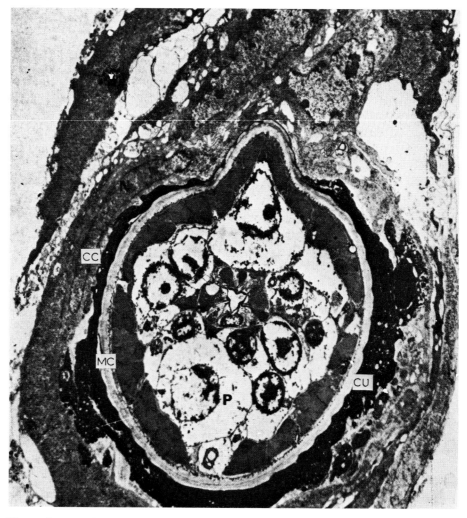

Fig. 1. Section through a melanotic capsule formed by the beetle *Diabrotica* around the nematode *Filipjevimermis leipsandra*. CC = cellular capsule; CU = cuticle of nematode; MC = melanized capsule material; P = pseudocoel of nematode. (Photograph by courtesy of Dr. George O. Poinar, Jr.)

Periplaneta americana (Rotheram and Crompton, 1972; Lackie and Rotheram, 1972). Following initial attachment of blood cells, the surface membrane of the parasite proliferates and forms a coat of microvilli below the layer of haemocytes; the capsule then disintegrates and does not reform.

Evasion of encapsulation by parasites in other arthropodan hosts has been less extensively studied, but some examples have been recorded. Crompton (1967) has attributed the non-encapsulation of the acanthocephalan *Polymorphus minutus* in the haemocoele of *Gammarus pulex* to the fact that the parasite invades in such a way as to maintain a continuous coat of host connective tissue around itself and therefore does not elicit a haemocytic response.

Little is known of the evasion of humoral encapsulation responses, although it must of course, occur. It is interesting to speculate, in view of the established genetic basis of suitability of mosquitoes as hosts for malarial and filarial parasites, that survival in these hosts may be due more to genetically determined lack of recognition by the host than to the operation of successful evasion stratagems by the parasites.

Notwithstanding the effectiveness of host responses, many parasites survive successfully in their arthropod hosts without, as far as it is known, the operation of subtle evasion devices. Parasites can presumably survive encapsulation if their metabolic requirements are low, as in infective metacercarial stages, or survive if, by their activity or rapidity of development, they prevent complete encapsulation of their bodies, as has been suggested is the case in nematodes and acanthocephalans. Bartlett and Ball (1966) made the interesting suggestion that in the evolution of parasitic insects there may be selection for a shorter developmental period in the host, in order to minimize the time during which inactive stages could be successfully encapsulated. Similarly, evolution of behavioural patterns may enable the parasite to infect immature hosts, in which responses are less effective. It must not be forgotten, however, that similar evolutionary stratagems may be adopted by the host as well; Briggs (1964) has suggested that, in the absence of effective cellular responses to the parasitoid *Nemerita canescens*, *Ephestia kuehniella* may resort to avoidance behavioural responses to prevent oviposition.

6.6. Responses in vertebrates

Vertebrate responses to parasitic invasion can be considered as deriving principally from three interacting and interrelated mechanisms, phagocytosis, inflammation and adaptive immunity. These mechanisms occur in all vertebrates, but reach a peak of efficiency in the homoiotherms. Components of the adaptive immune response are present in the most primitive vertebrates, the agnathan fishes, amphibia and reptiles. In birds and mammals the adaptive immune response is highly evolved and probably contributes to all host-responses, however mediated and effected. It is artificial, therefore, to consider any component of a response in isolation, but for convenience a brief description of each of the three mechanisms will be given here; further discussion will be found in later chapters.

Phagocytosis is the property of two major populations of cells, the mononuclear phagocytes and the polymorphonuclear granulocytes, which can phagocytose and digest particulate material. Phagocytic activity can be enhanced in a variety of ways by interaction with components of the immune response, e.g. opsonising antibodies and sensitized lymphocytes.

Inflammation involves "the primary mechanisms by which anti-parasitic properties of the host, particularly the phagocytic cells, are mobilized at the site of parasitic invasion in an attempt to erect a defence barrier, sequester and then destroy the parasite" (McCall, 1971). It is initiated by tissue injury and is closely linked to tissue repair processes and fibroblast activity. Two distinct forms of immunologically initiated inflammation occur, antibody-mediated (immediate) and cell-mediated (delayed) hypersensitivity. The interaction of components in inflammation is complex and the reader is referred elsewhere for details (Lepow and Ward, 1972).

Phagocytosis and inflammation can be considered as non-specific responses to the presence of "not-self" material or to tissue damage; they are probably the dominant responses when parasites invade abnormal hosts. Adaptive immune responses are specific responses to "not-self" and are the most important means by which an animal can limit infection by the parasites with which it is normally associated. Antibodies are a common component of immune responses to parasites, both irrelevant and protective (Sinclair, 1970) and may act directly as antiparasite agents as well as indirectly in enhancing phagocytic and inflammatory responses. Sensitized lymphocytes, the mediators of graft rejection and delayed hypersensitivity reactions, are known to be involved in many host responses elicited by parasites, but their role in protective immunity is, in many cases, controversial (Soulsby, 1970, 1972b; Larsh and Weatherly, 1974).

As the following chapters will point out, host responses operate in almost all of the environments that parasites occupy in the bodies of vertebrate hosts. In only a few environments will a parasite be exempt from responses; such immunologically privileged sites include the chambers of the eye and parts of the central nervous system. In each environment responses take characteristic forms depending upon the nature and construction of the tissues concerned, their vascularization and lymph drainage and their regenerative capacity. The responses elicited by invasion may show a wide spectrum of activity upon the parasites concerned, ranging from no visible effect at one extreme (irrelevant responses) to death and elimination at the other (protective responses). Between these extremes the host responses may affect parasites in various ways, e.g. limiting movement, interfering with feeding and metabolism, retarding growth and development or reducing reproductive potential. All of these effects will have protective value for the host if they limit parasite invasion and reduce the degree of damage sustained. However, protective value can be outweighed if the response itself is harmful to the host, as in the chronic inflammatory responses to the eggs of schistosomes (Warren, 1972) and the inflammatory responses made to larvae and adults of the nematode *Dictycaulus viviparus* in the lungs of cattle (Poynter and Selway, 1966). The host may also be harmed by the immunological consequences of both protective and irrelevant responses, e.g. the immunodepression associated with malarial and trypanosome infections (Phillips et al., 1974; Greenwood, 1974) the deposition of antigen-antibody complexes in malarial nephrosis (Voller, 1975), the potentiation of reaginic antibodies and allergic responses by helminth infections (Jarrett, 1972) and the possible stimulation of autoimmune responses (Mills, 1969; Voller, 1975). As far as the parasite is concerned, host responses will be disadvantageous only if they interfere, directly or indirectly, with the continuation of its species. Host responses can indeed be positively advantageous in certain circumstances, e.g. Dawes (Dawes and Hughes, 1964) has suggested that *Fasciola hepatica* may feed on the hyperplastic epithelium of inflamed bile ducts. In addition, responses that harm the host can be beneficial to the parasite if reproduction and continuation of the life cycle depend upon the host being eaten by another animal, as is the case in cyclophyllidean tapeworms utilizing herbivores as intermediate hosts (Smyth and Heath, 1970). Responses may also be advantageous to a parasite if they reduce the survival of other parasites competing for the same ecological niche (Schad, 1966).

It should be obvious from the above that there will in many cases be selection pressure on the host to moderate the severity of its responses to particular parasites. Equally there will be selection pressures on parasites to evolve mechanisms by which they elicit weaker responses or evade harmful responses and so allow longer survival in the host, or to evolve altered developmental patterns to allow reproduction before host responses become effective. Since detailed accounts of host responses to parasites in particular habitats will appear in later chapters of this volume and have, in addition, been covered in a number of recent reviews (Taliaferro and Stauber, 1969; Jackson et al., 1970; Jarrett and Urquhart, 1971; Soulsby, 1972a; Ogilvie and Jones, 1973; Mauel and Behin, 1974; Ogilvie, 1974a; Cohen and Sadun, 1976) they will be considered only briefly here and largely in relation to the means by which parasites escape some of the consequences of host responses.

Much of our information on these topics has come exclusively from experimental studies using laboratory strains of hosts and parasites. Extrapolation of the results of such studies to the situation in natural infections must therefore be made with considerable caution. In the majority of cases nothing is known about the involvement of host responses in natural infections and thus there is no information as to whether parasites survive because no response is elicited or because responses are successfully evaded.

6.6.1. Protozoa

Protozoan parasites present hosts with problems that are in many ways distinct from those posed by helminths. One important distinction is that protozoa, like bacteria and viruses, can multiply within the host; another is their ability to occupy intracellular habitats. Protective host responses must therefore be directed at limitation of parasite increase and must often overcome the difficulty of gaining access to parasites enclosed within the host's own cell membranes.

Of the three facets of the host response discussed earlier (p. 123) two, phagocytosis and adaptive immunity, are of prime importance in protection against protozoal infection. Protozoa are small enough to be ingested by phagocytic cells and it may be that in unnatural hosts blood and tissue forms are effectively destroyed in this way. In natural hosts phagocytosis appears to require cooperation with the immune system in order to be effective (Desowitz, 1970). Antibodies formed against the antigens of the parasite may act as opsonizing agents and render it susceptible to phagocytosis, or they may cause individual parasites to agglutinate with one another, again facilitating phagocytosis. However, phagocytosis does not necessarily result in the death of the organisms and some parasites, e.g. *Leishmania* and *Eimeria*, utilize phagocytic cells as vehicles for their developmental processes.

Antibody responses are a common consequence of protozoal infections, yet, as with helminth parasites, the majority of antibodies appear to have no significance in protection. Indeed, certain protozoa, notably the trypanosomes, stimulate the formation of large amounts of immunoglobulin (IgM) in their hosts, most probably as a result of non-specific activation of plasma cells (Urquhart et al., 1973). Protective antibodies have been demonstrated in a number of infections (Ogilvie, 1970) by passive transfer and by in vitro studies. Direct activity against the parasite is seen as agglutination, lysis, immobilization and neutralization (Cohen, 1974). Antibodies may also exert less directly obvious effects against the metabolism and reproduction of protozoa, one of the best known examples being the antibody "ablastin" which inhibits division of *Trypanosoma lewisi* (Taliaferro and Stauber, 1969). Unlike helminths, protozoa appear to be poor stimulators of reaginic antibody (Sadun, 1972); immediate-hypersensitivity reactions are therefore less common. Delayed hypersensitivity reactions are associated with a number of protozoal infections, particularly those caused by intracellular parasites, and it might be thought that activation of macrophages, a common correlate of delayed-hypersensitivity, would provide the host with an effective weapon against infection. However, although activated macrophages are known to function protectively against some protozoa, e.g. *Toxoplasma* (Soulsby, 1972) the nature of their involvement in other infections, particularly *Leishmania* (Bryceson, 1975) is complex.

Evasion of the host response has been best documented in certain mammalian trypanosomes which, though eliciting protective antibody responses, avoid the consequences by changing the surface antigens against which the response is directed (Vickerman, 1974). The precise mechanisms by which this is achieved are still obscure, but appear to involve phenotypic selection by host antibodies. Antigenic variation as a survival mechanism has also been demonstrated in malarial parasites of monkeys and rodents (Brown, 1974) and in rodent piroplasms (Phillips, 1971); it may well exist more widely. A number of protozoans, including trypanosomes, malarial parasites and piroplasms are known to release quantities of soluble antigens into the serum of the host and Wilson (1974) has made the interesting suggestion that such antigens may block or divert the protective response of the host, thus enhancing the survival of the parasite. As has already been mentioned, a number of protozoa exert profound immunodepressive effects upon their hosts and may therefore prolong their own survival, although there is the anomaly that trypanosomes depress the immune response of their hosts to other antigens whilst still eliciting anti-trypanosome antibodies and undergoing antigenic variation.

The adoption of the intracellular habit can perhaps be considered as a device for escaping host responses although it is equally possible that successful, persistent intracellular parasites such as *Toxoplasma* and *Leishmania* may be forms that have been able to inactivate or in some way tolerate intracellular destructive devices. Hirsch et al. (1974) described an apparent interference with normal lysosome activity in intracellular vacuoles containing *Toxoplasma* organisms; the survival of *Leishmania* amastigotes within activated macrophages is also known (Mauel and Behin, 1974). Movement into immunologically privileged sites as a means of survival appears to be rare in protozoa, although Ormerod and Venkatesan (1971) have suggested that the development of amastigotes of *Trypanosoma brucei* in the choroid plexus may represent such an escape mechanism.

Immune responses to protozoa are, like all immune responses, under genetic control (see Wakelin, 1975b), and protozoan parasites may therefore survive in individuals or strains of host species, other members of which mount effective, protective responses. One laboratory example has been investigated in some detail, namely the survival of *Leishmania donovani* in various strains of mice (Bradley, 1974). This aspect of parasite survival will be discussed more fully in relation to helminth parasites.

6.6.2. Helminths

Helminths are too large to be dealt with directly by phagocytic cells and responses therefore draw upon the inflammatory and immune armoury of the host. Inflammatory responses are the common reactions when helminths invade and establish for a short time in species other than their customary hosts and may be major determinants of host specificity, as is shown by the fact that infections can frequently be established in unnatural hosts when the inflammatory and immune responses of the host are suppressed (Wakelin, 1970). Behnke (1975) has shown that the failure of the mouse pinworm *Aspiculuris tetraptera* to mature in *Apodemus sylvaticus* (an unnatural host) is due to an early loss of larvae. This loss is prevented by limited treatment with the antiinflammatory, immunosuppressant drug cortisone acetate and the infection then persists and the worms mature. Behnke attributed the loss of worms to intestinal inflammation, although no mechanism for the loss was proposed. It is tempting to speculate that increased mucus production, a common intestinal response, might be a contributory factor in this and in similar cases of natural immunity, an idea which was first proposed more than thirty years ago (Ackert, 1942). Some evidence that inflammatory reactions in unnatural hosts are elicited by soluble products of the parasites comes from the work of Crandall (1965) who found that polymorphonuclear leucocytes of rats showed no chemotaxis to extracts of *Trichinella spiralis*, a natural parasite, but responded strongly to extracts of *Ascaris suum*, an unnatural parasite. These results lead to the conclusion that natural hosts have a reduced response to such factors, a view that conforms with the long-held view that, during evolution, parasites and host undergo mutual adjustment to lessen the intensity of responses ("adaptation tolerance" of Sprent, 1959).

Although in unnatural hosts inflammation frequently leads to elimination and/or death of the parasite, in natural hosts inflammation may occur without deleterious effects. In a number of tissue invading forms the inflammatory response results in encapsulation, with the parasite surviving for long periods within the capsule, e.g. larval *Trichinella spiralis*, cysticerci of *Taenia* sp, hydatid cysts of *Echinococcus granulosus* and the adults of *Paragonimus westermani* and *Onchocerca volvulus*. In all these responses there is a strong immunological component, but encapsulation is not necessarily dependent upon the immune response (Walls et al., 1973).

Immunologically-mediated inflammatory reactions, i.e. immediate- and delayed-hypersensitivity responses, are common consequences of helminth infections and have been extensively studied. Helminths are potent stimulators of reaginic antibodies (Ogilvie and Jones, 1969; Sadun, 1972) and infections are frequently associated with eosinophilia; delayed hypersensitivity reactions have also been demonstrated in many infections. In many cases there is no obvious protective role for hypersensitivity reactions, but evidence has been provided for participation in the immune expulsion of *Nippostrongylus brasiliensis* (Murray, 1972) and *Trichinella spiralis* (Larsh and Race, 1975); the latter workers have indeed suggested a wide involvement of delayed-hypersensitivity reactions in immunity to helminths.

Almost all helminth infections are known to stimulate the production of antibodies in their hosts but it would appear that the majority of these antibodies are quite irrelevant to the survival of the parasite concerned. That antibodies can protect hosts against helminths has been repeatedly demonstrated by passive transfer experiments, but the nature of their effect upon the parasite is obscure in all but a few cases. It is accepted that antibodies interfere with metabolism, stunt growth, reduce reproductive potential and cause physical damage, but only in *Nippostrongylus brasiliensis* is there convincing evidence for the role of antibody in producing such consequences in vivo (Ogilvie and Hockley, 1968; Lee, 1969), although it has been assumed in other infections, e.g. *Haemonchus placei* (Harness et al., 1973), *Trichinella spiralis* (Love et al., 1976), *Schistosoma mansoni* (Hockley and Smithers, 1970) and *Hymenolepis diminuta* (see Chapter 15) and has been demonstrated in vitro (Clegg and Smithers, 1972). Figs. 2 to 6 illustrate some examples of antibody-mediated damage to helminths. In systems where the components of protective immune responses have been analysed it is apparent that the action of antibody alone is not sufficient to bring about removal and death of the parasite, the cooperation of lymphoid and possibly other cell populations is also necessary (Ogilvie and Love, 1974; Wakelin, 1975a). This situation may be true of other systems also; for example, the destructive effects of antibody against *Schistosoma mansoni* are enhanced by the activities of leucocytes, particularly neutophils (Hockley and Smithers, 1970; Perez and Terry, 1973; Dean et al., 1975). Whether protective responses can be mediated entirely via cells with no involvement of antibody is not entirely clear; the most convincing experimental evidence has come from studies on the expulsion of *Trichostrongylus colubriformis* from guinea pigs (Dineen et al., 1968; Rothwell et al., 1974). Even in this case, however, successful passive transfer of immunity with serum has been recorded (Connan, 1972).

Despite the battery of responses that infection can elicit, it is a fact that helminths survive in their natural hosts, often for considerable periods, and thus it is obvious that there must be well developed mechanisms by which the parasite, at the level of the individual or of the species, can evade the harmful consequences of the host's responses. In evolutionary terms, evasion of the response may have been achieved by the loss of those functional antigens which stimulate the protective responses of the host, or, more likely, by modification, so that either the antigens become intrinsically less immunogenic (Dineen, 1963a and b) or come to resemble antigens possessed by the host, are recognized as "self" and elicit no response (Damian, 1964). The existence of protective host responses would exert a strong selection pressure on populations of parasites, favouring the emergence of less foreign i.e. less immunogenic forms. In the situation where responses to infection were so severe as to incapacitate or even kill the host, there would correspondingly be selection pressure on the host to modify its responsiveness. This mutual modification of host and parasite, or adaptation tolerance, has been discussed at length by Sprent (1959, 1969). Although the long-term survival of many helminths and the absence of harmful responses in their hosts appears to support the concept of adaptation tolerance, it cannot be a universal phenomenon since the transmission of many helminths is favoured by incapacitation of the host, however caused, and it is apparent, at least from laboratory studies, that many parasites evoke strong protective responses in their natural hosts.

Survival of individual parasites in the face of protective responses is achieved in a variety of ways, the best known, perhaps, being the acquisition

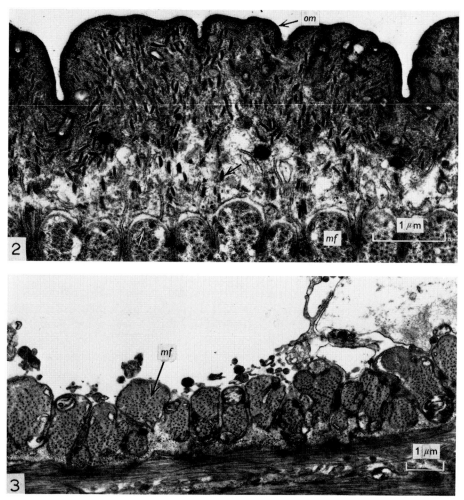

Figs. 2, 3 and 4. Sections through the tegument and underlying layers of *Schistosoma mansoni* recovered from hyperimmune monkeys.
Fig. 2. Tegument showing relatively normal structure.
Fig. 3. Damaged worm, showing loss of tegument and exposure of muscle fibres.

of host antigens by schistosomes (Smithers et al., 1969). By this means adult schistosomes can survive in immune hosts, whereas incoming schistosomula are destroyed before they can acquire host-antigens, a state of "concomitant immunity". Although host antigens have been identified in the bodies of a variety of helminths (Capron et al., 1968) schistosomes remain the only group known to successfully evade host responses in this way. Survival as a result of a different type of phenotypic adaptation is known to occur in *Nippostrongylus brasiliensis*, although the mechanism is not understood. Worms developing in immune rats given large challenge infections (Ogilvie and Hockey, 1968) or repeated low-level (trickle) infections (Jenkins and Phillipson, 1972a and b) show adaptation to their immune environment and become less immunogenic than normal worms (Jenkins, 1972). Nothing is known of the changes which give rise to reduced immunogenicity as far as host protection is concerned, but there is evidence of changes in the acetyl-cholinesterase isoenzyme pattern of adapted worms (Ogilvie, 1974b). Parasite survival at the level of the individual may also be enhanced if antigens of

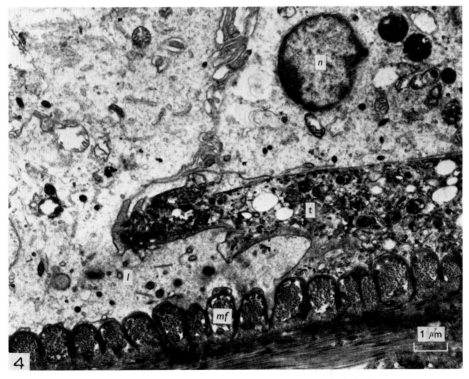

Fig. 4. Damaged worm with host leucocyte applied to and extending beneath damaged tegument.

l = leucocyte; mf = muscle fibre; n = nucleus of leucocyte; om = outer membrane of tegument; t = tegument. (Photographs reproduced by kind permission of Dr. D.J. Hockley and Cambridge University Press).

the parasite delay or divert the expression of the host response. There are a number of ways in which this might be achieved, for example by antigenic competition between irrelevant and functional antigens. The possibility that helminths prolong their survival by active immunosuppression of the host is raised by the work of Tanner and Faubert (1974), who have made the interesting observation that larval *T. spiralis* contain a factor that agglutinates host lymphocytes and which, if injected into mice prior to infection, results in the establishment of many more muscle larvae than in controls.

When parasite survival in the face of responses is considered in terms of the species rather than the individual, there are a number of strategems available to the parasite which are less sophisticated than the phenotypic adaptations described above. Since the important factor for the parasite is the reproduction of its species it may not matter that protective immune responses develop and eliminate an infection as long as a new generation has been produced. Hence rapidity of sexual maturation, as occurs for example in the nematodes *Nippostrongylus brasiliensis* and *Trichinella spiralis* and in many

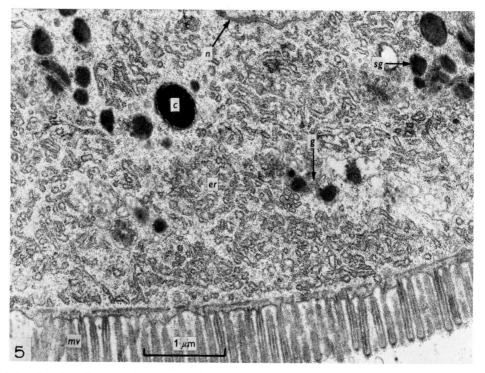

Figs. 5 and 6. Sections through intestinal cells of female *Nippostrongylus brasiliensis* recovered from infected rats.

Fig. 5. Worm recovered 10 days after infection before damage is evident.

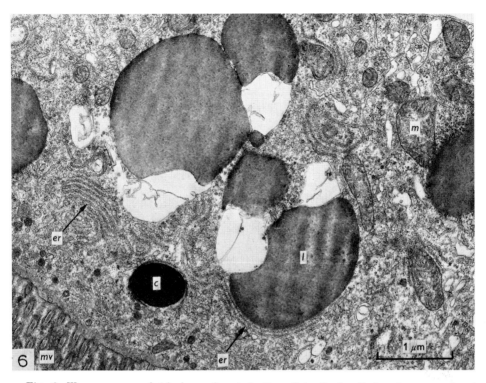

Fig. 6. Worm recovered 19 days after infection. Intestinal cell showing presence of large lipid droplets, whirls of rough endoplasmic reticulum and vacuolated areas. C = crystalline inclusion; er = endoplasmic reticulum; g = Golgi apparatus; l = lipid droplet; m = mitochondrion; mv = microvilli; n = nucleus; s.g. = secretory granule. (Photographs reproduced by kind permission of Professor D.L. Lee and Cambridge University Press).

avian intestinal digeneans, ensures species survival, even though the adult worms are subsequently expelled. Selection of parasite populations for this character would enhance their survival. Alternatively the consequences of host responses can be evaded if the responses generated by infection are effective only against juvenile stages (as in *Trichostrongylus colubriformis* (Dineen and Wagland, 1966) and *Nematospiroides dubius* (Bartlett and Ball, 1974)). This is a form of concomitant immunity which does not depend on masking of "not-self" by host antigens but may be due to differences in amount or importance of functional antigens between larvae and adults, or to differences in accessibility to the effectors of the response.

The work of Jenkins and Phillipson (1971, 1972a and b) with trickle infections of *Nippostrongylus brasiliensis* has emphasized an important point, which is that the development of host responses and the fate of parasites may be quite different when the level and rate of infection approximate more closely to the situation that exists in the field. It may be that in the wild, the population structures of host and parasite, and their ecological interactions, result in low-levels of infection being acquired over long periods of time. If this is the case then parasites may survive because they do not provide the host with an antigenic stimulus large enough to evoke a protective response. Experimental evidence for the operation of a threshold of this type in infections of *Trichuris muris* was obtained by Wakelin (1973) and Behnke and Wakelin (1973) whose survey data, on natural infections in wild mice, lend support to the idea that this may be a significant survival mechanism for a parasite which, in both laboratory and wild hosts is otherwise expelled before sexual maturity.

An interesting aspect of the survey data obtained by Behnke and Wakelin was that there was overdispersion of the parasites, with the majority of worms being found in a few mice. These mice were all adult females and it was suggested that the overdispersion resulted from a pregnancy- or lactation-induced suppression of host responses. Suppression of immunity to *T. muris* by lactation has been experimentally confirmed and is known to occur with a number of other nematodes (Selby and Wakelin, 1975). Lactating females will therefore function as a major means of parasite survival in certain situations, the parasite escaping the consequences of the host response because of physiological changes in the host. A corollary of this situation, which has significant consequences for parasite survival, is that there is a greater chance of new-born animals being exposed to infection at a time when their immunological competence may be incomplete and they may therefore develop a degree of unresponsiveness to the antigens of the parasite (Jarrett and Urquhart, 1971).

Overdispersion of parasites in populations of vertebrate hosts is well documented (Crofton, 1971; Bradley, 1972) and may have a variety of causes. One relevant and relatively neglected cause is the variability in immune responsiveness inherent in outbred populations of hosts. It is now established that there can be direct genetic control of responses to specific antigens and the possibility therefore exists that hosts may be genetically unresponsive to parasites, a possibility that was raised by Sprent (1959) before there was any experimental support for the idea. It is well known that there is considerable variation in the development of parasites within strains of hosts and, in one particular relationship, that of *Trichuris muris* in mice, the extent of development in strains of inbred mice may vary from expulsion after only 12 days of infection to maturation and prolonged survival for more than 35 days (Wakelin, 1975b). This variation arises from genetically determined strain differences in host response; a similar genetically determined variability exists between individual mice in outbred strains, some mice being quite incapable of mounting a protective response (Wakelin, 1975c). The importance of this means of parasite survival in the wild, in the face of effective immune responses from the majority of hosts, has been demonstrated for the cestode *Hymenolepis citelli* in *Peromyscus maniculatus* by Wassom et al. (1974). If no deleterious effects result from parasitism there should be no evolutionary selection against such unresponsive hosts and it is reasonable to suggest that they probably play an important role in the survival of a wide variety of helminth parasites.

6.7. REFERENCES

Ackert, J.E. (1942). "Natural resistance to helminthic infections." J. Parasitol. 28, 1—24.

Arnold, J.W. (1974). "The haemocytes of insects." In: "The Physiology of Insecta." 2nd Edn. (M. Rockstein, ed.) Vol. 5, pp. 202—254. Academic Press, New York.

Bartlett, A. and Ball, P.A.J. (1974). "The immune response of the mouse to larvae and adults of *Nematospiroides dubius.*" Int. J. Parasitol. 4, 463—470.

Bartlett, B.R. and Ball, J.C. (1966). "The evolution of host suitability in a polyphagous parasite with special reference to the role of parasite egg encapsulation". Ann. Entomol. Soc. Amer. 59, 42—45.

Basch, P.F. (1970). "Relationships of some larval strigeids and echinostomes (Trematoda): Hyperparasitism, antagonism and "immunity" to the snail host." Exp. Parasitol. 27, 193—216.

Beckett, E.B. and MacDonald, W.W. (1971). "The survival and development of subperiodic *Brugia malayi* and *B. pahangi* larvae in a selected strain of *Aedes aegypti.*" Trans. Roy. Soc. Trop. Med. Hyg. 65, 339—346.

Behnke, J.M. (1975). "The effect of immunosuppressive agents on infection with *Aspiculuris tetraptera* in abnormal hosts." Parasitology 71, xx.

Behnke, J.M. and Wakelin, D. (1973). "The survival of *Trichuris muris* in wild populations of its natural hosts." Parasitology 67, 157—164.

Boyden, S.V. (1963). "Cellular recognition of foreign matter." Int. Rev. Exp. Pathol. 2, 311—356.

Bradley, D.J. (1972). "Regulation of parasite populations. A general theory of the epidemiology and control of parasitic infections." Trans. Roy. Soc. Trop. Med. Hyg. 66, 697—708.

Bradley, D.J. (1974). "Genetic control of natural resistance to *Leishmania donovani* in mice." Nature, London 250, 353.

Briggs, J.D. (1964). "Immunological responses." In: "The Physiology of Insecta." 1st Edn. Vol. 3 (M. Rockstein, ed.) pp. 259—283. Academic Press, New York.

Bronskill, J.F. (1962). "Encapsulation of rhabditoid nematodes in mosquitoes." Can. J. Zool. 40, 1269—1275.

Brooks, W.M. (1969). "Molluscan immunity to metazoan parasites." In: "Immunity to Parasitic Animals." Vol. 1 (G.J. Jackson, R. Herman and I. Singer, eds.) pp. 149—171. North-Holland, Amsterdam.

Brown, K.N. (1974). "Antigenic variation and immunity to malaria." In: "Parasites in the Immunized Host: Mechanisms of Survival." (R. Porter and J. Knight, eds.) (Ciba Foundation Symp. 25) pp. 35—51. Associated Scientific Publishers, Amsterdam.

Bryceson, A.D.M. (1975). "Mechanisms of disease in leishmaniasis." In: "Pathogenic Processes in Parasitic Infections." (A.E.R. Taylor and R. Muller, eds.) (Symp. Brit. Soc. Parasitol. 13) pp. 85—100. Blackwell, Oxford.

Cantacuzene, J. (1916). "Production expérimentale d'hémo-agglutinines et de précipitines chez *Helix pomatia.*" C.R. Seances Soc. Biol. 79, 528—530.

Capron, A.R. (1970). "L'antigene parasitaire, structure et fonction." J. Parasitol. 56 (No. 4, Sect. II, 3), 515—521.

136

Capron, A., Biguet, J., Vernes, A. and Afchain, D. (1968). "Structure antigénique des helminthes. Aspectes immunologiques des relations hôte-parasite." Pathol. Biol. 16, 121—138.

Chapman, R.N. (1928). "The quantitative analysis of environmental factors." Ecology 9, 111—122.

Cheng, T.C. (1967). "Marine molluscs as hosts for symbioses." Advan. Mar. Biol. 5, 1—424.

Chernin, E. (1962). "The unusual life-history of *Daubaylia potomaca* (Nematoda: Cephalobidae) in *Australorbis glabratus* and in certain other fresh water snails." Parasitology 52, 459—481.

Clegg, J.A. and Smithers, S.R. (1972). "The effects of immune rhesus monkey serum on schistosomula of *Schistosoma mansoni* during cultivation in vitro." Int. J. Parasitol. 2, 79—98.

Cohen, S. (1974). "The immune response to parasites." In: "Parasites in the Immunized Host: Mechanisms of Survival." (R. Porter and J. Knight, eds.) (Ciba Foundation Symp. 25) pp. 3—20. Associated Scientific Publishers, Amsterdam.

Cohen, S. and Sadun, E.H. (1976). (eds.) "Immunology of Parasitic Infections." Blackwell, Oxford.

Connan, R.M. (1972). "Passive protection with homologous antiserum against *Trichostrongylus colubriformis* in the guinea pig." Immunology 23, 647—650.

Cort, W.W., Olivier, L. and Brackett, S. (1941). "The relation of physid and planorbid snails to the life cycle of the strigeid trematode *Cotylurus flabelliformis* (Faust, 1917)." J. Parasitol. 27, 437—448.

Crandall, R.B. (1965). "Chemotactic response of polymorphonuclear leukocytes to *Trichinella spiralis* and *Ascaris suum* extracts." J. Parasitol. 51, 397—404.

Crofton, H.D. (1971). "A quantitative approach to parasitism." Parasitology 62, 179—193.

Crompton, D.W.T. (1967). "Studies on the haemocytic reaction of *Gammarus* spp and its relationship to *Polymorphus minutus* (Acanthocephala)." Parasitology 57, 389—401.

Damian, R.T. (1964). "Molecular mimicry: antigen sharing by parasite and host and its consequences." Amer. Natur. 98, 129—149.

Dawes, B. and Hughes, D.L. (1964). "Fascioliasis: the invasive stages of *Fasciola hepatica* in mammalian hosts." Advan. Parasitol. 2, 97—168.

Dean, D.A., Wistar, R. and Chen, P. (1975). "Immune response of guinea pigs to *Schistosoma mansoni*. I. In vitro effects of antibody and neutrophils, eosinophils and macrophages on schistosomula." Amer. J. Trop. Med. Hyg. 24, 74—82.

Desowitz, R.S. (1970). "Antiparasitic mechanisms in parasitic infections." J. Parasitol. 56 (No. 4, Sect. II, 3), 521—525.

Despommier, D.D. and Muller, M. (1970). "The stichosome of *Trichinella spiralis*: its structure and function." J. Parasitol. 56 (No. 4, Section II, 1), 76.

Dineen, J.K. (1963a). "Immunological aspects of parasitism." Nature, London 197, 268—269.

Dineen, J.K. (1963b). "Antigenic relationship between host and parasite." Nature, London 197, 471—472.

Dineen, J.K. and Wagland, B.M. (1966). "The cellular transfer of immunity to *Trichostrongylus colubriformis* in an isogenic strain of guinea-pig. II. The relative susceptibility of the larval and adult stages of the parasite to immunological attack." Immunology 11, 47—57.

Dineen, J.K., Wagland, B.M. and Ronai, P.M. (1968). "The cellular transfer of immunity to *Trichostrongylus colubriformis* in an isogenic strain of guinea pig. III. The localization and functional activity of immune lymph node cells following syngeneic and allogeneic transfer." Immunology 15, 335—341.

Esslinger, J.H. (1962). "Behaviour of microfilariae of *Brugia pahangi* in *Anopheles quadrimaculatus.*" Amer. J. Trop. Med. Hyg. 11, 749—758.

Feng, S.Y. (1974). "Lysozyme-like activities in the hemolymph of *Crassotrea virginica.*" In: "Contemporary Topics in Immunobiology." Vol. 4 (E.L. Cooper, ed.) pp. 225—231. Plenum Press, New York.

Garnham, P.C.C. (1964). "Factors influencing the development of protozoa in their arthropodan hosts." In: "Host-Parasite Relationships in Invertebrate Hosts." (A.E.R. Taylor, ed.) (Symp. Brit. Soc. Parasitol. 2) pp. 33—50. Blackwell, Oxford.

Gingrich, R.E. (1964). "Acquired humoral immune response of the large milkweed bug, *Oncopeltis fasciatus* (Dallas) to injected materials." J. Insect. Physiol. 10, 179—194.

Good, R.A. and Papermaster, B.W. (1964). "Ontogeny and phylogeny of adaptive immunity." Advan. Immunol. 4, 1—115.

Götz, P. (1969). "Die Einkapselung von Parasiten in der Haemolymphe von Chironomus-Larven (Diptera)." Zool. Anz. 33, 610—617.

Götz, P. and Vey, A. (1974). "Humoral encapsulation in Diptera (Insecta): defence reactions of Chironomus larvae against fungi." Parasitology 68, 193—205.

Greenwood, B.N. (1974). "Immunosuppression in malaria and trypanosomiasis." In: "Parasites in the Immunized Host: Mechanisms of Survival." (R. Porter and J. Knight, eds.) (Ciba Foundation Symp. 25) pp. 136—146. Associated Scientific Publishers, Amsterdam.

Harness, E., Smith, K. and Bland, P. (1973). "Structural changes in the bovine nematode *Haemonchus placei* that may be associated with host immune response." Parasitology 66, 199—205.

Hirsch, J.G., Jones, T.C. and Len, L. (1974). "Interactions in vitro between *Toxoplasma gondii* and mouse cells." In: "Parasites in the Immunized Host: Mechanisms of Survival." (R. Porter and J. Knight, eds.) (Ciba Foundation Symp. 25) pp. 205—223. Associated Scientific Publishers, Amsterdam.

Hockley, D.J. and Smithers, S.R. (1970). "Damage to adult *Schistosoma mansoni* after transfer to a hyperimmune host." Parasitology 61, 95—100.

Hostetter, R.K. and Cooper, E.L. (1974). "Earthworm coelomocyte immunity." In: "Contemporary Topics in Immunobiology." Vol. 4 (E.L. Cooper, ed.) pp. 91—107. Plenum Press, New York.

Humphrey, J.H. and White, R.G. (1970). "Immunology for Students of Medicine." (3rd Edn.). Blackwell, Oxford.

Jackson, G.J., Herman, R. and Singer, I. (eds.) (1970). "Immunity to Parasitic Animals." Vol. 2. North-Holland, Amsterdam.

Jarrett, E.E.E. (1972). "Potentiation of reaginic (IgE) antibody to ovalbumin in the rat following sequential trematode and nematode infections." Immunology 22, 1099—1101.

Jarrett, E.E.E. and Urquhart, G.M. (1971). "Immunity to nematode infection." Int. Rev. Trop. Med. 4, 53—96.

Jenkins, D.C. (1972). "*Nippostrongylus brasiliensis.* Observations on the comparative immunogenicity of adult worms from primary and immune-adapted infections." Parasitology 65, 547—550.

Jenkins, D.C. and Phillipson, R.F. (1971). "The kinetics of repeated low-level infections of *Nippostrongylus brasiliensis* in the laboratory rat." Parasitology 62, 457—465.

Jenkins, D.C. and Phillipson, R.F. (1972a). "Increased establishment and longevity of *Nippostrongylus brasiliensis* in immune rats given repeated small infections." Int. J. Parasitol. 2, 105—111.

Jenkins, D.C. and Phillipson, R.F. (1972b). "Evidence that the nematode *Nippostrongylus brasiliensis* can adapt to and overcome the effects of host immunity." Int. J. Parasitol. 2, 353—359.

138

Kendall, S.B. (1964). "Some factors influencing the development and behaviour of trematodes in their molluscan hosts." In: "Host-Parasite Relationships in Invertebrate Hosts." (A.E.R. Taylor, ed.) (Symp. Brit. Soc. Parasitol. 2) pp. 51—73. Blackwell, Oxford.

Kent, N.H. (1963). "Seminar on immunity to parasitic helminths. V. Antigens." Exp. Parasitol. 13, 45—56.

Kinoti, G.K. (1971). "Observations on the infections of bulinid snails with *Schistosoma mattheei*. II. The mechanism of resistance to infection." Parasitology 62, 161—170.

Lackie, J.M. and Rotheram, S. (1972). "Observations on the envelope surrounding *Moniliformis dubius* (Acanthocephala) in the intermediate host *Periplaneta americana.*" Parasitology 65, 303—308.

Larsh, J.E. and Race, G.J. (1975). "Allergic inflammation as a hypothesis for the expulsion of worms from tissues: a review." Exp. Parasitol. 37, 251—266.

Larsh, J.E. and Weatherly, N.F. (1974). "Cell-mediated immunity in certain parasitic infections." Curr. Top. Microbiol. Immunol. 67, 113—137.

Lee, D.L. (1969). "Changes in adult *Nippostrongylus brasiliensis* during the development of immunity to this nematode in rats. I. Changes in ultrastructure." Parasitology 59, 29—39.

Lepow, I.H. and Ward, P.A. (eds.) (1972). "Inflammation: Mechanisms and Control." Academic Press, New York.

Lim, H-K. and Heyneman, D. (1972). "Intramolluscan inter-trematode antagonism: a review of factors influencing the host-parasite system and its possible role in biological control." Advan. Parasitol. 10, 191—268.

Love, R.J., Ogilvie, B.M. and McLaren, D.J. (1976). "The immune mechanism which expels the intestinal stage of *Trichinella spiralis* from rats." Immunology 30, 7—16.

Maramorosch, K. and Shope, R.E. (1975). (eds.) "Invertebrate Immunity." Academic Press, New York.

Mauel, J. and Behin, R. (1974). "Cell mediated and humoral immunity to protozoan infections." Transplant. Rev. 19, 121—146.

McCall, C.E. (1971). "Host-parasite interaction." In: "Principles of Pathobiology." (M.F. La Via and R.B. Hill, eds.) pp. 133—161. Oxford University Press, London.

Metchnikov, E. (1884). "Ueber eine Sprosspilzkrankheit der *Daphnien* Beitrag zur Lehre über den Kampf der Phagocyten gegen Krankheiterreger." Virchows Arch. Pathol. Anat. Physiol. 96, 177—195.

Michelson, E.H. (1963). "Development and specificity of miracidial immobilizing substances in extracts of the snail *Australorbis glabratus* exposed to certain agents." Ann. N.Y. Acad. Sci. 113, 486—491.

Mills, A.R. (1969). "A quantitative approach to the epidemiology of Onchocerciasis in West Africa." Trans. Roy. Soc. Trop. Med. Hyg. 63, 591—602.

Murray, M. (1972). "Immediate hypersensitivity effector mechanisms. II. In vivo reactions." In: "Immunity to Animal Parasites." (E.J.L. Soulsby, ed.) pp. 155—190. Academic Press, New York.

Nappi, A.J. (1973). "The role of melanization in the immune reaction of larvae of *Drosophila algonquin* against *Pseudeucoila bochei.*" Parasitology 66, 23—32.

Nappi, A.J. (1975). "Cellular immune reactions of larvae of *Drosophila algonquin.*" Parasitology 70, 189—194.

Newton, W.L. (1952). "The comparative tissue reaction of two strains of *Australorbis glabratus* to infection with *Schistosoma mansoni.*" J. Parasitol. 38, 362—366.

Ogilvie, B.M. (1970). "Immunoglobulin responses in parasitic infections." J. Parasitol. 56 (No. 4, Section II, 3), 525—534.

Ogilvie, B.M. (1974a). "Immunity to parasites (helminths and arthropods)." In: "Progress in Immunology II." Vol. 4 (L. Brent and J. Holborow, eds.) pp. 127—135. North-Holland, Amsterdam.

Ogilvie, B.M. (1974b). "Antigenic variation in the nematode *Nippostrongylus brasiliensis.*" In: "Parasites in the Immunized Host: Mechanisms of Survival." (R. Porter and J. Knight, eds.) pp. 81—100. Associated Scientific Publishers, Amsterdam.

Ogilvie, B.M. and Hockley, D.J. (1968). "Effects of immunity on *Nippostrongylus brasiliensis* adult worms. Reversible and irreversible changes in infectivity, reproduction and morphology." J. Parasitol. 54, 1073—1084.

Ogilvie, B.M. and Jones, V.E. (1969). "Reaginic antibodies and helminth infections." In: "Cellular and Humoral Mechanisms in Anaphylaxis and Allergy." (H.Z. Movat, ed.) pp. 13—22. Karger, Basel.

Ogilvie, B.M. and Jones, V.E. (1973). "Immunity in the parasitic relationship between helminths and hosts." Progr. Allergy 17, 93—144.

Ogilvie, B.M. and Love, R.J. (1974). "Cooperation between antibodies and cells in immunity to a nematode parasite." Transplant. Rev. 19, 147—168.

Ormerod, W.E. and Venkatesan, S. (1971). "The occult visceral phase of mammalian trypanosomes with special reference to the life cycle of *Trypanosoma (Trypanozoon) brucei.*" Trans. Roy. Soc. Trop. Med. Hyg. 65, 722—735.

Perez, H. and Terry, R.J. (1973). "The killing of adult *Schistosoma mansoni* in vitro in the presence of antisera to host antigenic determinants and peritoneal cells." Int. J. Parasitol. 3, 499—503.

Phillips, R.S. (1971). "Antigenic variation in *Babesia rodhaini* demonstrated by immunization with irradiated parasites." Parasitology 63, 315—322.

Phillips, R.S., Selby, G.R. and Wakelin, D. (1974). "The effect of *Plasmodium berghei* and *Trypanosoma brucei* infections on the immune expulsion of the nematode *Trichuris muris* from mice." Int. J. Parasitol. 4, 409—415.

Poinar, G.O. (1969). "Arthropod immunity to worms." In: "Immunity to Parasitic Animals." Vol. 1 (G.J. Jackson, R. Herman and I. Singer, eds.) pp. 173—211. North-Holland, Amsterdam.

Poinar, G.O. (1974). "Insect immunity to parasitic nematodes." In: "Contemporary Topics in Immunobiology." Vol. 4 (E.L. Cooper, ed.) pp. 167—178. Plenum Press, New York.

Poinar, G.O., Leutenegger, R. and Götz, P. (1968). "Ultrastructure of the formation of a melanotic capsule in *Diabrotica* (Coleoptera) in response to a parasitic nematode (Mermithidae)." J. Ultrastruct. Res. 25, 293—306.

Poynter, D. and Selway, S. (1966). "Diseases caused by lungworms." Helminth. Abstr. 35, 105—127.

Rabin, H. (1970). "Hemocytes, hemolymph and defence reactions in Crustaceans." J. Reticuloendothel. Soc. 7, 195—207.

Rotheram, S. (1967). "Immune surface of eggs of a parasitic insect." Nature, London 214, 700.

Rotheram, S. and Crompton, D.W.T. (1972). "Observations on the early relationship between *Moniliformis dubius* (Acanthocephala) and the haemocytes of the intermediate host *Periplaneta americana.*" Parasitology 64, 15—21.

Rothwell, T.L.W., Prichard, R.K. and Love, R.J. (1974). "Studies on the role of histamine and 5-hydroxytryptamine in immunity against the nematode *Trichostrongylus colubriformis.* I. In vivo and in vitro effects of the amines." Int. Arch. Allergy Appl. Immunol. 46, 1—13.

Sadun, E.H. (1972). "Homocytotropic antibody response to parasitic infections." In: "Immunity to Animal Parasites." (E.J.L. Soulsby, ed.) pp. 97—129. Academic Press, New York.

Salt, G. (1963). "The defence reactions of insects to metazoan parasites." Parasitology 53, 527—642.

Salt, G. (1969). "A note on the defence reactions of insects to Protozoa." Parasitology 59, 753—756.

Salt, G. (1970a). "The Cellular Defence Reactions of Insects." Cambridge Monographs in Experimental Biology, No. 16. Cambridge University Press, Cambridge.

Salt, G. (1970b). "Experimental studies in insect parasitism. XV. The means of resistance of a parasitoid larva." Proc. Roy. Soc. B. 176, 105—114.

Schad, G.A. (1966). "Immunity, competition and natural regulation of helminth populations." Amer. Natur. 100, 359—364.

Selby, G.R. and Wakelin, D. (1975). "Suppression of the immune response to *Trichuris muris* in lactating mice." Parasitology 71, 77—85.

Shapiro, M. (1969). "Immunity of insect hosts to insect parasites." In: "Immunity to Parasitic Animals." Vol. 1 (G.J. Jackson, R. Herman and I. Singer, eds.) pp. 211—228. North-Holland, Amsterdam.

Sher, A., Kusel, J.R., Perez, H. and Clegg, J.A. (1974). "Partial isolation of a membrane antigen which induces the formation of antibodies lethal to schistosomes cultured in vitro." Clin. Exp. Immunol. 18, 357—369.

Sinclair, I.J. (1970). "The relationship between circulating antibodies and immunity to helminthic infections." Advan. Parasitol. 8, 97—138.

Smithers, S.R., Terry, R.J. and Hockley, D.J. (1969). "Host antigens in schistosomiasis." Proc. Roy. Soc. B. 171, 483—494.

Smyth, J.D. and Heath, D.D. (1970). "Pathogenesis of larval cestodes in mammals." Helminth. Abstr. A. 39, 1—23.

Soulsby, E.J.L. (1963). "The nature and origin of functional antigens in helminth infections." Ann. N.Y. Acad. Sci. 113, 492—509.

Soulsby, E.J.L. (1970). "Cell mediated immunity in parasitic infections." J. Parasitol. 56 (No. 4, Section II, 3), 534—547.

Soulsby, E.J.L. (1972a). "Immunity to Animal Parasites." Academic Press, New York.

Soulsby, E.J.L. (1972b). "Cell mediated immunity responses in parasitic infections." In: "Immunity to Animal Parasites." (E.J.L. Soulsby, ed.) pp. 57—95. Academic Press, New York.

Sprent, J.F.A. (1959). "Parasitism, immunity and evolution." In: "The Evolution of Living Organisms." (G.S. Leeper, ed.) pp. 149—165. Melbourne University Press, Melbourne.

Sprent, J.F.A. (1969). "Evolutionary aspects of immunity in zooparasitic infections." In: "Immunity to Parasitic Animals." Vol. 1 (G.J. Jackson, R. Herman and I. Singer, eds.) pp. 3—62. North-Holland Publishing Co., Amsterdam.

Taliaferro, W.H. and Stauber, L.A. (1969). "Immunology of protozoan infections." In: "Research in Protozoology." Vol. 3 (T-T. Chen, ed.) pp. 506—564. Pergamon Press, Oxford.

Tanner, C.E. and Faubert, G. (1974). "Purification of a leukoagglutinating factor from *Trichinella spiralis* larvae." In: "Trichinellosis." (C.W. Kim, ed.) pp. 327—334. Intext Educational Publishers, New York.

Taylor, R.L. (1969). "A suggested role for the polyphenol-phenoloxidase system in invertebrate immunity." J. Invert. Pathol. 14, 427—428.

Tripp, M.R. (1963). "Cellular responses of molluscs." Ann. N.Y. Acad. Sci. 113, 467—474.

Tripp, M.R. (1974). "A final comment on invertebrate immunity." In: "Contemporary Topics in Immunobiology." Vol. 4 (E.L. Cooper, ed.) pp. 289—290. Plenum Press, New York.

Urquhart, G.M., Murray, M., Murray, P.K., Jennings, F.W. and Bate, E. (1973). "Immunosuppression in *Trypanosoma brucei* infections in rats and mice." Trans. Roy. Soc. Trop. Med. Hyg. 67, 528—535.

Vickerman, K. (1974). "Antigenic variation in African trypanosomes." In: "Parasites in the Immunized Host: Mechanisms of Survival." (R. Porter and J. Knight, eds.) (Ciba Foundation Symp. 25) pp. 53--80. Associated Scientific Publishers, Amsterdam.

Vickerman, K. and Luckins, A.G. (1969). "Localization of variable antigens in the surface coat of *Trypanosoma brucei* using ferritin-conjugated antibody." Nature, London 224, 1125—1126.

Vinson, S.B. (1972). "Factors involved in successful attack on *Heliothis virescens* by the parasitoid *Cardiochiles nigriceps.*" J. Invert. Pathol. 20, 118—123.

Vinson, S.B. (1974). "The role of the foreign surface and female parasitoid secretions on the immune response of an insect." Parasitology 68, 27—33.

Voller, A. (1975). "Aspects of the immunopathology of malaria." In: "Pathogenic Processes in Parasitic Infections." (A.E.R. Taylor and R. Muller, eds.) (Symp. Brit. Soc. Parasitol. 13) pp. 69—84. Blackwell Scientific Publications, Oxford.

Wakelin, D. (1970). "Studies on the immunity of albino mice to *Trichuris muris*. Suppression of immunity by cortisone acetate." Parasitology 60, 229—237.

Wakelin, D. (1973). "The stimulation of immunity to *Trichuris muris* in mice exposed to low-level infections." Parasitology 66, 181—189.

Wakelin, D. (1974). "The antigens of *Trichuris muris* which stimulate protective immunity in mice." Proc. IIIrd Int. Congr. Parasitol. 2, 1172.

Wakelin, D. (1975a). "Immune expulsion of *Trichuris muris* from mice during a primary infection: analysis of the components involved." Parasitology 70, 397—405.

Wakelin, D. (1975b). "Genetic control of immune responses to parasites: immunity to *Trichuris muris* in inbred and random-bred strains of mice." Parasitology 71, 51—60.

Wakelin, D. (1975c). "Genetic control of immune responses to parasites: selection for responsiveness and non-responsiveness to *Trichuris muris* in random-bred mice." Parasitology 71, 377—384.

Walls, R.S., Carter, R.L., Leuchars, E. and Davies, A.J.S. (1973). "The immunopathology of trichiniasis in T-cell deficient mice." Clin. Exp. Immunol. 13, 231—242.

Warren, K.S. (1972). "The immunopathogenesis of schistosomiasis: a multidisciplinary approach." Trans. Roy. Soc. Trop. Med. Hyg. 66, 417—432.

Wassom, D.L., DeWitt, C.W. and Grundmann, A.W. (1974). "Immunity to *Hymenolepis citelli* by *Peromyscus maniculatus:* genetic control and ecological implications." J. Parasitol. 60, 47—52.

Weathersby, A.B. (1960). "Further studies on exogenous development of malaria in the haemocoels of mosquitoes." Exp. Parasitol. 10, 211—213.

Whitcombe, R.F., Shapiro, M. and Granados, R.R. (1974). "Insect defence mechanisms against microorganisms and parasitoids." In: "The Physiology of Insects." 2nd Edn., Vol. 5 (M. Rockstein, ed.) pp. 447—536. Academic Press, New York.

Wilson, R.J.M. (1974). "Soluble antigens as blocking antigens." In: "Parasites in the Immunized Host: Mechanisms of Survival." (R. Porter and J. Knight, eds.) (Ciba Foundation Symp. 25) pp. 185—203. Associated Scientific Publishers, Amsterdam.

Winfield, G.F. (1932). "On the immunity of snails infested with the sporocysts of the strigeid *Cotylurus flabelliformis*, to the penetration of its cercariae." J. Parasitol. 19, 130—133.

Zelodon, R. and Monge, E. (1966). "Natural immunity of the bug *Triatoma infestans* to the protozoan *Trypanosoma rangeli.*" J. Invert. Pathol. 8, 420—424.

Ecological Aspects of Parasitology
Editor: C.R. Kennedy
© North-Holland Publishing Company, Amsterdam, 1976

CHAPTER 7

REPRODUCTION AND DISPERSAL

C.R. KENNEDY

Department of Biological Sciences, University of Exeter, Exeter, EX4 4PS (Great Britain)

CONTENTS

7.1. Introduction

Reproduction and the dissemination of the reproductive products are as essential features of the life cycles of parasites as they are of the life cycles of free living animals. The nature of the parasitic existence is such, however, that they may present particular problems to parasites. Reproduction may have to relate to the low levels of parasitic infections sometimes encountered, and hence to the low probability of encountering a reproductive partner. Fecundity must relate to, amongst other factors, the enormous mortality experienced in the course of many parasitic life cycles. The method and timing of the liberation of the reproductive products must relate both to the site of the parasite within its host and to the behaviour of its hosts. Dispersal in turn involves not merely dissemination in space, but escaping from and spreading away from the host individual in which reproduction took place. Since the stages effecting dispersal are also often the stages which infect the next host, they may have to accomplish both dissemination and host location. For many parasites, dissemination in space is less important than dissemination in time, so that they can await the return of suitable conditions for infection or even of the next host itself. None of these problems is unique to specific parasites: the combination of many or all of them is characteristic of parasitic life cycles.

Because dispersal is frequently achieved by the reproductive products, the nature and timing of reproduction and the nature and abilities of these products are closely related to the method and success of dispersal. This is especially the case when it is necessary for the parasite to disseminate in time, since the timing of the liberation of the reproductive products and their structure and metabolism may largely determine their life span. It is difficult therefore to discuss dispersal without prior discussion of the reproductive process. The data available on which to base such a discussion is, however, very limited. It consists essentially of estimates of fecundity, and scattered observations relating to the life span and activity of the dispersal stages. There is very little quantitative information on the reproductive process, including sex ratios, copulation frequency and hence egg production rates, or on distances covered by and mortality experienced during the dispersal phases. Since very few studies have had the investigation of parasite dispersal as their specific aim, the data may be particularly difficult to interpret in this context. Rather therefore than attempt to present a comprehensive survey of all this scattered information, this chapter will be concerned with outlining the strategy of the solutions to the above problems and providing selected specific examples.

7.2. Reproduction

7.2.1. Mating

Where free living species are of separate sexes, the need to find a reproductive partner presents problems to many animals, especially if they occur at low population densities and have poor mobility. These problems may also be encountered by many parasites. In bisexual infections of low density the probability of mating may be very low (MacDonald, 1965), and the extent of mating depends upon the size of the infection in or on any individual host. At the population level, however, the probability of any individual parasite mating must also relate to the degree of overdispersion of the parasite population throughout that of its host. Most, if not all, parasites are overdispersed throughout their host population (Crofton, 1971), and the greater the overdispersion, the greater is the mean number of parasites for a given incidence of infection. This means that a large number of parasites occurs in a small number of hosts, and in extreme cases most of the parasites may be concentrated into a very few host individuals. Within such individuals, the probability of mating by any parasite must be high. By contrast, if a parasite were randomly dispersed throughout its host population, the majority of hosts would contain only a single individual and so the probability of mating would be zero. The probability of mating is increased in some species by arrangements for permanent association following contact between two individuals, such as occurs in *Diplozoon paradoxum* and *Schistosoma* spp. when one partner will not develop fully to maturity in the absence of the other. Where parasites are hermaphroditic or protandrous the need to find a reproductive partner and the probability of mating may be less important considerations. Although hermaphroditic parasites appear always to be capable of cross-fertilisation if the need arises, they are also capable of self-fertilisation. Hermaphroditism is not an adaptation to parasitism as such, although it is clearly an advantage in infections by a single individual: it appears to be a feature of certain groups of animals, some of which, notably the platyhelminths, contain a large number of parasitic representatives. As far as is known, there are no elaborate reproductive behaviour patterns amongst parasites which increase the probability of contact between individuals such as are found amongst many species of free living animals.

7.2.2. Fecundity

Two basic strategies of egg production can be recognised in animals, although these should perhaps be more correctly regarded as extremes of a continuum. The two strategies result in species which are respectively 'r' and 'k' selected, and the application of this concept to parasitic animals is discussed in detail by Esch (in press) and Jennings and Calow (1975). At one extreme ('r' selection) animals produce very large numbers of eggs, which are in energetic terms cheap since little provision is made in the way of food reserves for the developing young. At the other extreme ('k' selection) only small numbers of eggs or young are produced, but these are provided with ample reserves and in many cases also given some form of parental protection. The first pattern, corresponding to 'r' selection, is much more frequently encountered amongst parasites than the second (Esch, in press; Jennings and Calow, 1975). High fecundity is generally held to be one of the most characteristic features of parasites, although it is not exclusive to them and many free-living animals, especially those having planktonic larvae, also exhibit it. This high fecundity is generally held to be associated with the heavy mortality encountered during the course of the life cycle, in both parasitic and free-living animals, and to ensure maintenance of the population if the probability of establishment and reproduction is low (Cole, 1954), although Jennings and Calow (1975) have suggested that high fecundity may be a consequence of the ready availability of energy to parasites and so the low probability of completing a life cycle may be a consequence and not a cause of high fecundity. This is presumed to be the case with most parasite species, although precise figures and details of life tables are very scarce indeed. Certainly the low probabilities of infection by *Schistosoma mansoni* larval stages (2.59×10^{-2} for miracidia and 6.9×10^{-7} for cercariae) suggest heavy larval mortality and are consistent with the high fecundity of the adults (a net reproductive rate of 11,493 in man) (Hairston, 1965).

Fecundity is in fact very difficult to estimate. In addition to sexual reproduction, many parasites, especially amongst the Protozoa and Digenea, are also capable of extensive asexual reproduction. This has the advantage that every parasite can in population terms be regarded as a reproductive female, so that heavy mortality in the larval stages is counteracted to a large extent by the ability of the few survivors to reproduce extensively. The perpetuation of existing genotypes in this way undoubtedly contributes towards the population isolation and strain formation found amongst so many parasites. The ability of parasites to reproduce asexually, however, further complicates the difficulties of estimating fecundity. Figures for egg or larval production are quoted frequently in the literature, but the fact that *Ascaris lumbricoides* can produce 200,000 eggs per day for a year or that one oocyst of *Eimeria tenella* can theoretically produce 2.5 million second generation merozoites only confirms that there must be a very heavy mortality at some stage in the life history and provides an idea of the reproductive potential of the species.

A realistic estimate of fecundity must also take into account the generation time of the parasite and the age at which it reproduces. The importance of generation time has been emphasised by Crofton (1963), who has shown that a nematode with a generation time of 3 days but producing only 5 eggs per female has a reproductive rate equivalent to that of *Haemonchus contortus*, which has a generation time of one month and produces 10,000 eggs per female. The importance of the age at which the parasite commences reproduction, and the relative lengths of the pre- and post-maturation spans has been demonstrated by Cole (1954). Egg production by *Hymenolepis nana* has been shown to relate not only to the age of the parasite but also to the number of parasites present (see also Chapter 2) and the nutrition of the host (Ghazal and Avery, 1974). A decline in the rate of egg production or of fecundity is also a common result of a host's immune response (Chapter 6). Values for egg production alone therefore may give little idea of the true reproductive potential of a parasite unless additional information on parasite population size and conditions and on generation time and length of life span are also provided, and this is very seldom the case. Species such as *Neoplectana glaseri* have a short generation time and low rate of egg production, whereas *Taenia saginata* has a long generation time, commences egg production whilst still young, and has a high rate of egg production. The tactics adopted by any particular species are those best suited to its particular life cycle, to its host(s) and the ecological conditions it encounters.

7.3. Release of reproductive products

7.3.1. Liberation from host

In the vast majority of free-living animals, reproductive products can be liberated directly into the surrounding medium. For endoparasites, however, the surrounding medium is the host, and the products then have to be released from this. The extent to which this presents a problem depends very largely upon the precise site occupied by the parasite within the host, since many sites may have no direct access to the surrounding medium. Release may take place in the following ways:

(a) From the body surface. Eggs may be released into the surrounding medium and then dispersed by wind or water currents (*Entobdella soleae*), or they may be released and then attached to hairs or feathers on the same host individual (*Pediculus humanus* and Mallophaga).

(b) From gills. Eggs may be released and become entangled in the gill filaments of the same fish, but more usually pass out of the fish with the ventilation current (*Gastrocotyle trachuri*).

(c) From the alimentary tract and bile duct. Eggs are passed into the lumen of the alimentary tract directly or enclosed in proglottids, and are then liberated with the faeces (*Fasciola hepatica* and *Hymenolepis diminuta*).

(d) From the vascular system. Eggs may be released into the blood stream in the region of the bladder and then pass through the bladder wall (*Scistosoma haematobium*) and out of the host with the urine. Microfilaria larvae and the infective stages of protozoans circulate in the blood, and are removed from it by the arthropod vector ingesting a meal of blood which contains the parasites (*Onchocerca volvulus*, *Trypanosoma brucei*, *Plasmodium vivax*).

(e) From the lungs and trachea. Eggs may be coughed up, swallowed and passed out with the host faeces (*Syngamus trachea*), or they may be retained in the uterus until the gravid parent migrates to the outside via the trachea and mouth, passes into the surrounding medium and releases the eggs (*Heronimus mollis*).

(f) From the body cavity. Eggs may be released by way of the genital ducts or abdominal pores (*Dictyocotyle coeliaca, Amphilina foliacea*), or they may be retained in the uterus until the gravid parent is discharged when the host spawns (*Acetodextra amiuri*), or the gravid parent may burst through the body wall of the host and release eggs as it decays in the surrounding medium (*Archigetes limnodrili*).

(g) From the eyes. Adults in the ocular sack release eggs which are washed from the host's eyes when it feeds (*Philophthalmus megalurus*).

(h) From tissue sites. Cercariae migrate actively through the tissues or passively with the circulatory system currents, and emerge through the mantle cavity, mantle pore or mantle surface.

Further details of methods of release are given in Chapters 8—20. It must be appreciated also that release is often an active process, occurring in response to specific stimuli provided by the host or by the physico-chemical changes in the habitat (see also Chapter 1).

7.3.2. Timing of reproduction and release of larvae

7.3.2.1. Timing of reproduction

Many parasites appear to be capable of a steady rate of egg production throughout the whole course of their reproductive life span. Many mammalian parasites, such as *Ascaris lumbricoides* and *Taenia saginata*, release eggs or proglottids containing eggs at a fairly constant daily rate for periods up to and exceeding a year, schistosomes for periods of up to 20 years and others such as *Nippostrongylus brasiliensis* for periods of up to a month. In such species the next host, whether another individual of the definitive host species or an intermediate host, is generally available for infection at all times, or the egg is able to survive outside the hosts for long periods in a resting stage. This ability to reproduce repeatedly over a long period not only increases the biotic potential of the species, but also assists the dispersal of the eggs (Cole, 1954). If eggs are released over a long period of time, then they are distributed widely in time, and the movements of the host throughout the period also ensure a wide distribution in space.

In many other parasites, egg production takes place at one time of year only, and extends over a period of a few days or weeks. The timing of production in such cases is related to the appearance of suitable climatic conditions for the eggs, the availability of the next host, or the behaviour of the parent host. Thus eggs of many fish crustaceans such as *Lernea cyprinacea* and monogenea such as *Dactylogyrus vastator* are produced only in spring when water temperatures are favourable for their development (Kennedy, 1970, 1975a), eggs of *Mazocraes alosae* are only produced when adult hosts aggregate for spawning (Bychovsky, 1957), eggs of *Polystoma integerrimum* are only produced when frogs spawn and in direct response to host reproductive hormones, and eggs of many fish proteocephalid cestodes are only produced in spring when large numbers of copepods, their next intermediate host, appear (Kennedy, 1975a). The timing of the reproductive period is always such that it occurs when the probability of contact with the next host is greatest.

7.3.2.2. Timing of release

Even though egg production may take place over a more or less extended period, egg or larval release may take place over a much more restricted period, and may in addition exhibit circadian rhythms. The timing again ensures the greatest probability of contact with the next host.

Eggs of *Heronimus mollis* accumulate in the uterus of the parasite during the hibernation period of the turtle, and are not liberated until hibernation has ceased in early summer (Crandall, 1960). Eggs of *Renicola* in shearwaters are also retained in the uterus of the parasite for most of the year, and specifically when the host is over the ocean, and are released only when the bird comes in shore (Wright, 1971). Larvae of *Dracunculus medinensis* are released only when the host comes into contact with water, when the intermediate host, a cyclopoid copepod, can be encountered.

Eggs passed into the surrounding medium in faeces are often released at one particular time of day, since defaecation itself often occurs at particular times only. The majority of the eggs of *Schistosoma haematobium* are released into the urine at mid-day or early afternoon, when the probability of its human host entering the water is highest in a tropical climate. By contrast, eggs of *Enterobius vermicularis* are released at night through the host's rectum, thus avoiding faecal contamination.

Rhythmical behavior is particularly evident and well documented in blood parasites, (Hawking, 1968: Worms, 1972). The appearance of microcilariae in the peripheral blood circulation of their vertebrate host exhibits a marked circadian periodicity, which is in every species related to the circadian rhythm of its vector's biting activity. Some species of *Plasmodium* show rhythms in reproduction rather than activity. Gametocysts are infective to mosquitos for a short period only, and the cycle of sexual multiplication is timed to ensure that this infective phase coincides with the time at which the vectors suck blood. These rhythms are seldom strictly circadian, although their periodicities are generally simple multiples of 24 h.

Release of cercariae from molluscs is also often periodic and at particular times of day. Thus cercariae of *Schistosoma mansoni* are released at mid-day, when the probability of encountering a human host in water is highest, whereas those of *S.douthitti* are released in the evening or night, when the probability of encountering a rodent is highest (Yamaguti, 1970).

Even when eggs are released from the host over an extended period, emergence of the larvae may take place at particular times of day only. Eggs of *Entobdella soleae* show an endogenous rhythm in hatching, and oncomiracidia are released just after dawn when the probability of encountering a resting sole is highest, (Kearn, 1973). By contrast, *E.hippoglossi* hatches only in the dark, which coincides with its host's resting period (Kearn, 1974).

It would appear that rhythmic behaviour in release of products and restricted periods of reproduction are related to increasing the probability of transmission and infection rather than to dispersal. Wide dispersal in both space and time is better accomplished by release of eggs continuously over a long period of time.

7.4. Dispersal

7.4.1. Theoretical considerations

Dispersal can be considered as the act of breaking up from an assembled state. Free living animals disperse away from the population centre and so avoid over-crowding, colonise new areas and extend the range of the species. Dispersal away from the reproductive stage and especially from the host in which it is contained is also necessary for parasites in order to avoid over-infection and to colonise new hosts. Since the stages that accomplish this dispersal are often also those that locate and infect the next host, it is important to realise that many of their activities are related to transmission rather than dissemination, and that the latter may be achieved almost incidentally. It is also important to separate considerations of dispersal away from an individual host from considerations of dissemination in space and range extension. The scales of this can be very different. Dissemination in space is often achieved principally, if not exclusively, by the movements of the host and so of its contained parasites. These generally cover much larger distances and extend over wider areas than the movements of the reproductive products or larvae of the parasites when free living. Dispersal away from the host may involve only very short distance movements of the free living stage, or in some cases no movement at all if eggs are deposited in one place and remain there until ingested by the next host. The very act of release ensures that it has dispersed from its parent host, and the time lag before it becomes infective to the next host reduces the probability that it will encounter the same host individual that contained its parent. Dispersal in space away from a host can be accomplished equally successfully by movements in a horizontal or vertical plane.

In addition to dispersal and dissemination in space, many parasites disperse in time. The reproductive products may have to survive unfavourable climatic conditions and await the return of favourable ones, or they may have to await the appearance of hosts of a new generation. Dispersal in time is generally accomplished by the incorporation of a resting stage into the life cycle. These extend its range in time so that it can await location and acquisition by a new host. If the resting stage is passed within an intermediate host, then the movements of this host will also serve to disseminate it and extend its range in space. Thus all parasites are dependent to a greater or lesser extent upon the movements of another animal, the host, for dissemination; a situation that contrasts sharply with that in free living animals which are dependent upon their own movements or those of the surrounding medium. The mobility of their hosts removes any necessity for the parasites themselves to be especially mobile in order to accomplish dispersal or dissemination, though not necessarily host location, and so many parasites have adopted a strategy of limited mobility which serves also to conserve energy resources.

7.4.2. Dispersal in space

7.4.2.1. By motile free living stages

Most interest in the motile free living stages of parasites has focused on the problems of host location and infection rather than dispersal. Very little information is actually available about their effectiveness in accomplishing dissemination in space, despite frequent reference to them as dispersal stages (e.g. Bychovsky, 1957) and unsupported assertions that they do so. The sparce data available on survival times, distances actually covered and the dispersion of infective larvae is also difficult to interpret in the context of dispersal. The fact that an individual larva is *capable* of travelling a particular distance has to be related to the situations in which the tactic or kinetic responses of many larvae cause them to aggregate in particular regions rather than disperse widely. Similarly, the fact that an individual larva is *capable* of surviving a particular time has to be related to the facts that the death rate of the larval population increases with time as energy stores are used up or senescence sets in, that their activity may be maximal in early life and decline thereafter, and that their infectivity depends on the age of the larva and so on the age structure of the larval population. Anderson and Whitfield (1975) have shown that cercariae of *Transversotrema patialensis* can survive up to 44 h at 24°C, but that 26 h is the period of 50% survival. Both activity and infectivity of the cercariae are also age dependent, and both drop to low levels many h before death (21 and 22 h, respectively). Thus wide dispersal by old larvae may be of little or no significance in view of their small numbers and reduced infectivity. Those that actually accomplish infection of the next host may well be those that have travelled only short distances.

Oncomiracidia larvae of monogeneans are in general short lived: up to 30 h at 7°C in *Entobdella*, and 9—14 h at 17°C, up to 10 h in *Diplozoon* and 4—6 h in *Discocotyle*. Their swimming speeds vary from 5 mm/s for *Entobdella* to 0.4 mm/s for *Discocotyle*. Assuming that they swam continuously in a straight line in water, then the maximum distances they could reach from the site of hatching would be 300 m for *Entobdella*, 100 m for *Diplozoon* and 10 m for *Discocotyle*. Such an event is extremely unlikely indeed, and it seems very probable that the larvae do not travel very far at all but infect their hosts very close to the site of egg deposition (Llewellyn, 1972). This must certainly be the case for species such as *Acanthocotyle lobianchi*, whose larvae possess no cilia and do not swim at all (Kearn, 1967), and for *Gyrodoctylus*, where infection is accomplished by transfer of adults from fish to fish following contact (Bychovsky, 1957: Malmberg, 1970).

Digenean miracidia are also generally short lived, and the degree of dispersal possible must relate, amongst other factors such as current, to the duration of their free living period. This in turn is dependent upon water temperature and pH, (Oliver and Short, 1956; Fowley, 1962), and the amount of endogenous food reserves, and an average span is 10—16 h (Erasmus, 1972). Miracidia of *Schistosoma mansoni* have an active life span of 8—9 h, can travel at a speed of about 2 mm/s, and can cover distances of 690—750 cm in their first h of life (Chernin and Dunavin, 1962). The same authors have shown that they can locate a host and therefore travel, distances of 86 cm in the horizontal plane and 33 cm in the vertical plane. The dispersion of miracidia in the water is not, however, random, and they tend to aggregate. Indeed the behaviour of most miracidia is such as to bring them together and relates to host location (Cable, 1972), and consists of a period of responding to physical stimuli, generally a positive phototaxis that takes them to the water surface, a period of random movements, and a period of chemotaxis or other directed responses to stimuli produced by snails. The behavioural sequence thus brings larvae into contact with a host, and leads to aggregation, and it is consequently very difficult to determine the extent to which they have actually dispersed away from the eggs or disseminated in space.

Cercariae are also short lived, the span depending again upon food reserves and temperature (Dutt and Srivastava, 1962), but also being characteristic of the species. In *Notocotylus* it may only be a matter of minutes, in schistosomes 8—16 h and in other furcocercariae 45—70 h (Erasmus, 1972). The distance covered in this time is very difficult to determine, as it depends so much upon the behaviour pattern of the cercaria as well as water currents and wave action. Not all cercariae are free living, and of those that are, not all are active or are swimmers. Even the pattern of swimming shows considerable variation. Some swim continuously, but with sudden and erratic changes in direction. Others alternate periods of swimming with periods of sinking, and the rate of sinking depends upon their structure and posture.

Cercaria hamata swims for periods of only 1—4 s, then sinks for periods of 20—107 s (Erasmus, 1972). Cercarial dispersal as a consequence of their own activity may be very poor indeed if cercariae penetrate the same host individual, or encyst very rapidly, or if in still waters are feeble swimmers. Whereas some species such as *Cercaria X* may be distributed uniformly in the water, the majority appear to aggregate. Indeed in one group, that of zygocercous cercariae, the cercariae aggregate in a rosette which is held together by mucus, and the aggregate does not swim around at all (Hendrickson and Kingston, 1974). The most detailed study of dispersal is that of *Schistosoma mansoni* cercariae. Radke et al. (1961) were able to show that they could travel up to 1600 m downstream of a release point. Cercarial densities were similar up to 600 m, but declined rapidly thereafter. However, infectivity declined with distance downstream, and presumably therefore ageing of the larvae, and the highest infection levels were found at only 33 m downstream. It would seem therefore that in general cercariae do not achieve a wide dissemination in space despite the ability of some individuals to be carried long distances by currents.

Nematode larvae, although longer lived than those of digeneans, do not appear to move as far. Most of the activity of the infective larvae of sheep and cattle nematodes takes place in, and in the immediate surroundings of, the host's faeces in which their eggs were deposited. They do not move far away from the faeces laterally, and their vertical migrations, although of major significance for infecting a new host, only cover very small distances, in the region of cm (Crofton, 1963). Even *Dictyocaulus* larvae which migrate to the sporangiophores of the fungus *Pilobolus* and are discharged with the sporangia only pass by this means onto the herbage surrounding the faecal pad (Robinson, 1962). This is probably of greater importance in ensuring transmission than in effecting dispersal. Although dispersal must depend to a large extent upon the climatic conditions and activity of the larvae, this latter is very variable, and different larval stages show different characteristics of behaviour pattern (Croll, 1970). Movements are in general non-linear, and locomotory patterns are dominated by the interaction of inherent asymmetry and reversals, (Cross, 1970, 1972). This makes it very difficult to estimate the distances actually covered away from the egg, and a species such as *Ancylostoma tubaeforme* may undergo extensive and continuous movement, yet after 15 min only have travelled 25 mm in a straight line (Croll, 1972). The larvae of most species move randomly in a continuous water film, but unidirectionally in a narrow channel of moisture such as is encountered on a grass blade (Crofton, 1954; Lee, 1965). Both the eggs and the larvae are aggregated and overdispersed. It seems therefore that the movements of nematode larvae are very local and that they achieve escape from the faecal mass, but they do not achieve wide dissemination.

Far less is known about the infective larvae of other groups of parasites.

Many are also short lived, including coracidia larvae of pseudophyllidean tapeworms (24 h) and infective stages of flagellates such as *Ichthyophthirius multifiliis* (33—48 h). Crustacean nauplii may live longer. None of these larvae are strong swimmers, and their dissemination must depend to a very large extent upon their longevity and water currents. The larvae of most arthropod parasites including fleas, lice and ticks also have limited powers of movement and cover only short distances.

7.4.2.2. By non-motile free living stages

The non-motile infective stages of many parasites are also often resting stages, and so accomplish dispersal in time as well as space. They occur at some stage of the life cycle in probably the majority of parasites, whether as eggs, cysts or encysted larvae.

Eggs, or cysts in the case of protozoans, are not all released from the host instantaneously even when reproduction and release are restricted to particular times. The movements of the host between each period of release achieves the distribution of the eggs in space, and the longer the reproductive period of the parasite, the greater the distance the host has covered during this time interval and so the wider the spread of the eggs.

The eggs or cysts are usually light and small, and so are easily moved by wind or water currents. Those of aquatic species are in general slightly denser than water, and so sink slowly. The shape, density and movement, however, may be such as to increase the likelihood of encountering the next host, and Jarecka (1961) has shown that the eggs of related species of cestode are of different structures and densities, such that some sink slowly and are more likely to encounter benthic copepods whilst others float and encounter planktonic copepods. The eggs of monogeneans sink at a rate of about 2 cm/ min, and those of *Gastrocotyle trachuri* take 1—4 days to reach the sea bottom at 50 m depth. Allowing for lateral drift with currents, they probably hatch within a few miles of where they were laid (Llewellyn, 1962). Other species may travel less far. Eggs of some species adhere to the gills of fish (Bychovsky, 1957), and of others to any hard substrate, so that eggs of *Acanthocotyle lobianchi* probably do not travel very far from the rays that laid them (Kearn, 1967). Digenean eggs probably also sink in a similar manner. Freshwater currents can carry eggs downstream, as acanthors of the acanthocephalan *Polymorphus minutus* have been found 2,500 m downstream of the source of infection (Hynes and Nicholas, 1963).

Eggs of terrestrial species seem to be carried much shorter distances. This is probably due to the fact that many are deposited in faeces, and can only be carried by wind when faeces have dried out, or by water trickling through and dispersing the faecal mass. Nematode eggs and *Eimeria* cysts may be carried laterally by wind, water and surface run-off, but the distances covered are generally a few metres at the most. Thus whilst eggs laid on land play only a small part in dispersal in space compared with the movements of the host that released them, those laid in water may disperse over larger areas as a result of current and tidal movements.

7.4.2.3. By hosts or vectors

This is probably the most important method of achieving dissemination in space, and as such it is perforce employed by all parasites. For those parasites which lack free living stages in their life cycle, including the sporozoans, trypanosomes and filarial nematodes, it is the only means of dispersal. It is also the most important method whereby new areas are colonised and the range of a parasite species extended.

Although in many cases it can be presumed that the movements of the host cover a wider area than the movements of the larvae or eggs, this is not necessarily so, as closer investigation may show that host movements are very local, and the life span of the parasite within the host may be much shorter than the period it can survive as a free living stage, especially if this is a resting stage. Many vectors are themselves relatively inactive or cover only short distances. Mites, ticks and fleas, which serve as vectors for some trypanosomes and filarial nematodes, may spend all or the greater part of their lives on a single host individual. Their transfer to another individual host may serve to disperse the parasite, but they do not disseminate it widely in space, and their role may be more important in transmission rather than dispersal. *Trypanosoma brucei* employs tsetse flies as vectors and humans as hosts. Although the flies can achieve speeds of 15 m.p.h., they are in fact relatively inactive, and probably do not spend more than 15 min in flight a day. They exist as very local populations and their dispersal is estimated at no more than 200 m per week (Gatehouse, 1972). The behaviour pattern of mosquitoes is heavily influenced by the choice and location of breeding sites. In the savannah — cultivated situation, breeding and feeding sites may be separated by 2—3 km, and the mosquitos range widely. In other situations, the mosquito may be restricted to a particular vegetation type, or in the urban situation may undertake unrestricted but short flights (Gillies, 1972). Even if the host or vector undertakes extensive migrations, as do many birds, a restricted period of egg release may result in the parasites being confined to a small area of the host's range.

Nevertheless, in many cases the movements of the host achieve a much wider range of dispersal than the parasites themselves can achieve when free living. The movements of nematode larvae are very local indeed when compared with the movements of their hosts, even when these exhibit a behavioural pattern that keeps them within a limited range and ensures they return to any specific area within a given time, as is the case with sheep flocks (Crofton, 1963). Where dispersal by hosts and by eggs have been compared, the former is generally found to be more effective. Acanthors of *Polymorphus minutus* spread only 2,500 m downstream of the source of infection, but infected *Gammarus pulex* were found 900 m upstream and 5,000 m downstream (Hynes and Nicholas, 1963). The movement of the ducks could be expected to be greater still, just as the movement of infected humans is greater than the 1,600 m that can be attained by *S. mansoni* cercariae (Radke et al., 1961).

7.4.3. Dispersal in time

Dispersal in time by the existence of resting stages in a life cycle is widespread amongst parasites. Such stages are long-lived, and generally encysted or encapsulated, resistant and non-feeding. They may survive for only periods of 1 month (acanthors of *Polymorphus minutus*), for about six months (oocysts of *Eimeria* and larvae of *Haemonchus contortus*), for up to 3.5 years (acanthors of *Moniliformis hirudinaceous*), or for up to 5 years (metacercariae of *Diplostomum spathaceum*). The period of survival depends also upon climatic factors, and both survival rate and infectivity vary with temperature and decline with time (e.g. Ciordia and Bizzell, 1963; Narain, 1965). They are especially common amongst terrestrial parasites, which face particular problems of survival in view of dehydration. They may be free living, or contained within another host. Their presence in a host both increases their dispersal in space as a result of the host's movements, and may ensure persistence of the parasite during a period when its other hosts disappear temporarily. Indeed, their major role appears to be to enable the parasites to survive periods of unfavourable conditions, whether climatic or related to the unavailability of a host. The ability of many trypanosomes to persist for long periods in their vertebrate host when their life span in their vector is so short and host/vector contact so spasmodic, may subserve this role, although resting stages sensu strictu do not exist for them. Amongst nematodes of sheep and cattle, it is the resting stages that survive the overwintering period of unfavourable climatic conditions, that relate the parasites of one year to another, and that relate infections of one host generation to those of the next (Crofton, 1963). *Nematodirus battus* overwinter as resting eggs, wheras *Ostertagia* spp. overwinter as inhibited larvae within a host. Resting stages show little or no endogenous activity, and so achieve little or no dispersal in space by their own efforts. They rely on the movements of the hosts and their habits for transmission.

7.4.4. Efficacy of dispersal

It is clear from the foregoing review that all parasites have some method of ensuring that progeny are dispersed away from their parent; whilst in a few cases such free living progeny may be distributed fairly widely in space as a result of their own activities or of wind or water currents, in the majority of species such movements are local in nature. Dissemination in space is achieved far more often by the movements of the host containing the parasites, and by release of the reproductive products over an extended period of time and hence space.

Despite the fact therefore, that many parasites have considerable potential for dispersal and dissemination, and may extend over a wide georgraphical range, the actual extent to which they achieve this potential may be much more limited in practice. The distribution of parasites is seldom homogenous or indeed contiguous throughout their range, and they tend to occur in discrete foci in local areas where conditions favour their dissemination and survival. This is to a large extent inevitable, since the persistence of a parasite depends not only upon its dispersal powers and ability to locate a new host, but also upon the persistence of all its hosts species in sufficient numbers and upon the correct climatic and physico-chemical conditions. This pattern of distribution is especially evident in schistosomes, where the discontinuous populations of molluscs serve as the focal centres (Wright, 1971). in many other parasitic diseases of man, such as trypanomiasis and malaria, and in littoral digeneans (James, 1968), where species populations are restricted not only to particular shores but also to particular zones and regions of the shores. Dispersal between foci relies upon the movements of hosts carrying the parasites, and in turn therefore upon the life span of the contained parasite. The extent to which this is actually achieved depends almost entirely upon the behaviour of the hosts. This may result in very local movements within a restricted area, as many animals remain within a home range or territory and may be unable to spread further to physical or behavioural barriers. Thus although host may be potentially able to disseminate parasites over large areas, they may in practice not do so, and the parasite may be restricted to particular foci despite the fact that vacant habitats apparently suitably in all respects may exist in close proximity. Colonisation of new habitats outside the foci must depend to a large extent upon chance invasions by parasitised hosts. Within the focal habitats, dispersal and dissemination may be very effective. A single example is sufficient to demonstrate this point. Slapton Ley is an isolated lake in South West England. The nearest natural similar freshwater body is about 100 km away. The lake is about 1500 years old. It is dominated by *Esox lucius* and *Perca fluviatilis*, and more recently by *Rutilus rutilus*. Despite this, despite the fact that the intermediate hosts are present, and despite the fact that it is within their geographical range, no nematodes maturing in fish and characteristic cestodes such as *Proteocephalus percae* or *Triaenophorus nodulosus* occur in the lake. *Ligula intestinalis* and *Tylodelphys clavata*, both widespread British parasites maturing in birds, only reached the lake 2 years ago, and then as a result of the chance arrival of a breeding pair of *Podiceps cristatus*. Within these two years, both species have spread throughout the entire lake so that they now attain infective levels of 30% and 70% respectively (Kennedy, 1975b). This amply demonstrates the reliance of parasites on chance host movements for extending range and spreading from one focus to another, but their efficacy of dispersal and dissemination within any focus.

7.5. Conclusion

It is clear from the foregoing account that although there is a considerable amount of information available on reproduction and on larval survival of individual parasites, this has seldom been examined or interpreted in relation to the parasite population as a whole or in a wider ecological context. The actual, as opposed to the potential, number of parasite progeny produced by a *population* has seldom been determined under natural conditions, and survivorship curves and life tables for most parasite populations are unknown. It may be possible in general terms to deduce the parasite strategy. Thus, the duration and timing of reproduction and release of reproductive products is closely related to the opportunities for infecting the next host: if this is available at all times of year then the products may be liberated over a long period and be capable of surviving for some time, whereas if the next host is available at one time of year and for a short period only, then the release of the reproductive products may take place over a short period only, and they may have a very short life span. Once liberated from the parent host, the behaviour of active free living parasite larvae appears to relate primarily to location of the next host, dissemination in space being accomplished more or less incidentally as a consequence of wind and water movements. The strategy of parasites with resting stages whether in or outside of a host appears to be to disperse in time, and by being capable of surviving for long periods to await the appearance of the next host. This type of parasite behaviour is very common and widespread since, as stated earlier, the mobility of their hosts removes any necessity for the parasites themselves to be especially mobile in order to accomplish dissemination or dispersal, and such immobility conserves the parasite's energy reserves. Nevertheless, understanding of the parasite's strategy, however important, is not in itself sufficient; it is essential that far more precise, detailed and quantitative data on survivorship and dispersal should be obtained. Only when such data is available can the population biology and ecology of parasites be thoroughly understood (Chapter 22), meaningful models constructed (Chapter 22) and the epidemiology of economically important parasites such as *Plasmodium* and *Trypanosoma* described.

7.6. REFERENCES

Anderson, R.M. and Whitfield, P.J. (1975). "Survival characteristics of the free-living cercarial population of the ectoparasitic digenean *Transversotrema patialensis* (Soporker, 1924)". Parasitology 70, 295—310.

Bychovsky, B.E. (1957). "Monogenetic Trematodes, their Classification and Phylogeny". Academy of Sciences, U.S.S.R.: Moscow.

Cable, R.M. (1972). "Behaviour of degenetic trematodes". In: "Behavioural Aspects of Parasite Transmission". (E.U. Canning and C.A. Wright, ed.). J. Linn. Soc. (Zool.) 51, Suppl. 1, 1—18.

Chernin, E. and Dunavan, C.A. (1962). "The influence of host-parasite dispersion upon the capacity of *Schistosoma mansoni* miracidia to infect *Australorbis glabratus*". Amer. J. trop.Med. Hyg. 11, 455—471.

Ciordia, H. and Bizzell, W.E. (1963). "The effects of various constant temperatures on the development of the free living stages of some nematode parasites of cattle". J. Parasit. 49, 60—63.

Cole, L.C. (1954). "The population consequences of life history phenomena". Quart. Rev. Biol. 29, 103—137.

Crandall, R.B. (1960). "The life history and affinities of the turtle lung fluke, *Heronimus chelydrae* MacCallum, 1902". J. Parasit. 46, 289—307.

Croll, N.A. (1970). "The behaviour of nematodes, their activity, senses and responses". Edward Arnold, London.

Croll, N.A. (1972). "Behaviour of larval nematodes". In: "Behavioural Aspects of Parasite Transmission". (E.U. Canning and C.A. Wright, ed.). J. Linn. Soc. (Zool.) 51, Suppl. 1, 31—52.

Crofton, H.D. (1954). "The vertical migration of infective larvae of strongyloid nematodes". J. Helminth. 28, 35—52.

Crofton, H.D. (1963). "Nematode parasite populations in sheep and on pasture". Commonw. Agric. Bur. Publ. 35, 1—104.

Crofton, H.D. (1971). "A quantitative approach to parasitism". Parasitology 62, 179—193.

Dutt, S.C. and Srivastava, H.D. (1962). "Biological studies on *Orientobilharzia dattai* (Dutt and Srivastava, 1952) Dutt and Srivastava, 1955 — a blood fluke of ruminants". Ind. J. Vet. Sci. 32, 216—228.

Erasmus, D.A. (1972). "The biology of trematodes". Edward Arnold, London.

Esch, G.W. (In press). "Parasitism and 'r' and 'k' selection".

Fowley, G. (1962). "Effect of temperature and pH on the longevity of *Schistosomatium douthitti* miracidia". Canad. J. Zool. 40, 615—620.

Gatehouse, A.G. (1972). "Host finding behaviour of tsetse flies". In: "Behavioural Aspects of Parasite Transmission". (E.U. Canning and C.A. Wright, ed.). J. Linn. Soc. (Zool.) 51, Suppl. 1, 83—95.

Ghazal, A.M. and Avery, R.A. (1974). "Population dynamics of *Hymenolepis nana* in mice: fecundity and the 'crowding effect' ". Parasitology 69, 403—416.

Gillies, M.T. (1972). "Some aspects of mosquito behaviour in relation to the transmission of parasites". In: "Behavioural Aspects of Parasite Transmission". (E.U. Canning and C.A. Wright, ed.). J. Linn. Soc. (Zool.) 51, Suppl. 1, 69—81.

Hairston, N.G. (1965). "On the mathematical analysis of schistosome populations". Bull. World Hlth. Org. 33, 45—62.

Hawking, F. (1968). "The 24-hour periodicity of microfilariae: biological mechanisms responsible for its production and control". Proc. roy. Soc. B 169, 59—76.

Hendrickson, G.L. and Kingston, N. (1974). "*Cercaria laramiensis* sp.n., a freshwater zygocercous cercaria from *Physa gyrina* Say, with a discussion of cercarial aggregation". J. Parasit. 60, 777—781.

Hynes, H.B.N. and Nicholas, W.L. (1963). "The importance of the acanthocephalan *Polymorphus minutus* as a parasite of the domestic ducks in the United Kingdom". J. Helminth. 37, 185—198.

Jarecka, L. (1961). "Morphological adaptations of tapeworm eggs and their importance in the life cycles". Acta Parasit. Polon. 9, 409—426.

James, B.L. (1968). "The occurrence of larval Digenea in ten species of intertidal proso-branch molluscs in Cardigan Bay". J. nat. Hist. 2, 329—343.

Jennings, J.B. and Calow, P. (1975). "The relationship between high fecundity and the evolution of entoparasitism". Oecologia 21, 109—115.

Kearn, G.C. (1967). "The life cycles and larval development of some acanthocotylids (Monogenea) from Plymouth rays". Parasitology 57, 157—167.

Kearn, G.C. (1973). "An endogenous circadian hatching rhythm in the monogenean skin parasite Entobdella soleae and its relationship to the activity rhythm of the host (Solea solea)". Parasitology 66, 101—122.

Kearn, G.C. (1974). "Nocturnal hatching in the skin parasite Entobdella hippoglossi from the halibut, Hippoglossus hippoglossus". Parasitology 68, 161—172.

Kennedy, C.R. (1970). "The population biology of helminths of British freshwater fish". In: "Aspects of Fish Parasitology". (A.E.R. Taylor and R. Muller, ed.). Symp. Brit. Soc. Parasit. 8, 145—149.

Kennedy, C.R. (1975a). "Ecological Animal Parsitology". Blackwell Scientific Publications, Oxford.

Kennedy, C.R. (1975b). "The natural history of Slapton Ley Nature Reserve VIII. The parasites of fish, with special reference to their use as a source of information about the aquatic community". Fld. Stud. 4, 177—189.

Llewellyn, J. (1962). "The life histories and population dynamics of monogenean gill parasites of Trachurus trachurus (L.)". J. mar. biol. Ass. U.K. 42, 587—600.

Llewellyn, J. (1972). "Behaviour of monogeneans". In: "Behavioural Aspects of Parasite Transmission". (E.U. Canning and C.A. Wright, ed.). J. Linn. Soc. (Zool.), 51, Suppl. 1, 19—30.

Lee, D.L. (1965). "The Physiology of Nematodes". Oliver and Boyd, London.

MacDonald, G. (1965). "The dynamics of helminth infections with special reference to schistosomes". Trans. R. Soc, trop. Med. Hyg. 59, 489—506.

Malmberg, G. (1970). "The excretory systems and the marginal hooks as a basis for the systematics of Gyrodactylus (Trematoda: Monogenea)". Ark. Zool. (2) 23, 1—235.

Narain, B. (1965). "Survival of the first stage larvae and infective larvae of Bunostomum trigonocephalus Rudolphi, 1808". Parasitology 55, 551—558.

Oliver, J.H. Jnr. and Short, R.B. (1956). "Longevity of miracidia of Schistosomatobium douthitti". Expl. Parasit. 5, 238—249.

Radke, M.G., Ritchie, L.S. and Rowan, W.B. (1961). "Effects of water velocities on worm burdens of animals exposed to Schistosoma mansoni cercariae released under laboratory and field conditions". Expl. Parasit. 11, 323—331.

Robinson, J. (1962). "Pilobolus spp. and the translation of the infective larvae of Dictyo-caulus viviparus from faeces to pasture". Nature, Lond. 193, 353—354.

Worms, M.J. (1972). "Circadian and seasonal rhythms in blood parasites". In: Behavioural Aspects of Parasite Transmission". (E.U. Canning and C.A. Wright, ed.). J. Linn. Soc. (Zool.) 51, Suppl. 1. 53—67.

Wright, C.A. (1971). "Flukes and snails". George Allen and Unwin, London.

Yamaguti, S. (1970). "On the periodicity of natural emergence of cercariae". H.D. Srivastava Comm. Vol., 485—492.

Part II

HABITATS

Ecological Aspects of Parasitology
Editor: C.R. Kennedy
© *North-Holland Publishing Company, Amsterdam, 1976*

CHAPTER 8

INTERACTION BETWEEN ARTHROPOD ECTOPARASITES AND WARM BLOODED HOSTS

D.R. ARTHUR

Department of Zoology, University of London King's College, London (Great Britain)

Contents

8.1. Introductory remarks

In presenting this contribution I have attempted to develop a point of view in respect of the intimacy between ectoparasitic arthropods and warm blooded hosts, particularly on aspects of feeding relationships. The skins of birds and mammals may act as "through ways" for the invasion of internal tissues by a number of parasites. In some, like schistosomes, this is achieved by direct entry, in others like *Plasmodium* and piroplasms this is through the intermediary of an arthropod vector. Many skin inhabiting arthropods, even though not necessarily implicated in transmitting pathogenic organisms, induce changes in the cellular pattern and mechanisms of both epidermal and dermal components. The degree to which this is achieved reflects the evolutionary history and intimacy of the associations.

Some arthropods like lice spend the whole of their life cycles on the hosts; all stages of ixodid ticks feed on hosts for several days before detaching to fall to the ground either to moult to succeeding stages or, as females, to lay eggs. Other arthropods such as fleas, tsetse flies and sand flies feed on vertebrate hosts only as adults or as in the case of blowflies as larvae. Larval forms of the warble fly and the bot fly arrive in the integument after a migratory period as endoparasites but most occupy this habitat either from eggs laid on the skin or from larvae deposited there. The environments presented by mammalian and avian skins are a consequence of the evolution of their possessors to meet the exigencies of the land habitat. They have evolved not only as proof against mechanical trauma but also to limit loss of water, electrolytes and other body constituents as well as to prevent the ingress of noxious and unwanted molecules from the external environment. To this end, the skins of birds and mammals are relatively impermeable, although most molecules pass through to some degree. The colonisation of land also further involves the mediation of sensation and the regulation of heat loss. In the latter connection moveable hairs and feathers are devices to assist insulation, although where these are much reduced, as in the new born or on the adoption of a less hairy form (as in man), additional problems of protection are posed. In general, this has been compensated for by an inverse relationship between epidermal thickness and hairiness. Thus the epidermis of a sheep is very thin: the hairy adult rat has a thinner epidermis than does the hairless new-born rat. Nevertheless skin temperature tends to vary more than body temperature: in cattle, body temperature ranges from 101-103°F and skin temperature from 95 to 105°F.

As an environment the skin may be broadly classified into the epidermis, consisting of stratified epithelium and an underlying dermis of three dimensional connective tissue in a ground substance and through which is distributed a range of physiological and dynamic cellular components, and elements of the vascular and nervous systems. Below the dermis is a fatty subcutaneous layer, the panniculus adiposus which, in most mammals, is separated from the rest of the body tissues by a flat sheet of striated muscle, the panniculus carnosus. This muscle layer is vestigeal in man.

8.2. The epidermal environment

Five cellular layers are recognized in the epidermis, these being (i) the most deeply seated stratum germinativum of columnar epithelial cells (ii) the stratum spinosum, or prickle cell layer, being a layer of irregular poly-hedral nucleate cells of several cells thick, (iii) a 2—4 layered stratum gran-ulosum resembling the stratum spinosum but flattened and showing a pro-gressive increase of keratohyalin granules, (iv) the stratum lucidum appearing as a thin transparent line and consisting of eleidin, a transformation product of keratohyalin and (v) the most superficial stratum corneum of several layers of flat, hexagonal, interdigitating cells tightly bound to their neigh-bous to form lamina. Over most of the body it is generally 10—20 cells thick, less in some parts than in others, as in the scrotum of man when compared with the back of the hand.

During epidermal cell differentiation certain proteins, which are insoluble in water, dilute acids, alkalis and organic solvents and resistant to trypsin and pepsin, build up within the cell cytoplasm. Their insolubility in the stratum corneum is attributed to the establishment of covalent cross links between the polypeptide chains, including sulphur containing amino-acids. They are linked together by disulphide bonds derived from the joining to-gether of two cysteine residues of adjacent chains to yield a single cystine. In man about 65—70% of the stratum corneum consists of such insoluble material. The hardened material, keratin, is a substance and not a definable protein material and keratinisation must be regarded not solely as being due to the cross linking of polypeptide chains but as a series of changes whereby groups of cells are rendered into a continuous coherent mass. In this respect keratinised epidermis differs from that of the vagina and buccal cavity, and in the epidermis the cells appear to be cemented together by a mucopoly-sacharide material to form a membrane. The horny layer desquamates at a fixed level possibly due to the distintegration of the mucopolysacharide cement, which is not influenced appreciably by external factors and cells are shed in clusters. The factors implicated in keratinisation are, as yet, not completely resolved: a matter reviewed by Ebling (1972).

On desiccation cornified epithelium, detached from the skin, becomes hard and brittle but remains pliable if it contains 10% water/wt. The water capacity of the cells of the stratum corneum may be determined by hygro-scopic water soluble materials, but a surface liquid film may play some part in controlling moisture loss. The main barrier to water loss in the epidermis is controversial: some workers allege that the rate-limiting membrane is two or three cells deep and about 10 μm thick, others that it is determined by a zone at the base of the horny layer. That the whole horny layer constitutes the barrier system is a further possibility. Despite this it has even been questioned whether the epidermal barrier, as such, exist, in view of the trans-epidermal passage of materials through epidermal intercellular canals and because epidermal cells are actively phagocytic.

The barrier properties are those of a diffusional resistance, having a purely physico-chemical basis, and not being dependent on the energy-requiring activities of living cells. The outward transepidermal water diffusion is a passive process, dependent only upon the ambient humidity (which determines the water concentration difference across the stratum corneum), the temperature of the stratum corneum and the thickness and integrity of the membrane itself. Perhaps the point should be made that since the skin is reconstituted constantly by accretion of cornified cells from below and by shedding of effete cells from the surface the stratum corneum does not constitute a static barrier, but its condition will vary from time to time. Moreover, it should be realised that there are regional variations in the skin of the same animal at any one time, some of which are discussed by Ebling (1972), and these may also alter with age. Their net effect is to produce a range of microenvironments over the body of an individual host in respect of such abiotic parameters as temperature, relative humidity and shade. A proportion of these microenvironments, too, will be within range of mechanical grooming by a host, a matter referred to later. All these in turn will influence the selection of sites for occupation by ectoparasites.

8.3. Coping with the epidermal environment

8.3.1. The epidermis as a food source

The high bulk content and concentration of keratin in the epidermis makes it a habitat in which highly specific requirements are demanded of those ectoparasitic inhabitants which colonise it successfully. The only alternative for their survival is to be able to utilise supplementary resources, such as blood and epithelial debris, and dependance on keratin is not exclusive. The most complete digestion of keratin takes place in *Tineola*, the clothes moth larva, and involves the enzyme keratinase under anaerobic conditions. This releases cystine which is probably reduced to cysteine by cystine reductase.

Cysteine is further reduced to hydrogen sulphide by cysteine desulphydrase and, since both are reducing agents, they will effect the breakdown of the disulphide bonds in the keratin, so assisting enzyme activity. Those insects capable of digesting keratin have a poor tracheal supply to the crop and the mid-gut. This is also poorly developed in mammal-infesting lice which, if generalisations can be drawn from some trichodect species, do not maintain reducing conditions in their digestive tracts. Thus it appears that the maintenance of highly reducing conditions may require poor tracheation but Waterhouse (1952) reports that this alone will not produce a negative redox potential. More recent work (Gilmour, 1965) suggests, however, that the low redox potential is a consequence of, rather than an essential condition for, keratin digestion. Bird-infesting lice regularly ingest feather particles (Martin, 1934; Nelson, 1971; Nelson and Murray, 1971) and the fluffy area of the feather is preferred. During digestion there is a continuous and vigorous circulation of digestive juices and partly digested products between the crop and the mid-gut (Waterhouse, 1952). These movements are probably more important than mechanical disintegration in the digestive process, for the teeth in the crop of both Amblycera and Ischnocera are poorly adapted for maceration. As a result of this circulation in avian mallophaga the pH in both parts of the gut is about 8.0 and the redox potential is approximately −200 mv. Partially digested material with disorganised structure is found in the mid-gut and a valvular arrangement in the crop prevents the passage of most feather particles into it. That avian mallophaga digest keratin is suggested by the ingestion of barbs and barbules and there are partially digested feathers in the crop. Moreover, the detection of free sulphydryl groups in both crop and mid-gut are indicative of at least mildly reducing conditions. Reduction of cystine does not proceed, however, beyond cysteine (Waterhouse, 1952) and in this is similar to dermestid larvae.

8.3.2. Alternative nutrient sources

Waterston (1926) indicated that bird lice are more catholic in the material they ingest than is provided by their living environment, for in the crop of *Falcolipeurus* he found small grains of quartz, mica, olivine, felspar, a lepidopterous scale, a seed coat and fungal spores. These were in addition to cast egg shells and lice cuticles as well as skin scurf and scales. Cannibalism is also reported in in vitro cultures of *Colpocephalum turbinatum* by Nelson (1971) but this has not been hitherto confirmed in vivo. Whilst some species can be fed in vitro on feathers for short periods, the evidence indicates that feeding is not exclusively on pure feather keratin and various accessory contaminants of feathers are likely to be ingested, unless they are supplied by symbionts known to be harboured by the lice (Ries, 1931). Many, like *Menacanthus*, have blood in the crop in addition to feather particles (Crutchfield and Hixson, 1934; Stockdale and Raun, 1965), probably as a result of chewing holes in the developing quill, whilst others, *Ricinus* spp, feed exclusively on blood of small passerine birds as a result of cutting holes through the epidermis (Clay, 1949). *Piagetiella*, which lives in the pouches of pelicans, probably subsists on mucus and blood, and its only association with the epidermis is to lay eggs on the pelican's feathers. A highly specialised feeding habit also occurs in *Actornithophilus patellatus* which inhabits the inside of the quill or shaft of the primary and secondary feathers of the curlew, and is rarely found elsewhere. It subsists on the pith of the shaft and all stages — eggs, nymphs and adults, may be present in large numbers in a single feather.

Digestion by mallophaga infesting mammals is a less thoroughly investigated field and on the available evidence the interchange between crop and midgut is a less conspicuous and general feature. It was observed by Waterhouse (1952) in *Heterodoxus* but was not detected in *Gliricola* or *Damalinia*. There is some physiological support for this view as the pH of the crop contents approximated to that of the ingested food. Reducing conditions are not encountered during the passage of food along the gut, and unlike avian mallophaga, no free sulphydryl groups are detected. Neither hair nor wool were detected regularly in the digestive tract of mammals infesting lice by Waterhouse (1952) and other workers, and it is presumed that due to their inability to digest keratin they subsist on epithelial debris and skin secretions.

In vitro rearing of cattle lice has been accomplished by Matthtyse (1944), using fresh cattle hair sprinkled with dried brewer's yeast and similar success has been achieved by Scott (1952) for the biting lice of sheep by feeding them on greasy wool, skin scurf and powdered yeast. Some gyropids e.g. *Gyropus ovalis* feed largely on epithelial debris, skin secretions and possibly epidermal cells and others like *Gliricola porcelli* are thought to take in sebum: Ewing (1924) observed that the latter thrust its head into the opening of the hair follicle and fed actively. Blood may be imbibed as in the Wallaby louse (*Heterodoxus*), where Waterhouse (1952) noted that the crop, mid-gut and rectum contained reddish-brown contents. Additionally, there was a substantial quantity of birefrigent material in the crop which appeared to be host skin and epithelial debris. Two interesting species in this connection are the louse associated with the African and Indian elephants, *Haematomyzus elephantis*, and that which lives on warthogs *(H. hopkinsi)*. They live within the skin folds of these hosts and their elongated mandibulate probosci penetrate the hosts' hides, when it is presumed that they then suck up blood or lymph by the pumping action of the pharynx.

Mallophaga probably have a common ancestry with the free living Psocoptera, which live in situations rich in organic debris, moulds and lichens and on which they feed. Accordingly it is not too surprising then to find that some psocopterids have been found in the fur of wild and domestic animals (e.g. Pearman, 1960) and on the plumage of birds (e.g. Mockford, 1967) inhabiting organically strewn nests, feeding or roosting areas. It is suggested by Mockford, that the skin and pelage of birds offer a constant warm temperature, probably high relative humidity and perhaps a food supply in the form of skin scurf and other debris. It is then not too far a cry evolutionarilly (Kellogg, 1902; Harrison, 1914; Webb, 1946; Calaby, 1970) and ecologically to the permanent lice population on these warm blooded vertebrates.

For the most part the mallophagan lice have not usually penetrated into or below the epidermis, largely because of their inability to completely penetrate this physico-chemical barrier. There are obviously exceptions to this as seen in the evolution of "piercing" mouthparts in the mallophagan

Trochilaecetes which enables it to reach the dermal blood supply which is its sole means of sustenance (Clay, 1949).

8.3.3. Host transfer

Since all stages of the louse, as far as is known, feed on the same materials, and egg laying and moulting from one instar to the next occur on the same host, invasion of other host species becomes largely irrelevant to them. The transfer from one host to another of the same species may then occur during aggregation of hosts in the nest as during feeding or in copulation. This makes for a high degree of host-specificity which can only have been maintained as a consequence of very long established relationships. There are many examples of this phenomenon and details of the host associations of mammalian lice, for example, are given by Hopkins (1949). It is likely, too, that during the evolutionary history of the hosts the lice also evolved, possibly less conspicuously and rapidly, but sufficiently so as to be useful indicators of host affinities (Eichler, 1948), although caution is required in making judgements (Clay, 1951, 1957). A need still exists, however, for effective transfer experiments.

The constancy of the physical parameters of the skin of most of the mammalian and avian hosts permits continuous breeding throughout the year, and theoretically a louse population could buid up to a large population, but on wild mammals, according to Hopkins (1949), the louse population is of variable size but often small. This may in part be due to active delousing or grooming on the part of the hosts. Anting in birds (confined to the passeriformes) has also been considered to be a delousing device and Simmons (1957, 1966) reports that after "anting" birds had numbers of dead and dying feather mites in their plumage. Possibly changes in the stability of the skin environment, as a consequence of preening and moulting of feathers or grooming and hair fall of the pelage may be key factors in preventing a large build up in lice populations, as impairment of these activities results in the rapid growth of parasite populations.

8.3.4. Louse distribution on individual hosts

While wild mammals seldom have more than three Phthiraptera species on a single host species, birds are frequently parasitised by four or more species. In such species, particularly among the Ischnocera, there is considerable structural variation which reflects the eco-evolutionary response of the lice to mechanical interference by the host, and which allows more than one parasite species to live without interspecific interference on the same host. Lice which occupy the head and the neck of the host are usually short or broad, have more rounded bodies and expanded heads and move relatively slowly, (e.g. *Philopterus, Saemundssonia, Anatoecus, Ibi-*

doecus): they are relatively safe from self-delousing by the birds as they are out of range of the hosts' beak. In contrast those (e.g. *Lipeurus, Columbicola, Degeeriella, Colpocephalum, Ferribia*) on the hosts' wings are usually elongated with narrow heads and are capable of rapid movement across the large feathers. Those occupying other body sites (e.g. *Menopon, Goniodes, Menacanthus, Coloceras*) are somewhat intermediate in shape and are generally able to move rapidly. Frequently too, there is a size relationship between species belonging to a single genus and larger hosts usually have larger lice. Other chemico-physical factors may be relevant in host distribution on the skin as suggested by the location of the louse *Lepidophthirus macrorhini* on the southern elephant seal. The host comes ashore twice a year for periods of from three to five weeks when the lice will reproduce at temperatures above 25°C. This they do on the hind flippers of the seal which are the warmest parts of the body both on land and in the sea. The Weddell seal, on the other hand, come on the ice daily and its louse, (*Antarctophthirus ogmorhini*) tolerates a wide range of temperature and can reproduce at temperatures between 5°—10°C. The particular sensitivity of all stages of lice, including the eggs, to high temperatures is well established and not infrequently under such conditions they will migrate to the shaded parts of the host.

8.3.5 Penetration of the stratum corneum by mites

Sarcoptid mites have exploited the epidermis to a greater degree than have lice in that they actively penetrate into the stratum corneum. Female *Sarcoptes* excavate a sloping burrow extending it daily by about 2 mm and depositing two or three eggs, to a total of 10—25, before dying in the burrow. In time these burrows are sloughed off as new layers are added to the horny layer. The larvae which emerge from the eggs wander from the burrow on to the skin surface where they form shallow pockets. Here they undergo three moults and attain maturity at 14—17 days after the eggs are laid. Copulation occurs in the pocket and the female then excavates a new burrow and the male dies soon after copulation. The total population in human scabies is usually small; about a dozen ovigerous females is the average number in adults and twenty in children. Rarely do infestations exceed fifty ovigerous females, which contrasts with Norwegian scabies, or crusted scabies, caused by the same species but in which the normal response of the host to the parasite is modified or impaired. Such a form of scabies usually occurs in mentally retarded patients, in those suffering from vitamin deficiencies, in cases of severe systemic disease, diabetes, leukaemia and where the immunological response is depressed. Sites of predilection are the hands, wrists and areas of flexural skin. Elbows, the feet and ankles, the penis and scrotum are frequently heavily infected; in children the palms and soles are also subject to attack. Rarely do the mites occur on the chest, back and head and the preferred sites are those with a low density of pilosebaceous follicles. The reasons for this are not known. Demodicid mites go a stage further than the sarcoptid mites by invading the human pilosebaceous follicles usually on face and neck. After copulation at the follicular orifice the female lays her eggs in the sebaceous gland, where subsequent development occurs. Here then are the beginnings of the invasion of the dermis and of a physiological response induced by the dermal components of the host.

8.4. The role of dermal components

Impingement of the dermis by ectoparasites induces dynamic changes in the synthesis of chemicals, in the spatial redistribution of cellular components and in their accelerated rates of production which may exercise effects on the epidermis. The dermis consists of a three-dimensional network of protein fibres and elastin embedded in a mucopolysaccharide ground substances containing dermatan sulphate (chondroitin sulphate B) and hyaluronic acid. The former is firmly associated with collagen and the latter with the extra fibrillar spaces. Fibroblasts are the most abundant cells in the loose connective tissue of the dermis, originating in mesenchyme cells and producing collagen. Their microstructure is indicative of active synthesis and secretion. Mast cells occur predominantly in loose connective tissue, particularly in the vicinity of small blood vessels and in serous membranes. These cells produce heparin and histamine and rodent mast cells also contain serotonin (5-hydroxy tryptamine); hyaluronic acid is also secreted. Functionally mast cells operate as defence mechanisms against cell injury and a close relationship exists between them and tissue resistance to numerous pathogens. Heparin delays or prevents the clotting of blood by an antithrombin or an antiprothrombin, prevents blood platelets agglutinating and clears the blood of fat. The histamine is involved in allergic and inflammatory reactions and the discharge of both chemicals causes profound hypersensitivity reactions. Macrophages or histiocytes are common cells in connective tissues and during inflammatory reactions increased numbers invade from local sites and from haematogeneous sources. In the latter instance they arise from monocytes that migrate from capillaires.

As a result of dermal feeding by ectoparasites, an inflammatory oedema results from plasma leaking from injured capillaires and certain of the leucocytes migrate through their walls. Diagnostically leucocytes are divided into granular forms (or polymorphs) and non-granular forms. The former are further divided into (i) acidophils or eosinophils (ii) neutrophils and (iii) basophils and the latter into (iv) lymphocytes and (v) monocytes. Eosinophils probably release histamine in allergic responses and it is reported that their frequency may increase in inflamed tissues, although pathologically they are not though to be important. Damaged tissue are invaded by neutrophils which migrate through the walls of dilated blood vessels into areas of tissue damage and usually a small accumulation of neutrophils in loose connective tissue is indicative of an acute inflammatory process. Basophils are probably involved in the release of histamine. Of the non-granular leucocytes lymphocytes are relatively common in loose connective tissue often in the form of nodules. They can leave the blood stream, migrate into tissues at the sites of inflammation and develop into monocytes and macrophages: antibody may be carried on the cell surface. The other non-granular type — the monocytes — are rare in normal tissues but on entering it may develop into macrophages which are cells with considerable phagocytic capabilities. Plasma cells are present in connective tissue but are far more numerous in haemopoietic tissue. They produce antibodies and develop from precursor cells at the sites of entry of new antigens and hence often occur at sites of inflammation.

8.5 Host location and probing

8.5.1 Requirements for dermal feeders

Most parasites feeding on the epidermis or on epidermal derivatives live there permanently and are transmitted from host to host by contact, whereas those which obtain their nutriment regularly from the blood and tissue fluids of the dermis leave their hosts on completion of feeding, which may be for minutes or days. Moreover, with the exception of the Anoplura, dermal feeding arthropodan parasites either incorporate free living stages into their life histories or detach from the host to carry out certain biological activities such as moulting or egg laying. Thus, in terms of satisfying ecological needs these parasites have to (i) locate the host (ii) identify a site for penetration, as for example, by probing (iii) penetrate the epidermis, (iv) produce a haemorrhage or blood pool, or locate a blood vessel (v) ingest the blood meal and (vi) on cessation of feeding withdraw the mouth parts.

8.5.2. Host location

Host location may involve a number of sensory components: fleas detect hosts within a range of a few cm and normally do not wander far from the nest of the host. In the absence of compound eyes it is presumed to be due to olfactory responses, although in *Ceratophyllus gallinae* the change to becoming negatively geotactic and positively phototactic after mating and their orientation towards light suggests that jumping is elicted when the light intensity is lowered as a result of a passing host. Unfed, eyeles ticks on attaining appropriate positions on vegetation are also stimulated to wave their first pair of legs preparatory to grasping as the shadow of the host passes over them. Here there may be further mediation by temperature and olfactory sensilla and a positive response to carbon dioxide concentration wafted by air currents. The sensitivity of fleas to air currents through the pygidial sensillum is now well known, and on stimulation random jumps occur serving either as a means of escape or of gaining access to the body of a host. The means of clinging to the hair has been described by Rothschild (1965). Whilst host selection in a number of ectoparasites is determined by the adult and a level of host specificity is demonstrated, most fleas are not strictly host-specific (Elton et al., 1931; Smit 1957a, 1957b). Accordingly, the composition of the ingested blood meal will vary and laboratory experiments suggest that this may affect their fertility. Fleas are normally thought to be nest-specific since their development from the free-living larva and pupa will depend to a great degree on the ecological conditions within the nest, of which the most variable appears to be the relative humidity. This may account for the greater success of fleas as parasites of mammals rather than of birds, whose nests are generally not only drier but also are more impermanent or less continuously occupied seasonally. The rabbit flea, *Spilopsyllus cuniculi*, is of particular interest in that the female's ovaries develop after the flea feeds on a pregnant rabbit or its new born young (Mead-Briggs and Rudge, 1960). Subsequently it was demonstrated experimentally by Rothschild and Ford (1966) that hormones released from the anterior lobe of the rabbit's pituitary gland regulate ovarian maturation, egg laying and regression in the flea. This process is highly specific as shown by the work of Exley, Ford and Rothschild (1965), Rothschild and Ford (1966, 1969) but how the host's hormones mediate on the rabbit flea is unknown, and it is likely that their effects operate through the flea's own hormonal system. It has also been postulated that a pheromone-like factor from the nestling rabbits stimulates both copulation and maturation in this flea. This association does appear to be unique among fleas, for in other species maturation will take place on castrated and adrenalextomised rats.

Vision is important in host location by some species of biting flies but within groups of species other integrated physiological or behavioural patterns are involved in host selection. In respect of tsetse flies, for example, this is reflected in the habitat preferences of their vertebrate hosts (Swynnerton, 1936) in the flight patterns of the flies (Nash, 1948) and in their blood meal choices as demonstrated by Weitz (1963) and Glasgow (1963).

8.5.3. Probing

Having obtained access to a host the selection of feeding sites depends on ambient temperature, skin thickness and the age and degree of starvation of the arthropod. From membrane experiments Galum (1971) noted that thin membranes may be necessary for feeding by arthropods with short probosces, such as lice of the order Anoplura or fleas, but that tsetse flies require thick membranes. Moreover, it was observed that such synthetic membranes as Paraffin M were not as effective in the induction of feeding as natural membranes like chick skin. Many haematophagous insects, such as tsetse flies and blackflies, probe before gorging and it is not unusual for some species, having pierced the skin, to withdraw their mouth parts and to try elsewhere. There is evidence that some blood feeding insects, such as *Rhodnius prolixus*, sample the food before gorging but why this should be necessary is not exactly clear. Friend and Smith (1972) consider that the repeated sampling by maxillary probing may either enable the food to come into contact with certain receptor cells or trial activity of the pharyngeal pump as the insect tests for the availability of fluid may be required. Similar probing occurs in tsetse flies. Whatever its significance it appears to be an important component of the initial phase of feeding. There must be some mechanism, too, for signalling that the mouth is suitably located in relation to the food source whether this be in a haemorrhagic pool or in a blood vessel. Much speculation has existed concerning the internal location of chemoreceptors that respond to chemicals on phagostimulatory activity: Von Gernet and Buerger (1966) have described sense organs in the cibarial pump of 22 mosquito species, Owen (1963) in the cibarium of the mosquito, *Culiseta inornata*, with possible sensory input for controlling sucking. Tawfik (1968) found two types of sensilla in the cibarium of *Cimex lectularius* and Friend and Smith (1972) in the epipharynx of *R. prolixus*. The host blood contains phagostimulants, as was first shown by Hosoi (1959), when he demonstrated that adenosine 5-phosphate provides the main stimulus for the gorging of *Culex pipiens*. Subsequently, certain nucleoside phosphates have been shown to act similarly for a number of blood feeding arthropods, and this discovery was used as a basis to evoke maximum uptake of artifical diets through membranes (Galun, 1971).

8.5.4. Penetration

In dermal feeders the initial process involves the penetration of the non-specific epidermal barrier. This is achieved by direct piercing stylets, by rasping or by blade like cutting mouth parts and there is much variation in the equipment used. This is followed by thrusting movements, sometimes accompanied by lysis, until the tips of the mouth parts are suitably located in the dermis. Inoculation of salivary secretions into the lesion follows and in some acari this is supplemented by regurgitation from the mid-gut (Arthur, 1965). The activity and functions of these secretions are not clear in most arthropods. Anticoagulins, once thought to be essential for mosquito feeding, are absent in as many species as those which possess them (Clement, 1963), whilst Lester and Lloyd (1928) showed that, though there is an anticoaglulin present in tsetse fly saliva, feeding can be successfully carried out after cutting the salivary duct without causing any host allergic responses. In rapidly feeding argasid ticks, there is evidence that the salivary secretion contains a potent anticoagulin and a proteolytic enzyme. In such species as *Ornithodorus savignyi* there is also a toxin of some potency, all of which suggests a destructive action on host tissues (Lavoipierre and Riek, 1955; Howell, 1966). Ixodid ticks, on the other hand, show some variation in the composition of their salivary secretions. Some, for example, *Hyalomma asiaticum*, and *Ixodes ricinus*, have salivary components which prevent blood clotting, others e.g. *Dermacentor andersoni* and *Boophilus microplus*, lack discernible anticoagulins, Fleas, mosquitoes, tsetse flies and argasid ticks are examples of arthropods which may be classified as temporary ectoparasites in that their engorging periods on the host are of the order of minutes rather than days. Ixodid ticks, however, feed continuously on hosts for days, and demand purchase for an extended period. During penetration the first trophic movements are the sweeping and cutting movements of the digits of the chelicarae, followed by the thrusting movements of the cheliceral shafts, and the backwardly toothed hypostome, through the stratum corneum. In a number of species this is accompanied by the ejaculation of a cement as a milky white secretion from the tick's salivary glands. Most of the cement forms a cone on the surface of the skin (Gregson, 1937, 1960) although there may be invasion of the outer skin layers and a flow around the penetrating mouth parts. When transilluminated the cement is seen to spread outwards in the host tissues as an opaque cloud. The piston-like action of the chelicerae pulls up the cement around the trophi from time to time and as it hardens rapidly a sheath forms around the embedded mouth parts. In some tick genera (*Amblyomma*, *Hyalomma*, and *Aponomma*) which insert their long mouth parts completely into the dermal tissues, the mouth parts are enveloped by cement, with little or no superficial deposition on the skin surface. In others (*Dermacentor*, *Boophilus* and *Rhipicephalus*) the cement is largely superficial and the mouth parts do not penetrate deeply into the dermis. In the genus *Haemiaphysalis* the inserted trophi come to lie just beneath the epidermis and the cement is predominantly superficial. Cement is not produced by some species of *Ixodes* and the mouth parts are inserted deeply into the dermis, where they are held in situ by transformed or compacted host tissues. Around the mouth parts of such species there is an abundance of collagen, suggesting that secure attachment may not be due solely to aggregation of fibres, but that there may even be active synthesis of collagen. Fibroblasts, generally believed to be responsible for secreting collagen are noticeably absent from around the mouth parts. Collagen is, however, a highly polymerized and relatively insoluble substance and the suggestion has been made that enzymes in the salivary secretion of *Ixodes* ticks may polymerize the collagen precursors in the skin.

8.6 Host reaction

Restriction or prevention of grooming in mice results in the increase of lice populations (*Polyplax serrata*) as demonstrated by Murray (1961) and Bell and Clifford (1964), and other methods of resistance then become demonstrable. Nelson et al. (1972) examined the histopathology of this situation in amputee white mice and found that over 12 weeks there was a two-phase reaction involving both the epidermis and the dermis. The initial four week phase was characterised by hyperplasia of the epidermis followed by reversion and accompanied by almost parallel increase in eosinophils, neutrophils and lymphocytes, to reach peak numbers at about the end of the second week. Thereafter eosinophils, and neutrophils declined to pre-invasive levels. Determinations of the patency of blood vessels at this period showed a reduced blood flow to the skin, which was reflected in the lowered capacity of the lice. This persisted to the eleventh week. The second phase covering eight weeks showed a renewal in epidermal thickness before thinning again, together with a secondary increase in lymphocytes and an increase in monocytes, mast cells and fibroblasts. The mast cells also released their granules. The evidence is similar to that indicated by Nelson and Bainborough (1963) for *Melophagus ovinus*, the sheep ked, and they suggested that the nature of the resistance was a reaction to hypersensitivity due to an extended period of arteriolar vasoconstriction, reducing the blood flow to the skin and its availability to the ectoparasite.

Resistance of the host to tick feeding has been claimed for a number of genera. Trager (1939), for example, has shown that one infestation of guinea pigs or rabbits with larvae of *Dermacentor andersoni* induces an acquired "immunity" which prevents subsequent batches of larvae from engorging. Repeated infestation with nymphs and adults results in a marked reduction in the amounts of blood inbibed by ticks of the later batches as demonstrated by Snow and Arthur (1966) by infesting guinea pigs with *Hyalomma anatolicum anatolicum*. The egg yields from females in which the blood intake is reduced as a result of repetitive feeding are considerably lower when compared with egg yields from females feeding on non-resistant hosts. This type of resistance then regulates the reproductive potential of the tick and operates as a negative feed-back mechanism. Under field conditions females of *Ixodes rubicundus* can also stimulate a resistance response in sheep, whereby ticks which attach are subsequently prevented either from completing engorgement or the rate of feeding is slowed down.

Larrivee et al (1964) described five stages of reactivity in guinea pigs exposed repeatedly to the bite of the flea *Ctenocephalides felis felis*. In the pre-sensitized skins (i.e. skin challenged in less than 5 days) the penetrating epipharynx passes through the epidermis and then the dermis, producing mild trauma in the skin together with slight infiltration of lymphocytes around the mouth parts. On occasions these workers reported cell fragmentation. Twenty four hours after the bite the mild lymphocytic infiltration is reduced and may be absent, but at no time was there the variety of components present in later allergic skin, nor was there evidence of abnormal macroscopic or microscopic changes. A stage of delayed skin reactivity occurs 5 days after initial exposure and persists for about four days. Skin examined 24 hours after the challenging bite at 5 days shows intense monocytic and lymphocytic infiltration of the dermis and hyperplasia of the epidermis. A third type of reactivity occurs when the challenging bite is given within 10-30 days after the initial exposure. The immediate reaction within 20 minutes of the challenge is one of predominantly eosinophilic response, to be replaced at 24 hours by an infiltration of mainly lymphocytes and monocytes. Any eosinpohilic infiltration is small at this time. The effect of challenge 50 days after the initial bite lasts about a month and is noteworthy by its immediate eosinophilic reaction and a mild delayed mononuclear response. Continued challenges result in diminution of response and this leads to the stage of non-reactivity where no skin reaction was noted and no cellular abnormalities were apparent. At this time fleas could still feed on the hosts. Comment on the absence of neutrophils in the histopathology of flea bites in the work of Larrivee et al (1964) has been made by Nelson et al. (1972) who note that they are the first cells to appear following mosquito bites in sensitized man (Goldman et al., 1952). In ixodid ticks with deeply inserted mouthparts and supported by cement, circumferential attachment is secure and rigorous host tissue destruction by salivary enzymes will take place at the tip of the hypostome. On the other hand, in superficially attaching genera, active digestion of surrounding tissues would weaken such an attachment. Accordingly such genera would be expected to have less active salivary secretions, to place less reliance on salivary anticoagulants to maintain blood flow, to rely on host tissue reactions for their source of nutriments and, on the release of such pharmacologically active host substances as histamines, to cause capillary dilatation, increased flow rate of blood and increased permeability of the capillaries. In other words, well adapted and established parasites do obtain their nutrient requirements by more subtle specific means as in *Boophilus microplus*, as has been observed by Tatchell and Moorhouse (1968). They noted that during feeding, except for the final stage of engorgement, small spaces filled with a regurgitate appear below the mouthparts. As the tick continues to feed up to 24 hours before detachment there is a progressive infiltration of neutrophils and lymphocytes around the mouth parts with some fibroblasts also being present. At all times the capillaries in the vicinity of the mouth parts are dilated and haemorrhagic and as feeding continues these changes extend to the deeper local blood vessels. The infiltration is accompanied by the disappearance of collagen fibres around the mouth parts in the dermis and the establishment of a cavity and a surrounding zone of heavily infiltrated collagen. The fluid filled cavity contains some leucocytes and haemorrhaged components from damaged blood vessels which could have deep-seated origins in the dermis. In some skin sections the widespread precipitation of the tissue nuclei, which occur in earlier in filtrations, below the mouth parts persist and extend the area of degeneration. Eosinophils also increase in number and are most numerous in the periphery of intense infiltrations. Monocytes and histocytes show intense phagocytic activity up until the deposition of further cement, to maintain the parasite in position, and then monocytes decrease in number.

Specific vascular damage results from the saliva of the tick while tissue damage is a consequence of the host response. Inducing a leukopenia in tick-infested dogs, by the use of nitrogen mustard, shows that polymorpho-nuclear leucocytes are rare around the mouth parts but heavy infiltrations characterise untreated tick-infested hosts. Towards the time of full engorge-ment well established cavities are present beneath and around the mouth parts of the latter, whereas in the former cavities are either absent or insigni-ficant and dermal collagen remains unaffected. Moreover, no secondary ce-ment is deposited by ticks in nitrogen mustard treated dogs (Tachell and Moorhouse, 1969), an indication of the breakdown of normal parasite-host responses.

Feeding by *I. ricinus* and other British species of this genus (Arthur, 1962; Stevens, 1968) and of *H. anatolicum* (Snow, 1967) produces hyper-plasia of the epidermis surrounding and extending beyond the mouthparts, which is not the case in *B. microplus*. Exposure of rats to the bites of imma-ture stages of *I. ricinus* leads to the destruction of the epidermis around the area of entry, with invasion by neutrophils: degenerating dermal fibroblasts are also present. Slight haemorrhages may occur immediately below the Malpighian layer, having their origin in the superficial capillaries, and there is evidence of pycnotic fibroblasts at the haemorrhagic edges; 9 min after at-tachment there is little change in the host tissue beyond compression of the connective tissue fibres and fibroblasts immediately around the mouthparts, and those further away are slightly re-orientated; 15 min after attachment the nuclei in contact with the mouthpart are pycnotic and more closely packed than in uninfected tissue.

The breakdown of tissues commences with those nearest the mouthparts whether these be muscle, adipose tissue or cartilage, and oedema is asso-ciated with histolysis as early as 3 hours after attachment. The oedematous fluid may contain a few red blood cells and remains of fibroblasts and con-nective tissue fibres. A frequent, but by no means invariable, feature of the wounds is the formation of a haemorrhage, and where this is extensive it may seep to the underlying tissues, but unless a dense capillary bed is pene-trated haemorrhages do not occur after short periods of attachment. Changes at the tip of the hypostome vary from negligible damage to extensive histo-pathological changes in the first 48 hours but thereafter the changes are in-variably marked. Nymphal feeding likewise produces marked effects in the skin after 48 hours. Tatchell and Moorhouse (1968) were unable to relate changes in either the frequency or appearance of mast cells to the feeding of the ticks, but in *I. ricinus* these cells migrate to the region penetrated by the trophi. They occur in greater numbers than in unaffected skin, extending to beneath the epidermis, where they do not normally occur. The cell membranes break down and the granules disperse to some distance from them, where they show either reduced staining reactions or appear as diffusely staining patches or clusters. These mast cells occur frequently in the uniformly staining area around the trophic zone and throughout attachment they, or their 'ghosts', and the released granules occur distal and lateral to this zone. Oedema is associated with the liberation of histamine from released mast cell granules and this possibly contributes to the increased flow of tissue fluids, although pharmacologically active components of the saliva could produce similar effects.

Spectrophotometric analysis of tick gut contents at varying intervals of tick feeding indicate that females of *I. ricinus* ingest nutrients other than blood at early periods of feeding and there may be ingestion of significant amounts of other tissue fluids during engorgement (Sutton and Arthur 1962). A subtle blend of blood and tissue fluid also seems to be necessary for the initiation of feeding in *Dermacentor andersoni.*

8.7. Conclusion

Many ticks are catholic in their choice of hosts, others are more rigidly limited. In some ticks there may be stadial specificity whereby larvae or nymphs may be limited to one group or species of host and adults to another. These limitations of host selectivity may, in part, be determined by ecological or behavioural associations between parasite and host or by more real restrictive specificity. Those ticks demonstrating greater host-specificity show a more stable equilibrium whereas in non-specific relationships mutual adaptation has less opportunity to evolve. In the latter situation the tick faces a broader spectrum of host tissues and responses and the problems inherent in host specificity would not restrict feeding. It must be appreciated, however, that even a host specific to a single tick species presents it with a formidable range of tissues ranging from the crossing dermal fibres on the back and flanks to the cartilage of the pinna of the ear. But in such a subtle association as exists between *Boophilus microplus* and cattle the importance of more specialised feeding microhabittats may be less critical, and greater exploitation of the whole skin surface may be possible.

The individual host protects itself from overwhelming parasitic challenges in a number of ways — grooming, moulting, delousing "anting" — and all are expressions of mechanical means of removing ectoparasites. Some areas of the bodies of mammals and birds are not accessible for physical removal by the host and, under certain circumstances, parasitic populations may be concentrated in such areas and as numbers increase, the area inhabited by the parasitic colony will increase. As the original area of skin becomes hypersensitive any attached parasites will stop, or reduce, feeding, which will be reflected in lower egg yields. This type of resistance then regulates the reproductive potential and operates as a negative feed back mechanism. Meanwhile extension of the parasite population beyond the original protected area will again bring mechanical control into play. Under field conditions females of *Ixodes rubicundus* can stimulate a resistance response in sheep whereby ticks which do succeed in attaching are subsequently prevented either from completing engorgement or the rate of feeding is slowed down. In the case of *Melophagus ovinus* and *Polyplax serrata* the acquired resistance is due to reduced blood flow in the host's skin, thus preventing them from obtaining blood, and suggests a reaction of hypersensitivity. Nelson et al (1972) suggest tentatively that the two phases of cellular reactivity observed in lice on mice represent two mechanisms. The first is considered to be an immune reaction, when resistance is initiated, the second as a non-immune reaction to chronic irritation, which serves to maintain the state of resistance until the lice population is again optimal.

The host response as a regulator of tick populations may operate in other ways. For example, the infestations of the ears of both wild and domestic animals by the immature stages of the two-host tick *Rhipicephalus evertsi* results in the secretion of wax by the hosts. The initial infestation by the larvae occurs in the folds of the inner parts of the pinna and in the external auditory meatus, where the larvae attach and feed. As a foreign body response the host produces wax from the wax secreting cells. On completion of feeding the larvae moult in situ and the emerging nymphs either re-attach on the same site or at some distance outwards to feed again. Frequently, the amount of secreted wax is sufficiently great to form a barrier between the nymphal mouth parts and the skin surface, so as to prevent complete feeding in a proportion of the nymphal population. And, since a threshold value of blood is necessary for metamorphosis to proceed, any ticks failing to attain this level will not develop. The very considerable aggregations of wax in the ears of wild Cape hares, for example, contain not only the cast skins of larvae but also the bodies of nymphs which have failed to attach.

In these, and other more subtle means, the host species is protected from excessive parasitic challenges, is able to regulate the reproductive capacity of the parasite in the long term, whilst the parasite, because of its reproductive potential, is still available in sufficient numbers to maintain the association. The successful host-parasite ecological relationship is therefore one where the innate resistance of the immune responses of the hosts interact to control, but not to overcome, the challenge of the parasite.

8.8. REFERENCES

Arthur, D.R. (1962). "Ticks and Disease". Pergamon Press, London. 1—445.

Arthur, D.R. (1965). "Feeding in ectoparasitic acari with special reference to ticks". Adv. Parasit. 3, 249—298.

Arthur, D.R. (1969). "Tick feeding and its implications". Adv. Parasit. 8, 275—292.

Bell, J.F. and Clifford, C.M. (1964). "Effects of limb disability on lousiness in mice. II Intersex grooming relationships". Expl. Parasit. 15, 340—349.

Calaby, J.H. (1970). "The Insects of Australia". (D.F. Waterhouse, ed.) Melbourne University Press, Melbourne.

Clay, T. (1949). "Piercing mouth parts in the biting lice (Mallophaga)". Nature, Lond., 164, 617—619.

Clay, T. (1951). "The Mallophaga as an aid to the classification of birds with special reference to the structure of feathers". Proc. X Ornith. Congress, June 1950.

Clay, T. (1957). "The Mallophaga of birds. First Symposium on Host Specificity among parasites of Vertebrates". Inst. Zool. Neuchatel, 120—155.

Clement, A.N. (1963). "The Physiology of Mosquitoes". Pergamon Press, London, 1—393.

Crutchfield C.M. and Hixson H. (1934). "Food habits of several species of Mallophaga (Poultry Lice) with special reference to blood consumption". Florida Entom. 26, 63—66.

Ebling, F.J.G. (1972). "Normal Skin" (Chp. 2, pp. 4—24, Vol. I) in Textbook of Dermatology by Rook, A., Wilkinson, D.S. and Ebling, F.J.G., Blackwell, Oxford. (I—1060; II—2118).

Eichler, W. (1937). "Einige Bemerkungen zur Ernahrung und Eiablage der Mallophagen". Sber. Ges. Naturf. Freunde. Berl. 80—111.

Eichler, W. (1948). "Some rules in ectoparasitism." Ann. Mag. nat. Hist. (Ser. 12). 1, 588—598.

Elton, C., Ford, E.B., Baker, J.R., and Gardener, A.D. (1931). "The Health and Parasites of a wild mouse population". Proc. zool. Soc., London, 657—721.

Ewing, H.E. (1924). "On the taxonomy, biology and distribution of the biting lice of the family Gyropidae". Proc. U.S. Nat. Mus., 63, Art. 20, 42 pp.

Exley, D., Ford, B. and Rothschild, M. (1965). "The rabbit flea Spilopsyllus caniculi Dale) as an indicator of hormones in the host." Proc. R. ent. Soc. Lond. C. 30, 32—36.

Friend, W.G. and Smith, J.J.B. (1972). "Feeding stimuli and techniques for studying the feeding of haematophagous arthropods under artificial conditions with special reference to Rhodnius prolixus". 241—256. In "Insect and Mite Nutrition". (J.G. Rodriquez, ed.) North-Holland, American Elsevier, Amsterdam.

Galun, R. (1971). "Recent developments in the biochemistry and feeding behaviour of haematophagous arthropods as applied to their mass rearing". Int. Atomic Energy Agency Pub. "Sterility Principle for Insect Control or Eradication". pp. 273—282.

Gilmour, D. (1965). "The Metabolism of Insects". Oliver and Boyd, Edinburgh.

Glasgow, J.P. (1963). "The Distribution and Abundance of Testes", Pergamon Press, London, 1—256.

Goldman L., Rockwell, E. and Richfield D. (1952). "Histopathological Studies on cutaneous reactions to bites of various arthropods". Amer. J. trop. Med. Hyg. 1, 514—525.

Gregson, J.D. (1937). "Studies on the rate of tick feeding in relation to disease." Proc. ent. Soc. Br. Columb. 33, 15—17.

Gregson, J.D. (1960). "Morphology and functioning of the mouth parts of Dermacentor andersoni Stiles". Acta. trop. 17, 48—79.

Harrison, L. (1914). "The Mallophaga as a possible clue to bird phylogeny." Aust. Zool. 1, 1—5.

Hopkins, G.H.E. (1949). "The host associations of the lice of mammals." Proc. zool. Soc. Lond. 119, 387—604.

Hosoi, T. (1959). "Identification of blood components which induce gorging of the mosquito". J. Insect. Physiol., 3, 191—218.

Howell, C.J. (1966). "Collection of salivary gland secretion from the argasid. *Ornithodoros savignyi* Audouin (1827) by the use of a Pharmacological Stimulant JIS." Afr. Vet. Med. Ass. 37, 236—239.

Kellogg, V.L. (1902). "Are the Mallophaga degenerate Psocids?" Psyche: 339—343.

Larrivee, D.H., Benjamini, E., Feingold, B.F. and Shimizu, M. (1964). "Histologic Studies of guinea pig skin: Different stages of allergic reactivity to flea bites." Expl. Parasit. 15, 491—602.

Lavoipierre, M.M.J. and Riek, R.F. (1955). "Observations on the feeding habits of argasid ticks and on the effects of their bites on laboratory animals, together with a note on production of coxal fluid by several of the species studied." Ann. trop. Med. Parasit., 49, 96—113.

Lester, H.M.O. and Lloyd, L. (1928) "Notes on the process of digestion in tsetse flies". Bull. ent. Res. 19, 39—60.

Martin, M. (1934). "Life History and habits of the Pigeon Louse *(Columbicola columbae)*". Can. Ent. 66, 6—16.

Matthyse, J.G. (1944). "Biology of the cattle biting louse and notes on cattle sucking lice". J. econ. Ent. 37, 436—442.

Mead—Briggs, A.R. and Rudge, A.J.B. (1960). "Breeding of the Rabbit Flea, *Spilopsyllus cuniculi* (Dale); requirement of a "factor" from a pregnant rabbit for ovarian maturation". Nature, Lond. 187, 1136—1137.

Mockford, E.L., (1967). "Some psocoptera from plumage of birds". Proc. ent. Soc. Wash., 69, 307—309.

Murray, M.D. (1961). "The ecology of the Louse *Polyplax serrata* (Burm.) on the mouse *Mus musculus* L." Aust. J. zool. 9, 1—13.

Nash, T.A.M. (1948). "Tsetse flies in British West Africa". London. H.M.S.O. 78 pp.

Nelson, B.C. (1971); "Successful rearing of *Colpocephalum turbinatum* (Pthiraotera)". Nature Lond. 232: 255.

Nelson, B.C. and Murray, M.D. (1971). "The distribution of Mallophaga on the domestic pigeon *(Columbia livia)*". Int. J. Parasit. 1: 21—29.

Nelson, W.A. and Bainborough, A.R. (1963). "Development in sheep of resistance to the Ked *Melophagus ovinus* (L.) III Histopathology of the skin as a clue to the nature of resistance". Expl. Parasit. 13, 118—127.

Nelson, W.A., Clifford, C.M., Bell. J.F. and Hestekin, B. (1972). *"Polyplax serrata*: Histopathology of the skin of louse—infested mice". Expl. Parasit. 31, 194—202..

Owen, W.B. (1963). "The contact chemoreceptor organs of the mosquito and their function in feeding behaviour". J. Insect. Physiol., 9, 73—87.

Pearman, J.V. (1960). "Some African Psocoptera found on rats." Entomologist, 93, 246—250.

Ries, E. (1931). "Die Symbiose der Lause und Federlinge" Z. Morph. Okol. Tiere 5 (20); 234—365.

Rothschild, M. (1965). "Fleas" Scient. Am. 213, 44—53.

Rothschild, M. and Ford, B. (1966). "Reproductive hormones of the host controlling the sexual cycle of the Rabbit flea *(Spilopsyllus cuniculi* Dale)" Proc. Int. Cong. Ent. 12. London (1964) 801—802.

Rothschild, M. and Ford, B. (1969). "Does a pheromone-like factor from the nestling rabbit stimulate impregnation and maturation in the rabbit flea?" Nature, Lond. 221, 1169—1170.

Scott, M.T. (1952) "Observations on the bionomics of the sheep body louse *(Damalinia ovis)*". Aust. J. agric. Res. 3, 60—7.

Simmons, K.E.L. (1957). "A review of the anting behaviour of passerine birds". Br. Birds. 50, 401—424.

Simmons, K.E.L. (1966). Anting and the problem of Self-Stimulation." J. zool. Lond. 149, 145—162.

Smit, F.G.A.M. (1957a). "The recorded distribution and hosts of Siphonaptera in Britain". Ent. Gaz. 8: 45—75.

Smit, F.G.A.M. (1957b). "Siphonaptera". Handbk. Ident. Br. Insects. 1 (16) 94 pp.

Snow, K.R. (1967). "Some aspects of the biology of *Hyalomma anatolicum* (Ixodoidea: Ixodidae)". Ph. D. Thesis, University of London, 1—409.

Snow, K.R. and Arthur, D.R. (1966). "Oviposition in *Hyalomma anatolicum anatolicum* (Koch, 1844) Ixodoidea: Ixodidae". Parasitology 56, 555—568.

Stevens, E. (1968). "Tick feeding in relation to disease transmission". Ph. D. Thesis, University of London, 1—210.

Stockdale, H.J. and Raun, E.D. (1965). "Biology of the Chicken Body Louse *Menacanthus stramineus*". Ann. ent. Soc. Amer. 58: 802—05.

Sutton, E. and Arthur, D.R. (1962). "Tick feeding in relation to disease transmission" in Aspects of Disease Transmission by ticks (D.R. Arthur, ed.). Symp. zool. Soc. Lond. No. 6, pp. 223—253.

Swynnerton, C.F.M. (1936). "The tsetse flies of East Africa. A first study of their ecology, with a view to their control." Trans. R. ent. Soc. Lond. 84: 1—579.

Tatchell, R.J. and Moorhouse, D.E. (1968). "The feeding processes of the cattle tick *Boophilus microplus* (Canestrini). II The sequence of host tissue changes." Parasitology, 58, 441—459.

Tatchell, R.J. and Moorhouse, D.E. (1969). "Neutrophils: their role in the formation of a tick feeding lesion". Science, N.Y.

Tawfik, M.S. (1968). "Feeding mechanisms and the forces involved in some blood sucking insects". Quaestiones Entomologicae, 4, 92—111.

Trager, W. (1939) "Acquired immunity to ticks". J. Parasit. 25, 57—81.

Von Gernet, G. and Buerger, G. (1966). "Labral and cibarial sense organs of some mosquitoes". Quaestiones Entomologicae, 2, 259—270.

Waterhouse, D.F. (1952). "Studies on the Digestion of wool by insects IX. Some features of digestion in chewing lice (Mallophaga) from bird and mammalian hosts." Aust. J. Biol. Sci. 6 (2) 256—275.

Waterston, J. (1926). "On the crop contents of certain Mallophaga". Proc. zool. Soc. Lond. 1926, 1017—1020.

Webb, J.E. (1946). "Spiracle structure as a guide to the phylogenetic relationships of the Anoplura (biting and sucking lice) with notes on the affinities of the mammalian hosts." Proc. zool. Soc. Lond. 116, 49—119.

Weitz, B. (1963). "The feeding habits of *Glossina*". Bull. Wld. Hlth. Org., 28, 711—729.

Ecological Aspects of Parasitology
Editor: C.R. Kennedy
© *North-Holland Publishing Company, Amsterdam, 1976*

CHAPTER 9

BODY SURFACE OF FISHES

Graham C. KEARN

School of Biological Sciences, University of East Anglia, Norwich NR4 7TJ
(Great Britain)

Contents

9.1. Introduction

Recent studies on fish skin have focused on two separate aspects in particular; first, the immunological properties of fish skin surfaces and secondly the role of fish skin secretions in the behaviour of fishes. These two aspects of fish biology are likely to be important in relation to skin parasitism, particularly the central problems of host specificity and how parasites identify their hosts. Inhabitants of the fish skin surface include protozoans, platyhelminths, leeches and crustaceans and range from epizoic organisms to highly specialized skin parasites. The limitations of space in a review of this kind preclude detailed coverage of such a wide range of organisms and because of this the parasitological part of this review is limited almost entirely to a consideration of the group which has received most attention in recent years and with which I am most familiar, namely monogenean (platyhelminth) parasites. It is hoped that this approach will illustrate more clearly the way in which fish skin structure and function and parasite biology are integrated.

9.2. Fish skin, with special reference to the epidermis and its secretions

The following account is mainly concerned with the structure and functions of the fish's epidermal cells with which parasites come into contact and with those properties of surface secretions of fishes which are likely to have a bearing on skin parasitism.

Accounts of the structure of the skin of both teleost and elasmobranch fishes have been given by Andrew (1959) and by Spearman (1973). The skin of the common sole is relatively unspecialized (Fig. 1), and will serve to illustrate the main features of the skin of bony fishes. Like the skin of other vertebrates it comprises an outer stratified epithelium, the epidermis, separated by a basement membrane from an inner layer or dermis. In the skin of tetrapod vertebrates the outer layers of the epidermis consist of squamous keratinized cells, whereas in fishes the superficial cells of the epidermis are non-keratinized and appear to be metabolically active. The surface of the epidermis is covered by a surface coat and Whitear (1970) presents evidence in support of the secretion of this surface coat by the outer layer of epidermal cells. Whitear regarded the surface coat as the mucus which is normally present on fish skin, but she saw a distinction between this and the "slipperiness", which assists escape from predators, and which she attributed to the more copious secretion of goblet mucous cells which lie in the epidermis of all the fishes which she studied.

Some fishes are known to undergo a process resembling ecdysis. In the South African fish *Agriopus*, it is the surface coat which is shed (Gilchrist, in Whitear, 1970), and Lester (1972) described the production of "slough", which is undoubtedly surface coat, in laboratory raised sticklebacks. It is possible that regular shedding of the surface coat in fishes is more widespread than hitherto supposed.

A phenomenon resembling ecdysis has been reported by Winn (1955) in four species of parrot fishes (Scaridae). A mucous envelope is produced by these fishes at night when they are resting on the bottom. The envelope has a hole in the mouth region and another behind the caudal fin, which permit the entrance and exit of the gill-ventilating current. At daybreak, or when the light is turned on in the laboratory, the fish breaks out of the envelope. It is not known whether these envelopes are derived from epidermal cells, from mucous cells or from both sources.

Fish skin mucus is reported to contain a variety of chemical substances, including nucleic acid, free protein, glycoprotein and mucopolysaccharides

(Wessler and Werner, 1957; Asakawa, 1970; Bremer, 1972). According to Asakawa, eel skin contains goblet cells producing a sialic acid—containing glycoprotein and columnar cells producing smaller quantities of sulphated mucopolysaccharide. Van Oosten (1957) referred to experiments which indicate that small quantities of nitrogenous waste products are excreted through the skin of goldfish and carp, and Kearn and Macdonald (in press) have recently identified small quantities of urea, ammonia and amino acids in the skin mucus of marine fishes.

However, the method employed by most workers to collect skin mucus from fishes involves scraping the skin, and Uskova et al. (1971) have shown that skin mucus collected in this way contains epithelial cells which are not present when the mucus is collected by blowing with a jet of air, and after hydrolysis, mucus collected by the scraping method contained 16 amino acids compared with 9 amino acids in mucus collected by the air jet method. Thus some of the substances identified in fish skin mucus may have been derived from stray epidermal cells harvested with the mucus.

Studies of the sugar and amino acid components of hydrolysed mucus have been undertaken (Lehtonen et al., 1966; Enomoto and Tomiyasu, 1960; Enomoto et al., 1960, 1961). Enomoto et al. (1960), comparing the eel *Astroconger myriaster* with the loach *Misgurnus anguillicaudatus*, found differences in the component sugars.

Using chromatographic methods Barry and O'Rourke (1959) were able to distinguish between samples of mucus from closely related fish species of the same genus. Recently, evidence has accumulated to show that fishes are able to recognize such odour differences and that fish secretions play an important part in interspecific and intraspecific chemical communication related to such activities as breeding, shoaling and escape from predation. Experiments performed by Todd et al. (reported by Bardach and Todd, 1970) have shown that in the yellow bullhead (*Ictalurus natalis*) skin mucus contains sufficient information to permit discrimination between two individual bullheads by another bullhead. However, they reported stronger responses to water in which the fish had been swimming than to skin mucus, indicating that other sources may contribute to individual odour.

According to Jakowska (1963) fishes respond to a variety of environmental and pathogenic agents by altering the quantity and the nature of their mucous secretions. Bardach and Todd (1970) describe evidence which indicates that following moderate stress in the bullhead, there is a change in the fish's secretions which can be detected by other bullheads. How much of this change can be attributed to skin mucus and how much to urine or to other sources is not known. They also give evidence that status in the social hierarchy is communicated to other bullheads and that a change in this status is detected by fishes in the community. The way in which these changes are controlled has received little attention, although there is some evidence for hormonal influence on size and numbers of mucous cells (e.g. Ogawa, 1970).

Reports of serum proteins in fish skin mucus may be of considerable parasitological significance. O'Rourke (1961) used an immuno-diffusion technique with anti-fish sera to demonstrate the presence of serum protein antigens in skin mucus from bass (*Morone labrax*) and cod (*Gadus callarias*), and Fletcher and Grant (1969) and Di Conza and Halliday (1971) detected immunoglobulin in the surface mucus of fishes. Fletcher and White (1973) found higher antibody titres in intestinal mucus from orally immunized plaice than in their sera, whereas parenteral immunization resulted in high serum antibody titres and low titres in intestinal and skin secretions. The implication is that antibodies in the mucous secretions of fishes may be derived not only from the serum but also by local synthesis at the mucous surfaces. Di Conza and Halliday (1971) found diffuse collections of lymphoid cells in both the skin and intestine of the catfish *Tachysurus australis*, and it was thought that these cells might be responsible for local antibody synthesis. Roberts et al. (1971) discovered cells which they described as eosinophilic granule cells in the epidermis of the plaice *Pleuronectes platessa* and they suggested that these cells might facilitate the transport of serum and lymph components through the skin. Hildemann (1962) has suggested that the epidermal secretion or "discus" produced by *Symphysodon discus* and eaten by the fry may contain antibodies against important pathogens. Thus the "discus" material may be analogous to the colostrum of mammals, giving immunity to the fry.

Another feature of fish epidermis which is important in relation to parasitism is its role in wound healing. Van Oosten (1957) has described the process whereby skin wounds are covered by migration of epidermal cells from the edge of the wound and by proliferation of these covering epidermal cells. He states that a covering epithelial layer forms much more rapidly in fishes than in warm blooded animals, but, after the wound is covered, further recovery in fishes is extremely slow. Berlin (1951) reported that in the loach (*Misgurnus fossilis*) deep skin wounds 25 to 100 mm^2 in area, involving both the epidermis and the dermis, were covered by migrating epidermis in as little as 24 to 36 h, and the epidermis had returned to its original thickness after 3—4 days. The dermis recovered much more slowly, damage still being detectable after 4 months. Lester (1972) found in the stickleback *Gasterosteus aculeatus leiurus* that a 20-cell wide band of epidermis, experimentally removed down to the dermis, was regenerated within half an hour at 15°C.

Conditions at the skin surface also depend on the behaviour and habitat of the fish. For example, the availability of oxygen close to the skin surface of a flatfish buried in the sediment may be considerably lower than at the surface of an actively swimming fish.

A comparison of the dermal layers of fishes is given by Andrew (1959) and a detailed account will not be given here. In the common sole *Solea solea* (Fig. 1) the dermis is thicker than the epidermis and contains fibrous

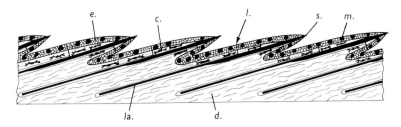

Fig. 1. Diagrammatic longitudinal section through the skin of the common sole (*Solea solea*). *c.*, Dermal chromatophore; *d.*, dermis; *e.*, epidermis; *l.*, weak line where skin fractures; *la.*, bony lamina of the scale; *m.*, epidermal mucus cell; *s.*, spine. (From Kearn, 1967b).

tissue, nerves, blood vessels, chromatophores (in the dorsal skin) and the overlapping scales which cover the upper and lower surfaces of the fish.

The skin of elasmobranch fishes also comprises epidermis and dermis but the denticles do not originate in the same way as the scales of bony fishes. The inner part of each denticle is manufactured by the dermis but the outer enamel is laid down by the epidermis. In the dogfish *Scyliorhinus canicula* the denticles project further above the skin surface than the scales of the sole and the hard projecting enamel surface is not covered by skin (Fig. 4). Denticles may be numerous and close together as in the dogfish, or few and widely separated as in the ventral skin of many rays.

9.3. Biology of skin parasites with special reference to monogeneans

9.3.1. Attachment

It would seem likely that the stresses imposed on the attachment organ of a skin parasite are related to the size of the parasite, its position on the skin (strong currents leave the gill cavity) and the behaviour and swimming speed of the host. With increasing size of the parasite, its resistance to water flow is likely to increase, requiring a more powerful attachment organ, and skin parasites on an active fish are likely to be exposed to more powerful water currents than parasites on a slow-moving or sedentary fish. Skin parasitic monogeneans orientate themselves in such a way that the attachment organ (haptor) is attached upstream with respect to water currents created when the host swims (e.g. *E. soleae*, see Kearn, 1963a, and *Gyrodactylus alexanderi*, see Lester, 1972). This attitude serves to minimize resistance to water flow.

The attachment organs of skin-parasitic monogeneans are varied (Fig. 2), and include the use of hooks, suckers and cement or combinations of these. The post-oncomiracidia of many skin-parasitic monogeneans and the skin-

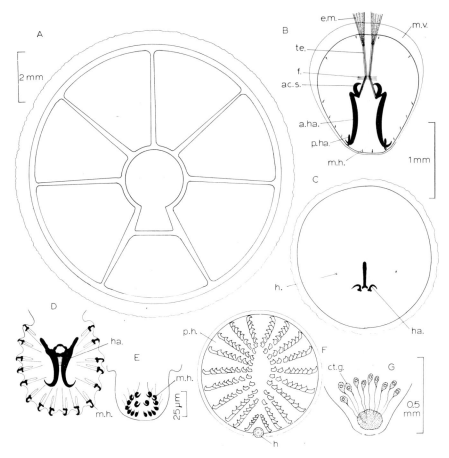

Fig. 2. Attachment organs of adult monogenean skin parasites. A. *Capsala martinieri*; B, *Entobdella soleae*; C, *Calceostoma calceostoma*; D, *Gyrodactylus* sp.; E, *Enoplocotyle minima*; F, *Acanthocotyle lobianchi*; G, *Leptocotyle minor* (semi-diagrammatic). Drawn from specimens in the author's collection. The following are drawn to the same scale: B and C; D and E; F and G. *ac.s.*, Accessory sclerite; *a.ha.*, anterior hamulus; *ct.g.*, cement glands; *e.m.*, extrinsic muscle, *f.*, transverse fibres; *h.*, haptor; *ha.*, hamulus; *m.h.*, marginal hooklet; *m.v.*, marginal valve; *p.h.*, pseudohaptor; *p.ha.*, posterior hamulus; *te.*, tendon.

inhabiting early larvae of some gill-parasitic monogeneans rely mostly on relatively small marginal hooklets (usually 14 or 16 in number) for attachment (see Kearn, 1967a, 1968). Some adult monogeneans also rely on marginal hooklets but these parasites are small. For example, living adult specimens of *Enoplocotyle minima* which I obtained from the skin of *Muraena helena* in Lisbon measured about 350 μm in length before flattening and possess 14 peripheral and 2 central marginal hooklets (Fig. 2E).

Increasing stress on the haptor has been met in gyrodactylids (Fig. 2D) by the acquisition of a pair of large hooks (hamuli or anchors). Lester (1972) however has shown that the marginal hooklets are used by the adult and have a major part to play in attachment. Lester found that the haptor will not become attached to clean glass but will adhere to the secretion left behind on the glass by the sticky glands in the head region. This may be an indication that suction is of little or no importance in the attachment of gyrodactylids.

Entobdellid monogeneans supplement their marginal hooklets with 2 pairs of hamuli (see Kearn, 1964 and Fig. 2B). In *E. soleae* the anteriormost pair of hamuli is associated functionally with a pair of accessory sclerites, which are derived from a persistently growing central pair of marginal hooklets. There is little doubt that the haptors of entobdellids are able to generate suction. *E. soleae* is able to attach to clean glass and is readily dislodged when the seal around the margin of the haptor is lifted at any point with a needle. Suction is created by means of a prominent pair of body muscles; each of these extrinsic muscles has a long tendon which enters the haptor and, after passing beneath transverse fibres which alter its direction, the tendon passes through a notch in the prop-like accessory sclerite and is inserted on the anterior end of the anterior hamulus (Fig. 3). Contraction of the muscles exerts on the anterior hamuli a lifting force, which, because the haptor has a water tight marginal seal and is pinned down peripherally by the marginal hooklets, leads to a drop in pressure in the sea-water-filled space between the haptor and the fish's skin. Lyons (1973) has described papillae with a possible proprioceptive function on the lower surface of the haptor.

Williams et al. (1973) studied preserved specimens of the benedeniid skin parasite *Pseudobenedenia nototheniae*. They recognized the importance of suction generated by haptoral musculature for attachment to the host's skin, but they claimed that this would only be necessary when the host swims, thereby creating strong currents which might tend to dislodge the parasites. They put forward the suggestion that "muscular gripping" by the rim of the haptor, together with surface tension forces between the marginal valve and the host's skin would be sufficient to attach the parasite to the resting host. However, surface tension forces are unlikely to operate in these circumstances, provided that the two opposed surfaces can easily be wetted.

Acanthocotylid monogeneans inhabiting the relatively denticle-free ventral side of rays, supplement their marginal hooklets not with hamuli but with a new attachment organ or pseudohaptor derived during larval development from the posterior region of the body (Kearn, 1967a). The pseudohaptor (Fig. 2F) appears to be a friction pad bearing radial rows of sclerites, but its mode of action is not yet fully understood.

Capsala martinieri from the skin of the oceanic sunfish, *Mola mola*, is a large monogenean (body length often exceeds 2 cm) and has well developed radial septa dividing up the haptor into muscular loculi (Fig. 2A). I was unable to find any hooks in specimens cleared in cedarwood oil.

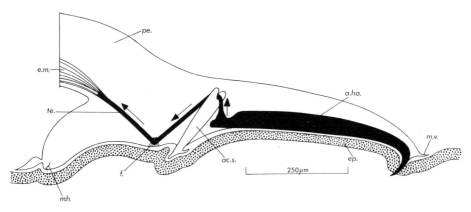

Fig. 3. A diagrammatic parasagittal section through the adhesive organ of *Entobdella soleae*. *ep.*, epidermis of fish; *pe.*, peduncle, joining haptor to body. Other lettering as in Fig. 2. The arrows show the direction of movement of the tendon when the extrinsic muscle contracts. The posterior hamulus has been omitted. (From Kearn, 1971).

Recent work by Kearn (1974a) has revealed the presence of gland cells with ducts which open on the ventral surface of the haptor both in oncomiracidia (Fig. 5) and in adult specimens of *E. soleae*. However, there is no evidence of an adhesive secretion at the sites of attachment of the parasites

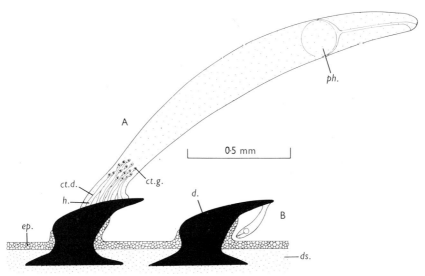

Fig. 4. (A) An adult specimen of *Leptocotyle minor*, and (B) a postlarval parasite, attached to the dogfish *Scyliorhinus canicula*. *ct.d.*, Duct of cement gland; *ct.g.*, cement gland; *d.*, dogfish denticle; *ds.*, dermis of the host; *ep.*, epidermis of the host; *h.*, haptor; *ph.*, pharynx. (From Kearn, 1965).

to glass, and the presence of similar glands opening elsewhere on the parasite's body surface indicates that they may have a different function. Gland cells which do have an adhesive function are associated with the haptor of some microbothriid skin parasites of elasmobranchs. *Leptocotyle minor* attaches itself to the denticles of its dogfish host (Kearn, 1965 and Figs. 2G, 4). Since these denticles are hard and not covered by soft skin tissue, hooks are unsuitable for attachment and are absent in adult parasites. An adhesive secretion is used to cement the parasite to the denticle.

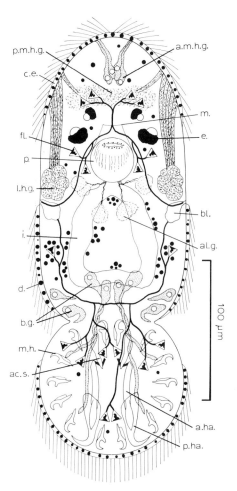

Fig. 5. The oncomiracidium of *Entobdella soleae* seen in ventral view. The body glands of the living larva are less conspicuous than 'the diagram indicates. *al.g.*, Glands associated with the alimentary canal; *a.m.h.g.*, anterior median head glands; *b.g.*, body glands; *bl.*, bladder; *c.e.*, ciliated epidermal cells; *d.*, refringent droplet; *e.*, eye; *fl.*, flame cell; *i.*, intestine; *l.h.g.*, lateral head glands; *m.*, mouth; *p.*, pharynx; *p.m.h.g.*, posterior median head glands. Other lettering as in Fig. 2. (From Kearn, 1974a).

Sticky secretions are widely used in monogeneans for temporary attachment. In the oncomiracidium of *E. soleae* the lateral head glands (Kearn, 1974a and Fig. 5), which communicate with the lateral borders of the head by means of long ducts, most probably provide temporary adhesion until the haptor is unfolded and brought into contact with the host's skin (Kearn, 1967b). The head region of the larva of *E. soleae* also contains medianly situated gland cells which have ducts opening close to those of the lateral head glands and which produce a similar secretion. These median glands may contribute to larval attachment, but their rapid growth during larval development indicates that their main role is to provide temporary adhesion for the adult during its leech-like movements (Kearn, 1974a).

Whitfield et al. (1975) studied specialized arm-like processes on the tail of a transversotrematid cercaria. These processes are used to make initial contact with the skin of the fish host, and at their distal extremity carry pad-like adhesive organs. They found that the adhesive organs did not incorporate discrete gland cells like the attachment organs of monogeneans described above, and that attachment was achieved by means of adhesive granules released by a specialized region of the epidermal syncytium.

9.3.2. Feeding

Feeding in *Entobdella soleae* has been studied in detail by Kearn (1963b). The parasite has no hard mouthparts, the mouth opening into a spacious cavity which houses a pharynx, the wall of which contains prominent gland cells (Fig. 6). The pharynx is everted through the mouth during feeding and placed on the host's skin so as to enclose a circular area of epidermis (Fig. 7). Presumably digestive secretions from the pharyngeal gland cells, and possibly also from oesophageal gland cells, are poured onto the enclosed skin and digested material is pumped back into the intestine by peristaltic contractions of the pharynx. Parasites feed for about 5 min at $19°C$ and then detach the pharynx and withdraw it into the head. Only epidermis is eroded at the feeding site, there being no evidence of damage to the dermis.

A more recent study (Kearn, 1974a) has shown that the pharyngeal gland cells in adult parasites are arranged in two rings, the outer ring comprising 47 (43—49) narrow gland cells and the inner ring 16 broader cells. The full complement of pharyngeal gland cells and the oesophageal gland cells are present in the oncomiracidium (Fig. 5), so that the larva appears to be fully equipped for feeding as soon as it alights on the host's skin. In the earlier study of feeding Kearn showed that the freshly cut surface of pharyngeal tissue in adult specimens of *E. soleae* would readily digest a gelatin layer. The pharyngeal gland cells are presumably the source of this proteolytic secretion, but it is not known whether other enzymes are produced or whether there are differences in the secretion of the cells of the inner and outer rings. The function of the oesophageal cells is unknown.

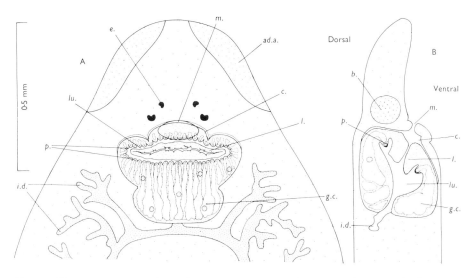

Fig. 6. The morphology of the feeding apparatus of *Entobdella soleae*. A, the retracted feeding organ as seen in dorsal or ventral view; B, a median sagittal section through the retracted feeding organ. *ad.a.*, Anterio adhesive area; *b.*, 'brain''; *c.*, cavity accommodating feeding organ; *e.*, eye; *g.c.*, pharyngeal gland cell; *i.d.*, intestinal diverticulum; *l.*, lip, tucked inside lumen of feeding organ; *lu.*, lumen of feeding organ; *m.*, mouth; *p.*, papilla. (From Kearn, 1963b).

Specimens of *E. soleae* remain attached to the glass bottom of a holding vessel for 2—6 days at 14—17°C (Kearn, 1967b), and they have been observed occasionally with the protruded pharynx attached to the glass (Kearn, unpublished observation). This indicates that a special stimulus from the skin of the sole is not required to evoke feeding behaviour in *E. soleae*.

There is no evidence that any monogenean skin parasite feeds on blood although polyopisthocotylinean gill parasites are known to do so (Llewellyn, 1954) and other skin parasites are blood feeders (e.g. the leech *Hemibdella soleae*, see Llewellyn, L.C., 1965). Epidermal feeding is known to occur in acanthocotylid monogeneans (Kearn, 1963b), in microbothriids (Kearn, 1965) and probably also in gyrodactylids (Kearn, 1963b; Lester, 1972). The

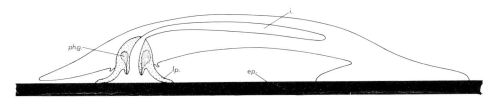

Fig. 7. A diagrammatic sagittal section through a specimen of *Entobdella soleae* in the act of feeding. *ep.*, Epidermis of fish; *i.*, intestine; *lp.*, pharyngeal lip; *ph.g.*, pharyngeal gland cell. (From Kearn, 1971).

microbothriid *Leptocotyle minor* gains access to the epidermis between the hard scales of the dogfish (Kearn, 1965). It is likely that epidermal feeding also takes place in the skin-inhabiting post-larval stages of monogenean parasites such as tetraonchids and dactylogyrids, which as adults live on the gills (Kearn, 1968).

Llewellyn, J. (1965) suggested that protomonogeneans evolved from free-living, ciliated, rhabdocoel-like platyhelminthes, which lived with the early sluggish vertebrates on the sea bottom. Since the unkeratinized epidermis of these fishes provided an abundant, exposed supply of living cells, it is not surprising that it was exploited by these early parasites. An important advantage of epidermis feeding would seem likely to be related to the regenerative properties of fish epidermis (p. 188). Epidermis removed by skin feeders will be rapidly replaced by the host and exposure of sub-epidermal host tissues to possible infection by micro-organisms will be minimized. It is also possible that dermal feeding might provoke more vigorous host responses. Freshly-caught soles carry 3 (1—9) adult speciments of *E. soleae* per infected fish, but in an enclosed laboratory tank the infestation may increase steadily until a single fish may carry up to 200 parasites (Kearn, 1971). Nevertheless, such heavily infected fishes often survive for many months and take food, indicating that the demand imposed on the fish by epidermal feeding in the natural environment is relatively small.

9.3.3. Breathing

Kearn (1962) found that *E. soleae* and *Acanthocotyle lobianchi*, which inhabit the lower surfaces of bottom-living flatfishes, perform undulating body movements. In *E. soleae* these movements draw water from the region behind the parasite and propel it forwards beneath the body and out in the head region. The rate and amplitude of the undulations in *E. soleae* increase when the oxygen content of the ambient water is lowered (at constant temperature), indicating that the body undulations may serve as breathing movements and that the oxygen available in the ambient sea water may fall to low levels. The oxygen content of marine sediments is known to be low over large areas of sea bottom (Perkins, 1957, Richards, 1957 in Kearn, 1962). Consequently when a sole is buried in the sediment during the hours of daylight the parasites attached to the sole's lower surface will be subjected to low ambient oxygen concentrations, whereas during the night when the fish is active the availability of oxygen is likely to be higher. There are indications that, in addition to changes in the rate and amplitude of undulation, *E. soleae* can adjust to changes in the availability of oxygen by changing the surface area and the thickness of its body.

Round-bodied fishes would seem less likely to subject their skin parasites to a wide range of ambient oxygen concentrations. The rigid body of *Leptocotyle minor*, which lives on the skin of the dogfish, *Scyliorhinus canicula*, has been seen to oscillate slowly backwards and forwards but these movements are feeble compared with the vigorous undulations of *E. soleae* and may contribute little to oxygen capture.

The relationship between the rate of body undulating movements and ambient oxygen concentration in *E. soleae* is an indication that the parasite possesses pathways for aerobic energy metabolism.

9.3.4. Host finding and host invasion

Most monogeneans lay eggs and in the majority of species these eggs fall to the sea or river bottom where they develop for a period of time varying in duration from a few days to many weeks (Llewellyn, 1972). On hatching, small ciliated larvae (oncomiracidia) emerge, which, according to Bychowsky (1957) have two behavioural phases, the first of which is characterized by positive phototaxis with a possible distributive function and the second by changes in the response to light and by readiness to attach to the host. Another possibility is that initial responses to light may ensure that larvae remain within or swim towards the region inhabited by the host, although the positive phototaxis determined in the laboratory for some monogenean larvae seems more likely to lead the parasite away from the host (see Llewellyn, 1972). There is evidence that some oncomiracidia have the ability to perceive water currents (see Llewellyn, 1972). Possible behavioural adaptations of this kind, which may promote proximity of host and infective stages of parasites, deserve further attention, but whether or not such adaptations occur it seems at first sight that the chances of slow moving oncomiracidia overtaking an actively swimming fish and making contact with its skin are likely to be low (see Llewellyn, 1972). However, in many fishes there are occasions when opportunities for these slow-moving animals to attach to the skin are greater. Many fishes have marked daily activity patterns, spending part of the day or night in mid-water or at the surface, and the remainder of the time resting or feeding on the bottom. Flatfishes spend most of their lives close to the bottom. Nevertheless, the common sole has a well-marked activity rhythm, swimming about and feeding on polychaetes in the sand at night and remaining immobile, partly buried in the sand, during the hours of daylight (see Kearn, 1973). Compared with round-bodied fishes much more of the body surface of a resting flatfish (at least half and possibly more of the body surface if the fish is partly buried) is in contact with the sediment and any parasite eggs it might contain. When active, even the relatively sluggish flatfish is capable of much higher cruising speeds than the oncomiracidia, and in spite of the fish's relatively large skin surface the fish would provide an elusive target for the oncomiracidia; any contact between host and parasite would most likely result from chance collisions. It is not surprising therefore that in *E. soleae* laboratory work indicates that most larvae hatch during the first 4 hours of daylight (Kearn,

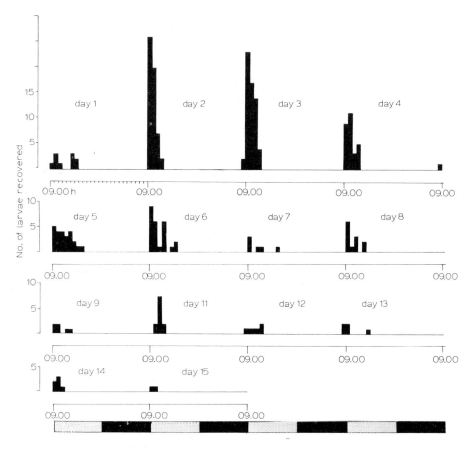

Fig. 8. Emergence of larvae of *Entobdella soleae* from a batch of about 400 eggs attached to the inner surfaces of a crystallizing vessel and exposed to alternating 12 h light and dark periods, indicated by bottom panel (light periods stippled, darkness black). No hatching on day 10. (From Kearn, 1973, to which reference should be made for further details).

1973 and Fig. 8), that is at the beginning of the sole's resting period. The related skin parasite *E. hippoglossi* hatches during darkness (Kearn, 1974b), indicating that its host, the halibut (*Hippoglossus hippoglossus*), has a night-time resting period.

In *E. soleae*, but curiously not in *E. hippoglossi*, there is a further safe-guard for the parasite. Fully-developed eggs of *E. soleae* can be stimulated to hatch at any time of day or night by mucus from the host's skin (Kearn, 1974c and Fig. 9), and work has shown that urea or ammonia provides a hatching stimulus (Kearn and Macdonald in the press). Urea is present in sole skin mucus in sufficient quantities to stimulate hatching.

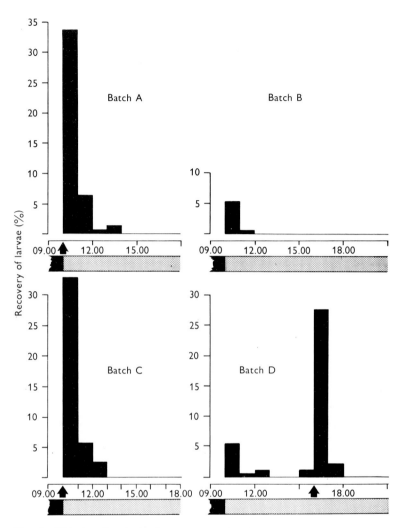

Fig. 9. Effects of sole body mucus on hatching in four batches of eggs of *Entobdella soleae*. Mucus was added at the times indicated by the arrows. No mucus was added to batch B (control). Eggs exposed to alternating 12 h light and dark periods (see bottom panel; light period stippled, dark period black). Number of larvae recovered is expressed as a percentage of the total number of eggs in the batch. (From Kearn, 1974c, to which reference should be made for further details).

Kearn (1974c) has shown that skin mucus from a wide range of fishes, including rays and plaice, readily induces hatching in *E. soleae*, and this lack of specificity would seem to be disadvantageous for the skin parasite since any species of fish resting for a sufficiently long time on the bottom would be likely to stimulate hatching in nearby eggs of *E. soleae*. One possible explanation is that non-host fishes with the habit of burying themselves in the sediment may be uncommon in areas where soles live, but present knowledge of the distribution of fishes on the sea bottom is fragmentary.

Kearn (1967b) also studied the behaviour of the oncomiracidium of *E. soleae* in relation to the skin of the sole and that of other fishes. It was shown that the larva prefers to settle on sole skin and that chemoperception is involved in host finding. This specificity of host selection contrasts sharply with the non-specific nature of host hatching factors. The specific substance (or substances) which permits the chemosensitive oncomiracidium to recognise the sole has not yet been identified.

Acanthocotyle lobianchi, a skin parasite of rays, employs a different host invasion strategy. Unlike the eggs of entobdellids, those of *Acanthocotyle* rarely hatch in the absence of the host, and eggs containing fully-developed larvae have been found to survive for up to 83 days at $13°C$ (Macdonald, 1974). When treated in the laboratory with ray mucus, or presumably when a ray settles on top of eggs on the sea bottom, hatching is immediate, the larva pushing off the operculum by extending its body. There are no cilia and the larva presumably relies on contacting the ray's skin with its extending body. Kearn and Macdonald (in the press) found evidence that the hatching factor in ray mucus is urea and that ammonia is ineffective as a hatching stimulant. Furthermore, the larvae of *A. lobianchi* fail to respond to low concentrations of urea which are sufficient to stimulate hatching in *E. soleae*. This finding is undoubtedly related to the fact that urea levels in ray mucus are significantly higher than those in sole mucus (see Kearn and Macdonald in the press); elasmobranch fishes maintain a high level of blood urea for osmoregulatory purposes and urea provides 80—90% of the total nitrogen excreted (see Nicol, 1960).

Gyrodactylid monogeneans are viviparous giving birth to individuals which are similar in body length to their parents. There is no ciliated larva and in my experience adult gyrodactylids fail to swim when allowed to fall freely through the water. Bychowsky (1957) came to the conclusion that *Gyrodactylus* spreads to new hosts when fishes come close enough for direct transfer of adult parasites, but he expressed some uncertainty about this possible mode of transfer, because direct contact between fishes is not common even in gregarious species. Malmberg (1970) divided a tank into 2 compartments by means of a barrier which permitted passage of parasites but not fishes, and he placed infected fishes on one side of the barrier and uninfected fishes on the other. After 4 weeks a single parasite was found on one of the uninfected fishes, but when an uninfected fish was placed with infected specimens parasites appeared on the uninfected fish in a few days. Khalil (1970) reported that fully mature specimens of *Macrogyrodactylus polypteri* were found on a previously uninfected host one hour after placing this fish in a tank with heavily infected fishes. Llewellyn, J. (personal communication) has found experimentally that an isancistrine monogenean readily transfers itself from the arms of one squid host to those of another specimen when the two squid touch.

My own unpublished observations on gyrodactylids on the guppy (*Lebistes reticulatus*) suggest another way in which the transfer could take place. When fishes were kept in a glass tank with a clean bottom healthy parasites were often found attached to the glass by the haptor. Some of these parasites would reattach readily to a freshly-killed fish or to a detached piece of fin. Guppies were frequently observed resting on the bottom (especially during the night), so that the opportunity would seem to arise for fishes to pick up parasites attached to the substrate.

In recent years there has been an improvement in our knowledge of the surface sensory receptors of parasites and some of these receptors may play an important part in host finding and in recognition of the host on contact. The head regions of gyrodactylids and of the oncomiracidium of *E. soleae* are well equipped with various kinds of ciliary receptor to which chemosensory and tangoreceptive functions have been tentatively ascribed (Lyons, 1972), and Whitfield et al. (1975) have made a detailed study of ciliary receptors at the tips of the "arms" of a transversotrematid cercaria. The arms are specialized organs for identifying and making initial attachment (see p. 194) to the skin of a fish and are held in such a position that their tips from which the receptors project are likely to make first contact with any object in the path of the sinking larva. These so-called mammiform receptors differ from the ciliary receptors so far described in monogeneans in that the single cilium of each receptor contains ciliary microtubules from several basal bodies. Lyons (1972) has pointed out the need for behavioural investigations and neurophysiological recording to confirm the supposed sensory nature of these organs in parasitic platyhelminths, although their small size makes recording difficult. A step in this direction has been taken by Wood (1974) who was able to throw some light on the function of hairlike receptors projecting between the shell valves of the glochidium larva of the bivalve mollusc *Anodonta*. The larva attaches itself to fish skin by closing the shell valves, thereby trapping a plug of skin between them. Using a micromanipulator Wood showed that valve closure occurred when the hairs were stimulated with a piece of fish skin, whereas a response rarely occurred to mechanical stimulation (except of a severe kind) by chemically inert materials. The indications are that the hairs are contact chemoreceptors. Wood also found that glochidia responded vigorously to fluid prepared by scraping mucus and epidermis from freshly-killed fishes and she found evidence that two small (dialysable) molecules are the active constituents of the fluid.

9.3.5. Host responses

The recent discovery of serum antibodies in fish skin mucus and the possibility that local synthesis of antibody might take place in fish skin (see

p. 188) have created renewed interest in the work of Nigrelli and Breder (1934) and Nigrelli (1935a,b) on the monogenean *Epibdella* (=*Benedenia*) *melleni*. Nigrelli and Breder studied the course of infection of a wide range of fishes in the New York Aquarium. Some fishes showed a natural immunity but others became heavily infected with the skin parasites but later developed a more or less complete resistance to further infection.

After treating moonfish (*Vomer setapinnis*) with "sol-argentum" to remove a primary infection around the eyes and on the dorsal surface it was found that re-infection took place on the lateral and ventral sides of the body and the parasites did not migrate to the area of the first infection. These observations indicate that some kind of localized skin immunity may develop (see p. 188).

Another noteworthy observation made by MacCallum (1927) and by Nigrelli and Breder (1934) concerns the localization of these skin parasites on the eyes of some of their hosts. This led Llewellyn (1957) to suggest that blood-borne antibodies may contribute to the mechanism of host specificity in monogeneans and that corneal-feeding parasites such as *B. melleni* may be able to establish themselves on a wide range of hosts because the cornea is non-vascularized. It is possible that the specialized corneal surface may provide a refuge for skin parasites at times when parasites are rejected by the rest of the skin surface of the host.

In a further series of experiments Nigrelli (1935a) immersed living parasites in skin mucus from a variety of fishes. Parasites in mucus from a susceptible host remained alive for 18—24 h (3 days if washed periodically), and those immersed in sea water lived for 3 days. Parasites placed in mucus from dogfish or rays, which do not become infected in aquaria, were mostly dead after 5 h and those placed in mucus from a normally susceptible host which had acquired immunity were dead in 8 h.

Kearn (1967b) found that specimens of *E. soleae* transferred experimentally from one sole to another survived for many months. However, parasites transferred to rays survive for only 2—8 days and those transferred to plaice for barely 24 h. One possible explanation is that the presence of the parasites may induce alien hosts to produce repellent substances. However, the survival time on rays is similar in duration to the period spent attached to glass, indicating that parasites may be unable to feed on ray skin so that the haptor is weakened and ultimately unable to remain attached.

More recently Lester (1972) has conducted experimental infections of the stickleback *Gasterosteus aculeatus leiurus* with the monogenean skin parasite *Gyrodactylus alexanderi*. When parasites were transferred to laboratory raised and previously uninfected fishes there was an initial increase in the numbers of parasites and then most of them disappeared over a period of 4 weeks. Fishes which were then reinfected within a few days of their recovery lost all their parasites within a week, whereas on fishes kept for 4 weeks before reinfection the population showed similar changes to those which followed the initial infection, i.e. first an increase in parasite numbers followed by a decrease over a further 3—4 week period. These observations indicate that a short-lived host response confers immunity for a period which is less than 4 weeks in duration. Lester also found that intramuscular injections of whole fluke antigen gave no protection and for this and other reasons he concluded that a local tissue response and not a humoral immune response was involved. He observed that the skin of the infected and uninfected hosts sheds a layer of mucoid material which he called "slough" and which most probably is epidermal surface coat (see p. 186), and he reports that flukes are often attached to the shed pieces of "slough". Lester did not make clear whether or not there is a difference between infected and uninfected hosts in the rate of shedding, which might account for the apparent acquired immunity, but he did record that surface coat associated with parasites has a greater density and is shed in larger sheets.

The nocturnal production of mucous envelopes by parrot fishes (p. 186), it may be related to parasitism. Many fishes have periods of activity followed by resting periods. When the host is active the chances of slow-swimming, short-lived larvae of skin parasites tracking down and making contact with the host would seem to be significantly smaller than when the host is resting, and selection would be expected to favour the emergence of infective stages at the beginning of the host's resting period (as in the monogenean *E. soleae*, see p. 197). The mucous envelope of the parrot fish is produced when the fish is resting, that is, at a time when the fish is likely to be most vulnerable to invading parasites, and it is conceivable that the envelope may prevent the attachment of parasites to the skin surface. If this explanation is correct it implies that there is some selective advantage for the fish in preventing the settlement of larval parasites. Winn and Bardach (1959) suggested that the mucous envelope may serve to deter predators such as nocturnal feeding moray eels which identify their prey by contact, but their evidence in support of this suggestion is by no means conclusive.

9.3.6. Predation

It has become increasingly apparent in recent years that many skin parasites of fishes (and sometimes gill parasites) are eaten by smaller fishes or by crustaceans, and in many instances a special relationship or cleaning symbiosis has evolved between certain host fishes and their cleaners (see Limbaugh, 1961; Feder, 1966; Hobson, 1971). It has also become apparent that predation on skin parasites is not confined to tropical seas but also occurs in temperate waters (McCutcheon and McCutcheon, 1964; Potts, 1973) and possibly also in freshwater (Tyler, 1963; Khalil, 1964; Spall, 1970). Hobson (1971) found large numbers of caligid copepods and gnathiid isopod larvae in the gut contents of the cleaner fish *Oxyjulis californica*, but although a monogenean (unidentified) and a leech occurred on the skin surfaces of fishes cleaned by *O. californica*, he found no evidence that they were taken. However the exoskeleton of crustaceans would be more recognizable in the gut contents of cleaner animals than the bodies of monogeneans, which, apart from relatively small hooks, are soft and probably easily digested. Tyler (1963), Khalil (1964) and Spall (1970) have observed fishes in aquaria picking off and eating gyrodactylid monogeneans from the skin of other fishes. Lester and Adams (1974) found hooks of *Gyrodactylus* in the intestine of immature sticklebacks, but since each of these fishes had been kept in isolation in the laboratory, the fishes had presumably eaten parasites which had become detached from their own bodies.

Adult skin parasites such as *Entobdella soleae* and *Acanthocotyle lobianchi* are found on the lower surfaces of their flatfish hosts and are encountered on their upper surfaces (Kearn, 1963a, 1967a). In *E. soleae* most young parasites alight on the upper surfaces but migrate to the lower surface before reaching sexual maturity (Kearn, 1963a). Such behaviour may have arisen as a result of selection pressure exerted by predators feeding on skin parasites. However, if this is so, then it is hard to see why similar selection pressures have not operated on *Acanthocotyle elegans*, which invades the lower surface of rays but migrates to the upper surface where sexual maturity is reached (Kearn, 1967a).

The evolution of protective coloration by some skin parasites has been referred to by Limbaugh (1961). In the oncomiracidium of *Capsala martinieri* from the skin of the ocean sunfish Kearn (1963c) found body pigment which may be the forerunner of pigment in the bodies of adult parasites (Kearn, 1963b), and it is conceivable that this may serve as camouflage. There are reports of sunfish being cleaned by fishes (Limbaugh, 1961) and by seabirds (Feder, 1966).

It is generally supposed that cleaning relationships have arisen because of the selective disadvantages to the host fishes of heavy parasite burdens. Such pressures may also have led to the production of mucous envelopes by parrot fishes as a means of reducing invasion by skin parasites (see above). Wyman and Ward (1972) regarded the removal of fungus from the skin as an important adaptive function of the relationship between two closely related fishes, young of *Etroplus maculatus* (the cleaners) and all age groups of *E. suratensis*. On the other hand Losey (1972) has questioned the assumption that removal of cleaners would lead to an increase in ectoparasites and has generated new interest in the adaptive significance of cleaning relationships. Wyman and Ward have thrown some light on the evolution of the cleaning relationship between *E. maculatus* and *E. suratensis*; it seems possible that intraspecific signals used when tending the young may have acquired an interspecific role facilitating contact between cleaner and host.

Acknowledgement

I would like to record my thanks to Cambridge University Press and to the Editors of Parasitology and The Journal of the Marine Biological Association of the United Kingdom for permission to republish Figs. 1, 4, 5, 6, 8 and 9, and to University of Toronto Press for permission to republish Figs. 3 and 7.

9.4. REFERENCES

Andrew, W. (1959). "Textbook of Comparative Histology". Oxford University Press, New York.

Asakawa, M. (1970). "Histochemical studies of the mucus on the epidermis of eel, *Anguilla japonica*". Bull. Jap. Soc. scient. Fish. 36, 83—87. (In Japanese with English Summary).

Bardach, J;E. and Todd, J.H. (1970). "Chemical communication in fish". In: "Advances in Chemoreception", Vol. 1. (J.W. Johnston, D.G. Moulton and A. Turk, ed.). pp. 205—240. Meredith Corporation, New York.

Barry, J.M. and O'Rourke, F.J. (1959). "Species specificity of fish mucus". Nature, Lond. 184, 2039.

Berlin, L.B. (1951). "Compensatory regeneration of the epidermis of the groundling". Dokl. Akad. Nauk SSSR. 80, 245—248. (In Russian).

Bremer, H. (1972). "Einige Untersuchungen zur Histochemie der sezernierenden Element der Teleostier-Epidermis." Acta Histochem. 43, 28—40.

Bychowsky, B.E. (1957). "Monogenetic Trematodes, their Classification and Phylogeny." Academy of Sciences, U.S.S.R., Moscow, Leningrad (In Russian). English translation by W.J. Hargis and P.C. Oustinoff (1961). American Institute of Biological Sciences, Washington.

Di Conza, J.J. and Halliday, W.J. (1971). "Relationship of catfish serum antibodies to immunoglobulin in mucus secretions." Aust. J. exp. Biol. med. Sci. 49, 517—519.

Enomoto, N., Nagao, T. and Tomiyasu, Y. (1960). "Studies on the external mucous substance of fishes—II. On the chemical properties of the mucous substance of conger, *Astroconger myriaster* Brevoort." Bull. Jap. Soc. scient. Fish. 26, 745—748. (In Japanese, English summary).

Enomoto, N., Nagao, T. and Tomiyasu, Y. (1961). "Studies on the external mucous substance of fishes—III. Identification of aminoacids in the mucous protein from some fishes by paper chromatography." Bull. Jap. Soc. scient. Fish. 27, 143—146. (In Japanese, English summary).

Enomoto, N. and Tomiyasu, Y. (1960). "Studies on the external mucous substance of fishes—I. On the chemical properties of the mucous substance of loach, *Misgurnus anguillicaudatus* Cantor." Bull. Jap. Soc. scient. Fish. 26, 739—744. (In Japanese, English summary).

Feder, H.M. (1966). "Cleaning symbiosis in the marine environment." In "Symbiosis". Vol. 1. (S.M. Henry, ed.) pp. 327—380. Academic Press, New York.

Fletcher, T.C. and Grant, P.T. (1969). "Immunoglobulins in the serum and mucus of the plaice, (*Pleuronectes platessa*)." Biochem. J. 115, 65P.

Fletcher, T.C. and White, A. (1973). "Antibody production in the plaice (*Pleuronectes platessa L.*) after oral and parenteral immunization with *Vibrio anguillarum* antigens." Aquaculture, 1, 417—428.

Hildemann, W.H. (1962). "Immunogenetic studies of poikilothermic animals." Am. Nat. 96, 195—204.

Hobson, E.S. (1971). "Cleaning symbiosis among California inshore fishes." Fishery Bull. Fish Wildl. Serv. U.S. 69, 491—523.

Jakowska, S. (1963). "Mucus secretion in fish— a note." Ann. N.Y. Acad. Sci. 106, 458—462.

Kearn, G.C. (1962). "Breathing movements in *Entobdella soleae* (Trematoda, Monogenea) from the skin of the common sole." J. mar. biol. Ass. U.K. 42, 93—104.

Kearn, G.C. (1963a). "The life cycle of the mongenean *Entobdella soleae*, a skin parasite of the common sole." Parasitology, 53, 253—263.

Kearn, G.C. (1963b). "Feeding in some monogenean skin parasites: *Entobdella soleae* on *Solea solea* and *Acanthocotyle* sp. on *Raja clavata.*" J. mar. biol. Ass. U.K. 43, 749—766.

Kearn, G.C. (1963c). "The oncomiracidium of *Capsala martinieri*, a monogenean parasite of the sun fish (*Mola mola*)." Parasitology, 53, 449—453.

Kearn, G.C. (1964). "The attachment of the monogenean *Entobdella soleae* to the skin of the common sole." Parasitology 54, 327—335.

Kearn, G.C. (1965). "The biology of *Leptocotyle minor*, a skin parasite of the dogfish, *Scyliorhinus canicula.*" Parasitology, 55, 473—480.

Kearn, G.C. (1967a). "The life cycles and larval development of some acanthocotylides (Monogenea) from Plymouth rays." Parasitology, 57, 157—167.

Kearn, G.C. (1967b). "Experiments on host-finding and host-specificity in the monogenean skin parasite *Entobdella soleae.*" Parasitology, 57, 585—605.

Kearn, G.C. (1968). "The development of the adhesive organs of some diplectanid, tetraonchid and dactylogyrid gill parasites (Monogenea)." Parasitology, 58, 149—163.

Kearn, G.C. (1971). "The physiology and behaviour of the monogenean skin parasite *Entobdella soleae* in relation to its host (*Solea solea*)." In: "Ecology and Physiology of Parasites, a Symposium" (A.M. Fallis, ed.), pp. 161—187. University of Toronto Press, Toronto.

Kearn, G.C. (1973). "An endogenous circadian hatching rhythm in the monogenean skin parasite *Entobdella soleae*, and its relationship to the activity rhythm of the host (*Solea solea*)." Parasitology, 66, 101—122.

Kearn, G.C. (1974a). "A comparative study of the glandular and excretory systems of the oncomiracidia of the monogenean skin parasites *Entobdella hippoglossi, E. diadema* and *E. soleae.*" Parasitology, 69, 257—269.

Kearn, G.C. (1974b). "Nocturnal hatching in the monogenean skin parasite *Entobdella hippoglossi* from the halibut *Hippoglossus hippoglossus.*" Parasitology, 68, 161—172.

Kearn, G.C. (1974c). "The effects of fish skin mucus on hatching in the monogenean parasite *Entobdella soleae* from the skin of the common sole, *Solea solea.*" Parasitology, 68, 173—188.

Kearn, G.C. and Macdonald, S. "The chemical nature of host hatching factors in the monogenean skin parasites Entobdella soleae and Acanthocotyle Lobianchi." Intr. J. Parasit. 6, in press.

Khalil, L.F. (1964). "On the biology of *Macrogyrodactylus polypteri* Malmberg, 1956, a monogenetic trematode on *Polypterus senegalus* in the Sudan". J. Helminth. 38, 219—222.

Khalil, L.F. (1970). "Further studies on *Macrogyrodactylus polypteri* a monogenean on the African freshwater fish, *Polypterus senegalus*". J. Helminth. 44, 329—348.

Lehtonen, A., Kärkkäinen, J. and Haahti, E. (1966). "Carbohydrate components in the epithelial mucin of hagfish, *Myxine glutinosa.*" Acta chem. scand. 20, 1456—1462.

Lester, R.J.G. (1972). "Attachment of *Gyrodactylus* to *Gasterosteus* and host response." J. Parasit. 58, 717—722.

Lester, R.J.G. and Adams, J.R. (1974). "*Gyrodactylus alexanderi*: reproduction, mortality, and effect on its host *Gasterosteus aculeatus.*" Can. J. Zool. 52, 827—833.

Limbaugh, C. (1961). "Cleaning symbiosis." Scient. Am. 205, 42—49.

Llewellyn, J. (1954). "Observations on the food and gut pigment of the Polyopisthocotylea (Trematoda: Monogenea)." Parasitology 44, 428—437.

Llewellyn, J. (1957). "Host-specificity in monogenetic trematodes." In: "First symposium on host-specificity among parasites of vertebrates." pp. 199—212. Neuchâtel.

Llewellyn, J. (1965). "The evolution of parasitic platyhelminthes". In: "Third Symposium of the British Society for Parasitology." pp. 47—78. Blackwell, Oxford.

Llewellyn, J. (1972). "Behaviour of monogeneans." In: "Behavioural aspects of parasite transmission." (E.U. Canning and C.A. Wright, ed.) pp. 19—30. Supplement No. 1 to the Zoological Journal of the Linnean Society, 51.

Llewellyn, L.C. (1965). "Some aspects of the biology of the marine leech Hemibdella soleae." Proc. zool. Soc. Lond. 145, 509—528.

Losey, G.S. (1972). "The ecological importance of cleaning symbiosis" Copeia, 4, 820—833.

Lyons, K.M. (1972). "Sense organs of monogeneans." In: "Behavioural aspects of parasite transmission." (E.U. Canning and C.A. Wright, ed.) pp. 181—199. Supplement No. 1 to the Zoological Journal of the Linnean Society, 51.

Lyons, K.M. (1973). "Scanning and transmission electron microscope studies on the sensory sucker papillae of the fish parasite Entobdella soleae (Mongenea)". Z. Zellforsch. mikrosk. Anat. 137, 471—480.

MacCallum, G.A. (1927). "A new ectoparasitic trematode, Epibdella melleni sp. nov." Zoopathologica 1, 291—300.

Macdonald, S. (1974). "Host skin mucus as a hatching stimulant in Acanthocotyle lobianchi, a monogenean from the skin of Raja spp." Parasitology, 68, 331—338.

Malmberg, G. (1970). "The excretory systems and the marginal hooks as a basis for the systematics of Gyrodactylus (Trematoda, Monogenea)." Ark. Zool. Band 23, 1—235.

McCutcheon, F.H. and McCutcheon, A.E. (1964). "Symbiotic behaviour among fishes from temperate ocean waters." Science, N.Y. 145, 948—949.

Nicol, J.A.C. (1960). "The Biology of Marine Animals." Sir Isaac Pitman, London.

Nigrelli, R.F. (1935a). "On the effect of fish mucus on Epibdella melleni, a monogenetic trematode of marine fishes". J. Parasit. 21, 438.

Nigrelli, R.F. (1935b). "Studies on the acquired immunity of the pompano, Trachinotus carolinus, to Epibdella melleni." J. Parasit. 21, 438—9.

Nigrelli, R.F. and Breder, C.M. (1934). "The susceptibility and immunity of certain marine fishes to Epibdella melleni, a monogenetic trematode." J. Parasit. 20, 259—269.

Ogawa, M. (1970). "Effects of prolactin on the epidermal mucous cells of the goldfish, Carassius auratus L." Can. J. Zool. 48, 501—503.

O'Rourke, F.J. (1961). "Presence of blood antigens in fish mucus and its possible parasitological significance." Nature, Lond. 189, 943.

Potts, G.W. (1973). "Cleaning symbiosis among British fish with special reference to Crenilabrus melops (Labridae)." J. mar. biol. Ass. U.K. 53, 1—10.

Roberts, R.J., Young, H. and Milne, J.A. (1971). "Studies on the skin of plaice (Pleuronectes platessa L.) 1. The structure and ultrastructure of normal plaice skin." J. Fish. Biol., 4, 87—98.

Spall, R.D. (1970). "Possible cases of cleaning symbiosis among freshwater fishes." Trans. Am. Fish. Soc. 99, 599—600.

Spearman, R.I.C. (1973). "The Integument". University Press, Cambridge.

Tyler, A.V. (1963). "A cleaning symbiosis between the rainwater fish, Lucania parva and the stickleback, Apeltes quadracus." Chesapeake Sci. 4, 105—106.

Uskova, Ye. T., Chaykovskaya, A.V. and Uskov, I.A. (1971). "Composition of amino acids in the mucus of fish skin". Hydrobiol. J. 7, 83—85.

Van Oosten, J. (1957). "The skin and scales." In: "The Physiology of Fishes", vol. 1. "Metabolism" (M.E. Brown ed.). pp. 207—244. Academic Press, New York.

Wessler, E. and Werner, I. (1957). "On the chemical composition of some mucous substances of fish." Acta chem. scand. 11, 1240—1247.

Whitear, M. (1970) "The skin surface of bony fishes." J. Zool. 160, 437—454.

Whitfield, P.J., Anderson, R.M. and Moloney, N.A. (1975). "The attachment of cercariae of an ectoparastic digenean, *Transversotrema patialensis*, to the fish host: behavioural and ultrastructural aspects." Parasitology, 70, 311—329.

Williams, I.C., Ellis, C. and Spaull, V.W. (1973). "The structure and mode of action of the posterior adhesive organ of *Pseudobenedenia nototheniae* Johnston, 1931 (Monogenea : Capsaloidea)." Parasitology, 66, 473—485.

Winn, H.E. (1955). "Formation of a mucous envelope at night by parrot fishes." Zoologica, N.Y. 40, 145—148.

Winn, H.E. and Bardach, J.E. (1959). "Differential food selection by moray eels and a possible role of the mucous envelope of parrot fishes in reduction of predation. Ecology 40, 296—298.

Wood, E.M. (1974). "Some mechanisms involved in host recognition and attachment of the glochidium larva of *Anodonta cygnea* (Mollusca : Bivalvia)." J. Zool. 173, 15—30.

Wyman, R.L. and Ward, J.A. (1972). "A cleaning symbiosis between the cichlid fishes *Etroplus maculatus* and *Etroplus suratensis*. I. Description and possible evolution." Copeia, 4, 834—838.

Ecological Aspects of Parasitology
Editor: C.R. Kennedy
© *North-Holland Publishing Company, Amsterdam, 1976*

CHAPTER 10

GILLS

C.H. FERNANDO [a] and C. HANEK [b]

Department of Biology, University of Waterloo [a], *Ontario (Canada) and Ministry of Agriculture and Fisheries* [b], *Nassau (Bahamas)*

Contents

10.1. Introduction

Gills or gill-like organs are found in a wide variety of aquatic animals. Many invading organisms are associated with gills but the present account will be restricted largely to the gills of teleost fishes. Our knowledge of the detailed structure and functioning of gills and their parasitofauna leaves much to be desired. However, recent studies on gill structure and function enable us to discuss the problems of this habitat in some depth in relation to gill inhabitating organisms.

The basic structure and functioning of fish gills have been well reviewed by Hughes and Morgan (1973). The gills are formed essentially of a much folded epithelium richly supplied with blood for respiratory exchanges. The respiratory surface is borne on the stiffened gill arch supports. The epithelium, which is considered to be "stratified", probably undergoes sloughing but the details of this are not clear. The area accessible to parasite infections is enormous and the readily available food supply in the exposed epithelium and the underlying blood vessels provides an ideal situation for parasitic infections.

The gill arches undergo movements while a current of water passes continuously over the respiratory surface. The water current is maintained by two pumps, the buccal and the opercular. The water currents and the "abrasive" movements of the gill arches necessitate a strong attaching apparatus in prospective parasites. Penetrative forms must be capable of adhesion or quick entry at the site of contact.

Parasitic infection is influenced by the ventilation volume and the dynamics of water flow. It is possible that active infective stages can reach the infection sites by limited positive movements in a suitable direction but it is generally assumed that the infective stages are carried passively to the infection site. It has been suggested that the copepod *Lernaeocera* reaches gill tissue by swimming against the ventilating current (see Kabata, 1970). The ability to infect the gill will then depend on adhesion, and penetration or grasping of the gill tissue. Site changes occur in active forms.

Since gills are the site of respiratory exchanges there is a high concentration of oxygen available to parasitic organisms. Carbon dioxide, other gases, and secretions from the gills may play a hitherto unknown role in attracting repelling or retaining parasites. Gills have a very wide spectrum of ectoparasites as compared with the skin and fins. Whether this richness of parasitic species is due in part to some chemical attractant in the gills needs study. Although the gills provide a very suitable site for parasitic infections, there are a number of factors acting against infection though our present knowledge of these is meagre. The abrasive effect of gill arch and opercular movements and secretions produced by the gills probably provide considerable protection against infection. After infection has occured a variety of host responses follow. The continuous water current interferes with the parasite's chances of reaching, attaching or penetrating and subsequent residence.

Gills are very susceptible to mechanical damage due to the exposed epithelial surfaces. Also, any interference with respiratory exchanges caused directly and lasting even for short periods can set up a chain reaction involving tissue damage and subsequent primary and secondary infections. There is considerable disagreement regarding the nature of associations of organisms of the gills. Banina (1969) for example considers the ciliate *Apiosoma* a parasite while Lom (1973) on the basis of ultrastructural studies considers this ciliate an ectocommensal.

10.2. Scope of present account

We shall begin with a brief introduction to the morphology of the gill microhabitat. Much remains to be learned about details of the structure of fish gills and their variability with age and in different species. Our remarks will thus be confined to general comments. The parasite spectrum of North American freshwater fishes has been compiled. This gives a measure of the suitability of the site for parasitic and other associations. We have dealt with some of the ways in which organisms attach themselves to gills or lodge after penetration. Recent work has elucidated some details of attachment but the dynamic aspects in relation to regenerating cells and host responses remain largely unexplored. Some comments on, and deductions from, the available data are made regarding the parasite spectrum, attachment, penetration and lodging of gill parasites.

Ecological aspects of gill parasites are discussed with special reference to spatial distribution. The interaction of parasites and host-response are briefly discussed. We have used the term parasite rather loosely as there is considerable disagreement regarding the nature of the association between gill inhabiting organisms and the gill.

10.3. Nature of microhabitat

Teleost fishes have four pairs of gills. They are curved posteriorly and each is supported by a series of bones, jointed flexibly. Each arch (Fig. 1) bears a large number of gill filaments or primary lamellae (Figs. 1, 2). These project laterally and posteriorly from the gill arch and are united by an interbranchial septem. On the upper and lower surface of each gill filament, projecting at right angles to the axis, is a row of closely packed leaf-like structures termed secondary lamellae (Figs. 2, 3). The secondary lamellae are delicate, flattened, structures, about 10 μ thick in *Gadus pollachius* according to Hughes and Grimstone (1965). Their asymmetric shape is related to the

212

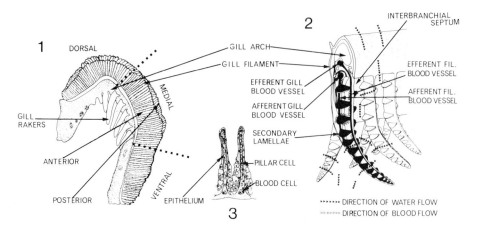

Fig. 1. Teleost gill arch showing divisions for recording spatial distribution of parasites.

Fig. 2. Diagram of portion of gill arch of a teleost fish illustrating the structure of the gill filaments and the position of the interbranchial septum. Direction of flow of water and blood indicated by arrows. (after Munshi and Singh, 1969).

Fig. 3. Section through two adjacent lamellae of the gill filament.

direction of water flow over the gills (Hughes, 1966). The secondary gill lamellae (Fig. 3) are lined externally by an epithelium which is more than one layer thick in parts. A basement membrane and an endothelium enclose a blood space. The wall of the lamella is usually only $1-3\ \mu$ thick. The endothelium consists of a series of loosely packed pillar cells which hold the epithelia of the two sides of the lamella together. The pillar cells are reinforced by strands of basement membrane material which run between them connecting the basement membranes of the two opposite surfaces. The spaces between the pillar cells are just large enough to allow the free passage of erythrocytes. One edge of each filament adjoins the buccal cavity and the other faces the operculum. An afferent blood vessel runs along the opercular edge and an efferent along the buccal edge. Blood flow is opposite to the direction of water flow across the lamellae. (Fig. 2). The blood supply to the gills has been well described and figured by Munshi and Singh (1968).

The non-respiratory portion of the gill arch bears a series of gill rakers often arranged in two rows like the teeth of a comb. They are directed opposite to the filaments and are widely spaced in predators and very closely spaced in plankton feeders. Intermediate spacing is common.

According to present knowledge (Hughes and Morgan, 1973; Hughes and Skelton, 1958; Skelton, 1970; Woskoboinikoff and Balabai, 1937), the flow of water over the gills is essentially continuous. This flow is maintained by two pumps, one in front (buccal) and one behind the gills (opercular). It is evident however, that all parts of the surface are not continuously or evenly exposed to the water current. Violent water movement over the gills occurs in "coughing". This helps to rid the gills of foreign particles.

Water flow over the gills provides the main source of infecting organisms. Since gills are designed for the exchange of respiratory gases, there is, in the living fish, a high concentration of oxygen immediately adjoining the active respiratory surface. This situation offers a favourable habitat for many protozoan and metazoan parasites.

The gills offer invading organisms a large surface area for attachment and the exposed epithelial respiratory surface of the gills provides suitable

TABLE 1

The ectoparasite spectrum of North American freshwater fishes. Only a few gill inhabitants are found on the other sites (skin and fins).

	Total ectoparasitic spectrum		Gill spectrum		% of total no. of species	Source
	No. genera	No. species	No. genera	No. species		
Fungi	6	?	?	?	uncertain	Hoffman (1967) Hoffman and Meyer (1974)
Protozoa	24	105	23	54	51	Hoffman (1967)
Monogenea	24	334	22	243	73	Hanek and Fernando (1972); Hanek et al. (1967); Mizelle and McDougal (1970)
Metacercariae	25	41	4	6	15	Hoffman (1967)
Copepoda	6	80	3	33	41	Kabata (1969) Roberts (1970)
Glochidia	?	about 500	?	about 500	almost 100	Pennak (1953)
Hirudinea	7	20	2	2	10	Hoffman (1967) Molnar et al. (1974)
Acarina	2	2	2	2	100	Mueller (1936) Tedla and Fernando (1970)
Total	94 plus glo- chidia	582 plus glo- chidia	58 plus glo- chidia	340 plus glo- chidia	58 plus glo- chidia	

material for direct ingestion. Invasive forms penetrate the epithelium and develop inter or intracellularly. The underlying blood supply can be tapped directly by parasites reaching capillaries or indirectly via food material reaching epithelial cells.

The movements of the individual gill arches, though restricted, and the opercular movements can be considered as abrasive to invading organisms on the gills and in the gill chamber.

Little is known about the rate of replacement of gill epithelial cells under normal or parasitized conditions. The viability of intracellular parasite life cycles will depend on the life span of epithelial cells. Sessile ectocommensals (ectoparasites) must also contend with sloughing of epithelial cells.

10.4. Parasite spectrum and mode of attachment

The parasite spectrum of North American freshwater fishes is given in Table 1. We have left out the endoparasitic groups although some nematodes do occur in the gill vascular system. Our compilation based on present records is certainly not comprehensive for all ectoparasites on fish gills. However, the table shows that a high proportion of Monogenea, Protozoa, Copepoda and glochidia occur on the gills.

Attachment of gill parasites to a particular site may be very short-lived or extend for the life span of the parasite. The mode of attachment may be mechanical involving only a grasp of the gill tissue as in the copepod genus *Ergasilus*. In others a portion of the parasite forms a more or less intimate junction with the host tissue. Active penetration of the gill tissue lodges the parasite intra or intercellularly with resulting host cell and tissue reaction. Even organisms whose bodies are outside the gill tissue may become partially or fully enclosed by gill tissue. For intracellular parasites and those attached to the outer cell surfaces the replacement time of epithelial cells and sloughing of portions of the epithelium will have important consequences.

10.4.1. Fungi

Secondary infection of diseased or damaged fish gills seems to be the rule in fungi. Reichenbach-Klinke and Elkan (1965) record a number of parasitic and saprophytic fungi on fish gills. Scott and O'Bier (1962) give an account of fungi associated with diseased fish and eggs. The active zoospores germinate on damaged or diseased fish gills and the mycelia gain access to the tissues.

Relatively few fungi have been reported from North American freshwater fish gills (Table 1).

10.4.2. Protozoa

A wide variety of Protozoa have been recorded on fish gills. Hoffman (1967) lists 54 species on North American freshwater fishes (Table 1). The genera *Costia, Ichthyopthirius, Trichodina, Myxosoma, Myxobolus* and *Henneguya* feature prominently among the species recorded.

The mode of attachment reflects the varying degrees of parasitism exhibited by Protozoa. Molnar (1971a) found *Cryptobia* attached to the gill epithelium by their longer flagellum. *Trichodina* and allied genera are only attached temporarily to any one site by an apparatus consisting of an adhesive ring and hooklets. More permanent attachment is achieved by other protozoans. Lom (1973) examined the mode of attachment in two sessile peritrichs, *Apiosoma piscicola* and *Epistylis lwoffi*. The terminal plate of the stalk of small colonies of *E. lwoffi* adheres to the host cells by a layer of fibrous material (Fig. 4). In *A. piscicola* the flattened, ciliated base (scopula) secretes a short, rudimentary stalk which is attached to the host cell membrane in the same way. Lom (loc. cit) on the basis of ultrastrucural studies concluded that there is a considerable pulling action on the host cell membrane by *Epistylis* but no serious damage is caused to the host cell. This association is ectocommensal. Lom and Lawler (1973) examined the attachment of two ectoparasitic dino-flagellates *Oodinium cyprinodontium* and *Amyloodinium* sp. The trophonts of *O. cypridodontum* attach to the gill filament by a ramified base bearing finger-like rhizoids. Host cell alterations at the points of contact were noted but no cell penetration was detected. In *Amyloodinium* sp. (Fig. 5), which lacks a chloroplast, the attachment is described as follows "At the base of the trophont is a gap in the thecal armour encircled by an osmophilic ring connected with an attachment plate bearing numerous filiform rhizoids. These rhizoids are embedded in the gill epithelium and inflict heavy damage on invaded cells. A long conspicuous stomopode is associated with the attachment plate; it has an axial fibrillar tube that passes through the osmiophilic ring deep into the trophont. This tube contains various types of vesicles and clove-like bodies with supposedly enzymatic function. The flagellar canal opens into a rudimentary sulcal fold at the base of the trophont. It is connected with the pustule and may also be the site of food ingestion." Besides surface attachment with inter or intracellular involvement many protozoans are found located inter and intracellularly in gill tissue.

10.4.3. Monogenea

Monogenea represent the most numerous group of true ectoparasites infecting the gills of fish. Two hundred and forty-three species or approximately 73% of the known North American Monogenea of freshwater fishes have been recorded on the gills. Among the 22 genera recorded *Dactylogyrus* (118

216

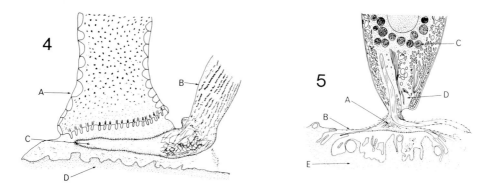

Fig. 4. *Apiosoma* using an *Epistylis* terminal platelet as an attachment site on gill; sche-
matized longitudinal section of their basal portions. A. *Apiosoma*; B. *Epistylis*; C. terminal platelet of *Epistylis* stalk; D. host cell. (redrawn from Lom, 1973).

Fig. 5. Schematic reconstruction of the basal portion of an attached *Amyloodinium* sp.
A. attachment plate; B. rhizoid; C. food vacuoles; D. flagellum; E. host cell. (after Lom
and Lawler, 1973).

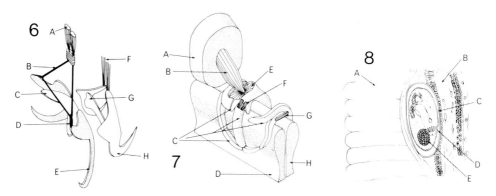

Fig. 6. Stereogram showing the principal adhesive apparatus of *Haliotrema balisticus*. A.
extrinsic adductor muscle; B. tendon; C. dorsal bar; D. fibrous loop; E. ventral anchor;
F. dorsal muscle; G. dorsal bar; H. dorsal anchor. (after Kearn, 1971).

Fig. 7. Stereogram of an adhesive organ of *Diclidophora* grasping secondary gill lamella of
host. A. peduncle; B. extrinsic suctorial muscle; C. sclerites, forming jaws; D. proximal
lamellae; E. diaphragm; F. intrinsic abductor muscle; G. intrinsic adductor muscle; H.
distal lamellae. (after Llewellyn, 1958).

Fig. 8. Metacercaria encysted in gill. A. gill filament; B. bone; C. metacercaria; D. circum-
oral spines; E. germinal cells. (after Martin, 1964).

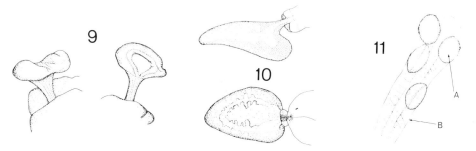

Fig. 9 and 10. Attachment organs of Lernaeopodidae: 9. bulla of *Salmincola californiensis;* 10. bulla of *S. salmonea.* (redrawn from Kabata and Cousens, 1972).

Fig. 11. Glochidia encysted in gill filament. A. glochidium; B. gill filament. (drawn from photograph in Hoffman and Meyer, 1974).

spp.), *Cleidodiscus* (20 spp.), *Urocleidius* (38 spp.) and *Actinocleidius* (21 spp.) figure most prominently.

Generally there are two attaching "complexes" in Monogenea. One is located at the anterior end and bears sucker-like attaching organs. This serves to attach the anterior end during feeding and locomotion. Bychowsky (1957) has described this anterior complex and classified it into two types according to whether or not they are connected to the mouth funnel and oral aperture. The second and main attaching complex is situated posteriorly. This complex consists of an adhesive disc (haptor) and different attaching organs thereon: Bychowsky (1957) distinguished haptors weakly or strongly delineated from the rest of the body, and primary and secondary haptors. The most primitive haptors weakly delineated from the body are found in the Protogyrodactylidae and the majority of Dactylogyridae. In contrast the Monocytolidae and the Capsalidae whose haptors are bound together by numerous connecting ridges are sharply delineated from the body and well developed for a sucker type of attachment. The attaching organs on the haptor can be divided into four basic types, viz. hook-shaped organs, supplementary haptors with chitinous armature, suckers and attaching valves. One species can possess from 1—3 types on its haptor.

Without exception the Monogenea are attached with their posterior end nearer the skeletal portion of the gill arch. In this way the attached end lies upstream of the gill ventilating current. There are, however, marked differences in the mechanism and mode of attachment even within a single genus (Llewellyn, 1958). Two specific examples will illustrate this. Kearn (1971) observed that *Haliotrema balisticus*, an ancyrocephalid on *Balistes capriscus*, was attached between the secondary gill lamellae of its host by a pair of ventral anchors directed ventrolaterally and a pair of dorsal anchors directed dorsally (Fig. 6). The ventral anchors, each of which pivots at the end of a ventral connecting bar, are operated by a pair of extrinsic muscles.

The tendonous regions of these muscles pass through fibrous loops attached to the ventral anchors and then join each other in the midline. Such an arrangement according to Kearn (1971) has a high mechanical advantage which may reduce the expenditure of energy required to maintain the anchors in position in the face of the strong gill ventilating current and may also preserve the mobility of the free region of the body. The dorsal anchors are not operated by muscles and tendons associated with the ventral anchors. Marginal hooklets and glands play only a subsidiary role in attachment. Llewellyn (1958) described the mode of attachment in all nine known species of *Diclidophora*. His findings can be briefly summarized as follows. The adhesive organs (jaws), formed of sclerites, are opened by the abductor muscle and applied to two or three secondary lamellae (Fig. 7). A preliminary grasp is taken closing the jaws with the intrinsic abductor muscles. Next the powerful extrinsic muscle contracts and the central part of the diaphragm is withdrawn into the peduncle. The marginal lips of the jaws act as valves to seal off the internal cavity of the adhesive organ and the suction pressure is set up in the cavity of the adhesive organ, the two jaws being freely hinged (the abductor muscle being relaxed). This pressure is converted into a clamping action whose force is proportional to the pull of the extrinsic muscle.

10.4.4. Digenetic metacercariae

Hoffman (1967) listed 41 species belonging to 25 genera of ectoparasitic metacercariae in North American freshwater fishes. Only six species have been recorded from gills (Table 1). The skin and fins are preferred sites for metacercariae probably because they can penetrate at these sites when the fish is stationary.

The first intermediate host of almost all Digenea are molluscs. After a series of developmental stages in the mollusc the cercariae are liberated and penetrate gill tissue. Martin (1964) described the life-cycle of *Pygidiopsoides spindalis* where the cercaria penetrates the gill of its host *Fundulus parvipinnis parvipinnis* and works its way to the bony support of the gill before encysting (Fig. 8). The metacercaria becomes enclosed in a cyst and the host may even react by enclosing the parasite in a thin layer of cartilage or bone.

10.4.5. Copepoda

Thirty-three species belonging to three genera of Copepoda have been reported on the gills of North American freshwater fishes. The genus *Ergasilus* is represented by 22 species while the general *Salmincola* and *Achtheres* are represented by 6 and 4 species respectively (Kabata, 1969; Roberts, 1970).

The parasitic females of *Ergasilus* grasp the gill filaments by the stout second antennae. Like the Monogenea they attach so as to face the ventilation current. The mouth parts are always within reach of the host epithelium. Einszporn (1965) showed that the first legs rasp off gill tissue and convey it to the oral cavity.

A unique mode of attachment is exhibited by the Lernaeopodidae. The way in which these parasites maintain their contact with the host can be seen as "A balanced compromise between security of attachment and the ability, however limited, to alter position on their host" (Kabata and Cousens, 1972). The attaching organ is a bulla which is a flattened sucker (Figs. 9, 10), permanently fused with the second maxilla. This dual appendage is contractile and enables the parasite to shift position within the radius of the appendage (Kabata and Cousens, 1972).

10.4.6. Unionidae (Mollusca)

According to Pennak (1953) there are over 500 species of Unionidae in North America. Although the parasitic stage is of comparatively short duration, it is an essential step in the life cycle. All Unionidae and a few other bivalves have glochidial larvae parasitic on fish (Table 1).

The parasitic period begins with the implantation of the glochidium. Contact of the sensory hairs inside the widely gaping valves of the glochidium with host tissue causes the valves to close vise-like on the gill tissue (Arey, 1921). The host reacts by enclosing the glochidium within the gill tissues (Fig. 11). Within the host tissue the glochidium undergoes a metamorphosis to a more complex juvenile with most of the adult organs (Howard and Anson, 1922). Having completed its metamorphosis the young mussel escapes apparently by its own efforts but perhaps aided by the local reaction in the host tissue.

Fuller (1974) mentions a rather interesting case where some Unionidae have such powerful hooks on their valves that they tear the gill tissue and are consequently confined to more resilient sites like skin and fins.

10.4.7. Acarina

Rarely found on gills, mite larvae have been reported in gill tissue by Mueller (1936). Tedla and Fernando (1970) found adult *Hydrozetes*, normally found on vegetation, on the gills of *Perca flavescens*. Some tissue reaction was noted at the site of attachment. They concluded that it was probably a case of incipient parasitism.

10.4.8. Hirudinea

There are only two records of leeches on fish gills in North American freshwater fishes. Mather (1948) found *Cytobranchus verilli* covering the gills of *Ictalurus punctatus* in Iowa and Molnar et al. (1974) found *Actinobdella trinagulata* on the gills of *Catostomus commersoni* in Ontario. The rarity of leeches on gills is probably due to the abrasive action of the gill and opercular movements.

10.5 Ecological aspects

10.5.1. Introduction

The systematics of gill infecting parasites have been studied extensively in Europe and North America particularily in the last thirty years (Bychowsky et al., 1962; Hoffman, 1967). There is, however, a paucity of literature on the ecology of gill parasites. Noble (1960) pointed out that the latter study would contribute to the better understanding of basic biological problems. One could postulate that the relative scarcity of ecological studies is due to this discipline being time consuming, costly and more liable to errors of interpretation.

We shall discuss the spatial distribution on gills as this casts some light on the heterogeneous nature of the habitat and then deal very briefly with parasite interactions and host response.

10.5.2. Spatial distribution

Early workers noted that some parasites were site specific. Cerfontaine (1896, 1898) was the first to demonstrate this phenomenon in *Diclidophora denticulata*, a monogenean on the gills of *Gadus virens*. More recently Suydam (1971) has reviewed previous accounts of spatial distribution in various Monogenea. All these workers defined the specific areas of attachment by dividing each gill arbitrarily into several regions. The parasite's position was then indicated with respect to these regions. However, almost all these studies were limited to single species of parasites and the majority of parasites studied were polyopisthocotylid monogeneans.

The gills of most fish species are infected with more than one species of parasite. Hanek (1973) in his study of the microecology of parasites infecting the gills of *Lepomis gibbosus* considered all parasites. Three hundred (25 monthly) specimens of *L. gibbosus* were examined from November 1971 to October 1972 in West Lake, Ontario. The gill parasite spectrum in this host consisted of seven species of Monogenea, three species of Copepoda and the glochidia of *Lampsilis radiata*. The gill arches were numbered and divided into sections in a generally acceptable manner for the spatial distribution census. Gills were numbered I—IV from anterior to posterior. The two faces of the hemibranchs were designated anterior and posterior and each hemibranch was divided into three subequal sections (dorsal, medial and ventral), thus giving six subequal parts for each gill and a total of 48 sections (Fig. 1). Numbers of specimens of each species of parasite collected on each section

TABLE 2

The spatial distribution of parasites on the gills of *Lepomis gibbosus* in West Lake Ontario. I–IV Gills: A, Anterior; P, Posterior; D, Dorsal; M, Medial and V, Ventral sections.

Parasites	No. para- sites	Inci- dence (%)	Inten- sity	Ante- rior (%)	Poste- rior (%)	Spatial distribution (Hemibranchs)									
						AD (%)	AM (%)	AV (%)	PD (%)	PM (%)	PV (%)	I (%)	II (%)	III (%)	IV (%)
Urocleidus ferox	28029	76.9	93.4	69	31	14	42	14	6	19	5	15	37	30	18
Actinocleidus gibbosus	4000	10.9	13.3	78	22	8	59	11	3	14	5	6	42	35	17
Actinocleidus recurvatus	1914	5.3	6.4	76	24	12	54	10	5	15	4	8	46	33	13
Urocleidus acer	634	1.7	2.1	82	18	13	41	28	2	14	2	5	65	25	5
Ergasilus caeruleus	1292	3.5	3.3	52	48	25	4	23	16	6	26	28	23	22	27
Totals/Means Monogenea	34978	96	116.6	66	34	13	39	14	7	20	7	17	35	29	19
Totals/Means Copepoda	1391	3.8	4.6	51	49	23	6	22	15	8	26	28	23	23	26
Totals/Means Glochidia	63	0.2	0.2	52	48	11	30	11	12	24	12	12	45	32	11

were recorded. Table 2 summarises the results.

The data does not include the individual numbers of the monogeneans *Cleidodiscus tobustis*, *Urocleidius attenuates*, *U. dispar*, Copepods *Achtheres ambloplitis*, *Ergasilus centrachidarum* and glochidia of *Lampsilis radiata*, whose combined incidence constituted only 1.6% of the total. As shown in Table 2 the dominant parasitic species, *Urocleidius ferox* reached an incidence of 77% while all seven Monogenea accounted for 96% of the parasite load. The Copepoda 3.8% and glochidia 0.2% accounted for only 4% of the parasites recorded. Site preference was evident in the Monogenea which were more abundant on the anterior face and medial sections. Although few glochidia were recorded the same seems to hold true for them too. Copepoda on the other hand exhibited an equal affinity for both faces of the hemibranch and a strong affinity for their dorsal and ventral sections. A detailed analysis indicated that all parasites found exhibited a strong affinity for particular gill arches. When considered synecologically Monogenea and glochidia occurred more often on gills II and III and less often on I and IV. *Ergasilus caerulus* lacked apparent affinity for any particular gill or gills.

We have concluded from the study that the "affinity" of Monogenea (and probably glochidia) for particular gill arches is influenced primarily by the direction and force of the ventilating current, the position of the hemibranches during the various phase of respiratory exchange and the crowding effect within the gill arches and their sections. The absence of any apparent affinity for particular gill arches in *Ergasilus caerulus* may be due to its strong attachment mechanism, which can change its site. However, Tedla and Fernando (1970) found that *Ergasilus confusus* prefers the second and third gill arch in *Perca flavescens*.

Molnar (1971b) found that site selection in *Dactylogyrus lamellatus* on the grass carp *Ctenopharyngdon idella* is dependent on the age of the fish but not on the developmental stage of the parasite.

10.5.3. Interaction of gill parasites

Wilson (1916) showed a negative correlation between glochidia and the copecod *Ergasilus caerulus* on the gills of the write carppie. Parasite interactions on the gills have been also demonstrated by many workers, e.g. Dogiel (1962), Kabata (1969) and Noble (1963). Infection by one parasite may increase the susceptibility of the gills to infection by a second parasite. On the other hand parasites may interfere mechanically with one another directly or indirectly by their effects on the gill tissue (Molnar, 1971).

10.5.4. Host response

The response of fish to gill parasitism is poorly understood. Acquired immunity, acquired resistance, specific allergy and "protection by infection" have been claimed by various authors. Molnar (1971b) has reviewed this subject in some detail for Monogenea, Kabata (1970) for Crustacea and Bauer (1961) generally for fish parasites. In general the tissue responses are hypertrophy, hyperplasia or cell proliferation.

10.6. Summary and discussion

Gills provide an ideal site for parasitic infection. They offer a large exposed epithelial surface well supplied with blood and adequately oxygenated. The ventilating current passing over the gills carries the infective stages to the gill surface directly, but active movements in reaching the gills and later site changes do occur. However, gill parasites are exposed to the abrasive action of the moving gill arches and the operculum and protective tissue responses. The strong water current necessitates some substantial means of attachment to withstand this hydrodynamic force. The large and varied parasite fauna found on the gills shows that many parasites have overcome these difficulties and are able to tap this rich source of food material. Parasites with well developed means of attachment are certainly at an advantage. This is seen in the Monogenea which are the most numerous parasites found upon fish gills. Infective forms capable of penetrating the epithelium on the face of the ventilating current also infect gills. These include Protozoa and metacercariae. Copepoda grasp the gill filaments or fix themselves by a highly specialized sucker to the gills. Glochidia grasp the gill filaments and the host surrounds it with tissue. Sessile Protozoa attach by a specialized plate to the gill epithelium or attach temporarily by various organelles.

According to Hughes and Morgan (1973) the degree of infection of the gills is probably directly related to the ventilation volume and the pattern of current flow over the gills. Experimental work by Paling (1968), Tedla and Fernando (1970) and Hanek (1973) support this contention. However, if the parasite is mobile on reaching the gills, like Ergasilidae and some Monogenea, the original "settling" pattern may change. If the ventilating current has a positive "value" to the parasite for any reason the original spatial pattern of infection will be maintained.

The spatial distribution of gill parasites must be studied using a more sophisticated approach utilizing high intensities for statistical analysis, a knowledge of ventilatory hydrodynamics and an appreciation of interaction between parasites.

The ecology of gill parasites in relation to their habitat will be understood when more detailed information is available on the many biological aspects so poorly known at present. Gills are highly specialized organs which are very susceptible to physiological stress, mechanical injury and parasite infections. The gill habitat is also heterogeneous from the parasites' point of view in the spatial sense. Gills have some similarities with the skin and fins on the one hand and the alimentary canal on the other. There are more parasites common to the skin and gills than to the gills and alimentary canal. Ectoparasites do not usually invade the alimentary canal though recently Paperna (1963) and Gussev and Fernando (1973) recorded Monogenea in the stomach and intestine of fish. One of the monogeneans had been recorded previously on gills.

10.7. Acknowledgements

We wish to thank Drs. J. Lom, Prague; Dr. Kalman Molnar, Budapest and N. N. Banina and O. N. Bauer, Leningrad; for kindly sending us literature and commenting on various aspects of this chapter.

We are also indebted to the editorial boards of the following journals for permission to reproduce illustrations: Journal of Zoology, London; Folia Parasitologia; Protistologia; Journal of Parasitology; Journal of the Marine Biological Association; Transactions of the American Microscopical Society; Journal of the Fisheries Research Board of Canada; THF Publications.

10.8. REFERENCES

Arey, L.B. (1921). "An experimental study on glochidia and the factors underlying encystment." J. Expl. Zool. 33, 463—499.

Banina, N.N. (1969). "Changeability of ciliates on the genus *Apiosoma.*" Folia Parasitologia (Praha) 16, 289—295.

Bauer, O.N. (1961). "Relationships between host fishes and their parasites. In: Parasitology of fishes." (Dogiel V.A. et al., ed.). English translation, Oliver and Boyd, Edinburgh and London, pp. 84—103.

Bychowsky, B.E. (1957). "Monogenoidea, their taxonomy and phylogeny." English translation, American Institute for Biological Sciences, Washington, D.C. 1961, pp. 1—509.

Bychowsky, B.E. (1962). "Key to parasites of freshwater fish of USSR." English translation by Israel Programme for Scientific translations O.T.S. No. 7764—11040, pp. 1—919.

Cerfontaine, P. (1896), "Contribution a l'étude des Octocotylides" I—III. Arch. Biol. 14, 497—560.

Cerfontaine, P. (1898). "Contribution a l'étude des Octocotylides" IV. Arch. Biol. 15, 301—328.

Dogiel, V.A. (1962). "General Parasitology." English translation, Oliver and Boyd, Edinburgh and London, pp. 1—516.

Einszporn, T. (1965). "Nutrition of *Ergasilus sieboldi* Nordman, II. The uptake of food and the food material." Acta Parasit. Polon 13, 373—380.

Fuller, S.L.H. (1974). "Clams and mussels" In: "Pollution ecology of freshwater invertebrates." (Hart, C.W. and Fuller, S.L.H. Ed.). Academic Press, N.Y., 215—273.

Gussev, A.V. and Fernando, C.H. (1973). "Dactylogyridae (Monogenoidea) from the stomach of fishes." Folia Parasitologica (Praha) 20, 207—212.

Hanek, G. (1973). "Micro-ecology and spatial distribution of the gill parasites infesting *Lepomis gibbosus* (L.) and *Ambloplitis rupestris* (Raf.) in the Bay of Quinte, Ontario." Ph. D. thesis, University of Waterloo.

Hanek, G. and Fernando, C.H. (1972). "Monogenetic trematodes from the Bay of Quinte area, Ontario IV Genus *Dactylogyrus* Diesing, 1859 with provisional host-parasite and parasite-host lists." Can. J. Zool. 50, 1313—1317.

Hanek, G. (1975). "New and previously known *Dactylogyrus* spp. from Southern Ontario fishes." J. Parasit. 61, 431—426.

Hoffman, G.L. (1967). "Parasites of North American freshwater fishes." Univ. California Press, Berkeley and Los Angeles, pp. 1—486.

Hoffman, G. L. and Meyer, F. P. (1974). "Parasites of freshwater fishes." T. H F. Publications, New Jersey, pp. 1—224.

Howard, A. D. and Anson, B. J. (1922). "Phases in the parasitism of the Unionidae." J. Parasit. 9, 68—82.

Hughes, G.M. (1966). "The dimensions of fish gills in relation to their function." J. exp. Biol. 45, 177—175.

Hughes, G.M. and Grimstone, A.V. (1965). The fine structure of the secondary lamellae of the gills of *Gadus pollachius*. Quart. J. microc. Sci. 106, 343—353.

Hughes, G.M. and Morgan, M. (1973). "The structure of fish gills in relation to their respiratory function." Biol. Rev. 48, 419—475.

Hughes, G.M. and Skelton, G. (1958). "The mechanism of gill ventilation in three freshwater teleosts." J. exp. Biol. 35, 807—823.

Kabata, Z. (1969). "Revision of the genus *Salmincola*, Wilson 1915 (Copepoda; Lernaepodidae). J. Fish. Res. Bd. Canada 26, 2987—3-41.

Kabata, Z. (1970). "Diseases of fishes, Book 1. Crustacea as enemies of fishes." T.H.F. Publications, New Jersey, pp. 1—171.

Kataba, Z. and Cousens, B. (1972). "The structure of the attachment organs of Lernaeopodidae (Crustacea: Copepoda)." J. Fish. Res. Bd. Canada 29, 1015—1023.

Kearn, G.C. (1971). "The attachment of the ancyropephalid monogenean *Haliotrema balisticus* to the gills of the trigger fish, *Balistes capriscus (=carolinensis)*." J. Parasit. 63, 157—162.

Llewellyn, J. (1958). "The adhesive mechanisms of monogenetic trematodes: the attachment of species of the Diclicophoridae to the gills of gadoid fishes." J. mar. biol. Ass. U.K. 37, 67—79.

Lom, J. (1973). "The mode of attachment and relation of the host in *Apiosoma piscicola* Blanchard and *Epistylis lwoffi* Faure-Fremiet, ectocommensals of freshwater fish." Folia Parasitologica (Praha), pp. 105—112.

Lom, J. and Lawler, A. R. (1973). "An ultrastructural study on the mode of attachment in dinoflagellates invading gills of Cyprinodontidae." Protistologica 9, 293—309.

Mather, C.K. (1948). "The leeches of the Okoboji region." M.Sc. thesis, University of Iowa.

Martin, W.E. (1964). "Life cycle of *Pygidiopsoides spindalis* Martin, 1951 (Heterophydae, Trematoda)." Trans. Amer. microsc. Soc. 83, 270—272.

Mizelle, J.D. and McDougal, H.D. (1970). "*Dactylogyrus* in North America. Key to species, host-parasite and parasite-host lists, localities, emendations and description of *D. kritskyi* sp. n.". Amer. Midl. Nat. 84, 444—462.

Molnar, K. (1971a). "Protozoan diseases of the fry of herbivorous fishes." Acta Vet. Acad. Sci. Hung. 21, 1—14.

Molnar, K. (1971b). "Studies on gill parasites of the grass carp *(Ctenopharyngdon idella)* caused by *Dactylogyrus lamallatus* Achmerov, 1952." Acta Vet. Acad. Sci. Hung. 21, 361—375.

Molnar, K., Hanek, G. and Fernando, C.H. (1974). "Parasites of fishes from Laurel Creek, Ontario." J. Fish. Biol. 6, 717—728.

226

Mueller, J. F. (1936). "Notes on some parasitic copepods and a mite, chiefly from Florida fresh water fishes." Amer. Midl. Nat. 17, 807—815.

Munshi, J.S.D. and Singh. B.N. (1968). "On the microcirculatory system of the gills of certain freshwater teleostean fishes." J. Zool. Lond. 154, 365—376.

Noble, E.R. (1960). "Fishes and their parasite-mix as objects of ecological studies." Ecology 41, 593—596.

Paling, J.E. (1968). "A method of estimating the relative volumes of water flowing over the different gills of freshwater fish." J. exp. Biol. 48, 533—544.

Paperna, I. (1963). *"Enterogyrus cichlidarum* n. gen. n. sp. a monogenetic trematode parasite in the intestine of a fish." Bull. Res. Counc. Israel (B), Zoology 11B, 183—187.

Pennak. R. W. (1953). "Fresh-water invertebrates of the United States." Ronald Press, New York, pp. 1—769.

Richenbach-Klinke, H. and Elkan, E. (1965). "The principal diseases of lower vertebrates." Academic Press, New York, pp. 1—600.

Roberts, L.S. (1970). *"Ergasilus* (Copepoda: Cyclopoida): Revision and key to species in North America." Trans. Amer. microsc. Soc. 89, 134—161.

Scott, W.W. and O'Bier, A.A. (1962). "Aquatic fungi associated with diseased fish and eggs." Prog. Fish Cult. 24, 3—15.

Skelton, G. (1970). "The regulation of breathing" In: "Fish Physiology", Vol. 4. (Hoar W.S. and Randall, D.J. Eds.), Academic Press, New York, pp. 293—359.

Suydam, E.L. (1971). "The micro-ecology of three species of Monogenetic trematodes of fishes from the Beaufort-Cape Hatteras area." Proc. helminth. Soc. Wash. 38, 240—246.

Tedla, S. and Fernando, C. H. (1969). "Observations on the glochidia of *Lampsilis radiata* (Gmelin) infesting yellow perch, *Perca flavescens* (Mitchell) in the Bay of Quinte, Lake Ontario." Can. J. Zool. 47, 705—712.

Tedla, S. and Fernando, C.H. (1970). "Some aspects of the ecology of the parasite fauna of the gills of yellow perch, *Perca flavescens.*" J. Fish. Res. Bd. Canada 27, 1045—1050.

Wilson, C.G. (1916). "Copepod parasites of freshwater fishes and their economic relation to mussel glochidia." Bull. U.S. Bureau Fish. 34, 331—374.

Woskoboinikoff, M.M. and Balabai, P.O. (1937). (Comparative experimental investigation on the respiratory apparatus of the bony fish II.). (Ukranian). Proc. Acad. Sci. Ukr. SSR 16, 77—127.

Ecological Aspects of Parasitology
Editor: C.R. Kennedy
© *North-Holland Publishing Company, Amsterdam, 1976*

CHAPTER 11

LUGS

J.H. ROSE

M.A.F.F. Central Veterinary Laboratory, Weybridge (Great Britain)

Contents

11.1. Introduction

The lungs and other parts of the respiratory tract of a wide range of animals have been exploited successfully as a habitat by both nematodes and digenetic trematodes. Amongst the nematodes, genera of the families Rhabdiasidae and Reticulariidae inhabit the lungs of snakes and lizards, respectively. The air sacs of birds are parasitised by genera of the Thelaziidae whilst the Syngamidae include genera parasitic in the respiratory systems of both birds and mammals. Genera of the Pseudaliidae occur in the lungs of both aquatic and land mammals whilst the Protostrongylidae, Metastrongylidae, Crenosomatidae, Filaroididae and Dictyocaulidae include many genera parasitic in the lungs of land mammals.

Digenetic trematodes also occur in a similar range of hosts. Genera of the family Plagiorchiidae inhabit the lungs of amphibia and reptiles. The latter are also parasitised by genera of the Heronimidae. The air sacs and trachea of birds provide a habitat for trematodes of the Cyclocoelidae. The genus Orchipedum, family Orchipedae, also inhabits the trachea of birds, while the Paragonimidae includes genera parasitic in the lungs of land mammals.

Other helminths have not exploited the lungs as a habitat to the same extent. Of the cestodes, hydatid cysts of *Echinococcus granulosus* may occur in the lungs but they also occur in the liver and other parts of the body of the mammalian intermediate host, and are clearly able to survive in a range of different habitats within the host with no particular pre-dilection for the lungs.

Protozoa are confined to the blood and lymphatic systems in the lungs. The problems of these habitats are considered in subsequent chapters. Detailed observations on the relationship between the host and parasite have been made on only a few of these lung parasites, mainly those infecting domestic animals or man himself, so that this chapter will be concerned essentially with work carried out on these parasites and their hosts.

11.2. Migration to the lungs

The first problem besetting parasites which live in the lungs and other parts of the respiratory tract is to migrate to their habitat. The routes of migration of several nematode genera have been studied in some detail and they are all similar. The infective larvae whether free-living, for example *Dictyocaulus filaria* and *Dictyocaulus viviparus*, or contained within an intermediate host, for example *Muellerius capillaris* and *Metastrongylus elongatus*, are ingested by the host while feeding. The larvae then pass down the alimentary canal to the small intestine. They pass via the wall of the intestine into the lymphatic system and to the heart, where they enter the circulatory system and are carried to the lungs. They rupture the capillaries to enter the lung parenchyma and alveoli and finally migrate to that part of the respiratory system which provides their habitat. Of the trematodes infecting the lungs, the most detailed studies have been made on species of *Paragonimus*. *P. westermani* metacercariae excyst in the small intestine and penetrate the intestinal wall to enter the peritoneal cavity. They then pass via the diaphragm into the pleural cavity and penetrate through the surface of the lungs. In the cat, the young trematodes after migrating into the abdominal cavity from the intestine then enter the wall of the abdominal cavity and, after a period of development, reappear in the abdominal cavity to continue their migration to the lungs (Yokogawa et al., 1962). According to Okura (1963) who studied the migratory route of *P. ohirai* in rats, the juvenile trematodes, after penetrating into the abdominal cavity, penetrate into the liver, where they may remain for 3 to 4 days, then continue on their migration to the lungs via the diaphragm and pleural cavity. *Paragoni-*

mus miyazakii larvae can follow one of two migratory routes. After reaching the abdominal cavity they may either migrate into the liver or into the abdominal wall, where they stay for several days before continuing their migration to the lungs (Yokogawa et al., 1962).

In order to complete these migrations the young parasites must be adapted to survive in a range of different environmental conditions which they encounter during their migrations within the body of the host.

Some of the parasites ingested by the host may fail to reach the respiratory system. Both Nickel (1962) and Michel (1968) found it difficult to infect young lambs with *D. filaria*. The latter worker infected both lambs and yearlings, some of which were autopsied 14 days after infection. The numbers of worms recovered represented only a small proportion of the infective larvae administered to the lambs. This would suggest that many larvae had failed to reach the lungs. Animals autopsied 30, 31 and 32 days after infection contained even less worms in their lungs indicating that worms had been lost from the lungs in the intervening period.

A number of workers have carried out experiments to ascertain the numbers of *P. westermani* which succeed in reaching the lungs. Yogore (1956) infected cats with 25 metacercariae each and on autopsy of 4 of them between 40 and 88% of the parasites were recovered from the lungs. Yokogawa (1965) described a series of experiments by Japanese workers using cats and dogs. In one group of dogs, each of which was given 10 to 20 metacercariae, an average of 68.1% of the metacercariae succeeded in reaching the lungs. Similar results were obtained in other experiments, indicating a high degree of success by the parasites in reaching the lungs. *P. westermani* has been found in other parts of the host's body e.g. the brain, liver, pericardium and stomach wall. Mature parasites have been found in some of these locations, but any eggs which they produce are unlikely to reach the external environment (Yokogawa, 1965). These parasites can be regarded as unsuccessful, as they fail to find the most favourable habitat within the host from which the eggs can pass out and continue the life-cycle.

11.3. Location in the respiratory system

Parasites of the respiratory system tend to occupy a particular niche within the system. Three niches can be recognised: the lung parenchyma and alveoli, the bronchioles and bronchi, and the trachea. The nematodes *Muellerius capillaris* and *Cystocaulus ocreatus* and the trematode *P. westermani* are found in the lung parenchyma and alveoli, although the host's tissues around them become modified to a considerable extent due to the presence of the parasite. *Metastrongylus elongatus*, *D. filaria* and *D. viviparus* are examples of lung-worms found in the bronchioles and bronchi of the host, while *Syngamus trachea* is parasitic in the trachea of numerous bird species including chickens, goslings and pheasants.

Some lung-worms inhabit the pulmonary artery, e.g. *Aelurostrongylus abstrusus* and *Angiostrongylus vasorum* which infect the cat and dog, respectively. The eggs and first stage larvae are found in the lung tissue, but Mackenzie (1960) also found young adults of *A. abstrusus* in squeeze preparations of lung tissue from a kitten. They appeared coiled in the lung parenchyma and were not in a bronchus or pulmonary artery. The problems associated with the pulmonary artery as a habitat will not be considered in this chapter.

Each of these habitats provide somewhat different problems for the parasites located therein.

11.4. Retention within the respiratory system

The bronchial tree with its lining of ciliated epithelium and mucus membrane is designed for the removal of foreign particles away from the lungs so that parasites in the bronchi and trachea must resist the flow of mucus if they are to remain within the host. The digenetic trematodes have a means of attachment to the host with their suckers, and nematodes of the family *Syngamidae* have a well developed buccal capsule with which they can attach themselves to the wall of the trachea. In some hosts, for example in the turkey and pheasant, the male worm of *S. trachea*, which is attached to the female, becomes embedded in the tissues of the host. Many of the other nematode species which inhabit the bronchi have no special means of attachment to the host, although the simple pointed head may penetrate the mucosa and give a certain amount of anchorage. Their own movements may assist in retaining them in the host or they may be so numerous as to block the small bronchi and render their expulsion more difficult. Many of them may be swept out of the bronchi and up the trachea, however, before reaching maturity. This was demonstrated by Michel (1969a) who infected lambs with *D. filaria* and collected worms expelled in the mucus by inserting a tracheotomy tube into the trachea with a rubber balloon attached to the external limb of the tube. Many worms were carried out of the lungs during the period between the 12th and 30th day after infection, so that expulsion started shortly after the developing worms reached the bronchial tree. The numbers of worms expelled from lambs pre-treated with cortisone, a known immuno-suppressant, were similar to those recovered from lambs not so treated, indicating that an immune response by the host was not responsible for the expulsion of the worms. They were expelled by the flow of mucus (Michel, 1969b).

Mackenzie (1959) working with experimental infections in pigs noted that the majority of *Metastrongylus elongatus* were expelled during the first few weeks following infection.

Detailed observations on the effect of the mucus flow in eliminating species other than *D. filaria* do not appear to have been made, but it seems likely that lung-worms which inhabit the bronchi and trachea and lack a means of attachment to the host will be affected similarly.

Parasites inhabiting the lung parenchyma and alveoli do not have to contend with the flow of mucus. Furthermore there is frequently a host reaction around the parasite forming a nodule or cyst which assists in retaining the parasite within the lungs. A similar type of reaction occurs around *Filaroides osleri* which is found in nodules in the walls of the trachea of the dog.

11.5. Metabolism

There appears to be little information available regarding the metabolism of parasites inhabiting the lung. The whole of the respiratory tract is rich in oxygen: for example alveolar air in man contains 14.2% O_2. The composition of the air in the bronchial tree varies during respiration but the oxygen level of both expired and inspired air is somewhat higher than that of the alveolar air, so that for many of the parasites within the respiratory system aerobic metabolism should be possible.

M. elongatus is said to be aerobic (Von Brand, 1972). It has a complete cytochrome system. Sato and Ozawa (1969) demonstrated the presence of ubiquinone, which is closely correlated with aerobic respiration, in the mitochondria of *M. elongatus*. A functioning citrate cycle has been identified in *D. viviparus*, but it does not appear to play a dominant role in the metabolism of the parasite as large amounts of succinate, acetate and proprionate have been identified in the worm (Vaatstra, 1969). This suggests that anaerobic metabolism can take place, although this nematode lives in the bronchi in the presence of a rich oxygen supply.

Parasites which are enclosed within a nodule or cyst within the lung tissues will not have such ready access to the rich oxygen supply within the lungs, and their metabolism may well be anaerobic. There is some evidence that *P. westermani*, which is enclosed within a cyst in the host's lungs, respires anaerobically, as α-ketoglutarate, citrate, malate, fumarate and succinate have been identified in it by Hamajima (1967) and it is known to develop a heavy oxygen debt.

11.6. Food supply

The feeding habits of lung parasites have not been investigated extensively, but the obvious source of food is the tissues of the host.

Syngamus trachea and *Cyathostoma bronchialis*, which inhabit the trachea of birds, have a well developed buccal cavity by means of which they can erode the cells of the mucosa. They also ingest the host's blood and the loss of blood may be so great as to cause anaemia.

The majority of the nematodes parasitic in the lungs lack a large buccal capsule, so probably feed by penetration and histolysis of the tissues or by puncturing the tissues. The migration of the young parasites from the circulatory system into the lung tissue ruptures capillaries and lung tissues, liberating blood and cells which can be utilised as food.

Those nematodes and trematodes which become enclosed within nodules or cysts formed by the host's tissues probably use the cells of these structures as part of their food supply. Oxyhaemoglobin has been found in *P. westermani* (Von Brand, 1972).

11.7. Reproduction

The problems facing lung parasites with regard to reproduction are those common to most helminth parasites within the host's body; to ensure that fertilization of the eggs takes place and that these reach the outside environment so that the life cycle can be continued. The difficulties involved in achieving these ends vary according to the part of the respiratory system occupied by the parasite.

The trematodes are hermaphrodite, so that self-fertilization is possible. On the basis of finding but a single specimen of *P. ohirai* containing eggs in a rat Yokogawa (1965) suggested that this species can be self-fertilizing, but the trematode was found in the pleural cavity, and not in the lungs. He considered, however, that within the lungs cross-fertilisation takes place, and this is also true for *P. westermani*. These trematodes are found in cysts which would seem to provide a barrier to two flukes coming together. However, experiments carried out by Omura (1960) would seem to indicate that a *cyst* is only formed in the lungs when two flukes are present within it. He transferred specimens of both *P. westermani* and *P. ohirai* from the lungs of one rat to the peritoneal cavity of another rat. When only one fluke was transferred, and subsequently migrated to the lungs, a cyst was never formed there, but when more than two adult flukes were transferred a cyst was formed in the lungs. In human infections, however, according to Yokogawa (1965) a worm cyst contains only one worm, although he suggests that as most examinations, whether at autopsy or by surgical excision, are made an appreciable time after infection, one worm may have migrated from the cyst. Yokogawa et al. (1960) suggested that copulation may occur while the flukes are migrating to the lungs.

Amongst nematodes, the problem of the two sexes meeting and copulating is particularly acute for those species which inhabit the lung tissues and become enclosed within a cyst or nodule. It is probable that many worms fail to copulate. Observations by Rose (1959) on lambs experimentally infected with *Muellerius capillaris* support this view. He found that many nodules contained but a single worm; the smallest nodules less than 1 mm in diameter contained immature fourth stage larvae and as the nodules increased in size so did the size and degree of development of the worm. In some nodules, however, more than one worm was present: indeed the largest nodules contained numerous worms and first stage larvae. There was no evidence that adult worms migrated through the lungs and it appeared that for copulation to occur worms of opposite sex needed to settle in the same part of the lung tissue at an early stage in their development; after breaking out of the capillaries and before the reaction of the host's tissues enclosed them within a nodule. In some lambs all of the nodules contained a single worm and no first stage larvae were ever recovered from the faeces of these animals. This would suggest that copulation had not taken place and that none of the worms had succeeded in completing their life cycle.

Parasites within the lung tissues produce eggs for a long time. Rose (1959) recovered first stage larvae of *M. capillaris* over a period of two years from experimentally infected lambs which were kept indoors under conditions which precluded any subsequent infection. Kassai (1962) recovered larvae of *Protostrongylus rufescens*, *M. capillaris*, *Cystocaulus ocreatus* and *Neostrongylus linearis* for periods of 28, 44, 65 and 67 months respectively, from the faeces of infected sheep. Specimens of *P. westermani* containing eggs were recovered from a dog which had been experimentally infected 6 years and 8 months previously (Yokogawa and Yoshimura, 1960), and Sugaro (1959) observed eggs in the faeces of dogs experimentally infected with 12 to 25 metacercariae of *P. westermani* for up to 300 days after the start of egg laying.

Parasites inhabiting the bronchial tree do not encounter the same barriers to copulation, as they are not isolated by the host's tissue reaction. A number of worms are frequently found together in the same bronchiole or bronchus, so that the chances of successful copulation should be reasonably good providing the worms remain within the lungs sufficiently long to become mature, but as noted earlier, worms can be eliminated from the lungs in the mucus before they have completed their development.

Syngamus trachea has overcome the problem of copulation by having the male worm permanently attached to the female. In pheasants the infective larvae develop into adults in the parabronchi of the lungs about 4 days after infection; they copulate on the following day; and by the seventh day start to migrate to the trachea. Male worms are firmly attached to the tracheal wall 11 days after infection (Fernando et al., 1971).

The expulsion of fertilized eggs or first stage larvae from the lungs is facilitated by the flow of mucus up the bronchial tree to the mouth, where they are swallowed and pass via the alimentary canal to the external environment in the faeces. In the case of *P. westermani* infections in humans, expulsion in the sputum is considered to be another pathway to the exterior.

Where the parasite is enclosed within a nodule or cyst, this must be broken down in some way to facilitate the transference of the eggs and first stage larvae into the bronchial tree and the flow of mucus. The precise way in which this is brought about is not clear. The eggs of nematodes enclosed within the tissues e.g. *M. capillaris*, hatch in situ, and the active movements of the first stage larvae combined with enzyme activity probably facilitate the breakdown of the nodules and liberation of the larvae into the bronchioles. According to Yokogawa (1919) the cyst around *P. westermani* has a small opening into the respiratory passage through which eggs can pass. Yokogawa (1965) refers to cases in which cysts containing only eggs have been found in humans infected with *P. westermani*. He suggests that the adult flukes may migrate from the cyst; this would be expected to rupture the cyst wall and so facilitate the release of the eggs.

11.8. Pathological changes in the tissues of the host

The pathological changes in the tissues of the host resulting from infection with lung parasites have been studied in most detail in domestic animals. They may adversely affect the parasites already present in the respiratory system or they may render part of it untenable as a habitat for parasites subsequently infecting the host.

Detailed observations have been made by a number of workers on the pathological changes in cattle infected with *D. viviparus*, in sheep infected with *D. filaria*, and in pigs infected with *M. elongatus*. All of these lungworms live in the bronchial tree of the host and the pathological changes which they invoke are broadly similar. These changes will not be considered in detail here, but, based on the work of Jarrett et al., (1960) with *Dictyocaulus viviparus*, reference will be made to the ways in which the parasite is affected by some of them.

The first pathological changes occur when the infective larvae migrate from the capillaries into the alveoli during the first few days after infection causing petechial haemorrhages which are soon resolved and do not produce any clinical effects. The developing worms then migrate into the bronchioles and bronchi and become surrounded by eosinophils. The subsequent blockage results in the collapse of the distal alveoli. If the infection is heavy, a marked increase in respiratory rate occurs at about days 10 to 14 after infection and coughing may become pronounced. This, together with the increased flow of mucus which may occur, can dislodge the lungworms and facilitate their expulsion from the host. A check to the disease may occur at 21 days or secondary complications may ensue. The alveoli become filled with fluid, resulting in pulmonary oedema, emphysema and secondary bacterial infections. Eggs and first stage larvae may be aspirated into the lungs and become surrounded by multi nucleate giant cells which try to engulf them, causing consolidation of lung tissue and giving rise to primary parasitic pneumonia. These eggs and larvae will, therefore, be unable to pass out of the host and so complete their life-cycle. The clinical symptoms become more pronounced, the respiratory rate increases and coughing becomes frequent so that further expulsion of the lungworms is likely to occur. The final effect may be the death of the host. This must be regarded as disadvantageous to the parasite as its life cycle will also be terminated.

Natural infections of sheep with *D. filaria* and pigs with *M. elongatus* do not generally cause such marked clinical effects, possibly because of the rate of worm loss already referred to, but the effects of these on the parasite are similar to those described above.

S. trachea causes no clinical symptoms in the turkey, but in chicks, goslings and pheasants it produces an accumulation of mucus in the trachea; dyspnoea and asphyxia occur in spasms and death may occur from asphyxia, emaciation and anaemia, thereby terminating the life of the gapeworm.

The pathological changes in the lungs caused by the migrating worms and in the trachea resulting from the attachment of the worms do not have any obvious ill effects on the parasite.

Lungworms which inhabit the parenchyma of the lungs evoke pathological changes which result in the formation of nodules and cysts around the parasites. Fairly detailed observations have been made on these changes in sheep due to infection with *M. capillaris* (Beresford-Jones, 1967). The migration of first stage larvae into the alveoli from the capillaries causes haemorrhages; the young worms become surrounded by lymphocytes and macrophages, forming a small nodule. The nodules around older worms consist of an inner zone of eosinophils and an outer zone of macrophages, eosinophils and lymphocytes. Dead and disintegrating worms may be present in some of the nodules, which may also be calcified. The nodules range in size from less than 1 mm to several cm in diameter. These latter may be confluent with one another and occupy considerable areas of lung tissue in older sheep. Some of them contain living adult worms together with eggs and first stage larvae (Rose, 1958; Beresford-Jones, 1967).

As already noted one of the effects of encapsulation is to isolate some individual worms and prevent copulation. The nodules themselves are unlikely to provide a suitable environment for subsequent infective larvae. The pathological changes do not generally give rise to any clinical disease. The host parasite relationship appears to be fairly well balanced. Infection of British Sheep with *M. capillaris* is widespread: Rose (1955) found 94.5% of healthy sheep sent to an abattoir in southern England infected with this lungworm.

The pathological changes resulting from *P. westermani* infection in man have been described in detail by Yokogawa (1965), who commented on the wide variations in the histopathological findings reported by different workers. These he considered were probably due to differences in the numbers of flukes in the lungs, the duration of the infection and the susceptibility of the individuals to the parasite. A detailed consideration of these findings is not appropriate here, apart from those which appear to affect the parasite. The parasites become enclosed within a cyst wall and death may take place within the cyst. Eggs within the lung tissue become enclosed within tubercles consisting of epitheliod cells, lymphoid cells, plasma cells, eosinophils and peripheral fibroplasts, and so are unable to pass out of the host and complete their life cycle.

The pathological changes resulting from infection of domestic animals and man are unfavourable to a greater or lesser extent to the parasite, either by their direct effects on it or by making part of the lung unsuitable as a habitat for parasites subsequently invading the host. It is not known whether this is true for the many parasites which infect wild animals, as detailed observations on the pathology in these host/parasite relations have not been made. It is possible that these relationships may be better balanced and the effects of the parasite on the host less harmful.

11.9. Immunity

The immune response of the host to infection with lung parasites has been investigated in only a few host parasite combinations, and then in varying degrees of detail. The most extensive investigations have again been made on lungworm infections in domestic animals. A number of examples have been selected to illustrate the way in which the immune response varies in its intensity in different-host parasite relations with regard to the effects on the parasite.

The immunity following infection of calves with *D. viviparus* has been studied by numerous workers and it is well established that an acquired resistance results from an initial infection. Infective larvae which are ingested by the immune host may fail to reach the lungs, or if they do so their development may be inhibited at the early fifth stage, or they may be eliminated or destroyed before they reach maturity (Michel, 1956). This acquired resistance does, however, decrease in intensity with time. Michel et al. (1965) infected each of a number of calves with 11,000 infective larvae on two occasions with an interval of 4 weeks between the two infections. Subsequently groups of calves were challenged after 3, 6, 12, 18 and 27 months. After each challenge calves from the group were autopsied 9, 10, 11, 29, 30 and 31 days after infection and examined for lungworms. By comparing the numbers of worms recovered from these and from control animals, it was evident that there was a strong resistance to the establishment of the challenge infection three months after immunisation, but 6 months after initial infection the resistance had decreased markedly and after 12 months it had decreased to such an extent that infective larvae of the challenge infection became established. The growth of these worms during the first 10 days was retarded, however, and there was a great reduction in the numbers of worms recovered between the 11th and 29th day after challenge. A large proportion of the worms which persisted after the 29th day failed to develop beyond the early 5th stage. Those worms which succeeded in developing beyond this stage did not grow as fast as worms in the control calves, and very few lung-worm larvae were recovered from their faeces.

Under natural conditions, calves which have been infected are not likely to be isolated from further infection unless they develop clinical disease, and a comparable loss of immunity is unlikely to occur. Calves at pasture which develop immunity following infection with *D. viviparus* can then be regarded as unfavourable hosts, in which infective larvae which are ingested subsequently are unlikely to be able to complete their life cycle success-fully. Michel (1956) considered that the age of the host did not have any marked effect on its susceptibility to infection.

Reference has already been made to the loss of *D. filaria* from lambs which is not related to an immune response by the host. An acquired im-munity does develop, however, following an initial infection. Kauzal (1934) was unable to re-infect lambs following an initial infection and similar re-sults were obtained by Michel (1956). He examined lambs, 7, 9 and 14 days after re-infection and found that only small numbers of larvae reached the lungs. In subsequent experiments it was shown that larvae reached the mes-enteric lymph glands (Michel and Sinclair, 1963) but were apparently unable to migrate further. By injecting larvae intravenously it was possible to bypass the mesenteric lymph glands so that larvae were able to reach the lungs, but the numbers of worms recovered from the lungs of resistant lambs declined sharply compared to the numbers of worms recovered from susceptible lambs. Fourth and fifth stage worms which reached the lungs were quickly destroyed in resistant animals. All of the developmental stages of *D. filaria* appeared capable of stimulating an immune response.

The acquired immunity resulting from infection persists for a considerable period of time. Wilson (1970) studied the duration of immunity to *D. filaria* in sheep and goats which had been infected experimentally with between 517 and 8000 infective larvae, and also in 4 naturally infected ewes. These animals were challenged at intervals from 1.5 to 46 months after infection. They were highly resistant to the establishment of infection for up to 46 months. Some worms reached the mesenteric lymph glands where they re-mained for 15 days, and it was suggested that they failed to migrate further. In some animals worms reached the lungs, but their growth was inhibited and they were eliminated before reaching sexual maturity. Generally, there-fore, sheep and goats which have developed immunity following an initial infection with *D. filaria* can be regarded as unfavourable hosts in which the life cycle will not be completed by infective larvae subsequently ingested. It is possible, however, for older sheep to develop patent infections. This was noted by Rose (1965) who observed light patent infections in ewes previously infected naturally at pasture, although it was not known whether re-infection had taken place or whether inhibited larvae present in the lungs had completed their development.

The immune responses of sheep to other species of lungworms has not been investigated in detail. Beresford-Jones (1967), after studying the pathological changes in lambs following infection with *M. capillaris*, suggested that an immune response was provoked which checked the development of the larvae in the fourth stage in the mesenteric lymph glands and lungs and inhibited the reproductive activity of adult worms. Under natural conditions, however, adult sheep frequently have patent infections of *M. capillaris*. Rose (1965) observed that adult sheep at pasture voided first stage larvae either continuously or intermittently for several years, and at autopsy after being in the pasture for five years, lungworms in all stages of development were seen in the lungs. From these observations it would seem that if an immune response is invoked it is not very strong, and that re-infection with *M. capillaris* is possible throughout the life of the sheep.

C. ocreatus, which occupies a similar niche within the lungs to *M. capillaris*, has been shown experimentally to invoke a strong immunity. Davtian and Panosian (1946) gave sheep an initial infection of 250—500 *C. ocreatus* larvae, and on challenge the rate of development of worms of the challenge infection was slowed down and the period during which the female worm produced eggs was reduced. With a larger initial infection of 3000—5000 larvae a stronger immunity developed which retarded the migration of larvae to the lungs, and some larvae became encysted in the wall of the intestine. It is difficult to envisage a situation under natural conditions when a sheep would ingest such large numbers of larvae, which are contained within the foot of the molluscan intermediate host within a short period of time. It is much more likely that small numbers of larvae will be ingested over a period of time, invoking only a light immunity so that sheep are likely to be susceptible to infection throughout their life.

The extent to which acquired resistance is evoked by pigs infected with *M. elongatus* has been investigated by a number of workers, but with somewhat differing results. Schwartz and Lucker (1935) infected young pigs with between 500 and 3500 larvae, then challenged them with 500 larvae. At autopsy the pigs which received the smallest initial infections contained many adult worms, suggesting that an immunity had not developed, whereas those animals which received the heaviest initial infection contained mainly juvenile worms without eggs, which was taken to indicate that an immunity had developed sufficient to limit the development of the worms. Jaggers (1965), however, found little evidence of any inhibition of development and growth of the lungworm resulting from an immune reaction, but both he and

Jaggers and Herbert (1968) reported that the immunity which developed resulted in the elimination of the existing lungworms. According to Dixon (1968), light infections can provoke immunity sufficient to limit the number of worms which become established in subsequent infection. He infected young pigs with 540 larvae of *M. elongatus* given as seven increasing doses over a period of 40 days, and subsequently challenged them with 20,000 larvae. Fewer lungworms were present in these pigs at autopsy than were present in controls.

A number of workers maintain that pigs exhibit age resistance. Hollo (1966) found that 25—54% of larvae developed in 2 months old pigs whereas in pigs aged 4.5 months only 10% developed. Polzenhagen et al. (1969) also noted a decrease in the level of infection with increase in age, and that the regular expulsion of worm eggs ceased when pigs reached the age of 8—9 months, irrespective of the age at which they had been infected. From the results of these various observations it would appear that the domestic pig only serves as a favourable host for *Metastrongylus* spp for a limited period.

Age resistance to infection with *S. trachea* has been reported by Euzéby et al. (1973). They infected four groups each of 15 White Leghorns aged 11 days, 5 weeks, 7 weeks and 12 weeks, with 200 *S. trachea* eggs. On the basis of faecal examination the 11 day old birds were completely susceptible to infection; there was 40% susceptibility at 5 weeks and only 7% susceptibility at 7 and 12 weeks of age. At autopsy, 30 days after infection, adult worms were present in 74% of the 11 days old birds, only two worms were found in one of the 12 weeks old birds and no worms were present in any of the others.

Acquired resistance following an initial experimental infection with *S. trachea* was demonstrated by Shikhobalova (1956). He found that following an initial experimental infection of chicks with 200—500 eggs, worms developing from a subsequent challenge infection were smaller in size and less numerous than those of the initial infection, and some worms were expelled from the chicks while still immature. He also reported that pullets were more readily infected than were cockerels. From these observations it would appear that only young domestic fowls are a suitable host for the development of *S. trachea*. It does not follow, however, that the results of these observations would be applicable to the other bird species, both domestic and wild, which can serve as hosts for *S. trachea*.

Information on immunity to lung trematodes is limited. Yokogawa (1965) described work on the immune response of dogs following infection with *P. westermani*. One dog was infected with 25 metacercariae; three months later it was challenged with 100 metacercariae. At autopsy 299 days later 80 adult worms were recovered, indicating that development of some of the worms of the second infection had taken place. A similar experiment on a second dog gave almost identical results. A third dog was given 40, 110 and 50 metacercariae respectively at intervals of 3 months. At autopsy 225 days later 167 adult worms were present, so that 83.5% of the total number of metacercariae had completed their development. From these results it was concluded that protective immunity had not developed and that the dog can provide a suitable host for the development of *P. westermani* irrespective of any previous infection.

In a review of the small amount of earlier work concerned with immunity Yokogawa et al. (1960) concluded from this work and from the observation that the developing stages and the adult flukes are in close proximity with the host's tissues that some acquired immunity probably developed in animals infected with lung flukes. Because of the complex life-cycle of *Paragonimus* species it is probable that the final host will only ingest small numbers of metacercariae at a time. These may well be too small to evoke an immune response by the final hosts sufficient to prevent reinfection, so they may well be susceptible to *Paragonimus* species throughout their lives.

Antibodies have been demonstrated in the host following infection with both lungworms and lungflukes but their effects, if any, on the parasite are not known.

11.10. REFERENCES

Beresford-Jones, W.P. (1967). "Observations on *Muellerius capillaris* (Muller, 1889) Cameron 1927. III Experimental infection of sheep". Res. Vet. Sci. 8, 272—279.

Davtian, E.A. and Panosian, M.A. (1946). "Immunity in sheep against hyperinfestation and reinfestation with *Cystocaulus ocreatus*". Gelm. Sb. Let Skrjabin 40, 96—103.

Dixon, J.B. (1968). "Immunity to *Metastrongylus apri*". Vet. Rec. 82, 417—418.

Euzéby, J., Gauthey, M. and Bordrez, C. (1973). "Studies on Syngamus infection. 2. Age resistance in fowl". Bull. Soc. Sci. vét. Méd. comp. Lyon. 75, 235—237.

Fernando, M.A., Stockdale, P.H.G. and Remmier, O. (1971). "The route of migration, development and pathogenesis of *Syngamus trachea* (Montagu 1811) Chapin, 1925, in pheasants". J. Parasit. 57, 107—116.

Hamajima, F. (1967). "Studies on the metabolism of the lung fluke genus *Paragonimus*. III Occurrence of organic acids in uterine eggs, larvae and adults". Jap. J. Parasitol. 16, 1—7.

Hollo, F. (1966). "Investigation on metastrongylosis in swine. III. The pathogenicity of lungworms in artificially infected animals". Acta. vet. hung. 16, 413—428.

Jaggers, S.E. (1965). "Studies on the resistance and immunity of swine to the lungworms *Metastrongylus* spp". Ph.D. thesis. University of Wales.

Jaggers, S.E. and Herbert, I.V. (1968). "Studies on the resistance of pigs to the lungworm *Metastrongylus* spp. Infections in minimal disease pigs from eight weeks of age". J. comp. Path. 78, 161—172.

Jarrett, W.F.H., Jennings, F.W., McIntyre, W.I.M., Mulligan, W., Sharp, N.C.C. and Urquhart, G.M. (1960). "Symposium on Husk. 1. The Disease Process". Vet. Rec. 72, 1066—1072.

Kassai, T. (1962). "Reproductive life span of *Cystocaulus*, *Muellerius* and *Neostrongylus* species in the lungs of sheep". Magy. Allator. Lap. 13, 9—13.

Kauzal, G. (1934). "Observations on the development of resistance to *Dictyocaulus filaria*". Aust. vet. J. 10, 110—111.

Mackenzie, A. (1959). "Studies on lungworm infection of pigs. III. The progressive pathology of experimental infections". Vet. Rec. 71, 209—214.

Mackenzie, A. (1960). "Pathological changes in lungworm infestation in two cats with special reference to changes in pulmonary arterial branches". Res. vet. Sci. 1, 255—259.

Michel, J.F. (1956). "Studies on host resistance to *Dictyocaulus* infection. II. Reinfection experiments with *D. filaria* in sheep". J. comp. Path. Therap. 66, 241—248.

Michel, J.F. (1968). " Experiments on the loss of *Dictyocaulus filaria* from the lungs of infected sheep. I. Reduction in worm numbers before patency". Folia parasit. Praha. 15, 309—316.

Michel, J.F. (1969a). "Experiments on the loss of *Dictyocaulus filaria* from the lungs of infected sheep. II. The demonstration of worms in the bronchial mucus". Folia parasit. Praha. 16, 57—65.

Michel, J.F. (1969b). "Experiments on the loss of *Dictyocaulus filaria* from the lungs of infected sheep. III. The effect of cortisone". Folia. parasit., Praha. 16, 121—128.

Michel, J.F. and Sinclair, I.J.B. (1963). "Host resistance to *Dictyocaulus filaria* infection". Parasitology 53, 7—8.

Michel, J.F., Mackenzie, A., Bracewell, C.D., Cornwell, R.L., Elliot, J., Herbert, Nancy C., Holman, H.H. and Sinclair, I.J.B. (1965). "Duration of the acquired resistance of calves to infection with *Dictyocaulus viviparus*", Res. vet. Sci. 6, 344—393.

Nickel, E.A. (1962). "Research on the course and effects of experimental *Dictyocaulus* infection in lambs and the productive power of Baermann's method in diagnosing *Dictyocaulus*". Ber. Münch. tierarztl. Wschr. 75, 2—6.

Okura, T. (1963). "Studies on the development of *Paragonimus ohirai*, Miyazaki, 1939, in the final host. 1. The route of migration of the larvae of *P. ohirai* in rats". Jap. J. Parasitol., 12, 57—67.

Omura, H. (1960). "Studies on host parasite relationships of lung flukes. Experiments of transplantation of adult *Paragonimus ohirai* and *Paragonimus westermani* into the peritoneal cavities and the subcutaneous tissues of rats". Jap. J. Parasitol. 9, 266—280.

Polzenhagen, M., Büchwalder, R. and Hiepe, T. (1969). "Lungworm infections in pigs. 2. Course and effects of *Metastrongylus* infection and the life span of *M. apri*". Mh. Vet. Med. 24, 847—850.

Rose, J.H. (1955). "The incidence of lungworms in sheep". Sanitarian, Lond. 63, 216—230.

Rose, J.H. (1958). "Site of development of the lungworm *Muellerius capillaris* in experimentally infected lambs". J. comp. Path. Therap. 68, 359—362.

Rose, J.H. (1959). "Experimental infection of lambs with *Muellerius capillaris*". J. comp. Path. Therap. 69, 414—422.

Rose, J.H. (1965). "Some observations on the transmission of lungworm infection in a flock of sheep at pasture". Res. vet. Sci. 6, 189—195.

Sato, M. and Ozawa, H. (1969). "Occurrence of ubiquinone and rhodoquinone in parasitic nematodes *Metastrongylus elongatus* and *Ascaris lumbricoides var suis*". J. Biochem., Tokyo, 65, 861—867.

Schwartz, B. and Lucker, J.T. (1935). "Experimental infections and super-infections of pigs with lungworms". J. Parasit. 21, 432.

Shikobolova, N.P. (1956). "Acquired immunity in chicks to *Syngamus* infection". Trudi Gelmint. Lab. acad Nauk. S.S.S.R. 8, 248—256.

Suguro, T. (1959). "Experimental study on the treatment of paragonomiasis". Jap. J. Parasit. 8, 518—522, 725—729.

Vaatstra, W.J. (1969). "Intermediary Metabolism of the Cattle Lungworm *Dictyocaulus viviparus*". Z. Physiol. Chem. 350, 701—709.

Von Brand, T. (1972). In "Parasitenphysiologie" 1—353. Gustav Fischer Verlag, Stuttgart.

Wilson, G.L. (1970). "The strength and duration of immunity to *Dictyocaulus filaria* infection in sheep and goats". Res. vet. Sci. 11, 7—17.

Yokogawa, M. (1965). In "Advances in Parasitology" (B. Dawes, ed.) pp. 9—158. Academic Press, London and New York.

Yokogawa, M. and Yoshimura, H. (1960). "Histopathological and parasitological examination of worm cysts which were removed surgically from the lungs of 16 cases of paragonimiasis". Jap. J. Parasitol. 9, 173—186.

Yokogawa, M., Yoshimura, H., Sano, M., Okura, T. and Tsuji, M. (1962). "The route of migration of the larva of *Paragonimus westermani* in the final hosts". J. Parasit. 48, 525—531.

Yokogawa, S. (1919). "Investigations on the lung distome (Japanese) 3. Report of the committee on the investigation of the endemic and infectious diseases". Government of Formosa 1—289.

Yokogawa, S., Cort, W.W. and Yokogawa, M. (1960). "Paragonimus and Paragonimiasis". Expl. Parasit. 10, 81—137, 139—205.

Yogore, M.G. (1956). "Studies on paragonimiasis in the Philippines". Thesis. The John Hopkins University. 1—148.

Ecological Aspects of Parasitology
Editor: C.R. Kennedy
© *North-Holland Publishing Company, Amsterdam, 1976*

CHAPTER 12

THE PERITONEAL CAVITY OF VERTEBRATES

M.J. HOWELL

Department of Zoology, Australian National University, P.O. Box 4, Canberra 2600,
A.C.T. (Australia)

Contents

12.1. Introduction

This chapter deals with the peritoneal cavity of vertebrates as a habitat for parasites, and is divided into three main sections. Initially, the physico-chemical properties of the peritoneal cavity are reviewed and this is followed by a brief survey of the parasites which occur in this site. The chapter concludes with a discussion of the parasitological problems posed by the habitat — so far as these can be specified at present — and the various ways in which these have been overcome by parasites.

12.2. Physicochemical properties of the peritoneal cavity

12.2.1. General remarks

The peritoneal cavity is a coelomic compartment of vertebrates which is lined by squamous, mesothelial cells (Bloom and Fawcett, 1962). It is not in direct contact with the external environment except by way of abdominal pores in some fishes and gonoducts in the majority of female vertebrates and a few male fishes (de Beer, 1951; Hoar, 1969). The evolution of a pleural cavity in birds and mammals means that the peritoneal cavity is relatively less extensive in these groups than in other vertebrates.

Most of the peritoneal cavity is occluded by the viscera. The remaining space is occupied by fluid (containing a number of free cells) which is largely distributed as a thin film over the viscera. The volume of fluid is limited and presumably related to the size of the animal although other factors can influence the amount present. Thus, in mice, volumes range from 0.02 ± 0.01 ml in males to 0.1 ± 0.05 ml in females; in the latter more fluid is present at prooestrus than at other stages of the oestrus cycle (Hartveit and Thunold, 1966). Fluid volumes of 0.33 to 5.58 ml in rabbits and 0.21 ± 0.32 ml in rats (Padawer and Gordon, 1956), and 15 to 45 ml in cattle (Tarkiewicz, 1956) have been reported. Fluid is relatively more copious in elasmobranchs than in other vertebrates (Bernard et al., 1966).

Somatic and visceral movements, the effects of gravity, and the activity of the diaphragm in mammals and lymph hearts in Amphibia and Reptilia, probably mix and circulate the peritoneal fluid (Yoffey and Courtice, 1970). These factors also, together with mechanisms controlling fluid balance in the body, probably have a bearing on the input of fluid into the cavity from blood vessels and tissue spaces associated with the viscera and body wall, and the output of fluid via the lymphatic system. Lymphatic vessels located on the peritoneal aspect of the right side of the diaphragm are principally involved in the absorption of fluid and particulate material (including cells and a variety of experimentally introduced particles) from the peritoneal cavity of mammals (Yoffey and Courtice, 1970). Approximately 25% of the total plasma volume of rats is absorbed by these vessels per hour; in rabbits the figure is about 10%. Proteins are absorbed more rapidly than particulate material.

Excessive fluid accumulation in the cavity is termed ascites. This is not apparently due to damaged diaphragm lymphatics, but rather to the inability of these vessels to cope with fluid that flows into the cavity when the gastrointestinal and hepatic lymphatic capacity is exceeded (Yoffey and Courtice, 1970).

TABLE 1

Physicochemical properties of plasma and peritoneal fluid in man and elasmobranchs *.

Property or constituent	Man		Elasmobranchs	
	Plasma	Peritoneal fluid	Plasma	Peritoneal fluid
Conductivity (mho × 1000)	11.4	13.4	—	—
pH	7.39	7.4	7.36 (6.45 [1])	5.8 (5.4 [1])
SG	1.027	1.012	—	—
Ash (%)	0.8	0.98	—	—
Solids (%)	8.6	2.5	—	—
Water (%)	93	97	—	—
Na (mEq/l)	138	138	259	246
K (mEq/l)	4.7	4.1	5.3	6.6
Ca (mEq/l)	5.0	4.0	4.9	3.4
Mg (mEq/l)	1.9	0.5	2.8	17.8
Cl (mEq/l)	102.4	109	236	274
Inorganic PO_4 (mg/100 ml)	4.0	4.0	—	—
Protein (g/100 ml)	7.2	2.1	3.35 [1]	0.132 [1]
Albumin (g/100 ml)	4.8	0.88	—	—
Globulin (g/100 ml)	2.3	0.81	—	—
Uric acid (mg/100 ml)	3.84	4.2	—	—
Total N (mg/100 ml)	1140	150	—	—
Ammonia N (mg/100 ml)	0.09	—	3.2 [1]	40.5 [1]
Urea N (mg/100 ml)	14.0	16.0	1179	1013
Non-protein N (mg/100 ml)	27.0	30.2	1090	870
Cholesterol (mg/100 ml)	190	60	—	—
Sugars (mg/100 ml)	81	114	—	—
CO_2 (mEq/l)	27.0	26.7	8.2	0.2
pCO_2 (mm Hg)	44.9	42.4 [2]	16.2 [2]	6.6 [2]
pO_2 (mm Hg)	39.4—93.0	28—40 [3]	—	—

* Values from Dittmer (1961) unless specified otherwise. Where available mean values are given; elsewhere, mid-points of ranges or single determinations are shown except for pO_2 where ranges are indicated.

[1] Bernard et al. (1966).

[2] Calculated from the Henderson-Hasselbalch equation (after Ludbrook, 1959) using values for pH and CO_2 concentration given in the table.

[3] Values for a range of mammals (von Brand, 1952; Del Monte, 1967).

12.2.2. Fluid other than cellular constituents

The properties of peritoneal fluid have not been studied extensively other than in man and elasmobranch fishes and this may be related to the difficulty of obtaining sufficient fluid for analysis. In man interest in its properties stems from the fact that levels of certain constituents may provide diagnostic clues to pathological conditions (Zhloba, 1968; Witte et al., 1972; McGowan et al., 1973).

Some data for man and elasmobranchs are given in Table 1. Apart from protein content the similarity between peritoneal fluid and plasma in man is striking. In elasmobranchs the situation is not so clear. It is of interest that although abdominal pores are present in these animals there are, nevertheless, marked differences between the composition of the fluid and sea water.

A few constituents of the fluid in other mammals and reptiles have been analysed by Dittmer (1961), Tarkiewicz (1956), 1962), Leone et al. (1963), Birkeland (1968a, b), Moline et al. (1972) and McGowan et al. (1973). With the exception of protein, their concentrations reflect plasma levels, but discrepancies between glucose and electrolyte concentrations have been noted (Birkeland, 1968a, b). Amino acid concentrations have not apparently been determined in normal animals. However, values for amino acid nitrogen in pleural fluid in man (Dittmer, 1961) suggest that amino acid levels comparable with those of plasma (3.5—6.0 mg/100 ml) would be present.

The osmotic pressure of extracellular fluids is generally considered to be close to that of plasma. Thus, that of the peritoneal fluid of marine elasmobranchs would be close to that of sea water (Δ = 1.8 to 2.2°C), while that of marine teleosts would be somewhat lower (Δ = 0.78°C) (Black, 1957). In other vertebrates, Δ probably lies within the range 0.5 to 0.6°C (Prosser and Brown, 1966).

Fluctuations in the properties of peritoneal fluid of normal animals have not been examined, but the changes that occur would almost certainly reflect plasma changes (see Chapters 18, 19). Some factors which are likely to affect its properties are therefore stage of development, phases of the reproductive cycle, dietary deficiencies and excesses, state of activity, cell populations present (see below) and changes in the external environment (particularly in poikilothermic animals). Disease, either indirectly (through plasma) or directly, may also bring about marked changes in the cavity (Del Monte, 1967; Drewes and McKee, 1967; Zhloba, 1968; Witte et al., 1972; McGowan et al., 1973). In rats, for example, oxygen tension is lowered from 30—40 mm Hg to 0 mm Hg within 8 days as a result of the growth of an ascites tumour (Del Monte, 1967).

Antigenic stimulation of mice at sites remote from the peritoneal cavity can lead to the presence of free and cytophilic antibody in the cavity (Lokaj and Klhůfek, 1968). Moreover, direct stimulation of these animals with antigens, particularly when they are incorporated in Freund's complete adjuvant, gives rise to ascites and high titres of antibody (Munoz, 1957; Anacker and Munoz, 1961; Banovitz et al., 1963). It would seem quite likely that antibody could reach the peritoneal cavity of all groups of vertebrates.

12.2.3. Cellular constituents

The peritoneal fluid is populated by a variety of cell types. These have been studied in some mammals but little is known about them in other vertebrates. The cells in mammals largely represent a mixture of white cells found also in the blood and lymph (Yoffey and Courtice, 1970). Since the blood of other vertebrates contains white cell types akin to those of mammals (Reichenbach-Klinke and Elkan, 1965) it seems probable that the cells in their peritoneal cavities will also be similar.

The cell types present in mammals are (i) mesothelial cells, derived from the lining of the cavity, and (ii) cells of both the lymphoid (lymphocytes) and myeloid (polymorphs, monocytes and macrophages) series, which are ultimately derived from stem cells in the bone marrow (Wu et al., 1968). There is conclusive evidence that monocytes are the precursors of macrophages (Van Furth, 1970; Van Furth et al., 1970) although it has been proposed that meso- and sub-mesothelial cells may also differentiate into macrophages (Mohr et al., 1971). Two types of lymphocytes, T and B cells, are recognised in mammals (Miller, 1972) and lymphocytes in the peritoneal cavity probably comprise a mixture of these.

Very few quantitative studies of the cells have been made. Padawer and Gordon (1956) showed there was wide variation both between and within a number of mammalian species in the number of cells present. For example, in rabbits there were 1.32×10^3 cells/mm^3, in CFl mice 119×10^3, and in DBA mice 103×10^3. These values lie outside the ranges for white cells in the blood of these species (Dittmer, 1961) and seem to bear little relationship to them. In rats, the number of cells/mm^3 was proportional to the reciprocal of the fluid volume, and an estimate of 20×10^6 cells for the entire cavity was made (Padawer and Gordon, 1956). Another estimate for rats put this figure at 8.2×10^6 (McLaughlin et al., 1972). For mice, values of 4×10^6 (Lokaj, 1968) and 8.13×10^6 (Chin and Hudson, 1972) have been reported.

Differential cell counts show that there are also wide variations both between and within species (Padawer and Gordon, 1956; Tarkiewicz, 1956; Forbes, 1966; Lokaj, 1968; Valtonen et al., 1968; McGowan and Davis, 1968, 1969; McGowan et al., 1968; Davis, 1973). Because various factors are known to affect the cell population (see below) this is, perhaps, not surprising. Mesothelial cell counts vary from 0 to 56%, lymphocytes from 25 to 68.2%, polymorphs from 3.25 to 46.5%, and monocytes and macrophages from 0.8 to 27.9%. Mast cells are present in low numbers in most mammals and a high percentage of eosinophils (30% or more) is characteristic of rats (Padawer and Gordon, 1956; Bosworth and Archer, 1961).

The cell population of mammals is not static. In normal animals, cells enter from the blood, rarely divide in the cavity, and probably reach the blood again via the lymphatics (Yoffey and Courtice, 1970). Although the degree of recirculation is still uncertain, the time required for the replacement of the entire population of macrophages in mice is 20—40 days (Van Furth, 1970). Proportions of the various cell types present depend on the age and sex of the individual, and may be influenced by pregnancy and the oestrus and menstrual cycles (Forbes, 1966; McGowan and Davis, 1968, 1969; McGowan et al., 1968; Jacques and Rüegg, 1970; Davis, 1973), direct or indirect antigenic stimulation or irritation (e.g. Litt, 1960a, b; Nelson and Boyden, 1963; Forbes, 1966; Lee and Schoen, 1968; Soldati et al., 1971; Fruhman, 1972; McLaughlin et al., 1972), exposure to cytotoxic drugs (Chin and Hudson, 1972), and radiation (McGowan and Davis, 1971). Cell division within the cavity increases in frequency following irritation and direct or indirect antigenic stimulation (Forbes, 1966; Mohr et al., 1971).

The functional properties of the lymphoid and myeloid cells of mammals are discussed in Chapter 6. These cells have received less attention in lower vertebrates but some of their properties are comparable (Nelstrop et al., 1968; Chiller et al., 1969a, b; Finn, 1970). Two main points can be made here with regard to peritoneal cells of mammals but which possibly apply

TABLE 2

Selected helminth parasites of the peritoneal cavity of vertebrates

Parasite	Taxonomic group		Stage in Life Cycle
Dictyocotyle coeliaca	Trematoda:	Monogenea	Adult
Stictodora lari	Trematoda:	Digenea	Larval; metacercaria
Ascocotyle mcintoshi	Trematoda:	Digenea	Larval; metacercaria
Coelomotrema antechinomes	Trematoda:	Digenea	Adult
Taenia crassiceps	Cestoda:	Cyclophyllidea	Larval; cysticercus
Taenia hydatigena	Cestoda:	Cyclophyllidea	Larval; cysticercus
Taenia pisiformis	Cestoda:	Cyclophyllidea	Larval; cysticercus
Mesocestoides corti	Cestoda:	Cyclophyllidea	Larval; tetrathyridium
Ligula intestinalis	Cestoda:	Pseudophyllidea	Larval; plerocercoid
Schistocephalus solidus	Cestoda:	Pseudophyllidea	Larval; plerocercoid
Diphyllobothrium spp.	Cestoda:	Pseudophyllidea	Larval; plerocercoid
Proteocephalus ambloplitis	Cestoda;	Proteocephalidea	Larval; plerocercoid
Nesolecithus africanus	Cestodaria:	Amphilinidea	Adult
Amphilina foliacea	Cestodaria:	Amphilinidea	Adult
Austramphilina elongata	Cestodaria:	Amphilinidea	Adult
Setaria labiatopapillosa	Nematoda:	Filaroidea	Adult
Dipetalonema proechimyis	Nematoda:	Filaroidea	Adult
Onchocerca gutturosa	Nematoda:	Onchocercoidea	Adult
Contracaecum multipapillatum	Nematoda:	Heterocheiloidea	Larval (3rd-stage)
Philometra cylindracea	Nematoda:	Dracunculoidea	Adult
Philometra sanguinea	Nematoda:	Dracunculoidea	Adult
Leptorhynchoides thecatus	Acanthocephala		Larval; cystacanth

* Key to characteristics
1 — Essentially free in the peritoneal cavity
2 — Encysted (enclosed by its own membranes) and/or encapsulated by host tissue
3 — Restricted to the peritoneal cavity

to other vertebrates as well. Firstly, they are sufficiently representative of lymphoid and myeloid cells elsewhere in the body since they have the potential to differentiate into a wide variety of apparently normal lymphoid and myeloid tissues (Goodman, 1963). Secondly, their ability to participate in immunological reactions, coupled with the ability of antibody to reach the peritoneal cavity, renders the site an immunologically hostile one for parasites.

12.3. Parasites of the peritoneal cavity

The peritoneal cavity of vertebrates has been exploited by a diverse array of parasites. That it surrounds the gut (a common portal of entry for parasites), contains a nutrient fluid, and provides security from dislodgement are

Host(s)	Characteristics *	References
Elasmobranch fish	1, 3, 6	Hunter and Kille (1950)
Teleost fish	2, 3, 7	Howell (1973)
Teleost fish	2, 3, 7	Leigh (1974)
Mammal	1, 3, 6	Angel (1970)
Mammal	1, 4, 7	Freeman (1962)
Mammal	2, 4, 7	Sweatman and Plummer (1957)
Mammal	1 or 2, 3, 6	Heath (1971)
Reptile; mammal	1, 4, 7	Specht and Voge (1965)
Teleost fish	2, 3, 7	Orr (1967)
Teleost fish	1, 3, 5	Orr et al. (1969)
Teleost fish	1 or 2, 4, 7	Halvorsen (1970)
Teleost fish	1, 3, 7	Fischer and Freeman (1969)
Teleost fish	1, 3, 5	Dönges and Harden (1966)
Ganoid fish	1, 3, 6	Bauer (1959)
Reptile	1, 3, 5	Baer (1951)
Mammal	1, 3, 7	Brengues and Gidel (1972)
Mammal	1, 4, 6	Everard et al. (1974)
Mammal	1, 4, 7	Eichler and Nelson (1971)
Teleost fish	2, 3, 7	Huizinga (1967)
Teleost fish	1, 4, 5	Molnár and Fernando (1975)
Teleost fish	1, 4, 5	Wierzbicki (1960)
Teleost fish	2, 3, 7	De Giusti (1939)

— Present in the peritoneal cavity and other sites
— Restricted to a single species of host
— Restricted to two or more related species of host
— Generally wide host range.

some factors which may have been involved in determining its suitability. However, it poses a number of problems for parasites which are considered further below.

The list of parasites in Table 2 is not an exhaustive compilation of species known from this site; a representative selection has been made on the basis of the various modes of infection and transmission known, life cycle stages present, and the existence of information other than that concerned directly with life cycles. The list does not include species such as *Echinococcus granulosus* and *E. multilocularis* which do not, or only rarely, occur naturally in the peritoneal cavity but which will develop in this site if inoculated (Lubinsky, 1960; D.D. Heath, 1970). Species such as *Fasciola hepatica* and *Spirometra* spp., which have only a transient association with the peritoneal cavity during movement to other sites, have also been excluded.

Adult and larval digenetic trematodes and nematodes, larval cestodes and acanthocephalans, and adult cestodarians and monogenetic trematodes — indeed representatives of all major groups of helminths — are listed in Table 2. To review their diverse biological characteristics, some of which are shown, is beyond the scope of this chapter. However, it is worth pointing out that there are no features known to be shared by the various species which set them off collectively from the parasites of other habitats.

12.4. Problems posed by the peritoneal cavity

12.4.1. Establishment of infection

There is no proof that the gonoducts of female vertebrates are used by helminth parasites as a means of reaching the peritoneal cavity. Thus, from the point of view of parasites the peritoneal cavity is an enclosed habitat and this raises the problem of access to the site. A variety of solutions to this problem have been made and these are summarised below. References, unless given, may be obtained from Table 2.

(a) Free living stages (cercariae) penetrate the skin of the host and move to the peritoneal cavity *(Stictodora lari)*.

(b) Free living stages are ingested and penetrate the gut wall *(Ascocotyle mcintoshi)*.

(c) Resistant stages in the external environment (eggs containing oncospheres) are ingested. These hatch in the intestine and the oncospheres pass via the gut wall, hepatic portal vein and liver to the peritoneal cavity *(Taenia* spp.).

(d) The host ingests prey infected with the preceding stage in the life cycle of the parasite. The parasite is released in the intestine and passes through the gut wall *(Ligula intestinalis, Schistocephalus solidus, Diphyllobothrium* spp., *Proteocephalus ambloplitis, Amphilina foliacea, Philometra* spp.,

Nesolecithus fasciatus, and *Contracaecum multipapillatum).* In some cases *(Diphyllobothrium* spp. and *Contracaecum multipapillatum)* a further act of predation may lead to the re-establishment of the parasite in the peritoneal cavity of the secondary predator (paratenic host). The same phenomenon is also seen with *Proteocephalus ambloplitis* but the plerocercoids, influenced by temperature and possibly by factors associated with the age of the host, subsequently move back through the gut wall into the lumen where they attain sexual maturity.

(e) An infected arthropod vector injects microfilariae into the host while taking a blood meal and these reach the peritoneal cavity, presumably via the blood stream *(Dipetalonema proechimyis* and *Setaria labiatopapillosa).*

(f) As for (e) but the microfilariae are presumed to move through the tissues *(Onchocerca gutturosa).*

(g) Transplacental and transmammary routes of infection are known for *Mesocestoides corti* (Hess, 1972), and there is a report of transplacental transmission of an unidentified species of *Setaria*, possibly *S. labiatopapillosa* (Fadzil, 1969).

(h) Artificial methods of infection have been established for *Taenia crassiceps* and *Mesocestoides corti.* Both species multiple asexually in the peritoneal cavity and they may be serially propagated by direct inoculation.

When tetrathyridia of *Mesocestoides corti* are fed to mice, they penetrate the gut wall and become established in the peritoneal cavity following a period of liver invasion. Although this experimental route of infection is known, the life cycle of this species has not been established with any certainty but it seems likely that route (d) above is involved. *M. corti* is also unusual in that tetrathyridia will multiply when placed in the peritoneal cavity of dogs but they can move from this site into the gut lumen (Eckert et al., 1969).

Stictodora lari (Howell, 1973), *Schistocephalus solidus* (Bråten, 1966) and *Taenia pisiformis* (Heath, 1971) can also be established in the peritoneal cavity by direct inoculation of their infective stages. This indicates that prior movement through the tissues is not essential for these species to resume their development and suggests that they have some means of identifying their habitat.

It is not known how the remaining species in Table 2 reach the peritoneal cavity. The cestodarians, like *Amphilina foliacea*, may use route (d). *Dictyocotyle coeliaca*, which has a free swimming oncomiracidium (Kearn, 1970; 1975), has the abdominal pores of its host as a potential site of entry. Angel (1970) proposed that since *Coelomotrema antechinomes* was recovered from female hosts only, the reproductive tract might be the route of infection.

The use of the peritoneal cavity as a habitat poses a number of other problems which are crucial to the establishment of an infection. These surround (a) the release of infective stages from eggs, ingested food or vectors at appropriate sites, or, in the case of free-living infective stages, making a direct encounter with the host; (b) the initiation of activity for movement through the tissues; and (c) the identification of the habitat when it is reached. These questions have received little attention.

Taenia pisiformis oncospheres hatch and activate under the influence of host enzymes and bile salts (Silverman, 1954; Heath, 1971), and it would seem likely that other parasites entering by the oral route are stimulated by the same or similar factors. Gland cells of various kinds, undoubtedly involved in facilitating movement through tissues, are found in the infective stages of many parasites but the chemical composition and functional properties of their secretions are unknown. In some species (e.g. *Diphyllobothrium* spp. plerocercoids and *Mesocestoides corti* tetrathyridia) gland cells are not particularly evident and how tissue migration is achieved is enigmatic.

Heath (1971) suggested that the gland cells of *T. pisiformis* oncospheres produce a mucopolysaccharidase for activation and penetration of the gut and a hyaluronidase to aid further movement through the tissues. However, Barker (1970) considered the secretions of this species had lubricant rather than lytic functions. Oncospheres of *T. pisiformis* possess hooks which may play a role in movement through tissues (Silverman and Maneely, 1955; Barker, 1970) and the larvae of *C. multipapillatum* possess a "boring tooth" which is possibly used for the same purpose (Huizinga, 1967).

The question as to whether the movement of parasites through tissues to the peritoneal cavity is a directional migration or whether the parasites become generally more widespread in the body of the host and only survive in the peritoneal cavity (and a few other sites) has not been resolved. Studies on *T. pisiformis* by Heath (1971), for example, could be interpreted either way.

12.4.2. Transmission to subsequent hosts; release of genital products

Parasites have evolved a variety of solutions to these problems which arise, like that of infection, from the fact that the peritoneal cavity is essentially an enclosed habitat. References, unless given, are listed in Table 2.

Predation is the means by which larval stages are transmitted. The parasite attains sexual maturity in the predator but there are some variations when paratenic hosts are involved (see above). Transplacental, transmammary and artificial modes of transmission of larval stages have been referred to above.

With adult parasites release of genital products (eggs except in nematodes which are viviparous) may take place in the following ways.

(a) Eggs are liberated into the peritoneal cavity and are presumed to escape via the abdominal pores *(Amphilina foliacea* and *Dictyocotyle coeliaca)* as well as the oviducts in female hosts *(D. coeliaca).*

(b) The parasite bores through the body wall and releases its eggs *(Nesolecithus africanus)*.

(c) The parasite possibly bores into the lungs from which the eggs reach the gut and pass in the faeces *(Austramphilina elongata)*.

(d) The female worm moves to the caudal fin of the fish host. The skin is perforated and the worm bursts (apparently due to osmotic pressure changes) releasing larvae *(Philometra sanguinea)*.

(e) Larvae are released into the peritoneal cavity and probably leave the host via the excretory ducts *(Philometra cylindracea)*. According to Molnár (1967), larvae and adult females of a related species, *P. abdominalis*, leave via the anus or the gonoducts.

(f) Larvae are released into the peritoneal cavity, reach the blood stream, and are transmitted by blood sucking arthropods *(Setaria labiatopapillosa* and *Dipetalonema proechimyis)*.

(g) Larvae are released into the peritoneal cavity, accumulate in tissue sites, and are transmitted by an arthropod vector *(Onchocerca gutturosa)*.

The route by which the eggs of *Coelomotrema antechinomes* (Table 2) are released is not known but it is possible they could escape via the oviducts (infected hosts were all female) or be disseminated when the host is eaten by a predator or dies of other causes.

Although much remains to be discovered in relation to the above phenomena, some information is available regarding the influence parasites have on the behaviour of their hosts, and the behaviour of the parasites, which probably facilitates transmission. Thus fish infected with *Schistocephalus solidus* and *Ligula intestinalis* behave in such a way as to render them more liable to predation than uninfected fish (Holmes and Bethel, 1972); a marked periodicity in microfilaremia occurs in *Dipetalonema proechimyis* infections (Everard et al., 1974), whose peaks may coincide with preferred feeding times of as yet unknown vectors; and microfilariae of *Onchocerca gutturosa* accumulate in the umbilical region of their cattle host, the preferred feeding site of their vector, *Simulium ornatum* (Eichler, 1971).

12.4.3. Problems for developing parasites

The nutritional properties and rate of turnover of peritoneal fluid, and the provision of security from dislodgement, would seem to be extremely favourable features of the peritoneal cavity for developing parasites. However, the following are considered to be the main problems faced by parasites of this site: (a) the nature of the habitat may be markedly different from that which the parasite experiences during the preceeding stage of its life cycle; (b) the parasite may bring about pronounced physicochemical changes in the cavity; (c) the environment is immunologically hostile. A fourth, which is more tentatively suggested, is that space and nutrients may be limited in availability. The last three in particular are probably closely interrelated but each will be considered separately. Whether fluctuations in the physicochemical conditions of the peritoneal cavity referred to in Section 12.2 pose significant problems for parasites, and how these fluctuations may in turn be modified by the presence of parasites, are aspects of the host-parasite relationship for which there is no information.

As judged by the biological diversity of the parasites listed in Table 2, the problems referred to above would assume greater or lesser degrees of significance and a variety of solutions to them have undoubtedly been made. However, details of the solutions are for the most part wanting.

(a) *Change of habitat.* On infection, the parasites listed in Table 2 have to adapt to a variety of physicochemical conditions that may differ markedly from those previously experienced. However, the physiological adjustments that are made have not been studied.

Experimental studies with some cestodes point to the existence of factors in the peritoneal cavity of some animals to which parasites are unable to adapt. Thus, *Taenia crassiceps* develops in mice but not in rabbits or guinea pigs when injected intraperitoneally (Freeman, 1962); *Mesocestoides corti* develops in mice, lizards, rats and dogs but not guinea pigs (Specht and Voge, 1965; Eckert et al., 1969); and *Schistocephalus solidus* develops in three, but not nine-spined sticklebacks (Bråten, 1966; Orr et al., 1969).

(b) *Adaptations to phsyicochemical changes during infection,* A spectrum of pathological effects accompany infections with peritoneal cavity parasites and these, together with defensive reactions of the host, would lead to changes in the physicochemical conditions of the habitat. Most of the available information is descriptive, relates to cestodes, and has been extensively reviewed by Williams (1967), Smyth (1969), Smyth and Heath (1970), Arme and Walkey (1970) and Weinmann (1970). More recent studies which provide clues to changes in the peritoneal cavity are those dealing with aspects of blood chemistry in a number of cestode infections (Izmagilova, 1968; Nemeth, 1972, 1973; Bundesen and Janssens, 1971; Enigk et al., 1972; Guttowa and Honowska, 1973).

Since the nature of the physicochemical changes in the peritoneal cavity that accompany infection have not been studied, it is impossible to define precisely the range of problems (apart from some related to defensive reactions of the host discussed separately below) which confront parasites and the solutions that have evolved. However, some adaptations which would clearly be of value in changing conditions have been demonstrated in a few species.

Schistocephalus solidus does not alter its internal pH in spite of pH changes over the range 4 to 8 in the external medium (Walkey and Davies, 1968). It was suggested that the calcareous corpuscles of the worms may buffer these pH changes. Thus, if such changes occur in vivo a relatively simple mechanism may operate to counteract them.

Cestodes are considered to be primarily anaerobic but they will consume oxygen when it is available (Smyth, 1969; von Brand, 1973). Those cyclo- and pseudophyllidean species in Table 2 appear to be no exception (Smyth, 1969; Weinbach and Eckert, 1969; von Brand, 1973; Köhler and Hansel- mann, 1974), and this facility may be of considerable advantage if fluctua- tions in oxygen tension take place.

Flexibility in the operation of metabolic pathways in relation to substrate availability has been shown in *M. corti* (R.L. Heath, 1970; R.L. Heath and Hart, 1970). When an exogenous source of purines and pyrimidines was present, these substances were incorporated into nucleic acid and there was no de novo synthesis of purines and pyrimidines; in their absence, de novo synthesis of pyrimidines (and probably purines) was initiated. Adaptations such as this enable the parasite to adjust to changes in substrate availability and make the most economic use of available resources.

Succinate is a major end product of carbohydrate metabolism in a number of cestodes listed in Table 2, but there are conflicting reports as to the in- fluence aerobic and anaerobic conditions have on its output (Reuter, 1967a, b, 1971; Köhler and Hanselmann, 1974; McManus, 1975). Its production suggests that the worms are capable of CO_2 fixation which is of possible ad- vantage if CO_2 levels in the peritoneal cavity rise. This property also enables $NADH + H^+$ formed during glycolysis to be reoxidised.

(c) *Avoidance of defensive reactions of the host.* As stressed earlier, the peritoneal cavity is an immunologically hostile environment. Thus a basic problem for a parasite is to avoid the defensive capabilities of its host. How- ever, this question has attracted little attention.

Taenia pisiformis (Heath, 1973a, b; Rickard, 1974) and *T. hydatigena* (Sweatman and Plummer, 1957; Gemmell, 1972a, b) provoke an immuno- logical response from their hosts which confers resistance to a challenge infection. However, this response is without apparent effect on the survival of established parasites; hence, a state of concomitant immunity (Smithers and Terry, 1969) exists.

Varela-Diaz et al. (1972) suggested that *T. hydatigena* cysticerci possessed two major antigenic determinants. One, designated L, elicits an adverse host reaction while another, N, is responsible for a host reaction favourable to the survival of the cysticercus. The relative degree of exposure of these antigens was considered to determine whether a given cysticercus survives or not. Support for the hypothesis was indirect and based on cysticercus survival following reciprocal, heterologous infections with *T. ovis* and *T. hydatigena* in lambs.

A similar type of mechanism was envisaged by Rickard (1974) who sug- gested, on the basis of indirect evidence, that *T. pisiformis* elicits the forma- tion of "blocking antibodies" which bind to, and prevent lethal antibodies and reactive cells from attacking, the cysticercus.

Both the above views are compatible with the "selective contaminative" hypothesis (Smithers and Terry, 1969). However, the "host antigens" are in these cases host-produced antibodies. More direct evidence for their existence is needed and some additional explanation seems necessary to account for the way invading parasites avoid non-specific host responses until such time as these antibodies are produced.

The cysts of *Stictodora lari* do not provoke an encapsulation response from their fish host for some weeks after infection (Leong and Howell, 1971; Howell, 1973). It has been shown (Howell, 1973) that the cyst wall is of parasite origin; it does not mimic host components, but it is coated by material synthesised by the fish. These results are also compatible with the "selective contaminative" hypothesis but a protective role for the host material remains to be demonstrated. If it is not protective, it has been suggested that spikes on the cyst wall may form an unsatisfactory substrate for the attachment of host cells. The occurrence of an encapsulation response some weeks after infection (coincident with infectivity to the definitive host) has been attributed to one or more of the following: (i) contamination of the cyst wall by material secreted by the parasite at this time, (ii) a reduction in the density of the spikes which makes cell attachment to the cyst wall possible, (iii) the manifestation of acquired immunity to the parasite.

S. lari may also possess a mechanism whereby the extent of provocation of the host is minimised. Cells lining the excretory bladder appear to synthesise and store protein throughout that period in the host when the cysts are not attacked by cells. Encapsulation coincides with the release of this material and its incorporation as an additional layer to the cyst wall (Leong and Howell, 1971). It is possible, therefore, that the bladder wall gland cells play an important part in the resorption and reutilisation of substances which, if passed out, could initiate a potentially damaging response from the host before development has been completed.

(d) *Space and nutrient availability*. The following examples show that there is as yet no conclusive evidence that space and nutrients are limited in availability to developing parasites, and it is thus uncertain whether these factors have posed major problems for parasites during the evolution of the host-parasite relationship.

The growth rate of *Schistocephalus solidus* plerocercoids is slower the greater the number present in a host (Meakins and Walkey, 1973) and the greater their size (McCaig and Hopkins, 1965; Sinha and Hopkins, 1967; Meakins and Walkey, 1973). The physical effects of the bulk of the plerocercoids and the amount of food they require may be responsible for these effects. However, some intrinsic properties of the plerocercoid which control its growth rate cannot be ruled out, since slower growth with increasing size was observed in vitro (McCaig and Hopkins, 1965) where space and nutrients were probably not limited. Further, plerocercoids become infective when they are relatively small (Hopkins and McCaig, 1963; Orr and Hopkins, 1969) and they may have the capacity to reduce their growth rate beyond that point. Field evidence indicates that *S. solidus* grows more rapidly in female than in male fish (Pennycuick, 1971) and this might be due to a diversion of host resources (which normally go into egg production) to the parasite.

Taenia crassiceps, which multiplies asexually in the peritoneal cavity, does not achieve a logarithmic growth rate; this indicates the existence of limiting factors. That available space is of some importance for this species is suggested by the facts that proliferation occurs more extensively in the peritoneal cavity rather than subcutaneously or within the pleural cavity (Freeman, 1962), and in female rather than male mice (Freeman, 1962; Culbreth et al., 1972) where available space is greater (Hartveit and Thunold, 1966). Genetic factors impose restrictions on the growth rate of this species. KBS and ORF strains of the parasite, differing in chromosome complement (Smith et al., 1972a) and other features (Fox et al., 1971; Smith et al., 1972b; Culbreth et al., 1972) have been characterised; cysticerci of the KBS strain (which possess two more chromosomes) multiply more slowly and do not grow as large as those of the ORF strain (Dorais and Esch, 1969; Beavers et al., 1971).

Although *Mesocestoides corti* grows exponentially during the first 50 days after infection (Specht and Voge, 1965; Novak, 1974b) it does not grow as rapidly as it is capable since cytostatic agents and splenectomy accelerate its growth rate (Novak and Lubinsky, 1973; Novak, 1974a). This information, coupled with the fact that *M. corti* possesses antigens capable of eliciting a protective host response (Kowalski and Thorson, 1972a, b), seems to indicate that immunological factors assume more significance than space or nutrients in limiting the growth rate of this species. In addition, it has been shown that the parasite grows faster in male than in female mice (Novak, 1972), a phenomenon related to hormonal differences (Novak, 1974b, 1975) and evidence that space is probably not a major limiting factor. In vitro studies (Voge and Coulombe, 1966) indicate that lack of oxygen and a pH outside the range 7.4—7.6 reduce the reproductive rate of *M. corti;* both these situations may well apply in vivo.

Certain features of a number of host-parasite relationships can be envisaged as having evolved in response to the limited availability of space and nutrients in the peritoneal cavity. These include (a) host factors which limit the intensity of infection and (b) parasite adaptations that involve (i) the use of food reserves acquired at some preceeding stage of the life cycle, during the establishment of infection, or from some other site later in the infection; (ii) a slow rate and/or minimal degree of developmental change; (iii) mechanisms which make efficient use of available nutrients, enhance space and nutrient availability, and offset the effects of compression due to limited space. Some of these are discussed below.

(i) *Intensity of infection.* Immunological responses of the host may limit the intensity of infection. For example the hosts of *Taenia hydatigena* and

T. pisiformis become immune to challenge following a primary infection (Sweatman, 1957; Heath, 1973a, b). Field data relating to plerocercoid infections in fish could be similarly interpreted (Powell and Chubb, 1966; Arme and Owen, 1968). Functional antigens are shared by several *Taenia* spp. (Gemmell, 1969; Wikerhauser et al., 1971). Although this phenomenon has not been demonstrated among species sharing the same host, it is feasible that some degree of reciprocal immunity could also impose limitations on intensity.

Other possible mechanisms that could operate are (a) age resistance; (b) intraspecific competition between parasites leading to the survival of one or a few specimens of a large infecting dose; (c) immunological responses which destroy a proportion of the developing parasite population; (d) a change in the ecology of the host which removes it from its source of infection. There are indications that (a) applies in *Taenia pisiformis* infections (Heath, 1971), (b) in *Schistocephalus solidus* infections (Hopkins and Smyth, 1951), and (d) in some *Ligula intestinalis* infections (Kamenski, 1964). The hypothesis of Varela-Diaz et al. (1972), discussed earlier in connection with the survival of *T. hydatigena* cysticerci, has possible relevance to (c).

(ii) *Utilisation of nutrients.* Very little is known regarding the nutritional requirements of peritoneal cavity parasites. Since all the cyclophyllidean species in Table 2 as well as *Schistocephalus solidus* plerocercoids can be cultivated in vitro through all or part of their peritoneal cavity phase of development (Voge, 1963; Taylor, 1963; Robinson et al., 1963; Voge and Seidel, 1968; Heath and Smyth, 1970; Heath, 1973c; McCaig and Hopkins, 1965; Sinha and Hopkins, 1967), an approach to this question is possible.

There seems to be no indication that the parasites rely on nutrients acquired at preceeding stages of their life cycles for their development, and they would thus appear to be dependent on their immediate hosts. Both *Taenia pisiformis* (Heath, 1971; Shield et al., 1973) and *Mesocestoides corti* (Specht and Voge, 1965; Specht and Widmer, 1972) undergo an initial period of development in the liver and this may help to offset limitations on nutrient availability in the peritoneal cavity.

Philometra cylindracea females complete their development only if they move into the serosa of the air bladder (to which the males are restricted) of their fish host, mate, and move back into the peritoneal cavity (Molnár and Fernando, 1975). Apart from the fact that this tissue invasion is essential for mating, it may enable the female worms to acquire a supply of nutrients which, on the return of the females to the peritoneal cavity, support the growth of developing larvae.

Schistocephalus solidus is apparently a more efficient energy converter than its host and may induce the host to consume more food than uninfected fish (Walkey and Meakins, 1970). There are also indications that it makes lower nutritional demands on its host in summer when the host is experiencing high metabolic demands (Davies and Walkey, 1966), and possesses enzymes adapted specifically for developmental processes in the fish which are distinct from those involved in maturation in the definitive host (Sinha and Hopkins, 1967). These characteristics would presumably enhance the availability of nutrients and the efficiency of their utilisation, and enable the parasite to balance its demands on the host against survival.

Adaptations involved in making efficient use of nutrients are clearly shown by the larval stages of cestodes. The fine structure of the tegument of the pseudophyllideans listed in Table 2 (Morris and Finnegan, 1969; Charles and Orr, 1968; Bråten, 1968; Grammelvedt, 1973; Andersen, 1975) and the bladder wall of *Taenia crassiceps* (Baron, 1968) shows many of the features of adult cestodes (see Smyth, 1969); the presence of microtriches clearly increases the surface area available for absorption. Active transport mechanisms have been demonstrated in larval *T. crassiceps* (Murrell, 1968; Haynes and Taylor, 1968; Haynes, 1970; Pappas et al., 1973a, b; Pappas and Read, 1973). These enable nutrients to be absorbed from solutions where their concentrations may be low, and they also prevent leakage of nutrients from parasite tissue when falls in their concentration in the external environment take place (Smyth, 1969).

The tegument of *Ligula intestinalis* is reported to possess amylase activity (Davydov and Kosenko, 1972) and thus membrane digestion (Smyth, 1972) may be an important adaptation for providing nutrients which might otherwise be inaccessible. There is evidence that macromolecules can be taken up directly by *Taenia crassiceps* cysticerci (Esch and Kuhn, 1971) but this has been challenged by Pappas and Read (1973).

Information on the extent of parasite dependence on the host for specific nutrients is very fragmentary. There is evidence that *Taenia hydatigena* is unable to synthesise cholesterol from acetate (Frayha, 1971; 1974). Hence, if this substance is an essential requirement it must be provided by the host. *T. crassiceps* cysticerci have been reported to have the ability to synthesise a number of amino acids (Taylor and Haynes, 1966). However, it is clear from more recent work (Pappas and Read, 1973) that the analytical procedures used by Taylor and Haynes (1966) were insensitive. Thus, the apparent absence of certain amino acids from the peritoneal fluid of infected animals may have been due to their low concentrations, and their presence in the cysticercus a reflection of the concentrating effect of its active transport mechanisms and not to any synthetic capabilities.

High levels of glycogen are present in many larval cestodes (Smyth, 1969), and in *Proteocephalus ambloplitis*, glycogen accounts for up to 68% of the dry weight (Marra and Esch, 1970). Although it is known that glycogenesis in *Taenia crassiceps* is stimulated by insulin (Esch, 1969) and glycogen reserves are maintained only in the presence of glucose (Taylor et al., 1966), and in *Mesocestoides corti*, C-14 from labelled glucose appears in glycogen (Köhler and Hanselmann, 1974), the extent of dependence on exogenous glucose for glycogenesis in vivo is far from clear. High levels of glycogen may be important, particularly in infections of long standing, in providing insurance against limitations on nutrient availability.

(iii) *Enhancement of space and nutrient availability*. Ascites accompanies experimental infections of mice with *Taenia crassiceps* (Freeman, 1962, 1964) and *Mesocestoides corti* (Specht and Widmer, 1972). While this is a pathological consequence of infection and eventually leads to the death of the host, it may, for a time, provide additional space and nutrients and facilitate further parasite development. A similar phenomenon, but to a less striking degree, could accompany infections with other parasites.

(iv) *Effects of compression*. The fluid-filled bladder of *Taenia* spp. may act as a hydroskeleton which offsets the effects of compression and protects the scolex. A glycoprotein in the cyst fluid of *T. hydatigena* is considered to be an osmotically active molecule involved in maintaining the turgidity of the cyst (Dixon et al., 1973).

In other species, possible adaptations to offset compression are cyst walls and general body shape which enable parasites to conform to the interstices of the convoluted viscera.

12.5. Conclusion

The peritoneal cavity of vertebrates is parasitised by a wide range of helminths. Knowledge of the physicochemical properties of the habitat, apart from in a few species of mammals, is negligible, and the biological characteristics of its parasites are in many cases virtually unknown. Hence, an analysis of the ecology of the parasites can only be superficial and somewhat speculative at the present time. It should also be stressed that any analysis is complicated by the fact that many characteristics of the parasites may not represent adaptations for life in the peritoneal cavity, but are perhaps more intimately involved with development in some other environment at a subsequent stage of the life cycle.

It is to be hoped that this review has succeeded in bringing together a somewhat scattered and diverse literature, highlighted the great range of helminth parasites which occur in the peritoneal cavity, and pointed to a number of areas where there is enormous scope for further work.

12.6. REFERENCES

Anacker, R.L. and Munoz, J. (1961). "Mouse antibody. Characteristics and properties of antibody in mouse peritoneal fluid." J. Immunol. 87, 426—432.

Andersen, K. (1975). "Comparison of surface topography of three species of *Diphyllobothrium* (Cestoda: Pseudophyllidea) by scanning electron microscopy." Int. J. Parasit. 5, 293—300.

Angel, L.M. (1970). "*Coelomotrema antechinomes* gen. et sp. nov., a prosthogonimid trematode from *Antechinomys* and *Antechinus* (Marsupialia: Dasyuridae) from Australia," Aust. J. Zool. 18, 119—124.

Arme, C. and Owen, W. (1968). "Occurrence and pathology of *Ligula intestinalis* infections in British fishes." J. Parasit. 54, 272—280.

Arme, C. and Walkey, M. (1970). In: "Aspects of Fish Parasitology." (A.E.R. Taylor and R. Muller, eds.) pp. 79—101. Blackwell, Oxford.

Baer, J.G. (1951). "The Ecology of Animal Parasites." University of Illinois Press.

Banovitz, J., Jordan, R.T. and Trapani, L. (1963). "Electrophoretic analysis of serum and pleuroperitoneal fluid proteins and antibody from guinea pigs." Nature, London 197, 704—705.

Barker, I.K. (1970). "The penetration of oncospheres of *Taenia pisiformis* into the intestine of the rabbit." Can. J. Zool. 48, 1329—1332.

Baron, P.J. (1968). "On the histology and ultrastructure of *Cysticercus longicollis*, the cysticercus of *Taenia crassiceps* Zeder, 1800 (Cestoda, Cyclophyllidea)." Parasitology 58, 497—513.

Bauer, O.N. (1959): In: "Parasites of Freshwater Fish and the Biological Basis for their Control." Israel Program for Scientific Translations (1962).

Beavers, P.E., Esch, G.W. and Kuhn, R.E. (1971). "Some effects of sub-lethal X-irradiation on reproduction and development in two strains of larval *Taenia crassiceps.*" Int. J. Parasit. 1, 235—239.

de Beer, G.R. (1951). "Vertebrate Zoology." Sidgwick and Jackson, London.

Bernard, G.R., Wynn, R.A. and Wynn, G.G. (1966). "Chemical anatomy of the pericardial and periviseral fluids of the sting ray, *Dasyatis americana.*" Biol. Bull. 130, 18—27.

Birkeland, R. (1968a). ["Comparative investigations on glucose concentrations in peripheral blood and peritoneal fluid in clinically healthy sheep and pigs."] Nord Vet. Med. 20, 146—154. (In Norwegian).

Birkeland, R., (1968b). "Comparative investigation on the concentration of calcium, magnesium, sodium and potassium in the blood plasma and peritoneal fluid of sheep and pig." Nord. Vet. Med. 20, 155—160.

Black, V.S. (1957). In: "The Physiology of Fishes." (M.E. Brown, ed.) pp. 163—205. Academic Press, New York.

Bloom, W. and Fawcett, D.W. (1962). "A Text Book of Histology." 8th Edn., Saunders, Philadelphia.

Bosworth, N. and Archer, G. T. (1961). "The eosinophil content of the peritoneal cavity of the rat." Aust. J. exp. Biol. med. Sci. 39, 165—169.

Bråten, T. (1966). "Host specificity in *Schistocephalus solidus.*" Parasitology 56, 657—664.

Bråten, T. (1968). "The fine structure of the tegument of *Diphyllobothrium latum* (L.). A comparison of the pleroceroid and adult stages." Ztschr. Parasitenk. 30, 104—112.

Brengues, J. and Gidel, R. (1972). "Recherches sur *Setaria labiatopapillosa* (Perroncito, 1882) en Afrique occidentale." Ann. Parasit. Hum. Comp. 47, 597—611.

Bundesen, P.G. and Janssens, P.A. (1971). "Biochemical tracing of parasitic infections. II. *Taenia pisiformis* in rabbits—a quantitative study." Int. J. Parasit. 1, 15—20.

Charles, G.H. and Orr, T.S.C. (1968). "Comparative fine structure of the outer tegument of *Ligula intestinalis* and *Schistocephalus solidus.*" Expl. Parasit. 22, 137—149.

Chiller, J.M., Hodgins, H.O., Chambers, V.C. and Weiser, R.S. (1969a). "Antibody response in rainbow trout *(Salmo gairdneri).* I. Immunocompetent cells in the spleen and anterior kidney." J. Immunol. 102, 1193—1201.

Chiller, J.M., Hodgins, H.O. and Weiser, R.S. (1969b). "Antibody response in rainbow trout *(Salmo gairdneri).* II. Studies on the kinetics of development of antibody-producing cells and on complement and natural hemolysin." J. Immunol. 102, 1202—1207.

Chin, K.N. and Hudson, G. (1972). "Quantitative study of the effects of cyclophosphamide on peritoneal cell populations." Acta Haematol. (Basel) 48, 239—244.

Culbreth, K.L., Esch, G.W. and Kuhn, R.E. (1972). "Growth and development of larval *Taenia crassiceps* (Cestoda). III. The relationship between larval biomass and the uptake and incorporation of ^{14}C-leucine." Expl. Parasit. 32, 272—281.

Davies, P.S. and Walkey, M. (1966). "The effect of body size and temperature upon oxygen consumption of the cestode *Schistocephalus solidus* (Müller)." Comp. Biochem. Physiol. 18, 415—425.

Davis, R.H. (1973). "Effect of birth on fetal mouse peritoneal fluid cellular composition." Proc. Soc. exp. Biol. Med. 142, 938—940.

Davydov, O.N. and Kosenko, L. Ya. (1972). [Membrane digestion of *Ligula intestinalis* plerocercoids."] Parazitologiya 6, 269—273. (In Russian).

De Giusti, D.L. (1939). "Further studies on the life cycle of *Leptorhynchoides thecatus.*" J. Parasit. 25, Suppl., 22.

Del Monte, U. (1967). "Changes in oxygen tension in Yoshida ascites hepatoma during growth." Proc. Soc. exp. Biol. Med. 125, 853—856.

Dittmer, D.S. ed. (1961). "Blood and Other Body Fluids." Federation of American Societies for Experimental Biology, Washington.

Dixon, S.N., Gibbons, R., Parker, J. and Sellwood, R. (1973). "Characteristics of a glycoprotein in the cyst fluid of *Cysticercus tenuicollis* from the goat." Int. J. Parasit. 3, 419—424.

Dönges, J. and Harden, W. (1966). "*Nesolecithus africanus* n. sp. (Cestodaria, Amphilinidea) aus dem coelom von *Gymnarchus niloticus* Cuvier, 1829 (Teleostei)." Ztschr. ParasitKde 28, 125—141.

Dorais, F.J. and Esch, G.W. (1969). "Growth rate of two *Taenia crassiceps* strains." Expl Parasit. 25, 395—398.

Drewes, P.A. and McKee, R.W. (1967). "Amino acid composition of ascitic fluid and blood plasma from mice bearing Ehrlich-Lettere tumour." Nature, London 213, 411—412.

Eckert, J., von Brand, T. and Voge, M. (1969). "Asexual multiplication of *Mesocestoides corti* (Cestoda) in the intestine of dogs and skunks." J. Parasit. 55, 241—249.

Eichler, D.A. (1971). "Studies on *Onchocerca guttorosa* (Neumann, 1910) and its development in *Simulium ornatum* (Meigen, 1818). II. Behaviour of *S.ornatum* in relation to the transmission of *O. guttorosa.*" J. Helminth. 45, 259—270.

Eichler, D.A. and Nelson, G.S. (1971). "Studies on *Onchocerca gutturosa* (Neumann, 1910) and its development in *Simulium ornatum* (Meigen, 1818). I. Observations of *O. gutturosa* in cattle in south-east England." J. Helminth. 45, 245—258.

Enigk, K., Feder, H. and Dey-Hazra, A. (1972). "Der Einfluss präpatenter Helmintheninfektionen auf dem Mineralstoff- und Enzymgehalt des Blutplasmas sowie den Mineralstoffgehalt der Leber beim Schwein." Ztschr. *Parasitenk.* 39, 323—338.

Esch, G.W. (1969). "*Taenia crassiceps:* insulin and carbohydrate metabolism in larval forms." Expl Parasit. 25, 210—216.

Esch, G.W. and Kuhn, R.E. (1971). "The uptake of ^{14}C-Chlorella protein by larval *Taenia crassiceps* (Cestoda)." Parasitology 62, 27—29.

Everard, C.O.R., Elisha, S.T. and Davies, J.B. (1974). "The biology of *Dipetalonema proechimycis* Esslinger, 1974 (Nematoda: Filaroidea) in Trinitrad." J. Parasit. 60, 556—558.

Fadzil, M. (1969). "*Setaria* sp. in a buffalo monster stillbirth." Kajian Vet. 2, 89—90.

Finn, J.P. (1970). "The protective mechanisms in diseases of fish." Vet. Bull. Weybridge 40, 873—886.

Fischer, H. and Freeman, R.S. (1969). "Penetration of parenteral plerocercoids of *Proteocephalus ambloplitis* (Leidy) into the gut of smallmouth bass." J. Parasit. 55, 766—774.

Forbes, I.J. (1966). "Mitosis in mouse peritoneal macrophages." J. Immunol. 96, 734–743.

Fox, L.L., Kuhn, R.E. and Esch, G.W. (1971). *"Taenia crassiceps:* Antigenic comparison of two larval strains." Expl Parasit. 29, 194–196.

Frayha, G.J. (1971). "Comparative metabolism of acetate in the taenüd tapeworms *Echinococcus granulosus, E. multilocularis* and *Taenia hydatigena."* Comp. Biochem. Physiol. 39B, 167–170.

Frayha, G.J. (1974). "Synthesis of certain cholesterol precursors by hydatid protoscolices of *Echinococcus granulosus* and cysticerci of *Taenia hydatigena."* Comp. Biochem. Physiol. 49B, 93–98.

Freeman, R.S. (1962). "Studies on the biology of *Taenia crassiceps* (Zeder, 1800) Rudolphi, 1810 (Cestoda)." Can. J. Zool. 40, 969–990.

Freeman, R.S. (1964). "Studies on responses of intermediate hosts to infection with *Taenia crassiceps* (Zeder, 1800) (Cestoda)." Can. J. Zool. 42, 367–385.

Fruhman, G.J. (1972). "The effects of zymosan upon the mast cell population of the peritoneal fluid." Proc. Soc. exp. Biol. Med. 141, 101–105.

Gemmell, M.A. (1969). "Immunological response of the mammalian host against tapeworm infections. XI. Antigen sharing among *Taenia pisiformis, T. hydatigena* and *T. ovis."* Expl Parasit. 26, 67–72.

Gemmell, M.A. (1972a). "Hydatidosis and cysticercosis. 4. Acquired resistance to *Taenia hydatigena* under conditions of strong infection pressure." Aust. Vet. J. 48, 26–28.

Gemmell, M.A. (1972b). "Hydatosis and cysticercosis. 5. Some problems of inducing resistance to *Taenia hydatigena* under conditions of strong infection pressure." Aust. Vet. J. 48, 29–31.

Goodman, J.W. (1963). "Transplantation of peritoneal fluid cells." Transplantation 1, 334–346.

Grammeltvedt, A.F. (1973). "Differentiation of the tegument and associated structures in *Diphyllobothrium dendriticum* ((Nitsch, 1824) (Cestoda: Pseudophyllidea). An electron microscopical study." Int. J. Parasit. 3, 321–327.

Guttowa, A. and Honowska, M. (1973). "Changes in the serum protein fractions in the course of *Ligula intestinalis* L. plerocercoid infestations in the bream, *Abramis brama* (L.)." Acta Parasit. Pol. 21, 107–114.

Halvorsen, O. (1970). "Studies on the helminth fauna of Norway. XV. On the taxonomy and biology of plerocercoids of *Diphyllobothrium* Cobbold, 1858 (Cestoda, Pseudophyllidea) from North–Western Europe." Nytt Mag. Zool. 18, 113–174.

Hartveit, F. and Thunold, S. (1966). "Peritoneal fluid volume and the oestrus cycle in mice." Nature, London 210, 1123–1125.

Haynes, W.D.G. (1970). *"Taenia crassiceps:* uptake of basic and aromatic amino acids and imino acids by larvae." Expl Parasit. 27, 256–264.

Haynes, W.D.G. and Taylor, A.E.R. (1968). "Studies on the absorption of amino acids by larval tapeworms (Cyclophyllidea: *Taenia crassiceps)."* Parasitology 58, 47–59.

Heath, D.D. (1970). "The development of *Echinococcus granulosus* larvae in laboratory animals." Parasitology 60, 449–456.

Heath, D.D. (1971). "The migration of oncospheres of *Taenia pisiformis, T. serialis* and *Echinococcus granulosos* within the intermediate host." Int. J. Parasit. 1, 145–152.

Heath, D.D. (1973a). "Resistance to *Taenia pisiformis* larvae in rabbits. I. Examination of the antigenically protective phase of larval development." Int. J. Parasit. 3, 485–489.

Heath, D.D. (1973b). "Resistance to *Taenia pisiformis* larvae in rabbits. II. Temporal relationships and the development phase affected." Int. J. Parasit. 3, 491–498.

Heath, D.D. (1973c). "An improved technique for the in vitro culture of taeniid larvae." Int. J. Parasit. 3, 481–484.

Heath, D.D. and Smyth, J.D. (1970). "In vitro cultivation of *Echinococcus granulosus,*

Taenia hydatigena, T. ovis, T. pisiformis and *T. serialis* from oncosphere to cystic larva." Parasitology 61, 329—343.

Heath, R.L. (1970). "Biosynthesis de novo of purines and pyrimidines in *Mesocestoides* (Cestoda). I." J. Parasit. 56, 98—102.

Heath, R.L. and Hart, J.L. (1970). "Biosynthesis de novo of purines and pyrimidines in *Mesocestoides* (Cestoda). II." J. Parasit. 56, 340—345.

Hess, E. (1972). "Transmission maternelle de tetrathyridia *(Mesocestoides*, Cyclophyllidea) chez la souris blanche." C.R. Hebd. Seanc. Acad. Sci. Series D 274, 596—599.

Hoar, W.S. (1969). In: "Fish Physiology." Vol. 3 (W.S. Hoar and D.J. Randall, eds.) pp. 1—72. Academic Press, New York and London.

Holmes, J.C. and Bethel, W.M. (1972). In: "Behavioural Aspects of Parasite Transmission." (E.U. Canning and C.A. Wright, eds.) pp. 123—149. Academic Press, New York and London.

Hopkins, C.A. and McCaig, M.L.O. (1963). "Studies on *Schistocephalus solidus*. I. The correlation of development in the plerocercoid with infectivity to the definitive host." Expl Parasit. 13, 235—243.

Hopkins, C.A. and Smyth, J.D. (1951). "Notes on the morphology and life history of *Schistocephalus solidus* (Cestoda: Diphyllobothriidae)." Parasitology 41, 283—291.

Howell, M.J. (1973). "The resistance of cysts of *Stictodora lari* (Trematoda: Heterophyidae) to encapsulation by cells of the fish host." Int. J. Parasit. 3, 653—659.

Hunter, G.C. and Kille, R.A. (1950). "Some observations on *Dictyocotyle coeliaca* Nybelin, 1941 (Monogenea)." J. Helminth. 24, 15—22.

Huizinga, H.W. (1967). "The life cycle of *Contracaecum multipapillatum* (von Drasche, 1882) Lucker, 1941 (Nematoda: Heterocheilidae)." J. Parasit. 53, 368—375.

Izmagilova, R.G. (1968). ["Changes in the vitamin A and C balance in helminthiases."] Materiali Seminaria-Soveshchaniya po Borbe s Gel'mintozanii sel'. — khoz. Zhivotnikh v Chimkente, Alma — Ata pp 73—74. (In Russian).

Jacques, R. and Rüegg, M. (1970). "Age and sex differences in mast cell count and histamine content." Agents and Actions 1, 144—147.

Kamenski, I.V. (1964). ["Study of the epizootiology of ligulid infections of fish in the Kakhovsk water reservoir."] Trudy vses. Inst. Gel'mint. 11, 62—70. (In Russian).

Kearn, G.C. (1970). "The oncomiracidium of the monocotylid monogeneans *Dictyocotyle coeliaca* and *Calicotyle kroyeri*." Parasitology 61, 153—160.

Kearn, G.C. (1975). "Hatching in the monogenean parasite *Dictyocotyle coeliaca* from the body cavity of *Raja naevis*." Parasitology 70, 87—93.

Köhler, P. and Hanselmann, K. (1974). "Anaerobic and aerobic energy metabolism in the larvae (tetrathyridia) of *Mesocestoides corti*." Expl. Parasit. 36, 178—188.

Kowalski, J.C. and Thorson, R.E. (1972a). "Protective immunity against tetrathyridia of *Mesocestoides corti* by passive transfer of serum in mice." J. Parasit. 58, 244—246.

Kowalski, J.C. and Thorson, R.E. (1972b). "Immunization of laboratory mice against tetrathyridia of *Mesocestoides corti* (Cestoda) using a secretory and excretory antigen and a soluble somatic antigen." J. Parasit. 58, 732—734.

Lee, S. and Schoen, I. (1968). "Eosinophilia of peritoneal fluid and peripheral blood associated with chronic peritoneal dialysis." Am. J. clin. Path. 47, 638—640.

Leigh, W.H. (1974). "Life history of *Ascocotyle mcintoshi* Price, 1936 (Trematoda: Heterophyidae)." J. Parasit. 60, 768—772.

Leone, E., Libonati, M. and Lutwak-Mann, C. (1963). "Enzymes in the uterine and cervical fluid and certain related tissue and body fluids of the rabbit." J. Endocrinol. 25, 551—552.

Leong, C.H.D. and Howell, M.J. (1971). "Formation and structure of the cyst wall of *Stictodora lari* (Trematoda: Heterophyidae)." Ztschr. ParasitKde 35, 340—350.

Litt, M. (1960a). "Studies on experimental eosinophilia. I. Repeated quantitation of peritoneal eosinophilia in guinea pigs by a method of peritoneal lavage." Blood, J. Haematol. 16, 1318—1329.

Litt, M. (1960b). "Studies on experimental eosinophilia. II. Induction of peritoneal eosinphilia by the transfer of tissues and tissue extracts.". Blood J. Haematol. 16, 1330—1337.

Lokaj, J. (1968). "The peritoneal cavity as a model in experimental immunology." Scr. Med. (Brno) 41, 169—176.

Lokaj, J. and Klhůfek, J. (1968). "Free and cell-bound proteins in the peritoneal cavity of normal and immunised mice." Scr. Med. (Brno) 41, 163—167.

Lubinsky, G. (1960). "The maintenance of *Echinococcus multilocularis sibiricensis* without the definitive host." Can. J. Zool. 38, 149—151.

Ludbrook, J. (1959). In: "A Symposium on pH and Blood Gas Measurement." (R.F. Woolmer, ed.) pp. 34—51, Churchill, London.

McCaig, M.L.O. and Hopkins, C.A. (1965). "Studies on *Schistocephalus solidus*. 3. The in vitro cultivation of the plerocercoid." Parasitology 55, 257—268.

McGowan, L. and Davis, R.H. (1968). "Cytologic changes in mouse peritoneal fluid between birth and sexual maturity." Proc. Soc. exp. Biol. Med. 129, 244—250.

McGowan, L. and Davis, R.H. (1969). "Effect of the mouse estrous cycle on peritoneal serous fluid cytology." Endocrinology 84, 175—177.

McGowan, L. and Davis, R.H. (1971) "Peritoneal fluid cytology in irradiation-induced ovarian tumours of mice." Obstet. Gynecol. 38, 125—135.

McGowan, L., Davis, R.H. and Bunnag, B. (1973). "The biochemical diagnosis of ovarian cancer." Am. J. Obstet. Gynecol. 116, 760—768.

McGowan, L., Davis, R.H. and Kressler, L.S. (1968). "The effect of pregnancy on the cytology of mouse peritoneal fluid." Proc. Soc. exp. Biol. Med. 128, 141—143.

McLaughlin, J.F., Ruddle, N.H. and Waksman, B.H. (1972). "Relationship between activation of peritoneal cells and their cytopathogenicity." Res. J. Reticuloendothel. Soc. 12, 293—304.

McManus, D.P. (1975). "Tricarboxylic acid cycle enzymes in the plerocercoid of *Ligula intestinalis* (Cestoda: Pseudophyllidea)." Ztschr. *Parasitenk.* 45, 319—322.

Marra, M. and Esch, G.W. (1970). "Distribution of carbohydrates in adults and larvae of *Proteocephalus ambloplitis* (Leidy, 1887)." J. Parasit. 56, 398—400.

Meakins, R.H. and Walkey, M. (1973). "Aspects of in vivo growth of the plerocercoid stages of *Schistocephalus solidus.*" Parasitology 67, 133—141.

Miller, J.F.A.P. (1972). "Lymphocyte interactions in antibody responses." Int. Rev. Cytol. 33, 77—130.

Mohr, W., Beneke, G. and Murr, L. (1971). "Proliferation der Zellsysteme in Cavum peritonei. I. Proliferation von Mesothelzellen, submesothelialen Bindgewebszellen, Endothelzellen und freien Zellen der Peritonealfluessigkeit induziert durch Phytohaemagglutinin." Beitr. Pathol. 143, 345—359.

Moline, J., Lavandier, M., Roulleier, A., Corcos, D. and Baudouin, J. (1972). "Mesure simultanée et comparative des pO₂, pCO₂ et pH dan le sang artériel et dans le compartiment séreux: Etude clinique et expérimentale." Pathol. Biol. 20, 263—276.

Molnár, K. (1967). "Morphology and development of *Philometra abdominalis* Nybelin, 1928." Acta Vet. Acad. Sci. Hung. 17, 293—300.

Molnár, K. and Fernando, C.H. (1975). "Morphology and development of *Philometra cylindracea* (Ward and Magath, 1916) (Nematoda: Philometridae)." J. Helminth. 49, 19—24.

Morris, G.P. and Finnegan, C.V. (1969). "Studies of the differentiating plerocercoid cuticle of *Schistocephalus solidus*. II. The ultrastructural examination of cuticle development." Can. J. Zool. 47, 957—964.

Munoz, J. (1957). "Production in mice of large volumes of ascites fluid containing antibodies." Proc. Soc. exp. Biol. Med. 95, 757—759.

Murrell, K.D. (1968). "Respiration studies and glucose absorption kinetics of *Taenia crassiceps* larvae." J. Parasit. 54, 1147—1150.

Nelson, D.S. and Boyden, S.V. (1963). "The loss of macrophages from peritoneal exudates following the injection of antigens into guinea pigs with delayed-type hypersensitivity." Immunology 6, 264—275.

Nelstrop, A.E., Taylor, G. and Collard, P. (1968). "Studies on phagocytosis. III. Antigen clearance studies in invertebrates and poikilothermic vertebrates." Immunology 14, 347—356.

Nemeth, I. (1972). "Immunological study of rabbit cysticercosis. V. Characterization of the antibody response to experimental infection with *Cysticercus pisiformis* (Bloch, 1780)." Acta Vet. Acad. Sci. Hung. 22, 377—408.

Nemeth, I. (1973). "Immunological study of rabbit cysticercosis. VI. Isolation of specific antibodies from serum of rabbits infected with the larvae of *Taenia pisiformis* (Bloch, 1780) Gmelin, 1790." Acta Vet. Acad. Sci. Hung. 23, 13—23.

Novak, M. (1972). "Quantitative studies on the growth and multiplication of tetrathyridia of *Mesocestoides corti* Hoeppli, 1925 (Cestoda: Cyclophyllidea) in rodents." Can. J. Zool. 50, 1189—1196.

Novak, M. (1974a). "Acceleration of the growth of tetrathyridial populations of *Mesocestoides corti* (Cestoda: Cyclophyllidea) by splenectomy." Int. J. Parasit. 4, 165—168.

Novak, M. (1974b). "Effect of sex hormones on the growth and multiplication of tetrathyridia of *Mesocestoides corti* (Cestoda: Cyclophyllidea) in mice." Int. J. Parasit. 4, 371—374.

Novak, M. (1975). "Gonadectomy, sex hormones and the growth of tetrathyridial populations of *Mesocestoides corti* (Cestoda: Cyclophyllidea) in mice." Int. J. Parasit. 5, 269—274.

Novak, M. and Lubinsky, G. (1973). "Acceleration of the growth of populations of and of the multiplication of tetrathyridia of *Mesocestoides corti* Hoeppli, 1925 (Cestoda: Cyclophyllidea) by some cytostatic agents." Can. J. Zool. 51, 83—90.

Orr, T.S.C. (1967). "Distribution of the plerocercoid of *Ligula intestinalis.*" J. Zool. 153, 91—97.

Orr, T.S.C. and Hopkins, C.A. (1969). "Maintenance of *Schistocephalus solidus* in the laboratory with observations on the rate of growth of, and proglottid formation in, the plerocercoid." J. Fish. Res. Bd. Can. 26, 741—752.

Orr, T.S.C., Hopkins, C.A. and Charles, G.H. (1969). "Host specificity and rejection of *Schistocephalus solidus.*" Parasitology 59, 683—690.

Padawer, J. and Gordon, A.S. (1956). "Cellular elements in the peritoneal fluid of some mammals." Anat. Rec. 124, 209—222.

Pappas, P.W. and Read, C.P. (1973). "Permeability and membranes transport in the larvae of *Taenia crassiceps.*" Parasitology 66, 33—42.

Pappas, P.W., Uglem, G.L. and Read, C.P. (1973a). "*Taenia crassiceps:* absorption of hexoses and partial characterization of Na^+-dependent glucose absorption by larvae." Expl Parasit. 33, 127—137.

Pappas, P.W., Uglem, G.L. and Read, C.P. (1973b). "Mechanisms and specificity of amino acid transport in *Taenia crassiceps* larvae (Cestoda)." Int. J. Parasit. 3, 641—651.

Pennycuick, L. (1971). "Differences in the parasite infections in three-spined sticklebacks (*Gasterosteus aculeatus* L.) of different sex, age and size." Parasitology 63, 407—418.

Powell, A.M. and Chubb, J.C. (1966). "A decline in the occurrence of *Diphyllobothrium* plerocercoids in the trout *Salmo trutta* L. of Llyn Padarn, Caernarvonshire." Nature, London. 211, 439.

Prosser, C.L. and Brown, F.A. (1966). "Comparative Animal Physiology." 2nd Edn., Saunders, Philadelphia.

Reichenbach-Klinke, H. and Elkan, E. (1965). "The Principal Diseases of Lower Vertebrates." Academic Press, New York and London.

Reuter, J. (1967a). "Studies on plerocercoids of *Diphyllobothrium dendriticum*. II. The dependence of lactic and succinic acid excretion on the gas phase." Acta Acad. Abo. B 27 (9) 7 pp.

Reuter, J. (1967b). "Studies on plerocercoids of *Diphyllobothrium dendriticum*. III. The oxygen uptake and the excretion of lactic and succinic acid in media with different osmotic concentrations." Acta Acad. Abo. B 27 (10) 10 pp.

Reuter, J. (1971). "Studies on plerocercoids of *Diphyllobothrium dendriticum*. V. Rates of acid production, change in weight, and depletion of glycogen during in vitro cultivation at different temperatures." Acta Acad. Abo. B 31 (10), 4 pp.

Rickard, M.D. (1974). "Hypothesis for the long term survival of *Taenia pisiformis* cysticerci in rabbits." Ztschr. ParasitKde 44, 203—209.

Robinson, D.L.H., Silverman, P.H. and Pearce, A.R. (1963). "The culture of *Taenia crassiceps* in vitro." Trans. roy. Soc. trop. Med. Hyg. 57, 238.

Shield, J.M., Heath, D.D. and Smyth, J.D. (1973). "Light microscope studies of the early development of *Taenia pisiformis* cysticerci." Int. J. Parasit. 3, 471—480.

Silverman, P.H. (1954). "Studies on the biology of some tapeworms of the genus *Taenia*. I. Factors affecting hatching and activation of taeniid ova, and some criteria of their viability." Ann. trop. Med. Parasit. 48, 207—215.

Silverman, P.H. and Maneely, R.B. (1955). "Studies on the biology of some tapeworms of the genus *Taenia*. III. The role of the secreting gland of the hexacanth embryo in the penetration of the intestinal mucosa of the intermediate host, and some of its histochemical reactions." Ann. trop. Med. Parasit. 49, 326—330.

Sinha, D.P. and Hopkins, C.A. (1967). "Studies on *Schistocephalus solidus*. 4. The effect of temperature on growth and maturation in vitro." Parasitology 57, 555—566.

Smith, J.K., Esch, G.W. and Kuhn, R.E. (1972a). "Growth and development of larval *Taenia crassiceps* (Cestoda). I. Aneuploidy in the anomalous ORF strain." Int. J. Parasit. 2, 261—263.

Smith, J.K., Parrish, M., Esch, G.W. and Kuhn, R.E. (1972b). "Growth and development of larval *Taenia crassiceps* (Cestoda)—II. RNA and DNA synthesis in the ORF and KBS strains determined by autoradiography." Int. J. Parasit. 2, 383—389.

Smithers, S.R. and Terry, R.J. (1969). "The immunology of schistosomiasis." Adv. Parasit. 7, 41—93.

Smyth, J.D. (1969). "The Physiology of Cestodes." Oliver and Boyd, Edinburgh.

Smyth, J.D. (1972). In: "Functional Aspects of Parasite surfaces." (A.E.R. Taylor and R. Muller, eds.) pp. 41—70. Blackwell, Oxford.

Smyth, J.D. and Heath, D.D. (1970). "Pathogenesis of larval cestodes in mammals." Helminth. Abstr. 39, 1—23.

Soldati, M., Intini, C., Isetta, A.M. and Ghione, M. (1971). "Further studies on the experimental infection of the mouse with trachoma agents." Rev. Int. Trach. 47, 31—41.

Specht, D. and Voge, M. (1965). "Asexual multiplication of *Mesocestoides* tetrathyridia in laboratory animals." J. Parasit. 51, 268—272.

Specht, D. and Widmer, E.A. (1972). "Response of mouse liver to infection with tetrathyridia of *Mesocestoides* (Cestoda)." J. Parasit. 58, 431—437.

Sweatman, G.K. (1957). "Acquired immunity in lambs infected with *Taenia hydatigena* Pallas, 1766." Can. J. comp. Med. 21, 65—71.

Sweatman, G.K. and Plummer, P.J.G. (1957). "The biology and pathology of the tapeworm *Taenia hydatigena* in domestic and wild hosts." Can. J. Zool. 35, 93—109.

Tarkiewicz, S. (1956). "Dane do znjajomosci plyna ortrzewnego u bydla." Ann. Univ. Mariae-Curie Sklodowska Sect. DD — Vet. Sci. 11, 315—335.

Tarkiewicz, S. (1962). "Badania porownawcze stezenia bialka ogolnego i zawartosci frakcju bialkowych w surowicy krwi i plynie otrzewnowycm bydlia." Ann. Univ. Mariae—Curie Sklodowska Sect. DD — Vet. Sci. 17, 235—242.

268

Taylor, A.E.R. (1963). "Maintenance of larval *Taenia crassiceps* (Cestoda: Cyclophylli-dea) in a chemically defined medium." Expl. Parasit. 14, 304—310.

Taylor, A.E.R. and Haynes, W.D.G. (1966). "Studies on the metabolism of larval tape-worms (Cyclophyllidea: *Taenia crassiceps*). I. Amino acid composition before and after in vitro culture." Expl. Parasit. 18, 327—331.

Taylor, A.E.R., McCabe, M. and Longmuir, I.S. (1966). "Studies on the metabolism of larval tapeworms (Cyclophyllidea: *Taenia crassiceps*). II. Respiration, glycogen utiliza-tion, and lactic acid production during culture in a chemically defined medium." Expl. Parasit. 19, 269—275.

Van Furth, R. (1970). In: "Mononuclear Phagocytes." (R. Van Furth, ed.) pp. 151—165. Blackwell Scientific Publications, Oxford and Edinburgh.

Van Furth, R., Hirsch, J.G. and Fedorko, M.E. (1970). "Morphology and peroxidase cy-tochemistry of mouse promonocytes, monocytes and macrophages." J. exp. Med. 132, 794—812.

Valtonen, E.J., Koivuniemi, A. and Mantere, R. (1968). "Cellular composition of normal human peritoneal fluid." Ann. Med. Intern. Fenn. 57, 125—127.

Varela-Diaz, V.M., Gemmell, M.A. and Williams, J.F. (1972). *"Taenia hydatigena* and *T. ovis*: Antigen sharing. XII. Immunological responses of the mammalian host against tapeworm infections." Expl Parasit. 32, 96—101

Voge, M. (1963). "Maintenance in vitro of *Taenia crassiceps* cysticerci." J. Parasit. 49, Suppl., 59—60.

Voge, M. and Coulombe, L.S. (1966). "Growth and asexual multiplication in vitro of *Mesocestoides* tetrathyridia." Am. J. trop. Med. Hyg. 15, 902—907.

Voge, M. and Seidel, J.A. (1968). "Continuous growth in vitro of *Mesocestoides* (Cestoda) from oncosphere to fully developed tetrathyridium." J. Parasit. 54, 269—271.

Von Brand, T. (1952). "Chemical Physiology of Endoparasitic Animals." Academic Press, New York and London.

Von Brand, T. (1973). "Biochemistry of Parasites." 2nd Edn., Academic Press, New York and London.

Walkey, M. and Davies, P.S. (1968). "Effects of pH on oxygen consumption of the cestode *Schistocephalus solidus.*" Expl Parasit. 22, 201—206.

Walkey, M. and Meakins, R.H. (1970). "An attempt to balance the energy budget of a host-parasite system." J. Fish. Biol. 2, 361—372.

Weinbach, E.C. and Eckert, J. (1969). "Respiration of the larvae (tetrathyridia) of *Meso-cestoides corti.*" Expl Parasit 24, 54—62.

Weinman, C.J. (1970). In: "Immunity to Parasitic Animals." Vol. 2. (G.J. Jackson, R. Herman and I. Singer, eds.) pp. 1021—1059. Appleton-Century-Crofts, New York.

Wierzbicki, K. (1960). "Philometrosis of crucian carp." Acta Parasit. Pol. 8, 181—194.

Wikerhauser, T., Zuković, M. and Dzakula, N. (1971). *"Taenia saginata* and *T. hydatige-na:* intramuscular caccination of calves with oncospheres." Expl Parasit. 30, 36—40.

Williams, H.H. (1967). "Helminth diseases of fish." Helminth. Abstr. 36, 261—295.

Witte, M.H., Witte, C. L., Davis, W.M., Cole, W.R. and Dumont, A.E. (1972). "Peritoneal transudate: A diagnostic clue to portal system obstruction in patients with intra-abdo-minal neoplasms or peritonitis." J. Am. med. Ass. 221, 1380—1383.

Wu, A.M., Till, J.E., Siminovitch, L. and McCulloch, E.A. (1968). "Cytological evidence for a relationship between normal hematopoietic colony-forming cells and cells of the lymphoid system." J. exp. Med. 127, 455—464.

Yoffey, J.M. and Courtice, F.C. (1970). "Lymphatics, Lymph and the Lymphomyeloid Complex." Academic Press, New York and London.

Zhloba, A.F. (1968). ["Sugar, amylase and aldolase content in the exudate of the ab-dominal cavity during acute appendicitis."] Khirurgiya 44, 86—89. (In Polish).

Ecological Aspects of Parasitology
Editor: C.R. Kennedy
© *North-Holland Publishing Company, Amsterdam, 1976*

CHAPTER 13

MUSCULATURE

Dickson DESPOMMIER

Division of Tropical Medicine, School of Public Health, College of Physicians and Surgeons, Columbia University, 630 West 168th Street, New York, N.Y. 10032 (U.S.A.)

Contents

13.1. Introduction

Motility is a biological phenomenon shared by all phyla of animals. While some animals rely solely upon flagella or cilia for their movement, most organisms have some kind of intracellular contractile protein system which they use for locomotion.

Musculature represents an advanced system of highly organized contractile proteins and related structures and is utilized by all organisms from the primitive echinoderms through the vertebrates.

In vertebrates, musculature has become differentiated into three varieties of cells; smooth, striated cardiac, and striated skeletal.

Musculature has been selected as a niche by many forms of parasitic life (see Garnham, 1973 for a review of the protozoan and helminthic parasites inhabiting mammalian muscle tissues), eg. viruses, bacteria, fungi, protozoans and helminths. However, even though the striated skeletal muscles may constitute as much as 40% of the wet weight of the mammalian host (Lockhart, 1973), surprisingly few parasites infect only this tissue. *Sarcocystis sp.*, a protozoan, and *Trichinella spiralis*, a nematode, represent the two most commonly encountered exclusive striated muscle parasites.

A description of the ecological balance that has been established between the larva of *Trichinella spiralis* and the striated skeletal muscle cell of mammals will follow. The main reasons for discussing this balanced system are:

1) Much information exists concerning the normal structure and function of the striated skeletal muscle cell, while a smaller, yet significant, body of information is available concerning the interactions of *T. spiralis* with its host cell.

2) The larva of *T. spiralis* lives intracellularly and differs from all other intracellular parasites in that it does not replicate in its host cell.

3) The infected host cell remains intact and alive throughout the infection period, which, in some instances, may be for the life of the host.

In the ensuing discussion, emphasis will be placed upon parasite-induced niche modification.

13.2. *Trichinella spiralis* — Striated Muscle Cell Interactions

13.2.1. The concept of the intracellular niche

The life cycle of *T. spiralis* is unlike that of most other helminths in that the same animal functions as both the definitive and intermediate host. While the adult stage lives in the small intestine, the infective larva lives intracellularly in a modified skeletal muscle cell (Ribas-Mujal and Rivera-Pomar, 1968).

Prior to larval invasion, the striated muscle cell, as will be seen, represents a rather variable niche into which the larva of *T. spiralis* must enter.

Since a niche is defined as the entire milieu in which an organism lives (Odum, 1971), it would be impossible to present, in comprehensive fashion, all of the physical, chemical, physiological, and other environmental factors that characterize the striated muscle cell (see Bourne, ed., 1973, Structure and Function of Muscle, Vols. 1—4 for a thorough review of normal and abnormal muscles). However, some general aspects of the striated muscle cell will be presented in order to point out those features of the niche that contribute to its variability. These features will be compared with the specific changes that have been induced at various stages of the intracellular infection.

13.2.2. The pre-infection niche

13.2.2.1. Physical environment

The striated skeletal muscle cell (often referred to as a fiber but, in reality, a multinucleated cell) can be as long as 30 cm (Lockhart and Brandt, 1938) and 0.01—0.1 mm in diameter (Lockhart, 1973). Muscle cells rarely exist apart from one another; they are commonly found in bundles surrounded by a connective tissue sheath.

Upon microscopical examination, each muscle cell is seen to possess randomly spaced peripheral nuclei and a highly ordered array of contractile filaments (actin and myosin), interrupted at regular intervals by a dark line, Z band, which lies perpendicular to the filaments. Over 50% of the total protein of the striated muscle cell is composed of contractile filaments. The region between two Z bands is the sarcomere, which is the basic contractile unit of the cell. The actin (thin) filaments are attached to the Z bands, while the myosin (thick) filaments remain unattached when the sarcomere is relaxed. The highly ordered array of actin and myosin in the relaxed muscle cell, as revealed by X-ray crystallography, forms a liquid crystal (April, 1975).

Non-filamentous subcellular components include mitochondria, glycogen, two varieties of smooth membrane systems (sarcoplasmic reticulum and transverse tubules) and a few ribosomes. The scarcity of ribosomes reflects the low rate of protein synthesis found in the striated muscle cell.

Throughout the contraction cycle, many changes in the physical environment take place, some of which may play a role in the initial stages of infection with *T. spiralis*.

The most obvious feature during contraction is the general shortening of the muscle cell by the interdigitation of the actin and myosin filaments. It is reasonable to assume that at this time the non-filamentous components become somewhat distorted in their cellular locations relative to the resting state. The contraction cycle is triggered by a nerve impulse, thereby eliciting changes in the electrical environment as well. The resting potential in most striated muscle cells is about −90 MV, while the excitation of the muscle fiber results in a +30 MV potential or a net change of 120 MV over a 20 millisecond interval (Nastuk and Hodgkin, 1950).

13.2.2.2. Chemical environment

The chemical composition of the striated muscle cell is as unique as its physical makeup suggests.

So well studied are those components of the striated muscle cell involved with contraction, that one needs only to consult a modern text of cell biology or physiology for a fairly complete listing. A few key biochemical constituents deserve reiteration, however, since their fate during the early stages of the intracellular infection has been recently elucidated (Stewart and Read, 1974).

These "marker" components include myosin, actin, creatine, creatine phosphate, and myoglobin. In addition, the striated muscle cell contains high levels of ATP.

Like all other cells the activity of the muscle cell determines the levels of metabolic by-products produced, as well as the kinds and levels of nutrients taken up. However, since the striated muscle cell is capable of anaerobic as well as aerobic metabolism, the secretion of lactate also varies.

13.2.2.3. The physiological environment

The physiological environment is the sum of the physical and chemical interactions resulting in a contraction-rest cycle.

As with the chemical and physical environment, the physiological aspects of muscle contraction have been well described (Zierlers, 1974).

13.2.2.4. The extracellular environment

Each striated muscle cell is surrounded by a membrane (sarcolemma) on top of which lies a non-membranous glycocalyx. Each striated muscle cell has direct contact with the extracellular fluid by its transverse tubule system that forms a complicated network of smooth membranes which run perpendicular to the contractile filaments.

Each bundle of striated muscle cells is surrounded by a variety of cellular elements, which include fibroblasts and satellite cells, while capillaries, arterioles and venules constitute adjacent organs. The interactions of these cells and organs with each striated muscle cell are incompletely understood, with the exception of the capillary. Apparently, the fibroblast contributes collagen to the general system of muscle bundles, the collagen functioning as a kind of intercellular "glue". The satellite cells are important during muscle regeneration, but there is still much debate as to their specific role in that process (Murray, 1973).

In summary, the striated muscle cell represents a unique pre-infection niche in that (1) its physical composition is highly ordered and, at rest, resembles a liquid crystal, (2) its chemical composition is restricted in the variety of biochemical components when compared to most other specialized mammalian cells, and (3) its metabolism can vary greatly depending upon the activity of the host.

In the discussion to follow, the ways in which the parasite brings about the stabilization of this changing niche will be elucidated.

13.2.3. The post-infection niche

The intracellular phase of *T. spiralis* infection has been somewhat arbitrarily divided into 8 stages, based upon functional, as well as morphological, considerations (Table 1). These stages are: (1) recognition; (2) attachment; (3) penetration; (4) migration; (5) initial growth phase; (6) lag phase; (7) exponential growth phase; and (8) maturation.

13.2.3.1. Recognition

The schematic drawings and relevant micrographs in Figs. 1—9 represent the proposed sequence of events leading to the invasion of a muscle cell.

No experimental evidence is available to explain how the newborn larva of *T. spiralis* finds its way from the thoracic duct (Harley and Gallicchico, 1971) into the blood stream, and finally out of the capillary and into a striated muscle cell.

However, a recent unpublished study has revealed some of the morphological events leading to the entry of the larva into the muscle cell (Lewis and Despommier, unpublished data).

Apparently, the first event following exit from the capillary into the extracellular space is an alignment of the larva parallel to the sarcolemma (Figs. 1, 3). Whether the larva is attracted to a specific surface area of the muscle cell has not yet been determined, but it is probable that the initial alignment of the larva is, at least in part, dictated by the spatial arrangement of the surrounding muscle cells.

TABLE I

Stages in the life cycle of the intracellular larva of *Trichinella spiralis*

Days after penetration	Life cycle stage	Site of stage
0	Recognition	Extracellular, perhaps even in capillary
0	Attachment	Glycocalyx
0	Penetration	Sarcolemma
0—3	Migration	Intracellular adjacent to sarcolemma
0—1	Initial growth	Intracellular, central portion of muscle cell
2—3	Lag phase	Intracellular
4—10	Exponential growth	Intracellular, inside Nurse cell cytoplasm
11—19	Exponential growth	Inside Nurse cell surrounded by double membrane
20	Mature larva	Inside Nurse cell surrounded by double membrane

274

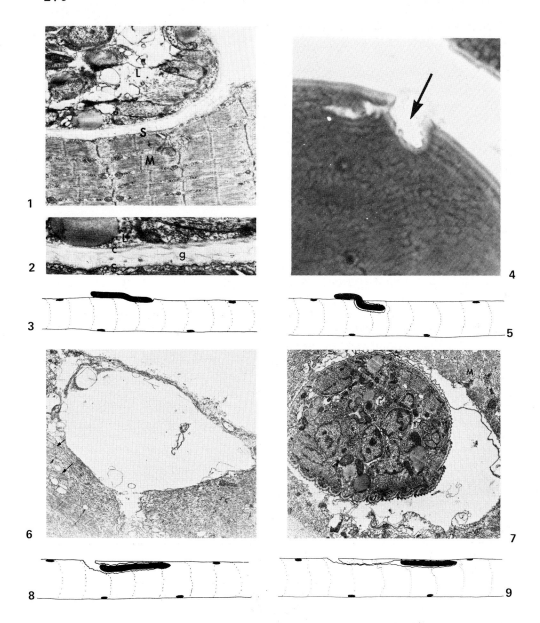

Fig. 1. The larva (L) as viewed at the ultrastructural level is aligned with the sarcolemma (S) during the attachment phase. Note indentation of the muscle cell (M). ×12,200.

Fig. 2. A high magnification EM view of the larva (L), sarcolemma (S), and glycocalyx (g). The glycocalyx follows the contour of the larval cuticle (C). ×27,000.

Fig. 3. Schematic representation of the attachment phase.

13.2.3.2. Attachment

There is little experimental evidence for a specific attachment process analogous to, for instance, the absorption of a virus particle to receptors on the surface of its host cell. However, it is clear from ultrastructural studies and preliminary in vitro experiments that the larva of *T. spiralis* becomes closely associated with the glycocalyx prior to penetration (Figs. 1, 2).

Apparently, the larva causes an indentation in the sarcolemma and appears to become attached to the glycocalyx (Fig. 1). In vitro studies employing mouse embryo muscle tissue have revealed that the larvae attach to the sarcolemma and are retained on the cell surface during changes of medium (Miranda, personal communication). Since larvae never entered the mouse embryo muscle cells, it is possible that the attachment observed in vitro was artifactual.

It is unlikely that the attachment of the larva to the glycocalyx represents a cell-specific recognition process, since larvae have been recovered from a variety of tissues other than muscle (Mauss and Otto, 1942).

13.2.3.3 Penetration

While all of the events of the penetration phase have yet to be documented, a distinct, ragged-edged hole in the sarcolemma (Fig. 4) represents indirect evidence that penetration of the fiber has taken place.

Data (personal observation) collected from experiments employing mice injected intramuscularly with newborn larvae (Despommier et al., 1975) suggest that penetration occurs rapidly, taking no longer than 10 minutes for a complete entrance of the larva into the muscle cell.

From the torn appearance of the hole that is made by the larva in the sarcolemma it is tempting to speculate that penetration is solely a mechanical process, but it is possible that hydrolytic enzymes of various classes may also play a role (Read, 1970). In any case, it is clear that the host cell does not participate actively in penetration by a phagocytic-like

---

Fig. 4. A ragged-edged hole (arrow) in the infected muscle cell resulting from the penetration of the cell by the larva. (Phase-contrast). × 1700.

Fig. 5. Schematic representation of the penetration phase.

Fig. 6. An ultrastructural view of the hole made by a penetrating larva. Note the disarray of myofilaments (arrows) adjacent to the site of penetration. × 15,000.

Fig. 7. An EM view of a cross section of a larva (L) during the early migration phase. Muscle, (M). × 17,000.

Fig. 8. Schematic representation of the early migration phase.

Fig. 9. Schematic representation of the late migration phase.

mechanism, since no host membranes surround the larva during this early phase of the infection (Despommier, 1975).

13.2.3.4. Migration

The larva, once inside the muscle cell, moves away from the site of penetration and is found just underneath the sarcolemma (Figs. 7, 8, 9). It is not known how far from the site of penetration the larva migrates, nor is it known for how long the larva is capable of migration. However, data collected from experiments with synchronized muscle infections (Despommier et al., 1975) suggest that migration is complete on the 3rd day after penetration. At this time, no observable physical changes have occurred in that portion of the muscle cell not in direct contact with the larva (Despommier, 1975), while a local disruption of contractile filaments is noted immediately adjacent to the larva (Fig. 8).

The reason for larval migration away from the site of penetration may relate to "leakiness" of the muscle cell which would presumably result from entry and migration, since a distinct ovoid lumen remains behind the worm during its migration (Fig. 6). The larva may be seeking a more stable area of cytoplasm. Eventually, however, the lumen collapses, and, in effect, seals off the larva from the original entry site (Fig. 9).

It is probable that in the region of the muscle cell affected by larval migration a normal contraction cycle could not take place because of the disruption of Z bands, but appropriate studies have not been carried out in order to test this possibility.

13.2.3.5. Initial growth phase

During the early stages of infection (i.e. penetration and migration), the larva doubles its volume (Despommier et al., 1975). However, from day 1 through day 3 after penetration, larval growth ceases (Fig. 10).

It has been speculated that during this phase of the infection, the muscle cell contributes little in the way of larval nutrition (Despommier, 1975), the larva deriving most of its nutrients from endogenous lipid and glycogen deposits. This notion is based upon ultrastructural observations which indicate that the newborn larva contains much lipid and glycogen in various cell types prior to infection.

During the first two days of infection no structural changes in the striated muscle cell, other than local disruption of contractile filaments around the larva, can be observed. Similarly there are no morphological changes in any of the surrounding cellular elements, nor is there any inflammatory process evident during the worm's initial growth phase.

13.2.3.6. Lag phase

The larva pauses in its growth beginning on the 2nd day after entry into the muscle cell (Despommier et al., 1975), and remains this way until day 4

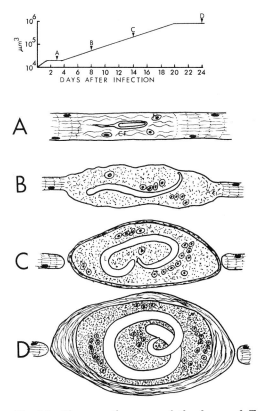

Fig. 10. The growth curve of the larva of *Trichinella spiralis* is depicted in the upper graph. Larval growth is expressed in μm^3/larva/day. Points A, B, C, and D are represented by the schematic drawings below the graph.

A. 3 days after penetration, the muscle cell has begun to convert to a Nurse cell. Contractile filaments are in dissarray (CF), and the nuclei (N) have enlarged and moved out of their peripheral location.

B. On day 8, the remnants of actin-Z band complexes (AZ) are seen in the polar regions, while the central portion of the cell has been fully converted to Nurse cell cytoplasm.

C. 14 days after infection, the larva (L) has begun to coil, and the nuclei (N) have completed enlargement and have increased in numbers.

D. The mature *T. spiralis* — Nurse cell complex.

(Fig 10A). Throughout the lag phase, a great number of physical and chemical changes are initiated in the muscle cell.

Physical changes include disarrangement of contractile filaments, nuclear enlargement and migration, and an increase in the overall volume of the infected portion of muscle cell (Faaske and Themann, 1961; Ribas-Mujal and Rivera-Pomar, 1968; Bäckwinkle and Themann, 1972; Purkerson and Despommier, 1974; Despommier, 1975).

Since numerous drastic alterations are initiated by the larva in its physical and chemical environment during the lag phase, it is obvious that the niche originally presented to the larva upon entering the muscle cell is unsatisfactory for the maintenance of larval growth.

Changes in the niche are invoked either just prior to or during a temporary cessation in larval development, and it is tempting to speculate that perhaps some metabolic product(s) elaborated during the initial growth phase or early lag phase interacts with the infected muscle cell much as a virus instructs its infected cell (Ribas-Mujal and Rivera-Pomar, 1968). It would be of importance in understanding the modification process to establish that such parasite products are indeed responsible for the restructuring of the niche.

As will be seen, the process of niche modification is carried to conclusion during the exponential phase, and thereby results in a host cell quite different in structure and function from the original striated muscle cell.

13.2.3.7. Exponential growth

The physical changes that occur in the infected muscle cell have been well characterized at the ultrastructural level (Faaske and Themann, 1961; Ribas-Mujal and Rivera-Pomar, 1968; Bruce, 1970; Bäckwinkle and Themann, 1972; Teppema et al., 1973; Purkerson and Despommier, 1974; Despommier, 1975).

The larva achieves its maximum size and completes its morphogenesis into an infective larva between days 5—19 after penetrating the striated muscle cell (Figs. 10B, C, D). Larval volume increases about 40% per day during this time, growing at an exponential rate (Despommier et al., 1975).

The larva becomes infective to a new host on about days 14—16 after infection (Villella, 1970), although certain larval morphogenic processes have yet to be completed (Wu, 1955; Kozek and Crandall, 1974; Despommier, personal observation).

The infected muscle cell on day 4 retains some of the characteristics of an uninfected striated muscle cell in that it still contains actin and myosin filaments, but their spatial arrangement is now much disrupted and by day 10 there are no further traces of them, either morphologically (Despommier, 1975) or chemically (Stewart and Read, 1974).

Histochemical evidence of an elevation in alkaline phosphatase levels in infected muscle cells 4—5 days after penetration has been reported (Bullock, 1953). The rate of incorporation of certain amino acids (i.e. tryptophan, methionine, and proline) is also increased from the 4th day after invasion (Stewart and Read, 1972; Stewart, 1973).

All of the "marker" substances associated with the normal striated muscle cell are gone by day 10 after penetration (Stewart and Read, 1974), while new biochemical constituents can now be found in or around the infected cell. For instance, collagen (Ritterson, 1966) is synthesized by the infected cell (Stewart and Read, 1972) and deposited on its external surface beginning on about day 10 after infection (Teppema et al., 1974; Despommier, 1975).

The level of acid phosphatase drastically increases during this remodelling period (Maier and Zaiman, 1966) suggesting an increase in general lysosomal activity. It is interesting to note that lysosomal enzymes are also induced as the result of non-infectious striated muscle cell disorders (Arnold et al., 1965).

The increased rate of uptake of a variety of amino acids indicates that niche modification probably involves more than a reutilization of existing amino acids obtained from autophagy, as had been originally proposed by Ribas-Mujal and Rivera-Pomar (1968).

When the striated muscle cell can no longer be recognized as a muscle cell, either structurally (Fig. 10B) or biochemically, it is termed the Nurse cell. Justification for the use of this term is based on the fact that all nutrients obtained by the larva from the extracellular environment must first pass through the Nurse cell cytoplasm, hence the infected cell "nurses" the larva during its growth phase (Purkerson and Despommier, 1974; and Despommier, 1975).

From days 4 to 10 after penetration, the larva exists in the cytoplasmic milieu of the Nurse cell, but thereafter until the worm is consumed by another host it is surrounded by a double membrane supplied by the Nurse cell (Purkerson and Despommier, 1974; Despommier, 1975). The significance of these membranes is not clear, however, since exponential growth of the larva has already begun some time before the membranes are present.

Of particular interest are the profound changes noted in the nuclei of the infected cell during this time. The nuclei undergo enlargement (Drachman and Tunchay, 1965), and become randomly distributed (Ehrhardt, 1896) throughout the cytoplasm (Figs. 10A—D). Nuclear structure is also altered greatly, so that the nucleoplasm appears sparse and the nucleolus becomes well defined and granular (Ribas-Mujal and Rivera-Pomar, 1968; Purkerson and Despommier, 1974; Despommier, 1975). Nuclear enlargement is complete on about day 9 (unpublished data).

Nuclear enlargement is accompanied by an increase in the number of nuclei during the twenty day infection period (Drachman and Tunchay, 1965). Usually, about 10 striated muscle cell nuclei are contained within the area of cytoplasm destined to become a Nurse cell (Despommier, 1975). By 20 days, there can be as many as 90 nuclei (unpublished data derived from serial sections of Nurse cells), while an average of about 40 nuclei is typically observed (Despommier, loc. cit.). However, the rate of nuclear division and the temporal nature of the division cycle have yet to be determined. Autoradiographic studies indicate that incorporation of H^3-thymidine occurs in nuclei of infected cells (Gabryel and Gustowska, 1967). Stewart and Read (1973) have reported that a 200% increase in the incorporation rate for H^3-

thymidine occurs from day 1—20 in infected tissue as compared to normal muscle tissue, but the possibility that at least some of the H^3-thymidine was incorporated by the lymphoid cells present in the accompanying inflammatory process could not be excluded.

13.2.3.8. Maturation

Larval and Nurse cell growth ceases on day 20 after penetration (Fig. 10). The infection is now stabilized and can remain in this configuration for many months or even years, depending upon the ability of the host to calcify the Nurse cell (Pagenstecher, 1865).

While the growth pattern for both parasite and host cell is complete, the system continues to exhibit an increased uptake rate of various amino acids (Hanks and Stoner, 1958; Stoner and Hanks, 1955; Stewart and Read, 1972; Stewart, 1973), suggesting that the turnover rate of at least some parasite proteins is rather high.

The metabolic rate (i.e. O_2 consumption), on the other hand, has been reported to be lower in infected muscle than in normal muscle (Karpiak et al., 1963), and correlates with data that suggest that mitochondria from infected tissue do not complete the breakdown of glucose (Michejda et al, 1972).

The rate of uptake of proline, a precursor for collagen, also remains high after larval growth ceases (Stewart and Read, 1972). The hypertrophied gly-

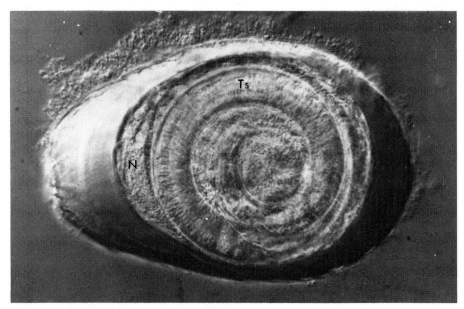

Fig. 11. A *Trichinella spiralis* (Ts) — Nurse cell (N) complex from a mouse infected 36 days previously. Isolation was accomplished using a 0.25% trypsin solution for 0.5 h at 37°C. (Nomarski interference photograph courtesy of Eric V. Grave). ×700.

cocalyx surrounding each Nurse cell continues to thicken after larval maturation (Teppema et al., 1974; Despommier, 1975) and is consistent with the above biochemical data.

The larva-Nurse cell complex (Fig. 11) can be isolated from normal host tissues by either tearing out individual Nurse cells by microdissection (Ritterson, 1966), or by a large scale isolation procedure employing homogenization of infected tissue in 0.25% trypsin at 37°C for 0.5 hours (Feldman, personal communication). The fact that the mature larva-Nurse cell complex can be isolated from other host tissues gives evidence for its anatomical independence from surrounding normal tissues and suggests that it would be possible to carry out a variety of relevant biochemical and morphological studies using the isolated complex as the starting material.

13.3. Conclusions

From an ecological standpoint, mammalian muscle tissue represents a niche that is unsuitable for the majority of parasitic organisms. This is probably a function of its variable physical, chemical, and physiological environment. Therefore, the only parasites that routinely infect this tissue are those capable of changing the niche.

The degree of niche modification is usually linked with deleterious effects to the host. Extreme examples of niche modification in muscle cells are *Trypanasoma cruzi* infection, in which the muscle cell serves directly as a source of nutrients for the replicating parasites (Sanabria, 1964), and *Nosema sp.* in a variety of crustaceans. These parasites, however, destroy their host cells (Weidner, 1970). In contrast, some parasitic organisms such as *Toxoplasma gondii* (Frenkel, 1953) or *Sarcocystis sp.* (Garnham, 1973) assume a benign posture in muscle, and lead to a limited amount of muscle destruction resulting from the encystment process.

The last examples are representative of a balanced host-parasite relationship, since the parasites remain alive while no further host damage occurs after encystment. In contrast, the *T. spiralis* — Nurse cell complex is one of the few examples of a balanced ecological association at the intracellular level in which no host cell destruction occurs.

It is a commonly held theory that host-parasite relationships that have existed for only a short time on the evolutionary time scale tend to be more harmful for the host than those that had evolved earlier. If this view is valid, then one may assume that the muscle phase of *T. spiralis* infection fits best into the latter group of host-parasite interactions. The fact that all mammals with the exception of the Chinese hamster (Ritterson, 1959) tolerate the intracellular larva strongly supports this concept.

In a balanced relationship between host and parasite, both specific (eg. immunological) and non-specific (eg. inflammatory) host responses serve to control the amount of damage the host incurs through the infection. Similarly, the parasite must constantly find new ways of avoiding these responses through natural selective mechanisms if it is to continue in that host (Sprent, 1959; Dobson, 1972). For this reason fascinating forms of adaptations have been developed by a number of diverse parasites, ranging from antigenic variation by African trypanosomes (Franke, 1905; Gray, 1970), to incorporation of host constituents into its own tegument by the Schistosome spp. infecting man (Smithers and Terry, 1969). Although they were not mentioned in the description of the *T. spiralis* — Nurse cell complex, host protective immune responses play a central role in maintenance of the balance of the relationship.

During the development of the intracellular larva, protective antigens are synthesized, stored, and, to a limited degree, secreted (Despommier, 1974) presumably into the cellular matrix of the Nurse cell. Eventually they come into contact with immuno-competent cells, whose protective responses are later directed against a newly acquired infection (Despommier, 1971).

Therefore, the intracellular larva regulates, by the host defense mechanisms, the degree to which the host becomes reinfected, thereby increasing the chances that its ecosystem (i.e. the host) will not be destroyed by reinfection. In addition, host protective immune responses function to reduce intraspecific competition for nutrients obtained for each intracellular larva by the Nurse cell from the extracellular environment.

There are many similarities between the *Trichinella spiralis* — Nurse cell complex and a variety of virus infections, as already noted by Ribas-Mujal and Rivera-Pomar (1968). A major common feature lies in the fact that each gains control over the metabolism of its host cell, inducing changes that make the niche conform to the needs of the parasite. Niche modification represents not so much an adaptive response by the parasite as one by the host cell. The parasite, however, is restricted in its ability to elicit changes in the environment by the genetic capabilities of that host cell. In the case of at least some viruses, the growth of the virus is coupled with host cell division, and results in an increase in the niche for the virus, but the process may lead to the death of the host. In contrast, *T. spiralis* induces growth of the Nurse cell and presumably an increase in the amount of active genetic material in each Nurse cell, but the Nurse cell remains non-proliferative.

In virus infections the mechanisms that signal host cell transformation are now generally understood to be under direct or indirect control of the viral genome. Whether this is the case with *T. spiralis* remains to be tested.

The mammalian striated muscle cell has been shown to be capable of reorganization into a specialized cell whose sole function appears to be that of ensuring the survival of its nematode parasite. The modification of this host cell into the Nurse cell and the development and growth of the larva presumably occurs through a series of feed-back mechanisms operating at crucial points throughout the infection. Since the nature of these specific feed-back steps has not been characterized, however, it would be of little value to speculate about them at this time. Suffice it to say that with the eventual elucidation of these critical exchanges of information between the parasite and its host, the basis for investigations of the molecular ecology of this interesting association will be established.

13.4. REFERENCES

April, E.W. (1975). "The Myofilament Lattice: Studies on Isolated Fibers IV. Lattice equilibria in striated muscle". J. Mechanochem. Cell Motility. 3 (in press).

Arnold, M.A., P.H. Weswig O.H. Muth; and J.E. Oldfield (1965). "Relationship of 5' Nucleotidase Activity in Lamb Muscle to Selenium Levels Administered to this Lamb." Proc. Soc. Exp. Biol. Med. 118, 75—76.

Bäckwinkel, K.D.; and H. Themann. (1972). "Elektronmicroskopische Untersuchungen über die Pathomorphologie der Trichinellose." Beitr. Pathol. 146, 259—271.

Bourne, G.H., ed. (1973). "The Structure and Function of Muscle. Vols. 1—4." Academic Press, New York and London.

Bruce, R.G. (1970). "The Structure and Composition of the Capsule of Trichinella spiralis in Host Muscle." Parasitology 60, 223—227.

Bullock, W.L. (1953). "Phosphatases in Experimental Trichinella spiralis Infections in the Rat." Expl. Parasit. 12, 150—162.

Despommier, D. (1971). "Immunogenicity of the Newborn Larva of Trichinella spiralis." J. Parasit. 57, 531—535.

Despommier, D. (1974). In: "3rd International Congress of Parasitology". Vol. 2, pp. 1163—1164.

Despommier, D. (1975). "Adaptive Changes in Muscle Fibers Infected with Trichinella spiralis." Am. J. Path. 78, 477—496.

Despommier, D.D., L. Aron and L. Turgeon. (1975). "Trichinella spiralis" Growth of the Intracellular (muscle) Larva." Expl. Parasit. 37, 108—116.

Dobson, C. (1972). In: "Immunity to Animal Parasites". (E.J.L. Soulsby, ed.). pp. 191—222. Academic Press, New York and London.

Drachman, D.A.; and T.O. Tunchay. (1965). "The Remote Myopathy of Trichinosis." Nemology. 15, 1127—1135.

Ehrhardt, O. (1896). "Zur Kenntniss der Muskelveränderungen bei der Trichinase des Menschen." Beitr. Path. Anat. 20, 1—42.

Faaske, E. and H. Themann. (1961). "Elektronmicroskopische Befunde an der Muskelfaser nach Trichinbefall." Virchow Arch. Path. Anat. 334, 459—474.

Franke, E. (1905). "Über Trypanosomentherapie." München Med. Wsch., 52, 2059—2060.

Frenkel, J.K. (1953). "Host, Strain, and Treatment Variation as Factors in the Pathogenesis of Toxoplasmosis." Amer. J. Trop. Med. Hyg. 2, 390—411.

Gabryel, P. and L. Gustowska. (1967). "Veränderungen der quergestreiften Muskelfosern in fruhen Stadium einer Trichinal spiralis-Infection." Gegenbaur. Morph. Jahrb. 111, 174—180.

Garnham, P.C.C. (1973). In: "The Structure and Function of Muscle." Vol. 4, Pharmacology and Disease. (G.H. Bourne, ed.) pp. 239—286. Academic Press, New York and London.

Gray, A.R. (1970). In: "The African Trypanosomiases". (H.W. Mulligan, ed.) pp. 113—116. Wiley-Interscience, New York.

Hanks, L.V. and R.D. Stoner. (1958). "Incorporation of DL-Tryosine-2-C-14 and DL-Tryptophan-2-C-14 by Encysted *Trichinella spiralis* Larvae." Expl. Parasit. 7, 82—89.

Harley, J.P.; and V. Gallicchico. (1971). "*Trichinella spiralis*: Migration of Larvae in the Rat." Expl. Parasit. 30, 11—12.

Karpiak, S.E., A. Kozai and M. Kryzyanowski. (1963). "Changes in the Metabolism of the Skeletal Muscles of Guinea Pigs Caused by the Invasion of *Trichinella spiralis*. I. Influence of the Invasion on the Carbohydrate Metabolism of Muscles." Wiad. Parazyt. 9, 435—446.

Kozek, W.J. and C.A. Carndall. (1974). In: "Trichinellosis". (C. Kim, ed.), pp. 231—238. Intext Educational Publishers, New York.

Lockhart, R.D. (1973). In: "The Structure and Function of Muscle. Vol. I. Structure. Part I". (G.H. Bourne, ed.). pp. 1—22. Academic Press, New York and London.

Lockhart, R.D. and P.W. Brandt. (1938). "Proceedings of the Society of Great Britain and Ireland." In: J. Anat. 72, 470.

Maier, D.M.; and H. Zaiman. (1966). "The Development of Lysosomes in Rat Skeletal Muscle in Trichinous Myosites." J. Histochem. Cytochem. 14, 396—400.

Mauss, E.A. and G.F. Otto. (1942). "The Occurance of *Trichinella spiralis* Larvae in Tissues other than Skeletal Muscles." J. Lab. Clin. Med. 27, 1384—1387.

Michejda, J.W. and K. Boczon. (1972). "Changes in Bioenergetics of Skeletal Muscle Mitochondria During Experimental Trichinosis in the Rat." Expl. Parasit. 3, 161—171.

Murray, R.M. (1973). In: "Structure and Function of Muscle. Vol. I., Structure, Part I". (G.H. Bourne, ed.). pp. 237—299. Academic Press, New York and London.

Nastuk, W.L. and A.L. Hodgkin (1950). "The Electrical Conductivity of Single Muscle Fibers." J. Cell. Comp. Physiol. 35, 39—73.

Odum, E.D. (1971). "Fundamentals of Ecology." Saunders, Philadelphia.

Pagenstecher, H.A. (1865). "Die Trichinen." pp. 116. Engelmann Publisher, Leipzig.

Purkerson, M. and D. Despommier. (1974). In: "Trichinellosis". (C. Kim, ed.) pp. 7—23. Intext Educational Publishers, New York.

Read, C.P. (1970). In: "Trichinosis in Man and Animals". (S.E. Gould ed.) pp. 91—101. C.C. Thomas, Springfield.

Ribas-Mujal, D. and J.M. Rivera-Pomar. (1968). "Biological Significance of the Early Structural Alterations in Skeletal Muscle Fibers Infected by *Trichinella spiralis*." Virchs. Arch. Abt. A. Path. Anat. 345, 154—168.

Ritterson, A.L. (1959). "Innate Resistance of Species of Hamsters to *Trichinella spiralis* and its Reverse by Cortisone." J. Infect. Dis. 105, 253—266.

Ritterson, A.L. (1966). "Nature of the Cyst of *Trichinella spiralis*." J. Parasit. 52, 157—161.

Sanabria, A. (1964). "Ultrastructure of *Trypanosoma cruzi* in Mouse Myocardium. II. Crithidial and Leishmanian Forms." Expl. Parasit. 15, 125—137.

Smithers, S.R. and R.J. Terry. (1969). In: "Advances in Parasitology, Vol. 7". (B. Dawes, ed.). pp. 41—93. Academic Press, New York and London.

Sprent, J.F.A. (1959). In: "The Evolution of Living Organisms". (G.S. Leeper, ed.). pp. 149—165. Melbourne Univ. Press. Melbourne.

Stewart, G.L. (1973). "Studies on Chemical Pathology in Trichinosis." Thesis. Rice University, Houston, Texas, U.S.A.

Stewart, G.L. and C.P. Read. (1972). "Some Aspects of Cyst Synthesis in Mouse Trichinosis." J. Parasit. 58, 1061—1064.

Stewart, G.L. and C.P. Read. (1973). "Desoxyribonucleic Acid Metabolism in Mouse Trichinosis." J. Parasit. 59, 164—167.

Stewart, G.L. and C.P. Read. (1974). "Studies on Biochemical Changes in Trichinosis. I. Changes in Myoglobin, Free Creatine, Phosphocreatine, and Two Protein Fractions in Mouse Muscle." J. Parasit. 60, 996—1000.

Stoner, R.D. and L.V. Hanks. (1955). "Incorporation of C^{14}-Labeled Amino Acids by *Trichinella spiralis.*" Expl. Parasit. 4, 435—444.

Teppema, J.S., J.E. Robinson; and E.J. Ruitenberg. (1973). "Ultrastructural Aspects of Capsule Formation in *Trichinella spiralis* Infections in Rats." Parasitology 66, 291—296.

Villella, J.B. (1970). In: "Trichinosis in Man and Animals". (S.E. Gould, ed.), pp. 19—60. C.C. Thomas, Springfield.

Weidner, E. (1970). "Ultrastructural Study of Microsporidian Development. I. *Nosema sp.* (Sprague, 1965) In: *Calbinectes sapidus* Rathburn." Z. Zellforsch. Mikrosk. Anat. 105, 33—54.

Wu, L.W. (1955). "The Development of the Stichosome and Associated Structures in *Trichinella spiralis.*" Can. J. Zool. 33, 440—466.

Zierlers, K.L. (1974). In: "Medical Physiology". (V.B. Mountcastle, ed.). pp. 77—120. C.V. Mosby Co., St. Louis.

Ecological Aspects of Parasitology
Editor: C.R. Kennedy
© *North-Holland Publishing Company, Amsterdam, 1976*

CHAPTER 14

LIVER AND OTHER DIGESTIVE ORGANS

Thomas C. CHENG

Institute for Pathobiology, Center for Health Sciences, Lehigh University, Bethlehem, Pennsylvania 18015 (U.S.A.)

Contents

14.1. Introduction

Ecologically, the liver and other digestive organs are either permanent or transient habitats for some parasites. Furthermore, the parasite may be either the mature organism or some developmental stage during its life cycle. Although there are more organs directly or indirectly involved in digestion in vertebrates other than the liver, pancreas, and gall bladder, this review will be concerned only with these organs.

14.2. The Liver

The vertebrate liver is both an exocrinic and an endocrinic organ. The exocrinic secretion is bile. Bile is comprised of bile pigment, bile salts, cholesterol, certain crystalloids, and some protein. Although this fluid is not known to influence the establishment of parasites in the organ of its origin, i.e., the liver, its presence in the intestine is known to influence or directly effect the establishment and subsequent development of certain parasitic organisms. In brief, among protozoans, Pratt (1937) and Itagaki (1954) have demonstrated that bile stimulates the excystation of *Eimeria tenella* in chickens. Similarly, Lotze and Leek (1960), Farr and Doran (1962), and Jackson (1962) have demonstrated that bile in combination with other substances stimulates the excystation of a number of species of *Eimeria*. The exact role of bile, however, requires further investigation, although it has been suggested that the bile salts alter the surface of the stieda body and enhance the action of the pancreatic enzymes. Subsequent to this, the bile salts and enzymes enter the oocysts and activate the sporozoites (Farr and Doran, 1962). It is noted that in some instances, such as in the case of *E. arloingi* of sheep, carbon dioxide is an essential part of the complex stimulus required for excystation (Jackson, 1962).

In the case of helminth parasites, the activation of the oncosphere of taeniid cestodes after release from the embryophore as a result of the action of digestive enzymes is dependent on the presence of bile. This has been demonstrated by Leonard and Leonard (1941) in the case of *Taenia pisiformis*, by Silverman (1954) in the case of *Taenia saginata*, and by Berberian (1957) in the case of *Echinococcus granulosus*.

Bile is also known to play a role in stimulating the evagination of the cestode scolex. Furhermore, it acts synergistically with digestive enzymes, especially trypsin, to cause the rupturing of the cyst membranes (De Waele, 1934; Read, 1955; Rothman, 1959a). In fact, except for *Taenia taeniaeformis*, bile appears to be essential for scolex evagination (Smyth and Haslewood, 1963).

There is some suggestion that the growth of *Hymenolepis diminuta* in the mammalian gut is also influenced by bile (Goodchild, 1958, 1960, 1961a, b; Goodchild and Harrison, 1961), although the supporting evidence has been questioned by Rothman (1959b), Smyth (1963), and Smyth and Haslewood (1963).

In the case of digenetic trematodes, there is some evidence that suggests that bile either enhances or is required for the excystation of certain metacercariae. Specifically, Oshima et al. (1958a, b) have shown experimentally that the time required for the excystation of *Paragonimus westermani* metacercariae is shortened from 3 hours to 30 minutes when 0.5% sodium cholate or sodium deoxycholate, both bile salts, are present. Kobayashi (1959) has reported that the metacercaria of *Metagonimus yokogawai* does not excyst in pepsin or pancreatin but the addition of pig bile causes excystation. Wikerhauser (1960) has demonstrated that the highest rate of excystation of the metacercaria of *Fasciola hepatica* occurs after pepsin treatment followed by trypsin with the addition of 20% ox bile. Also, Dixon (1966) has shown that the emergence of *F. hepatica* metacercariae is triggered by bile.

As a result of an elaborate series of studies involving testing the lytic action of bile from a number of species of animals as well as constituents of bile on *Echinococcus granulosus*, Smyth and Haslewood (1963) have suggested that specific bile salts may act as selective biochemical agents in determining host specificity by lysing or otherwise eliminating parasites in incompatible hosts.

According to Ham (1950), the exocrinic function of the vertebrate liver is not of sufficient importance to justify the large size of this organ, and consequently he has assumed that the endocrinic function of the liver is more important. It is well known that the hepatic parenchymal cells play many roles including the receipt, synthesis, and storage of nutrients. Furthermore, because of its anatomical location between the blood vessels of the hepatic portal system and the general circulation, it serves not only as a concentrating center for nutrients and other substances carried in the blood but also for species of blood-borne parasites. For example, it is well known that the sporozoites of *Plasmodium* spp. disappear from the general circulation shortly after their introduction by the mosquito vector, and invade hepatic parenchymal cells as well as other types of cells of the reticuloendothelial system to initiate the exoerythrocytic phase of their life cycles. Although the exact attraction, if such exists, between the sporozoites and hepatic cells remains unknown, it would not be surprising if the concentration of sporozoites, which enter and metamorphose into merozoites, is to a great extent due to the location of the liver in the circulatory system.

The liver is a chemically rich organ. It is a site of glycogen synthesis and storage as well as being the site where deamination of amino acids occurs. The nitrogenous moeity is then coupled with carbon dioxide to form urea which is circulated to the kidneys for excretion. Albumins, certain globulins, fibrinogen, and other proteins concerned with blood coagulation are synthesized in liver cells. These are but a few of the chemical constituents and functions of the liver. Consequently, it is not surprising to find a number of reports of parasites ingesting hepatic cells.

Electron microscopical studies on the exoerythrocytic merozoites of certain *Plasmodium* spp. and related organisms have given insights into the mechanisms involved in the uptake of nutrients by exoerythrocytic merozoites (Hepler et al., 1966; Aikawa et al., 1966, 1967; Sterling and DeGiusti, 1972; and others). It is evident from such studies that in most instances the host cell's cytoplasm is ingested by the parasite by employing a cytostome (Fig. 1). This, however, need not always be the case since the cytostome of such species as *Plasmodium fallax* is inactive (Hepler et al., 1966). Irrespective of the mechanism involved in the uptake of the host cell's cytoplasm, it is apparent that the hepatic cells in which the exoerythrocytic merozoites of many species occur provide their required nutrients. What remains to a large extent unanswered is, what specific nutrients do hepatic cells provide for these intracellular parasites? One approach that could lead to the resolution of this question is to examine what nutrients are re-

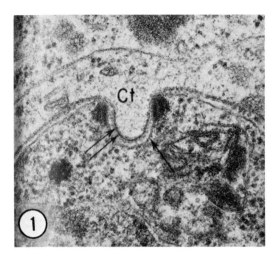

Fig. 1. Electron micrograph showing the cytostome (Ct) of a merozoite of *Plasmodium elongatum* in a host cell. Note the occurrence of a double membrane, with one (single arrow) originating from the wall of the cytostome and the other (double arrow) from the host cell. ×60,000. (After Aikawa et al., 1967).

quired for the in vitro cultivation of these parasites in defined media. Although much of the work that has been performed on in vitro cultivation of *Plasmodium* spp. has been directed at either the erythrocytic forms or the sexual stages that occur in the mosquito host (see Taylor and Baker, 1968, for review), there has been considerable progress made in culturing exoerythrocytic schizonts. Unfortunately the basic technique that has been employed has involved maintaining the parasites in tissues, and these have been almost exclusively avian malaria parasites in chick embryo tissues (see Pipkin and Jensen, 1958; Huff, 1964; Zuckerman, 1966, for reviews). Furthermore, since the exoerythrocytic stages of mammalian malaria parasites occur in liver cells and a totally satisfactory method for culturing these cells has yet to be achieved, the successful culturing of these parasites has also not been attained. Thus, definition of the chemical contributions of hepatic cells to intracellular protozoan parasites in the form of nutrients remains to be elucidated.

It is well known that the hepatic tissue of the host is invaded by the species of *Entamoeba*, especially *E. histolytica*, in advanced cases of amoebiasis. Thus the liver must be considered as a suitable habitat for these amoebae. While living in the liver, trophozoites of the virulent strains of *E. histolytica* ingest hepatic cells, and in so doing cause abscesses. This has caused the question to be raised as to whether the virulent strains are somehow different from the avirulent strains from the standpoint of their ability to ingest and utilize the contents of host cells, among other differences. This topic has been reviewed by Neal (1966). Studies designed to answer this question have not provided a clear-cut answer. For example, although Shaffer and Iralu (1963) have demonstrated differences in the ability of strains of *E. histolytica* to lyse erythrocytes of human and ox origin and Bragg and Reeves (1962) have reported that glucose metabolism in the Laredo strain is different from that in other strains, Neal (1960) was unable to find any difference in proteolytic activity of cell-free extracts prepared from different strains. However, Jarumilinta and Maegraith (1961) have demonstrated that a carboxypeptidase occurs in avirulent strains but not in virulent strains.

The temperature of the host's liver, i.e., the body temperature, was thought to be an ecological factor that governed the suitability of this habitat for amoebae. This was based exclusively on in vitro studies. Specifically, *E. histolytica* will only survive at $37°C$ with a tolerance of $± 1-2°C$ (McConnachie, 1955). However, the truism of this concept has been marred by the discovery of the so-called Laredo strain of *E. histolytica*. This strain was isolated from man and is morphologically indistinguishable from other *E. histolytica* strains. However, it wil grow in vitro at any temperature between $20°$ and $37°C$ (Siddiqui, 1963; Entner and Most, 1965). Richards et al. (1965) and Entner and Most (1965) have also reported additional strains which will grow between $25°$ and $37°C$. Thus, it is now questionable if the in vivo ambient temperature is as important as originally thought for the survival of parasitic amoebae, at least for some species.

Other than amoebiasis, fascioliasis is probably the most common parasitic disease involving the liver. Actually the monoecious adult of *Fasciola* spp. resides in the common bile duct of the mammalian host; however, young adults that have penetrated the surfaces of the hepatic lobules migrate through the parenchyma and eventually reach the common bile duct. There is some evidence that the hepatic environment can only support the successful passage of a limited number of flukes. Specifically, Ross (1964) has demonstrated that when cattle are fed *Fasciola hepatica* metacercarial cysts in numbers increasing from 200 to 1300, there is no increase in the percentage of infection as the number of parasites fed is increased. Furthermore, some young adult worms were inhibited in the damaged parenchyma, particularly in the ventral hepatic lobe where a preferential migration of the parasites occurs. This, as stated, suggests that the number of worms which can successfully migrate to the bile duct is in some yet undetermined manner influenced by the hepatic parenchyma. Also supporting this concept is the finding by Ross (1965) that the number of worms reaching the bile ducts is greatly reduced when the number of metacercarial cysts fed to each cow is increased to 2,500. Moreover, when large numbers of cysts, i.e., from 5,000 to 10,000, are fed to a cow, many immature flukes can be found trapped in the hepatic parenchyma. The finding of trapped and disintegrating flukes in the liver has also been reported by Ross (1967) in 8 to 12-week-old calves each of which had been fed 200—15,000 cysts.

The topic of pathogenesis in the liver due to *F. hepatica* and other liver flukes has been reviewed by Pantelouris (1965), Sinclair (1967), and others and need not be reiterated here. However, it is of interest to note that irrespective of the host species, young adult worms undergo tremendous growth while migrating through the liver (Urquhart, 1956; Lagrange and Gutmann, 1961; Dawes, 1962a, b; Thorpe, 1965; Ross et al., 1966). This, of course, implies that the hepatic cells are providing the essential nourishment but exactly what this consitutes remains to be elucidated. Nevertheless, it is known that young flukes feed on hepatic parenchymal cells by employing their oral suckers to pinch and burst host cells, and the resulting cellular debris is aspirated into the pharynx and ingested (Dawes, 1961). The liver is believed to include one or more nutrients not present in the host's abdominal cavity, since Dawes and Hughes (1964) have reported that young *F. hepatica* that have penetrated through the definitive host's alimentary tract but for some reason do not penetrate the hepatic capsule but remain in the abdominal cavity become stunted.

In addition to the direct ingestion of cells, helminth parasites in the liver may absorb nutrients. For example, Lewert and Lee (1955) have demonstrated, by employing cytochemistry, that the hepatic cells in the vicinity of young *Taenia taeniaeformis* lack glycogen; however, after the larvae become encapsulated by a well-formed connective tissue tunic, the glycogen content of the adjacent liver cells reverts back to normal. Similarly, Kuwamura (1958) has demonstrated that there is glycogen depletion in the hepatic parenchyma of rabbits infected with *Clonorchis sinensis*, Kublitskene and Goldbergiene (1966) have demonstrated a similar phenomenon in rabbits and guinea pigs harboring *Fasciola hepatica*, and Mostafa et al. (1965) have shown depletion of glycogen in hepatic cells in the vicinity of *Fasciola gigantica*. The depletion of glycogen, or course, need not mean that the parasites are directly utilizing the carbohydrate. It may reflect the disruption of carbohydrate metabolism in the cells. However, Cheng and Snyder (1962, 1963) have demonstrated that the reduction of glycogen in digestive gland cells of molluscs parasitized by trematode sporocysts can be attributed to the degradation of this polysaccharide to glucose, which is transferred from the host cells to the parasites. Whether this efficient method of carbohydrate utilization occurs in helminths in vertebrate livers remains to be elucidated.

In relation to the pathology of hepatic fascioliasis, it suffices to state that the basic pattern is very similar irrespective of the species of trematode involved. However, there are differences in severity, which, in part, are correlated with the number of worms involved. The similar pattern involves (1) the initiation of traumatic hepatitis as the young flukes penetrate the capsule. This can be appreciated by the occurrence of minute hemorrhagic spots and lines in the capsule which extend 1 or 2 mm into the parenchyma.

(2) Upon entry of the worm into the parenchyma, the lesion associated with it is grey-white or creamy in colour. It consists of compressed parenchymal cells, some of which are necrotic. (3) The trail following the migratory path of the worm is filled with debris of liver cells, neutrophils, erythrocytes, eosinophils, lymphocytes, and macrophages. The wall of such a trail is comprised of degenerate liver cells infiltrated with macrophages, eosinophils, and mononuclear cells. Dawes (1961) has suggested that this pathology may represent in part a toxic reaction to waste products excreted by the parasites. (4) Finally, in the later stages, the tracks become more extensive and fibrinous peritonitis may occur. A discussion of the pathology associated with bile ducts is beyond the scope of this chapter.

Relative to the pathoecology of fascioliasis, it is of interest to note that secondary bacterial infections may occur as a result of changes in the liver. Usually the organisms involved are already present in the liver but they may be introduced with the trematode. In most cases, small restricted abscesses develop, become encapsulated, and eventually disappear. However, the serious infectious necrotic hepatitis or Black disease may develop in parasitized cattle and sheep. The most comprehensive study to date on this disease is still that of Turner (1930) who has demonstrated that the lesions caused by fascioliasis provide a suitable environment for the rapid exsporulation and subsequent multiplication of the spores of *Clostridium oedematiens* lying dormant in the liver. This bacterial disease leads to hepatic necrosis, toxemia, and early death of the host. Thus, the development of infectious necrotic hepatitis may be thought of as a pathological condition triggered by alterations in the hepatic ecological niche by fascioliasis.

Another well known role of the liver as a habitat for helminth parasites involves the anisakid nematodes and fish livers. Although these nematodes can occur on the alimentary tracts, extending from the stomach to the posterior intestine, gonads, muscles, mesenteries, and peritoneum, body cavities, and connective tissues of fishes, they commonly occur in the hepatic parenchyma. Those anisakids occurring in the liver are almost invariably larvae. The literature pertaining to the biology of anisakids and the pathogenic alterations in fish resulting from their presence have been reviewed by Margolis (1970) and Cheng (1976) and need not be reiterated here. In brief, it is known that the sizes and weights of parasitized livers as well as their fat contents are inversely proportional to the number of *Contracaecum aduncum* present (see Margolis, 1970, for the primary literature) (Fig. 2). The livers of fish harboring larvae of *Anisakis* sp. and *Terranova decipiens* are similarly effected. Thus, ecologically, this group of nematodes causes the alteration of their habitats. The cause of these alterations still remains essentially speculative. There are two schools of thought on this matter. Because she found most of the *Contracaecum* larvae in cod situated beneath the hepatic capsule, Bazikalova (1932) postulated that the damage inflicted on the liver.is caused by toxins produced by the parasite. This idea

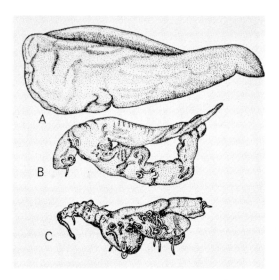

Fig. 2. Relative sizes of three specimens of livers of the Baltic cod, *Gadus morhua,* each of which measured 67 cm long. A. Nonparasitized liver. B. and C. Livers parasitized with 45 and 62 *Contracaecum aduncum,* respectively. (After Petrushevsky and Shulman, 1958).

was supported by Guiart (1938); however, Bauer (1958) and Dogiel (1962) have expressed the opinion that liver atrophy in gadid fishes is due to mechanical damage inflicted by the parasites.

There is another aspect of piscian hepatic anisakiasis that is interesting. Even if the larval nematodes occur in the hepatic parenchyma, only some are encapsulated (Guiart, 1938). In other words, the liver environment reacts differently to the same *Contracaecum* larvae. The cause of this remains undetermined.

The occurrence of parasites in the liver may only be transient. Such parasites may normally require passage through this habitat as a prelude to further maturation as in the case of *Fasciola hepatica* and *F. gigantica.* On the other hand, certain larval nematodes that enter incompatible hosts may cause what has been designated as larva migrans. If this involves the migration of the larvae through visceral organs, the condition is referred to as visceral larva migrans. This syndrome was originally recognized by Beaver et al. (1952) and the resulting histopathological lesions have been reported from the human liver (Beaver et al., 1952; Dent et al., 1956) and in experimental infection studies involving mice (Smith and Beaver, 1953). It is now known that visceral larva migrans in humans can be caused by helminths of animal origin including *Ascaris suum* and *Capillaria hepatica;* however, it is most commonly caused by *Toxocara canis,* the dog ascarid. Heavy and repeated infections with human parasites such as *Ascaris lumbricoides* and *Strongyloides stercoralis* can produce a similar pathological condition during the period of larval migration.

Visceral larva migrans generally results from the ingestion of the eggs of the responsible nematode and the resulting histopathological alterations are in the form of a trail of reaction cells and in the case of older infections, the parasite becomes encapsulated and is eventually destroyed. Ecologically, this means that the liver is not a suitable habitat for these nematode larvae but the complete understanding of the causes of the incompatibility is hidden in the maze that comprises the challenging problem of what factors govern and influence host-parasite compatibility and incompatibility. It is noted, however, that there is some evidence that antibodies play an important role in the eventual elimination of the larvae involved in visceral larval migrans (Huntley and Moreland, 1963; Huntley et al., 1965; Hogarth-Scott et al., 1969; Patterson et al., 1975).

The liver of certain mammals serves as a suitable habitat for the hydatid cysts of the cyclophyllidean cestodes *Echinococcus granulosus*, *E. multilocularis*, and related species since it is most commonly at this site that the cyst becomes lodged and develops, at least in man and certain other mammals. It is the oncosphere that enters the liver via a blood vessels and develops into a hydatid larva. This implies that the host's liver is providing the growth requirements and the necessary morphogenetic triggering factors. These, however, remain unknown. Furthermore, there is no direct evidence relating to how hydatid cysts take up nutrients, although it is assumed that it is by absorption or active transport across the surface.

A number of other species of protozoan, helminth, and arthropod parasites have been reported from vertebrate livers and associated ducts; however, these reports consist of no more than the descriptions of the parasites and their occurrences and consequently no ecological information can yet be derived from the available information.

14.3. The Gall Bladder

The gall bladder is the storage site for bile. It is a musculomembranous sac which may be suspended by a mesentery from the liver or may be more securely anchored to that organ. The role of bile in influencing or governing the establishment of parasites in the host's alimentary tract has been reviewed. In addition, it can serve as a habitat for certain symbionts. For example, this author has found large numbers of the ciliated protozoan *Haptophrya michiganensis* in the gall bladder of the salamander *Plethedon cinerus* and also the related *H. virginiensis* in the gall bladder of the dusky salamander, *Desmognathus fuscus*, in Giles County, Virginia. These protozoa were originally reported from the intestines of amphibia. Unfortunately, nothing more is known about the relationship between these symbionts and their habitats.

One of the more popular cestodes from the standpoint of serving as a model for cytochemical and physiological studies in recent years is *Hymeno-lepis microstoma*. This cestode inhabits the common bile duct of the mouse rather than the intestine proper. Most of these studies, however, have dealt with the parasite itself rather than from an ecological standpoint, i.e., examining the interaction of the parasite with its environment. Nevertheless, some tentative conclusions of an ecological nature can be drawn from the cytochemical studies by Bogitsh (1963).

The chemical composition and the function of the helminth tegument has been examined extensively since the advent of electron microscopy and cytochemistry, although most of the available information does not pertain to the relationship between the parasite and its habitat. On the other hand, Kruidenier (1948) has demonstrated that the tegument of *Paragonimus kellicotti* is a polymerized mucoprotein and Baer (1951) has suggested that this polymerization is undoubtedly influenced to a certain degree by the effect of the host's fluids on the tegument. Monné (1959) has shown the presence of acid mucopolysaccharides in the external portions of the tegument of several cestodes. He attributes to these substances the role of host enzyme inhibitors, which presumably protect the helminth from being digested. This idea has been adopted by von Brand (1952).

Bogitsh (1963) has demonstrated, by employing the periodic acid-Schiff technique, that the tegument of *H. microstoma* includes a protein-carbo-hydrate complex and is probably of mucoprotein type. Since the intensity of the Schiff reaction can be correlated with the hexose values, it is reasonable to assume that the intense PAS reaction signifies the occurrence of a more highly polymerized mucopolysaccharide. In view of this, Bogitsh has concluded that the high degree of polymerization encountered in *H. micro-stoma* tegument reflects the habitat of the worm, i.e., the bile duct. It also appears likely that the protective function of the tegument of *H. microstoma* is accomplished by means of a highly resistant protein carbohydrate complex fortified by a thin external layer of acid mucopolysaccharide. Furthermore, it has been suggested that these complex molecules may also serve to protect the parasite against bile and bile salts.

The presence and distribution of both acid and alkaline phosphatases in *H. microstoma* is quite uniform except for the scolex. It has been suggested by Bullock (1949, 1958) that in acanthocephalans the presence of cuticular alkaline phosphatase is concerned with the absorption of nutrients or with the elimination of fatty acids. Erasmus (1957a, b) has suggested that the acid and alkaline phosphatases in the cestode tegument may be involved in the release of phosphate ions and that these may be correlated with the phos-phorylated passage of substances through the tegument or in some manner related to carbohydrate metabolism. Numerous authors have adopted the hypothesis that these enzymes are involved in uptake, although there is some evidence to the contrary (Phifer, 1960). Nevertheless, in view of what is known about uptake by helminth surfaces (see Pappas and Read, 1975, for review), there can be no doubt that *H. microstoma* acquires its nutrients from the bile duct contents through its body surface.

14.4. The Pancreas

The pancreas is the major salivary gland of the abdomen. It is cradled in the curvature of the duodenum and its secretions are conveyed to the duodenum through the pancreatic duct. At least nine enzymes, in addition to water, bicarbonate, and salts, constitute the two litres of pancreatic juice secreted in 24 h in large mammals. Among the enzymes are the proteinases trypsinogen, chymotrypsinogen, and carboxypeptidase. In addition, there is a pancreatic amylase, lipase and lecithinase. There are also such hydrolases as ribonuclease and deoxyribonuclease.

In addition to the exocrinic function of the pancreas, cells of the islets of Langerhans comprise the endocrinic portion of the gland. These cells secrete insulin and glucagon into the capillary beds of the gland.

Very little is known about the pancreas as a habitat for parasites. In fact, only a few parasites have been reported from this gland. Probably the best known parasite associated with the pancreas is *Eurytrema pancreaticum*. This digenetic trematode actually does not live in the pancreas but in the pancreatic duct. Furthermore, it is not restricted to this habitat since it has also been reported from the gall bladder, bile duct, and intestine of several species of mammals. Although originally reported from sheep in Japan, it has since been found in *Bos taurus*, *B. indicus*, *Sus scrofa domestica*, *Camelus bactrianus*, *Buffelus bubalis*, *Capra hircus*, *Lepus coreanus*, and a human in Asia.

In addition to *E. pancreaticum*, several related species of the same genus occur in the pancreatic ducts of mammals (Yamaguti, 1958). It is noted, however, that one species, *E. coelomaticum*, does live in the pancreas, rather than the pancreatic duct, of cows and camels. Nothing is known about the dynamic relationship between this parasite and its environment.

The pancreas may also be indirectly involved in parasitism. For example, it is known that terminal hypoglycemia occurs in hosts parasitized by pathogenic African trypanosomes. This condition, however, cannot be attributed to the high rate of glucose consumption by the parasites, i.e., the exhaustion of the carbohydrate reserves in the host, because the level of blood glucose can be temporarily elevated by the injection of adrenaline (Regendanz, 1929; Krijgsman, 1933) and brought back to the normal level by the administration of trypanocidal drugs (Scheff, 1932). Such findings suggest the involvement of the suprarenals, the pancreas, and/or the thyroid (Bellelli and Caraffa, 1956; Lippi and Benedetto, 1958). In other words, some yet undefined function of the pancreas (and/or the other glands mentioned) acts synergistically with African trypanosomiasis in producing the hypoglycemia. Thus, ecologically, the pancreas may play a role in terminating the parasitism by rendering the blood an unsuitable habitat for the flagellates by contributing to the hypoglycemic condition.

In some instances of schistosomiasis, the eggs will infiltrate the pancreas (Seife and Lisa, 1950). This, however, cannot be responsible for the diabetic condition known to occur in some cases since the diabetes ceases rapidly after treatment with anthelmintics (Day, 1924; Erfan, 1933).

14.5. Conclusion

In conclusion, it is evident from the foregoing account that information pertaining to the role of the vertebrate liver and other digestive organs as ecological niches for parasites is still rudimentary. Nevertheless, the recognition that the hosts' internal organs serve as habitats for endoparasites and that dynamic interchange does occur and that such habitats do influence the structure and physiology of the parasites is growing, and will undoubtedly stimulate further studies on parasitism from the ecological standpoint.

Acknowledgment

Some of the information included in this chapter was obtained as a result of research supported by a contract (223-73-222) from the Food and Drug Administration, U.S. Department of Health, Education and Welfare.

14.6. References

Aikawa, M., Hepler, P.K., Huff, C.G. and Sprinz, H. (1966). "The feeding mechanism of avian malarial parasites". J. Cell Biol. 28, 355—373.

Aikawa, M., Huff, C.G. and Sprinz, H. (1967). "Fine structure of the asexual stages of *Plasmodium elongatum*". J. Cell Biol. 34, 229—249.

Baer, J.G. (1951). "Ecology of Animal Parasites". Univ. Illinois Press, Urbana, Ill.

Bauer, O.N. (1958). "Relationships between host fishes and their parasites". In "Parasitology of Fishes" (V.A. Dogiel, G.K. Petrushevsky, and Y.I. Polyansky, eds.). pp. 90—108. English translation by Z. Kabata. Oliver and Boyd, Edinburgh and London.

Bazikalova, A. (1932). "Data on the parasitology of Murman fishes". Sborn. Nauchn.-Promyslov. Rabot na Murmane. Snabtechnizdat, Moscow and Leningrad. pp. 136—153. (In Russian).

Beaver, P.C., Snyder, C.H., Carrera, G.M., Dent, J.H. and Lafferty, J.W. (1952). "Chronic eosinophilia due to visceral larva migrans", Pediatrics 9, 7—19.

Belleli, L. and Caraffa, V. (1956). "Castellanosi (tripanosomiasi) sperimentale della cavia da *Castellanella evansi (Trypanosoma evansi)* comportamento del tasso glicemico". G. Mal. Infet. Parasit. 8, 521—523.

Berberian, D.A. (1957). "Host specificity and the effect of digestive juices on ova of *E. granulosus*". 10th Rep. Orient Hospital, Beirut, Lebanon. pp. 33—43.

Bogitsh, B.J. (1963). "Histochemical studies on *Hymenolepis microstoma* (Cestoda: Hymenolepididae)". J. Parasit. 49, 989—997.

Bragg, P.D. and Reeves, R.E. (1962). "Pathways of glucose dissimilation in the Laredo strain of *Entamoeba histolytica*". Expl. Parasit. 12, 393—400.

Bullock, W.L. (1949). "Histochemical studies on the Acanthocephala I. The distribution of lipase and phosphatase". J. Morph. 84, 185—200.

Bullock, W.L. (1958). "Histochemical studies on the Acanthocephala III. Comparative histochemistry of alkaline glycerophosphatase". Expl. Parasit. 7, 51—58.

Cheng, T.C. (1976). "The natural histpry of anisakiasis in animals." J. Milk Food Tech. 39, 32—460.

Cheng, T.C. and Snyder, R.W. Jr. (1962). "Studies on host-parasite relationships between larval trematodes and their hosts. I. A review. II. The utilization of the host's glycogen by the intramolluscan larvae of *Glypthelmins pennsylvaniensis* Cheng, and associated phenomena". Trans. Am. Microsc. Soc. 81, 209—228.

Cheng, T.C. and Snyder, R.W. Jr. (1963). "Studies on host-parasite relationships between larval trematodes and their hosts. IV. A histochemical determination of glucose and its role in the metabolism of molluscan host and parasite". Trans. Am. Microsc. Soc. 82, 343—346.

Dawes, B. (1961). "On the early stages of *Fasciola hepatica* penetrating into the liver of an experimental host, the mouse: a histological picture". J. Helminth. R.T. Leiper, supplement, 41—52.

Dawes, B. (1962a). "On the growth and maturation of *Fasciola hepatica* L. in the mouse". J. Helminth. 36, 11—38.

Dawes, B. (1962b). "Additional notes on the growth of *Fasciola hepatica* L. in the mouse, with some remarks about recent researches in Belgium". J. Helminth. 36, 259—268.

Dawes, B. and Hughes, D.L. (1964). "Fascioliasis: the invasive stages of *Fasciola hepatica* in mammalian hosts". Adv. Parasit. 2, 97—168.

Day, H.B. (1924). "The etiology of Egyptian splenomegaly and hepatic cirrhosis". Trans. Roy. Soc. Trop. Med. Hyg. 18, 121—130.

DeWaele, A. (1934). "Study of the function of the bile in the evagination of the cysticerci of cestodes". Ann. Parasit. Hum. Comp. 12, 492—510.

Dent, J.H., Nichols, R.L., Beaver, P.C., Carrera, G.M. and Staggers, R.J. (1956). "Visceral larva migrans, with a case report". Am. J. Pathol. 32, 777—803.

Dixon, K.E. (1966). "The physiology of excystment of the metacercaria of *Fasciola hepatica* L." Parasitology 56, 431—456.

Dogiel, V.A. (1962). "General Parasitology". (English translation by Z. Kabata). Oliver and Boyd. Edinburgh and London.

Entner, N. and Most, H. (1965). "Genetics of *Entamoeba*. Characterization of two new parasitic strains which grow at room temperature (and at $37°$)". J. Protozool. 12, 10—13.

Erfan, M. (1933). "Bilharziasis and diabetes mellitus". J. Trop. Med. Hyg. 36, 348—349.

Erasmus, D.A. (1957a). "Studies on phosphatase systems of cestodes I. Studies on *Taenia pisiformis* (cysticercus and adult)". Parasitology 47, 70—80.

Erasmus, D.A. (1975b). "Studies on phosphatase systems of cestodes II. Studies on *Cysticercus tennicollis* and *Moniezia expansa* (adult)". Parasitology 47, 81—91.

Farr, M.M. and Doran, D.J. (1962). "Excystation of the poultry coccidium, *Eimeria acervulina*". J. Protozool. 9, 154—161.

Goodchild, C.G. (1958). "Growth and maturation of the cestode *Hymenolepis diminuta* in bileless hosts". J. Parasit. 44, 352—362.

Goodchild, C.G. (1960). "Effects of starvation and lack of bile upon growth, egg production, and egg inability in established rat tapeworms, *Hymenolepis diminuta*". J. Parasit. 46, 615—623.

Goodchild, C.G. (1961a). "Carbohydrate contents of the tapeworm *Hymenolepis diminuta* from normal, bileless, and starved rats". J. Parasit. 47, 401—405.

Goodchild, C.G. (1961b). "Protein contents of the tapeworm *Hymenolepis diminuta* from normal, bileless, and starved rats". J. Parasit. 47, 830—832.

Goodchild, C.G. and Harrison, D.L. (1961). "The growth of the rat tapeworm *Hymenolepis diminuta* during the first five days in the final host". J. Parasit. 47, 819—829.

Guiart, J. (1938). "Etude parasitologique et épidémiologique de quelques poissons de mer". Bull. Inst. Oceanogr. Monaco No. 755, 1—15.

Ham, A.W. (1950). "Histology". Lippincott, Philadelphia.

Hepler, P.K., Huff, C.G. and Sprinz, H. (1966). "The fine structure of the exoerythrocitic stages of *Plasmodium fallax*". J. Cell Biol. 30, 333—358.

Hogarth-Scott, R.S., Johnansson, S.G.O. and Bennich, H. (1969). "Antibodies to *Toxocara* in the sera of visceral larva migrans patients: the significance of raised levels of IgE." Clin. Exp. Immunol. 5, 619—625.

Huff, C.G. (1964). "Cultivation of the exoerythrocytic stages of malarial parasites". Am. J. Trop. Med. Hyg. 13, 171—177.

Huntley, C.C. and Moreland, A. (1963). "Gel diffusion studies with *Toxocara* and *Ascaris* extracts". Am. J. Trop. Med. Hyg. 12, 204—208.

Huntley, C.C., Costas, M.C. and Lyerly, A. (1965). "Visceral larva migrans syndrome: clinical characteristics and immunologic studies in 51 patients". Pediatrics 36, 523—536.

Itagaki, K. (1954). "Further investigation on the mechanism of coccidial infection of fowl.". J. Fac. Agri. Tottori. Univ. 2, 37—53.

Jackson, A.R.B. (1962). "Excystation of *Eimeria arloingi* (Marotel, 1905): stimuli from the host sheep". Nature 194, 847—849.

Jarumilinta, R. and Maegraith, B.G. (1961). "The patterns of some proteolytic enzymes of *Entamoeba histolytica* and *Acanthamoeba* sp. II. The action of *E. histolytica* and *Acanthamoeba* sp. on various synthetic substrates". Ann. Trop. Med. Parasit. 55, 518—528.

Kobayashi, A. (1959). "Studies on excystation of the metacercaria of *Metagonimus yokogawai*". Acta Schol. Med. Gifu 7, 822—828.

Krijgsman, B.J. (1933). "Biologische Untersuchungen über das System: Wirtstier-Parasit. III. Telil: Das Verhalten der Blutproteine und IV Telil: Das Verhalten des Blutzuckers Während der Entwicklung von *Trypanosoma evansi* in Mans und Ratt". Z. Parasitenk. 6, 1—22.

Kublitskene, O. and Goldbergiene, M. (1966). "Histochemical and biochemical studies on metabolism in experimental fascioliasis". Acta Parasitol. Litu. 6, 98—99. (In Russian).

Kruidenier, F.J. (1948). "Metachromatic determination of mucoprotein distribution in *Paragonimus kellicotti*". J. Parasit. 34 (Sect. 2), 22.

Kuwamura, T. (1958). "Study on experimental clonorchiasis, especially on the histochemical change in the liver". Shikoku Igaku Zassi, 12, 28—57.

Lagrange, E. and Gutmann, A. (1961). "Sur l'infestation experimentale de la souris par *Fasciola hepatica*". Riv. Parasit. 22, 93—101.

Leonard, A.B. and Leonard, A.E. (1941). "The intestinal phase of the resistance of rabbits to the larvae of *Taenia pisiformis*". J. Parasit. 2, 375—378.

Lewert, R.M. and Lee, C.O. (1955). "Studies on the passage of helminth larvae through host tissues. III The effects of *Taenia pisiformis* on the rat liver as shown by histochemical techniques". J. Infect. Dis. 97, 177—186.

Lippi, M. and Benedetto, A. (1958). "Variazione della glicemia in corso di tripanosomiasi sperimentale nella cavia de *Trypanosoma brucei*". Arch. Ital. Sci. Med. Trop. Parassitol. 39, 446—451.

Lotze, J.C. and Leek, R.G. (1960). "Some factors involved in excystation of the sporozoites of three species of sheep coccidia". J. Parasit. 46 (Suppl.), 46.

Margolis, L. (1970). "Nematode diseases of marine fishes". In "A Symposium on Diseases of Fishes and Shellfishes". (S.F. Snieszko, ed.) pp. 190—209. Spec. Publ. No. 5, Am. Fisher. Soc., Washington, D.C.

McConnachie, E.W. (1955). "Studies on *Entamoeba invadens* Rodhain 1934 in vitro and its relationship to some other species of *Entamoeba*". Parasitology 45, 452—481.

Monné, L. (1959). "On the external cuticles of various helminths and their role in the host-parasite relationship: a histochemical study". Arch. Zool. 12, 343—358.

Mostafa, A.M.B., Moustafa, I.H., Soliman, M.K. and El-Amrousi, S. (1965). "Histochemical studies of glycogen concentration in the liver of buffaloes affected with *Fasciola gigantica* in comparison with normal ones". Vet. Med. J. Giza 10, 183—190.

Neal, R.A. (1960). "Enzymic proteolysis by *Entamoeba histolytica:* biochemical characteristics and relationship with invasiveness". Parasitology 50, 531—550.

Neal, R.A. (1966). "Experimental studies on *Entamoeba* with reference to speciation". Adv. Parasit. 4, 1—51.

Oshima, T., Yoshida, Y. and Kihata, M. (1958a). "Studies on the excystation of the metacercariae of *Paragonimus westermani*. 1. Especially on the effect of bile salt". Bull. Inst. Publ. Health Tokyo 7, 256—269.

Oshima, T., Yoshida, Y. and Kihata, M. (1958b). "Studies on the excystation of the metacercariae of *Paragonimus westermani*. 2. Influence of pepsin pretreatment on the effect of bile salts". Bull. Inst. Publ. Health Tokyo 7, 270—274.

Pantelouris, E.M. (1965). "The Common Liver Fluke, *Fasciola hepatica* L". Pergamon, Oxford.

Pappas, P.W. and Read, C.P. (1975). "Membrane transport in helminth parasites: a review". Expl. Parasit. 37, 469—530.

Patterson, R., Huntley, C.C., Roberts, M. and Irons, J.S. (1975). "Visceral larva migrans: immunoglobulins, precipitating antibodies and detection of IgG and IgM antibodies against *Ascaris* antigen". Am. J. Trop. Med. Hyg. 24, 465—470.

Phifer, K. (1960). "Permeation and membrane transport in animal parasites: further observations on the uptake of glucose by *Hymenolepis diminuta*". J. Parasit. 46, 137—144.

Pipkin, A.C. and Jensen, D.V. (1958). "Avian embryos and tissue culture in the study of parasitic Protozoa". Expl. Parasit. 7, 491—530.

Pratt, I. (1937). "Excystation of coccidia, *Eimeria tenella*". J. Parasit. 23, 426—427.

Read, C.P. (1955). "Intestinal physiology and the host-parasite relationship". In "Some Physiological Aspects and Consequences of Parasitism". (W.H. Cole, ed.). pp. 27—43. Rutgers Univ. Press, New Brunswick, New Jersey.

Regendanz, P. (1929). "Der Blutzucker bei Trypanosomeninfektionen". Arch. Schiffs. Trop. Hyg. 33, 242—251.

Richards, C.S., Goldman, M. and Cannon, L.T. (1965). "Cultivation at $25°C$ and behavior in hypotonic media of strains of *Entamoeba histolytica*". J. Parasit. 52 (Sect. 2), 45.

Ross, J.G. (1964). "Experimental infection of cattle with *Fasciola hepatica*". Proc. 1st Int. Congr. Parasit. Rome 2, 885—886.

Ross, J.G. (1965). "Experimental infections of cattle with *Fasciola hepatica:* a comparison of low and high infection rates". Nature 208—907.

Ross, J.G. (1967). "Experimental infection of cattle with *Fasciola hepatica*. High level single infections in calves". J. Helminth. 41, 217—222.

Ross, J.G., Todd, J.R. and Dow, C. (1966). "Single experimental infection of calves with the liver fluke *Fasciola hepatica* (Linnaeus 1758)." J. Comp. Path. Therap. 76, 67—81.

302

Rothman, A.H. (1959a). "Studies on the excystment of tapeworms". Expl. Parasit. 8, 336—364.

Rothman, A.H. (1959b). "The role of bile salts in the biology of tapeworms. II. Further observations on the effects of bile salts on metabolism". J. Parasit. 45, 379—383.

Scheff, G. (1932). "Ueber den intermediaren stoffwechsel der mit Trypanosomen infezierten Meerschweinchen". Biochem. Z. 248, 168—180.

Seife, M. and Lisa, J.R. (1950). "Diabetes mellitus and pylephlebitic abcess of the liver resulting from *Schistosoma mansoni* infestation". Am. J. Trop. Med. 30, 769—772.

Shaffer, J.G. and Iralu, V. (1963). "The selective ability of strains of *Entamoeba histolytica* to hemolyse red blood cells". Am. J. Trop. Med. Hyg. 12, 315—317.

Siddiqui, W.A. (1963). "Comparative studies of effect of temperature on three species of *Entamoeba*". J. Protozool. 10, 480.

Silverman, P.H. (1954). "Studies on the biology of some tapeworms of the genus *Taenia*. I. Factors affecting hatching and activation of taeniid ova, and some criteria of their viability". Ann. Trop. Med. Parasitol. 48, 207—215.

Sinclair, K.B. (1967). "Pathogenesis of *Fasciola* and other liver flukes". Helminth. Abstr. 36, 115—134.

Smith, M.H.D. and Beaver, P.C. (1953). "Persistence and distribution of *Toxocara* larvae in the tissues of children and mice". Pediatrics 12, 491—497.

Smyth, J.D. (1963). "Biology of Cestode Life Cycles". Comm. Bur. Agri. Tech. Bull. No. 34.

Smyth, J.D. and Haslewood, G.A.D. (1963). "The biochemistry of bile as a factor in determining host specificity in intestinal parasites, with particular reference to *Echinococcus granulosus*". Ann. N.Y. Acad. Sci. 113, 234—260.

Sterling, C.R. and DeGiusti, D.L. (1972). "Ultrastructural aspects of schizogony, mature schizonts, and merozoites of *Haemoproteus metchnikovi*". J. Parasit. 58, 641—652.

Taylor, A.E.R. and Baker, J.R. (1968). "The Cultivation of Parasites In vitro". Blackwell, Oxford and Edinburgh.

Thorpe, E. (1965). "The pathology and experimental fascioliasis in the albino rat". J. Comp. Pathol. 75, 39—44.

Turner, A.W. (1930). "Black disease (infectious necrotic hepatitis) of sheep in Australia". Bull. Coun. Sci. Ind. Res., Melbourne, No. 46, 1—139.

Urquhart, G.M. (1956). "The pathology of experimental fascioliasis in the rabbit". J. Path. Bact. 71, 301—316.

von Brand, T. (1952). "Chemical Physiology of Endoparasitic Animals". Academic Press, New York.

Wikerhauser, T. (1960). "A rapid method for determining the biability of *Fasciola hepatica* metacercariae". Am. J. Vet. Res. 21, 895—898.

Yamaguti, S. (1958). "Systema Helminthum. Vol. 1. The Digenetic Trematodes of Vertebrates". Interscience, New York.

Zuckerman, A. (1966). "Propagation of parasitic protozoa in tissue culture and avian embryos". Ann. N.Y. Acad. Sci. 139, 24—38.

Ecological Aspects of Parasitology
Editor: C.R. Kennedy
© *North-Holland Publishing Company, Amsterdam, 1976*

CHAPTER 15

INTESTINE

A.D. BEFUS [a] and R.B. PODESTA [b]

*Department of Medicine [a], McMaster University, Hamilton, Ontario L8S 4J9 (Canada)
and Department of Zoology [b], University of Toronto, Toronto, Ontario M5S1A1
(Canada)*

Contents

15.1. Introduction

An accurate description of the microecology of the intestinal lumen is important because prevailing concepts concerning the luminal "biocoenose" profoundly influence the types of questions asked about the host-parasite relationship. A prime difficulty facing parasitologists is keeping abreast with the vast proliferation of methodology and data on the structure and function of the gastrointestinal tract in health and disease. Detailed descriptions of the luminal environment have recently appeared in the parasitological literature (Crompton, 1973; Mettrick and Podesta, 1974) and coverage of other aspects of the gastrointestinal tract of interest to parasitologists have appeared-(Allen and Snary, 1972; Argenzio and Stevens, 1975; Waldram, 1975). Only limited coverage of recent advances in mucosal resistance have been given in parasitological literature (Dobson, 1972; Murray, 1972), but a number of comprehensive reviews have been published elsewhere (Dayton et al., 1971; Tomasi and Grey, 1972; Brandtzaeg, 1973; Mestecky and Lawton, 1974; Bienenstock, 1974).

Herein we direct the reader to pertinent literature and describe in as much detail as space limitations allow, major problems encountered by metazoan parasites in the intestinal microcosm. We will face this task by first, considering problems associated with luminal physical-chemical characteristics and, second, the problems of host immune responses. The interests of the authors dictate that we will deal primarily with helminths in the mammalian gut. A further objective is to pose some unsolved questions regarding the host-parasite relationship in the hope that this will stimulate the interest of the researcher.

15.2. Problems associated with luminal physical-chemical characteristics

Due to the complex nature of intestinal structure and function it is not surprising that parasitologists struggle to control variables while attempting to deal with problems concerning the intestinal lumen as a discrete habitat. Nevertheless, several laboratories have succeeded in isolating some of the more important variables affecting various aspects of the biology of metazoan parasites (e.g. see Crompton, 1973; Mettrick and Podesta, 1974; Podesta and Mettrick, 1974a; Castro et al., 1974a; Komuniecki and Roberts, 1975). There are therefore a number of recent accounts of the advances and reviews on the luminal environment as it affects the biology of metazoan parasites and further review at this time is unnecessary. Instead a novel approach to two problems associated with the luminal physical-chemical characteristics will be dealt with, including the problems of differential host and parasite surface area and competition for nutrients and a discussion of the adaptations of metazoan parasites to life in an environment with excessive levels of carbon dioxide.

15.2.1. The effect of unstirred water layers on surface area considerations

The function of the microvilli covering the external surface of platyhelminths has been a matter of discussion for some time. Since microvilli are found only in epithelial layers involved in absorption and secretion, their first function would appear to be one of increasing the available surface area. Estimates of the increase in surface area due to the microvilli covering cestode parasites vary from a factor of 2.3 to 12.4 (see Berger and Mettrick, 1971). Using the results of Mettrick and Cannon (1970), the mature strobila of *Hymenolepis diminuta* would have an anatomical surface area of 101 cm^2 per 100 mg dry weight, assuming an average increase in surface area due to the microvilli of 2.6 (Berger and Mettrick, 1971).

However, the increase in surface area due to the microvilli in cestodes is small in comparison to the increase in the small intestine, which has prompted speculation on other functions of the microvilli in cestodes (Berger and Mettrick, 1971). The anatomical surface area of the rat intestine, for example, varies from between 1,226 cm^2 100 mg^{-1} dry weight in the proximal small bowel to 696 cm^2 100 mg^{-1} in the distal ileum (Wilson and Dietschy, 1974; Westergaard and Dietschy, 1974). Per unit weight, therefore, the anatomical surface area of the rat intestine exceeds that of the rat tapeworm, *H. diminuta*, by a factor of 7—12. Since both surfaces compete for the same substrate present in the intestinal lumen, the question arises as to how the tapeworms are so successful in maintaining their nutritional status in the intestine while competing with the mucosa for the same nutrient source. The answer may lie in the structure of water adjacent to the mucosal and worm surface.

Overlying all biological membranes there exists a hydrodynamic layer or lamellae of water through which the movement of solute molecules is determined only by diffusional forces. This unstirred layer of water may exert a large part of the total resistance encountered by solutes passing from the bulk extracellular water phase into the intracellular compartment. The properties of water leading to its structure adjacent to membranes and the effects of lamellar water on transport processes have been recognized for some time (see Wilson and Dietschy, 1974; Westergaard and Dietschy, 1974). In the small intestine, the aqueous diffusion barrier may cause pronounced alterations in permeability coefficients and in the case of fatty acids and steroids the unstirred layer, and not the cell membrane, is rate-limiting for intestinal absorption (Westergaard and Dietschy, 1974).

The unstirred layer also markedly alters the effective or physiological surface area in the small bowel. Although the anatomical surface area is quite large (see above) the effective surface area of the small intestine of rats covers only the tips of the villi, due to the unstirred aqueous layer effect, and is equal to only 8.2—9.5 cm^2 100 mg^{-1}, or about 1/100—1/200 the anatomical surface area (Wilson and Dietschy, 1974; Westergaard and Dietschy, 1974).

If we assume an unstirred water layer coating *H. diminuta*, the effective surface area would be 39 cm^2 100 mg^{-1}, or approximately 1/3 of the anatomical surface area. Therefore, one obtains the interesting result that, per unit weight, the effective surface area of the worm is greater than that of the intestine by a factor of 3—4. This is consistent with the fact that the rate of transport per unit weight of electrolytes, water and nonelectrolytes by *H.*

diminuta and other cestodes is as great or greater than that in the small intestine (Podesta and Mettrick, 1976a).

It is apparent, therefore, that during the evolution of the cestodes selection pressures did not favor the development of an elaborate morphological design to increase the surface area. In contrast to the intestine, which serves multiple functions in absorption, digestion and secretion, the cestode surface structure is probably designed primarily for absorption, the other functions such as digestion and secretion being largely taken over by the host's intestine. This is consistent with the structure of other epithelial types such as the gall bladder, urinary bladder, kidney tubules, malpighian tubules, salt glands and others, which are primarily transporting epithelia and are relatively flat surfaces covered only with microvilli but lacking villi and mucosal folds.

The unstirred layer has other implications as well. It has been demonstrated that this lamellar water markedly alters active transport kinetics for various substrates. A thick unstirred layer, for example, results in artifactually high K_t values as well as altering J_{max} (Wilson and Dietschy, 1974). Since these kinetic parameters are significantly distorted by unstirred layers, then it is apparent that many of the conclusions drawn on the basis of these data for transport in platyhelminths must be reevaluated. For example, a thick unstirred layer may be the reason for the high K_t values (>25 mM) obtained for glucose transport by *H. diminuta* in vivo (Podesta and Mettrick, 1974b).

15.2.2. Anaerobes in an aerobic environment

A considerable body of evidence has accumulated in support of a predominantly anaerobic energy metabolism in metazoan parasites (Saz, 1971, 1972; Hochachka and Mustafa, 1972; Hochachka et al., 1973; Harpur, 1974a). As adults, most of these helminth parasites inhabit the lumen of the gastrointestinal tract of vertebrate animals. Their use of anaerobic metabolic pathways to generate ATP has been considered explicitly or implicitly as a logical extension of the belief that the lumen of the intestine is anoxic. However, recent evidence has shown that the intestinal lumen is not anoxic but contains a fluid phase with a pO_2 of 40—50 mm Hg (Hamilton et al., 1968; Podesta and Mettrick, 1974a). The aerobic nature of the lumen is further supported by the observation that the haemolymph of *Ascaris* has a pO_2 of at least 26 mm Hg (Harpur, 1974b). These results suggest that metazoan parasites invading the intestine of their host are not faced with anoxic conditions. Why then do they sacrifice an efficient aerobic means of generating metabolic energy in favor of a less efficient anaerobic energy metabolism? The use by parasites of anaerobic metabolic pathways to generate energy must therefore be related to some environmental factor other than the availability of oxygen.

Probably the most unique condition encountered by a parasite of the intestine is the high tensions of carbon dioxide and the concomitant acidic pH of most of the intestinal lumen. In the postprandial intestine acid chyme from the stomach reacts with bicarbonate-rich fluids, such as bile or secretions by the pancreas and Brunner's glands, producing a luminal pCO_2 which may exceed 600 mm Hg (see Mettrick and Podesta, 1974). There are a number of cases of animals living in an environment of very low pH, e.g. *Thermoplasma acidophila*, a mycoplasma-like organism that grows at pH 2 (Hsung and Haug, 1975), but these environments do not have a bicarbonate base so that the system is not complicated by high concentrations of a highly permeable molecule like free CO_2. Animals can also tolerate wide ranges of tissue CO_2, such as diving vertebrates or animals under anaesthesia. However, no other organisms are known to cope with the excessive levels of free CO_2 that are generated in the intestinal lumen.

The magnitude of the problem can be demonstrated by a few simple calculations. The reaction system of interest is the CO_2–HCO_3^- system:

$$CO_2 + H_2O \rightleftharpoons H_2CO_3 \rightleftharpoons HCO_3^- + H^+$$

The high luminal pCO_2 means that CO_2 will diffuse down a gradient into the tissues of the parasites where it will shift the poise of the CO_2–HCO_3^- reaction system towards the production of H^+ and HCO_3^-. As a first approximation, we will assume a permeability of 12.25 μmoles CO_2 h^{-1} cm^{-2} per mm Hg (Green et al., 1970; Gros and Moll, 1974), and using a conservative estimate of luminal pCO_2 of 500 mm Hg, the amount of CO_2 entering the tissues of a 16 day old *H. diminuta* having an effective surface area of 23 cm^2 (calculated from data of Mettrick and Cannon, 1970), would be 140.8 μmoles CO_2 h^{-1} per worm. Since for every mole of CO_2 entering the tissue a mole of H^+ and HCO_3^- will be generated, this means that each worm will accumulate 140.8 μmoles H^+ (equivalent to pH 0.85) and HCO_3^- per hour in the postprandial intestine. Even if it is assumed that the worm tissue concentration of the $CO_2 : HCO_3^-$ system is in a steady state condition with the luminal pCO_2, at 100 mm Hg luminal pCO_2 the worm tissue concentrations of CO_2, H^+ and HCO_3^- would be 3 mM or pH 2.52. The worm tissue values at 500 or 600 mm Hg luminal pCO_2 would be 15 mM CO_2 or pH 1.82 and 18 mM CO_2 or pH 1.74, respectively. This is the basis of the problem, assuming that the worms do not do anything with the CO_2 once it enters their tissue. If this is the case, the pH of their tissue fluids would be strongly acidic. If the tissue pH is indeed acidic, how does the metabolic machinery function and if, on the other hand, the tissue pH lies in the neutral range, how do the worms counter the effect of a huge gradient of CO_2 between their tissues and the intestinal lumen?

One obvious way to regulate their tissue pH would be to secrete H^+ ions and recent studies have demonstrated an active H^+ secretory mechanism in *H. diminuta* (Podesta and Mettrick, 1975, 1976a). However, the maximum rate of H^+ secretion by *H. diminuta* is 62.3 μmoles h^{-1} g wet weight^{-1} or in terms of a 16 day old worm, 27.16 μmoles h^{-1} per worm (R.B. Podesta, unpublished). This is obviously insignificant when compared to the amount of H^+ accumulated via the hydration of ambient CO_2 within the tissues of the worm.

Another way to counter the acidifying effect of CO_2 accumulation would be to accumulate buffer anions, and in particular, HCO_3^- by secreting H^+ ions and, at least in cestodes, additional buffer anions may be derived from the calcareous corpuscles. These corpuscles are small amorphous concretions consisting primarily of calcium phosphate and calcium carbonate (Bachra et al., 1963, 1965). The worms may mobilize the phosphate and carbonate for buffering purposes by absorbing a hypertonic absorbate, thus increasing the amount of calcium which may be dissolved in their tissue fluids (Podesta and Mettrick, 1976a). However, the maximum total calcium content of *H. diminuta* is only 2.38 μmoles (von Brand, 1966) and if all the calcium is bound to phosphate and carbonate, release of the buffer anions could take up only 3.17 μmoles H^+.

One further way of handling the buffer problem would be to use the CO_2 in carboxylating reactions before it is hydrated (i.e. the carboxylation of phosphoenolpyruvate to form oxaloacetate). However, the total amount of organic acids which may be secreted as end products of such reactions amounts to only 85.7 μmoles h^{-1} in *H. diminuta* (Von Brand, 1966). Even if all the acids are derived from carboxylation reactions, this would remove only 85.7 μmoles of CO_2/h/worm and a reduction in H^+ accumulation of the same amount. Thus, in *H. diminuta* at least, it would appear that the above mechanisms, including H^+ secretion, anion accumulation and carboxylation, are insufficient means for buffering the tissue pH in the postprandial intestine. It is possible only to conclude, based on the present state of knowledge, that *H. diminuta* cannot exist in the small bowel. The perplexing problem is, of course, that it does!

The evidence, other than that of the levels of pCO_2 attained in the lumen, supporting the magnitude of the buffering problem is scarce. Webster and Wilson (1970) sampled the fluid in the protonephridial canals of *H. diminuta* and found that it has a pH as low as 4.5 and a pCO_2 of at least 120 mm Hg, which is much higher than the level of pCO_2 maintained in the fluids of other organisms, excepting the luminal fluid of the intestine. Similarly, Harpur (1974b) found that fluid from the pseudocoele of *Ascaris* contained a pCO_2 of at least 150 mm Hg, which is again much higher than that found in the extracellular fluids of other animals. How *Ascaris* could accumulate anions is uncertain as, unlike cestodes, they do not contain calcareous corpuscles. Perhaps HCO_3^- is accumulated by a greater rate of H^+ secretion or by absorbing OH^-. However, the main point is that the worms in the lumen of the intestine do appear to contain high levels of buffer anions and free CO_2. Both of these undoubtedly function in, or are a result of, the buffering problems associated with a habitat containing what are usually considered as narcotizing or even lethal levels of CO_2.

Other possibilities for handling the excessive CO_2 also exist. The worms could bind the H^+ and HCO_3^- to tissue proteins. This would serve to partially regulate the tissue pH and, as shown with other protein—CO_2 systems (Gros and Moll, 1974), serve as a means of facilitated diffusion of CO_2 through the tissues to the excretory system. The haem proteins in *Ascaris* are a prime candidate for this function. Perhaps a large gradient of H^+ between the tissues of the worm and the intestinal lumen is maintained by a passive Donnan potential as suggested for *Thermoplasma* (Hsung and Haug, 1975). On the other hand, the proton conductance of the epithelial membranes in luminal dwelling helminths may be very high. This means that the concentration of H^+ would always be very close to electrochemical equilibrium across, say, the brush border of the cestode tegument. If this is the case, it would be unusual since membranes are thought to have a very low conductance to H^+ and it could give considerable support to the hypothesis of high proton conductance of excitable tissues (see Rehm et al., 1973). Although there are a number of possible explanations, none have been explored in any detail in helminth tissues.

As a consequence of the above buffering problem, which lends support to its details, we must return to the question of anaerobiosis in an aerobic environment. What are the metabolic consequences of increasing the levels of CO_2 and concentrations of buffer anions in the worm tissues? There are more than an ample number of reviews considering the biochemical pathways to succinate formation (see above) and space limitations do not allow a full reiteration of these pathways. However, as shown by Hochachka and Mustafa (1972), the point at which the aerobic and anaerobic pathways diverge is following the formation of phosphoenolpyruvate (PEP). In the aerobic system, PEP is reduced to pyruvate under the influence of pyruvate kinase (PK) while in the anaerobic system, PEP is carboxylated to form oxaloacetate under the influence of PEP carboxykinase (PEPCK). As queried by Saz (1972), why does PK compete so successfully for PEP in the aerobic system but is so unsuccessful in its competition with PEPCK for substrate PEP in the helminth system? Saz's (1971, 1972) answer to this question is based on his and other observations that the activity of PK is very low or negigible in anaerobic helminth tissues while PEPCK activity is high. However, the method used to assay PK activity in these studies has been questioned (Köhler and Hanselmann, 1974) and it appears that *H. diminuta* and *Mesocestoides corti* have considerable PK activity (Köhler and Hanselmann, 1974).

In marine bivalves a decrease in pH leads to an inhibition of pyruvate kinase with a concomitant activation of PEP carboxykinase (Hochachka and Mustafa, 1972). Since substantial acidification of tissues of marine bivalves occurs under anaerobic conditions, this effect of pH, combined with the inhibition of pyruvate kinase activity by alanine, has been credited as the major determinant controlling the formation of succinate during periods of anoxia (Hochachka and Mustafa, 1972). However, this hypothesis has been

310

criticized on the grounds that the anaerobic pathways must be turned on first in order to produce the organic acids which are thought to acidify the tissues (Livingstone and Bayne, 1974).

It is therefore an open question as to what controls the formation of oxaloacetate or pyruvate from PEP. However, it is well known that increased concentrations of the CO_2–HCO_3^- system inhibit the succinate-cytochrome c reductase component of the electron transport chain (Kasbekar, 1966). By simple mass action a high tissue pCO_2 may shunt PEP into the formation of oxaloacetate via the carboxylation reaction. Further, increased concentrations of buffer anions, including phosphate and bicarbonate, can by themselves inhibit the activity of PK (Black and Henderson, 1972). It is therefore possible that the use of anaerobic energy metabolism in helminths is not related to the question of the presence or absence of oxygen in the intestinal lumen, but is secondary to the large gradient of CO_2 across their tissues and the lumen and the subsequent buffering problems which arise (Podesta and Mettrick, 1976a). In the postprandial intestine, when the luminal pCO_2 is raised, the increased concentrations of tissue anions and CO_2 would inhibit PK and stimulate the carboxylation of PEP. Hours after the meal taken by the host, when the pCO_2 in the lumen is presumably much lower, the worms could resort to other pathways with appropriate changes in tissue buffer anions and pCO_2. There is increasing evidence that the worms have the complete metabolic machinery for aerobic energy metabolism coupled to molecular oxygen (Podesta and Mettrick, 1974a; Rahaman and Meisner, 1973) and, as suggested by Mounib and Eisan (1972), the retention of both aerobic and anaerobic systems offers considerable versatility, especially to a parasite whose life cycle involves free-living and parasitic stages.

15.3. Intestinal immunology

All organisms, to some degree, influence the environment which they inhabit and have evolved to cope with, if not utilize, the alterations they induce. Parasites, and especially those inhabiting the intestine, are not exceptional in this regard and some fascinating examples of the pathophysiological alterations induced and strategies adopted by hosts and parasites for coping with these alterations have been elucidated (see Symons, 1969; Gaafar, 1971; Podesta and Mettrick, 1976b). One alteration in the environment which free-living organisms do not encounter but with which all parasites, to a greater or lesser extent, must cope, is the host's immune response (see Chapter 6).

It has been accepted that the host's immune response is potentially lethal to parasites which inhabit the blood or other tissues. But, for parasites which inhabit the intestinal lumen, and particularly those which do not have a tissue migration in the same host prior to intestinal residence, or do not cause

obvious mucosal damage, the existence of host immune responses is only beginning to be recognized. In part this slow recognition has been caused by
the previously widely held, but erroneous, concept that macromolecular uptake does not occur in the normal adult intestine and therefore no antigenic
substances cross the undamaged mucosa (Heyneman, 1963). Furthermore,
the intestinal environment has been considered unfavourable for the action
of antibody or other immunological components. Early in the century it was
clear that intestinal resistance to cholera and shigellosis could exist in the absence of systemic immunity, but not until the late 1940's was local antibody
implicated in this resistance (Tomasi, 1971). With the identification of the
five classes of serum immunoglobulin and the subsequent discovery of the
striking abundance of IgA in mucosal secretions in the early 1960's, the components of intestinal resistance began to unravel (Tomasi and Grey, 1972).
The significance of these components in eucaryotic parasitic infections is
poorly understood and even less is known of how intestinal parasites cope
with them. As an accurate description of the environment is essential in ecological studies we provide details of the immunological components in the
intestine.

15.3.1. The gut-associated lymphoid system (GALS)

The intestine is continually exposed to a variety of antigens and it is well
established that potentially immunogenic quantities are absorbed intact
(Walker and Isselbacher, 1974). Therefore, it is not surprising that there is an
elaborate gut-associated lymphoid system (GALS) composed of lymphoid
cells, (a) dispersed in the epithelium, lamina propria and lymph of the thoracic duct or (b) aggregated in the Peyer's patches, appendix and mesenteric
lymph nodes (Guy-Grand et al., 1974). Other immunologically important
cells such as basophils, eosinophils, macrophages, mast cells, monocytes and
neutrophils also occur in the intestinal wall.

The central lymphoid organs, the thymus and the avian bursa of Fabricius
— in general responsible for cellular and humoral components of immunity
respectively — are derived from the embryonic gut. Therefore, in the search
for the mammalian equivalent of the bursa, the GALS, especially the lymphoid aggregates, the Peyer's patches and appendix, has been scrutinized closely (Cooper and Lawton, 1972). These lymphoid aggregates are covered by a
specialized epithelium which has unique pinocytotic capabilities (Bockman
and Cooper, 1973) and apparently samples the antigenic make-up of the
lumen; some antigens accumulate in the Peyer's patches while in transit to
the draining lymph nodes (Carter and Collins, 1974). Below the specialized
epithelium the structure of Peyer's patches is complex but regions which
may be bursal equivalents have been identified (Waksman, 1973).

Recent observations on large dividing cells (blasts) in the GALS have provided much insight into the characteristics of the system (Bienenstock, 1974;

Guy-Grand et al., 1974; Mestecky and Lawton, 1974). Thymus-independent (B) blasts apparently sensitized by antigens in the Peyer's patches migrate in the lymph through the mesenteric lymph nodes to the thoracic duct and enter the circulation, from whence many home to the lamina propria of the intestine and become predominantly IgA producing cells. Some blasts enter other lymphoid tissues, such as the spleen, where they may differentiate into IgG, IgM or IgA containing cells. During the course of the migration to the intestinal lamina propria the blasts initially acquire surface IgA and, subsequently, intracellular IgA; presumably components of maturation process which results in the homing specificity of the cells. As IgE producing cells can be relatively abundant in the intestinal lamina propria (see below) they may have a similar maturation and migration sequence. Thymus-dependent (T) blasts from the mesenteric lymph node and thoracic duct which home to the GALS, are found not only in the lamina propria, but also in the epithelium. In addition to being sites of initial antigen accumulation and of precursor cells, Peyer's patches may contain specific antibody forming cells (Veldkamp et al., 1973) and functional T cell populations (Levin et al., 1974).

The epithelium, which does not overlie the lymphoid aggregates, is often thought to be composed almost exclusively of digestive-absorptive columnar epithelial cells, but in fact contains a large, scattered population of lymphocytes representing about 10—15% of the total epithelial cell population. These so-called thelio-lymphocytes resemble other lymphocytes but a portion contains granular inclusions which can be liberated, releasing histamine. The derivation and function of these cells are not understood but they may act as sentinel cells in the epithelium, probably sensitized by IgE (Bienenstock, 1974). It seems highly probable that investigations of this large population of thelio-lymphocytes, lying immediately adjacent to luminal parasites, will be rewarding especially as IgE responses are well known in parasitic infections.

15.3.2. Immunoglobulins in the intestine

As stated above IgA is normally the predominant immunoglobulin in mucosal secretions, a fact which at first sight is surprising as IgG is the predominant immunoglobulin in the serum. Preferential local production and secretion of IgA (Brandtzaeg, 1973) as well as its enhanced survival characteristics in the proteolytic luminal environment (Fubara and Freter, 1972) contribute to its predominance in the intestinal contents.

In man about 85% of serum IgA is a 7S monomeric form but in the intestinal secretions polymeric IgA, specifically an 11S dimer, predominantes. This dimer contains, in addition to the monomers, a J (joining) chain (IgA_2J) which was thought to function in the polymerization. Recent evidence indicates that it is for the specific combination with secretory component (SC)

(Eskeland and Brandtzaeg, 1974). SC is synthesized in mucosal epithelial cells (Brandtzaeg, 1974) and combines with IgA_2J to form IgA_2JSC, the major form of IgA in mucosal secretions. The functions of SC are incompletely understood but it confers resistance to proteolysis (Lindh, 1975). It may also be involved in the homing of B blasts to the lamina propria (Guy-Grand et al., 1974; Mestecky and Lawton, 1974) and the specific transport of IgA_2JSC into the intestinal lumen (Brandtzaeg, 1974).

In many animals other than man (notably cats, dogs, guinea pigs, mice and rats) monomeric IgA is uncommon even in the serum; IgA_2J is the predominant form of serum IgA in these animals (Heremans, 1974). Despite the complexity of the system and the species differences, the important knowledge to be considered by parasitologists is that much of the IgA in the intestinal lumen is of a different origin, structure and probably function to serum IgA.

IgM is often the second most abundant immunoglobulin in the intestine (Brandtzaeg, 1973) and, like IgA, has some affinity for SC, as 60—70% of secreted IgM contains SC (Eskeland and Brandtzaeg, 1974; Weicker and Underdown, 1975). IgM normally exists as a pentamer with an associated J chain; the latter may explain the affinity for SC. In cases of IgA deficiency in man there are significant increases in IgM producing cells adjacent to mucosal surfaces, providing further support that IgM is important in intestinal immune responses.

In contrast to IgA and IgM, which can bind SC, IgG in the intestinal lumen is indistinguishable from that in the serum and is highly susceptible to proteolysis. Normally there is little production of IgG in the intestinal wall, unlike IgA and IgM, hence most IgG in the lumen must originate by leakage from the serum or the large extravascular pool. However, in chronic infections or diseases such as ulcerative colitis, local increases in IgG producing cells occur (Brandtzaeg et al., 1974). Increased local production of IgG, plus increased leakage due to enhanced vascular permeability (see below), result in increased IgG in the lumen. Brandtzaeg (1973) termed this the second line of mucosal defence, as large amounts of IgG can enter the gut in a short period of time to augment or replace the first line of defence: the IgA and IgM immunoglobulins which are specifically adapted for action in the intestinal lumen.

To our knowledge IgD and IgE have not been detected in the intestinal lumen, although cells containing these immunoglobulins occur in the lamina propria. In fact, the lamina propria below epithelial surfaces is a major site of IgE producing cells. The functions of IgD are unknown; IgE appears to be important in the leakage of serum antibody into the gut lumen and has received considerable attention from parasitologists (Dobson, 1972; Murray, 1972).

15.3.3. Some characteristics of gastrointestinal immune responses

We have described briefly the elaborate GALS and the present under-standing of immunoglobulins in the intestine; these components must be important in intestinal immunity and hence the integrity of the intestinal microcosm (Tomasi and Grey, 1972). Studies comparing different routes of immunization have established that following intestinal antigen adminis-tration there are local antibody (Dayton et al., 1971) and cellular responses (Morag et al., 1974; Spencer et al., 1974) which are distinct from systemic responses. Furthermore, following systemic immunization responses are different to those following intestinal immunization.

Resistance to reinfection with microbes often correlates with levels of local IgA antibody rather than with serum antibody. Moreover, Fubara and Freter (1973) protected naive mice from cholera infection by using intestin-al IgA from immunized mice. Pierre and Reynolds (1974) established pro-tection against cholera using serum IgG antibodies, but Tomasi and Grey (1972, p. 131) cautioned that such experiments using passively-transferred serum antibody for protection against intestinal infection do not provide clear answers as to the significance of this antibody in active immunity. Therefore, whether active immunity to cholera, and probably many other intestinal infections, resides in intestinal IgA, serum IgG acting in the intes-tine, or both, remains to be determined.

Although considerable information has accumulated about immune re-sponses to intestinal infection, there is little understanding of the actual protective mechanisms. The best understood mechanism is the inhibition of adherence of bacteria to the mucosal surface by IgA antibodies (Gibbons, 1974; Genco et al., 1974). Therefore, antibody can be directly protective against infection, but typically circulating antibody simply provides recog-nition of the antigen initiating other effectors of immunity such as the com-plement system or phagocytosis. Do such reactions occur in the gut?

Although complement may enter the gut as do other serum proteins, intestinal secretions are considered to be anticomplementary. Furthermore, unaltered IgA from the serum or secretions does not fix complement, although Boackle et al. (1974) have shown that aggregates of IgA fix comple-ment. Reports that IgA in secretions can act synergistically with complement and lysozyme to be bacteriolytic (Adinolfi et al., 1966; Burdon, 1973; Hill and Porter, 1974) suggest that complement, particularly in collaboration with lysozyme, is important in intestinal immunity and requires further in-vestigation.

Phagocytic cells enter the intestinal lumen but previously this was consid-ered only to be a mechanism of cell disposal. Recently, however, Bellamy and Nielsen (1974) established that following homologous challenge of the intestine of immunized animals there was a massive emigration of neutro-phils into the lumen and, moreover, that this phenomenon could be passively transferred with immune serum. It seems unlikely that these cells are important as phagocytes in the lumen as peristaltic flow would quickly dispose of neutralized or agglutinated microbes; more probably their function, once in the lumen, is to release their granules which contain the antimicrobial enzymes i.e., lipases, peroxidases, lysozyme and many others (Baggiolini, 1972).

Neutrophils may not be the only source of these enzymes in the intestine. Paneth cells, located at the base of the crypts, contain lysozyme which can be released into the lumen (Erlandsen et al., 1974) or be involved in intracellular digestion of intestinal micro-organisms phagocytosed by these cells (Erlandsen and Chase, 1972). These sessile cells may be most important in resistance to infections which do not induce a massive invasion of other cells containing antimicrobial enzymes.

15.3.4. Immune responses to intestinal parasites

There are many detailed reviews of immune responses to intestinal parasites (Soulsby, 1972; Ogilvie and Jones, 1973; Larsh and Weatherly, 1974, 1975; Ogilvie and Love, 1974). Herein we present only a simplified model of the responses which ultimately lead to the rejection of some parasites (Fig. 1). On the other hand some members of parasite populations and many species of parasites cope with these responses successfully. The model incorporates not only components which have been investigated by parasitologists but also components documented by other disciplines and which must apply in immune responses to intestinal parasites as well.

As outlined above, antigen from the lumen can accumulate in local Peyer's patches where precursor cells probably become sensitized. Antigen also may enter the circulation directly (Bienenstock, 1974) and sensitize cells in other lymphoid tissues. Sensitized blast cells differentiate while migrating via the lymph and circulation to the intestinal wall, where they proliferate. B blasts differentiate into plasma cells in the lamina propria producing antibodies which are predominantly IgA, and perhaps IgE. The latter is well known in helminthic infections (Ogilvie and Jones, 1973), but IgA is poorly known, although some reports exist (e.g. Dobson, 1972; Crandall and Crandall, 1972; Zinneman and Kaplan, 1972; Crandall et al., 1974; Befus, 1976). The established importance of IgA in procaryotic infections necessitates further investigation of this aspect of eucaryotic infections. In passive protection studies, IgG_1 antibody is of major importance in resistance to *Nippostrongylus brasiliensis*, but, recalling the caution of Tomasi and Grey (1972, see above), other antibody may be important in active immunity. Antibodies must act in two fundamental ways: (a) by entering the lumen and acting directly on the parasites and (b) by recruiting and/or stimulating other components of the immune system (Fig. 1). Perhaps IgA antibodies act directly on the parasites, whereas antibodies of other classes act indirectly.

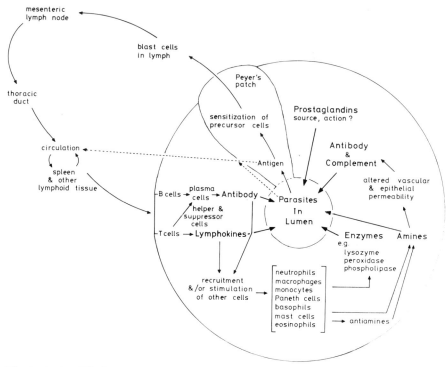

Fig. 1. A simplified model of the immunological events which may lead to the expulsion of parasites from the intestine (see text). Antigen(s) released from the parasite crosses the specialized epithelium overlying the Peyer's patches, cells become sensitized and undergo a complex migration returning to the intestinal wall; antigen may also enter the circulation directly and sensitize cells in lymphoid areas other than the Peyer's patches. In the intestinal wall sensitized cells complete their differentiation to effector cells whose products may act directly on the parasites or alternately influence other components of the immune system which can affect the parasites and/or regulate the immune responses. Direct action on the parasites may involve: antibody, lymphokines, enzymes, amines, complement, prostaglandins and other, as yet unknown, components.

T cells are essential in the expulsion of many parasites (Ogilvie and Jones, 1973). Recent studies on the GALS in parasitic infections which incorporate T cell responses include: the transfer of immunity with cells (Liburd et al., 1973; Selby and Wakelin, 1973; Ogilvie and Love, 1974); the proliferation of cells (Dobson and Soulsby, 1974); and the homing of cells (Dineen et al., 1968; Parrott and Ferguson, 1974). However, the mode of action of T cells in parasite infections is unknown. In accordance with what is known of T cell function, their action could be to regulate B cell responses or to produce lymphokines which would act either directly upon the parasites or on other immunologically important cells (Fig. 1).

The major action of antibodies and lymphokines may not be directly on the parasites but rather on the recruitment and/or stimulation of various cells (Fig 1), as basophils, eosinophils, macrophages, mast cells, monocytes and neutrophils have been implicated in immunity to parasites (Ismail and Tanner, 1972; Murray, 1972; Rothwell and Dineen, 1972; Larsh and Race, 1975). Other cells such as the Paneth cells (see above) require investigation also. Products from all these cells, which might be the effectors of protective immunity, can be placed into one of two categories, namely biogenic amines or antimicrobial enzymes. Amines such as histamine and 5-hydroxytrypt-amine could act directly on the parasites (Rothwell et al., 1974) or alter permeability (see Murray, 1972), enhancing the passage of antibody and other serum components such as complement into the lumen. Further, re-cent evidence that complement may be associated with intestinal parasites (Befus, 1976) suggests a role for complement in intestinal immunity. The action of the antimicrobial enzyme systems, e.g. lysozyme, peroxidases, lipases (Baggiolini, 1972), is probably directly on the parasites (Fig. 1). Striking elevations in levels of intestinal peroxidase (Castro et al., 1974b) and phospholipase (Ottolenghi, 1973; Larsh et al., 1974) in helminth infected animals strongly suggest that these enzymes play a role in intestinal ecology.

Local increases in amines and antimicrobial enzymes, however, can cause tissue damage and some mechanism to minimize such damage, yet allow effi-cient immune responses, is essential. Of the properties attributed to eosino-phils (Hubscher and Eisen, 1973), their antihistamine action, if delicately controlled, appears to be an excellent candidate for a mechanism of mini-mizing tissue damage. It also may help to explain the occurrence of eosino-philia in so many parasitic infections characterized by amine-mediated re-sponses.

Dineen et al. (1974) provided evidence of another component in the ex-pulsion of parasites, namely prostaglandins. It is known that these com-pounds have a number of important functions throughout the body and the authors suggest that they could act directly on the parasites (Fig. 1), or in-directly by influencing other components of the immune system.

In all, there are a complex series of integrated immune responses which ultimately lead to rejection of the parasites; direct action on the parasite may involve antibodies, lymphokines, amines, antimicrobial enzymes and prostaglandins (Fig. 1), and it seems unlikely that the list is complete. Apart from helping one to view the host's immune responses to intestinal infection as a complex integrated system rather than as a simplified two or three step mechanism (see Ogilvie and Jones, 1973), consideration of Fig. 1 makes it painfully clear that there is little information on the actual mechanisms which result in rejection. Immune responses result in parasite damage (Ogilvie and Jones, 1973; Befus and Threadgold, 1975), but whether rejec-tion is due to a progressive reduction in fitness resulting from continued immunological attack or due to an abrupt paralytic, or comparable, action on the parasites is unknown.

15.3.5. Survival despite host immune responses

In the absence of evidence to the contrary, it is probable that immune responses, albeit not necessarily protective, occur against most, if not all, intestinal parasites. Hence the fundamental question is, how do parasites survive despite these responses? Protective responses are often density-dependent, viz. there are thresholds in a host (Wakelin, Chapter 6 above) below which rejection does not occur, or occurs more slowly than at greater parasite densities (Hopkins and Stallard, 1974; Ogilvie and Love, 1974; Befus, 1975). Therefore, the strategy of a species may be to minimize the possibility of overpopulation in an individual host. Other mechanisms adopted by parasites to cope with host's immune responses have been reviewed extensively (Porter and Knight, 1974), although little is known about mechanisms employed by intestinal parasites. Presumably intestinal parasites employ fundamentally similar strategies to those of parasites inhabiting other environments (see Chapters 17 and 18).

With mucosal-associated procaryotic infections there are two interesting strategies for survival which may apply to eucaryotic intestinal infections. The first is the active depression of immune responses, e.g. specific cleavage of IgA (Plaut et al., 1974), depression of monocyte chemotaxis (Kleinerman et al., 1974) and inactivation of complement components and chemotactic and phagocytic factors (Schultz and Miller, 1974). The pathophysiological alterations induced by parasites may depress the effectiveness of intestinal immune responses. The second strategy is to utilize some of the host responses as survival mechanisms. Centifanto and Kaufman (1971) showed that *Herpes simplex* "sensitized" by IgG antibodies were insusceptible to neutralization by IgA. Rubinstein et al. (1974) enhanced or inhibited tumor growth depending upon the class, combination of classes or concentration of antibodies transferred to mice. Perhaps some of the immunoglobulins detected on the surface of *Hymenolepis* spp. by Befus (1976) are relatively host protective antibodies and some relatively worm protective. Whether the worms survive or not may depend upon the interaction of these different antibodies. The activity of IgA as a blocking antibody has been reviewed (Bienenstock, 1974) and Walker and Isselbacher (1974) have suggested that IgA antibodies depress the uptake of antigens from the intestinal lumen; a mechanism which could induce a partially tolerant state, permitting parasite survival. There are many exciting possibilities for mechanisms of parasite survival despite, or because of, host immune responses.

15.4. Conclusions — the physiology of the immune response

Although considerable advances have been made to further our understanding of the histopathology and pathophysiology of helminth infections (Symons, 1969; Gaafar, 1971; Brandborg, 1973; Gallagher et al., 1971; Scofield, 1974; Podesta and Mettrick, 1976b) and the role of the host's immune response in parasite expulsion (see previous section and Chapter 6), very little information is available uniting these two fields. Are there changes in the microflora and intestinal structure and function in helminth infections which alter the potential of the immune response? For example, Murray (1972) has suggested that entry of antibody into the lumen of the intestine occurs via the tight junctions which are broken down and separated in the flat mucosa of intestines infected with *Nippostrongylus*. However, in other diseases characterized by flat mucosa, such as sprue, the remaining mucosa is less permeable than normal (Fordtran et al., 1967). The permeability of the mucosa remains unaltered in cholera, in which there is a copious active secretion of fluid and a marked immune response (Lifson et al., 1972). The mucosa of rats infected with *H. diminuta* is also less permeable to passively transported solutes and water than controls (Podesta and Mettrick, 1976b). These permeability studies suggest that the tight junctions are intact in these diseases and therefore as is well known the separation of the cells at the tight junctions is not necessary for antibody to traverse the intestinal mucosa.

Although advantageous in terms of surface area (see above), the unstirred layer coating membranes may also be a disadvantage to luminal dwelling parasites. It has been suggested that serum-derived antobody (and indeed secretory antibody and other components of the immune response such as complement) may exert its effect while in this unstirred layer at the mucosal surface and, in a parasitological sense, at the surface of the parasites; a position in which antibody may be protected from rapid destruction by proteolytic enzymes (Pierce and Reynolds, 1974).

The major question facing parasitologists, however, is what are the underlying physiological or biochemical changes in the parasite induced by the immune response leading to the expulsion of the parasites? The rather limited knowledge in this respect has been reviewed (Ogilvie and Jones; Ogilvie and Love, 1974). It is our belief that in this very basic question lies a most promising future for parasitology.

15.5 REFERENCES

Adinolfi, M., Glynn, A.A., Lindsay, M. and Milne, C.M. (1966). "Serological Properties of γA Antibodies to *Escherichia coli* Present in Human Colostrum." Immunology 10, 517—526.

Allen, A. and Snary, D. (1972). "The Structure and Function of Gastric Mucus." Gut 13, 666—672.

Argenzio, R.O. and Stevens, C.E. (1975). "Cyclic Changes in Ionic Composition of Digesta in the Equine Intestinal Tract." Am. J. Physiol. 228, 1224—1230.

Bachra, B.N., Trautz, O.R. and Simon, S.L. (1963). "Precipitation of Calcium Carbonates and Phosphates. I. Spontaneous Precipitation of Calcium Carbonates and Phosphates Under Physiological Conditions." Arch. Biochem, Biophys, 103, 124—138.

Bachra, B.N., Trautz, O.R. and Simon, S.L. (1965). "Precipitation of Calcium Carbonates and Phosphates. III. The Effect of Magnesium and Fluoride Ions on the Spontaneous Precipitation of Calcium Carbonates and Phosphates." Arch. Oral Biol. 10, 731—738.

Baggiolini, M. (1972). "The Enzymes of the Granules of Polymorphonuclear Leukocytes and Their Functions." Enzyme 13, 132—160.

Befus, A.D. (1975). "Secondary Infections of *Hymenolepis diminuta* in Mice: Effects of Varying Worm Burdens in Primary and Secondary Infections." Parasitology 71, 61—75.

Befus, A.D. (1976). "Hymenolepis diminuta and H. microstoma: Mouse Immunoglobulins Binding to the Tegumental Surface." Expl. Parasit. (In press).

Befus, A.D. and Threadgold, L.T. (1975). "Possible Immunological Damage to the Tegument of *Hymenolepis diminuta* in Mice and Rats." Parasitology 71, 525—534.

Bellamy, J.E.C. and Nielsen, N.O. (1974). "Immune-Mediated Emigration of Neutrophils into the Lumen of the Small Intestine." Inf. Immun. 9, 615—619.

Berger, J. and Mettrick, D.F. (1971). "Microtrichial Polymorphism Among Hymenolepid Tapeworms as Seen by Scanning Electron Microscopy." Trans. Amer. Microsc. Soc. 90, 393—403.

Bienenstock, J. (1974). "The Physiology of the Local Immune Response and the Gastrointestinal Tract." Prog. Immunol. II 4, 197—207.

Black, J.A. and Henderson, M.H. (1972). "Activation and Inhibition of Human Erythrocyte Pyruvate Kinase by Organic Phosphates, Amino Acids, Dipeptides and Anions." Biochem. Biophys. Acta 284, 115—127.

Boackle, R.J., Pruitt, K.M. and Mestecky, J. (1974). "The Interactions of Human Complement with Interfacially Aggregated Preparations of Human Secretory IgA." Immunochemistry II, 543—548.

Bockman, D.E. and Cooper, M.D. (1973). "Pinocytosis by Epithelium Associated with Lymphoid Follicles in the Bursa of Fabricius, Appendix, and Peyer's Patches. An Electron Microscopic Study." Am. J. Anat. 136, 455—478.

Brand, T. von. (1966). "Biochemistry of Parasites." Academic Press, London and New York.

Brandborg, L.L. (1973). In: "Gastrointestinal Disease." (M.H. Sleisenger and J.S. Fordtran, eds) pp. 989—1014. Saunders, Philadelphia, London, Toronto.

Brandtzaeg, P. (1973). "Structure, Synthesis and External Transfer of Mucosal Immunoglobulins." Ann. Immunol. (Inst. Past.) 124C, 417—438.

Brandtzaeg, P. (1974). "Mucosal and Glandular Distribution of Immunoglobulin Components: Differential Localization of Free and Bound SC in Secretory Epithelial Cells." J. Immunol. 112, 1553—1559.

Brandtzaeg, P., Baklion, K., Fausa, O. and Hoel, P.S. (1974). "Immunohistochemical Characterization of Local Immunoglobulin Formation in Ulcerative Colitis." Gastroenterology 66, 1123—1136.

Burdon, D.W. (1973). "The Bactericidal Action of Immunoglobulin A." J. Med. Microbiol. 6, 131—139.

Carter, P.B. and Collins, F.M. (1974). "The Route of Enteric Infection in Normal Mice." J. Exp. Med. 139, 1189—1203.

Castro, G.A., Johnson, L.R., Copeland, E.M. and Dudrick, S.J. (1974a). "Development of Enteric Parasites in Parenterally Fed Rats." Proc. Soc. Exp. Biol. Med. 146, 703—706.

Castro, G.A., Roy, S.A. and Stockstill, R.D. (1974b). *"Trichinella spiralis*: Peroxidase Activity in Isolated Cells from the Rat Intestine." Expl. Parasit. 36, 307—315.

Centifanto, Y.M. and Kaufman, H.E. (1971). In: "The Secretory Immunologic System." (D.H. Dayton, Jr., P.A. Small, Jr., R.M. Chanock, H.E. Kaufman and T.B. Tomasi, Jr., eds.) pp. 331—340. United States Government Printing Office, Washington, D.C.

Cooper, M.D. and Lawton, A.R. (1972). In: "Contemporary Topics in Immunobiology." (M.G. Hanna, Jr., ed.) pp. 49—68. Plenum Press, New York and London.

Crandall, R.B. and Crandall, C.A. (1972). *"Trichinella spiralis*: Immunologic Response to Infection in Mice." Expl. Parasit. 31, 378—398.

Crandall, R.B., Crandall, C.A. and Franco, J.A. (1974). *"Heligmosomoides polygyrus* (=*Nematospiroides dubius*): Humoral and Intestinal Immunologic Responses to Infection in Mice." Expl. Parasit. 35, 275—287.

Crompton, D.W.T. (1973). "The Sites Occupied by Some Parasitic Helminths in the Alimentary Tract of Vertebrates." Biol. Rev. 48, 27—83.

Crompton, D.W.T. and Nesheim, M.C. (1970). "Lipid, Bile Acid, Water and Dry Matter Content of the Intestinal Tract of Domestic Ducks with Reference to the Habitat of *Polymorphus minutus* (Acanthocephala)." J. Exp. Biol. 52, 437—445.

Dayton, D.H., Jr., Small, P.A., Jr., Chanock, R.M., Kaufman, H.E. and Tomasi, T.B., Jr. (1971). "The Secretory Immunologic System." United States Government Printing Office, Washington.

Dineen, J.K., Ronai, P.M. and Wagland, B.M. (1968). "The Cellular Transfer of Immunity to *Trichostrongylus colubriformis* in an Isogenic Strain of Guinea-Pig. IV. The Localization of Immune Lymphocytes in Small Intestine in Infected and Non-Infected Guinea-Pigs." Immunology 15, 671—679.

Dineen, J.K., Kelly, J.D., Goodrich, B.S. and Smith, I.D. (1974). "Expulsion of *Nippostrongylus brasiliensis* from the Small Intestine of the Rat by Prostaglandin-Like Factors from Ram Semen." Int. Arch. Allergy 46, 360—374.

Dobson, C. (1972). In: "Immunity to Animal Parasites." (E.J.L. Soulsby, ed.) pp. 191—222. Academic Press, New York and London.

Dobson, C. and Soulsby, E.J.L. (1974). "Lymphoid Cell Kinetics in Guinea Pigs Infected with *Trichostrongylus colubriformis*: Tritiated Thymidine Uptake in Gut and Allied Lymphoid Tissue, Humoral IgE and Hemagglutinating Antibody Responses, Delayed Hypersensitivity Reactions, and In Vitro Lymphocyte Transformations during Primary Infections." Expl. Parasit. 35, 16—34.

Erlandsen, S.L. and Chase, D.G. (1972). "Paneth Cell Function: Phagocytosis and Intracellular Digestion of Intestinal Microorganisms. I. *Hexamita muris*." J. Ultrast. Res. 41, 296—318.

Erlandsen, S.L., Parsons, J.A. and Taylor, T.D. (1974). "Ultrastructural Immunocytochemical Localization of Lysozyme in the Paneth Cells of Man." J. Histochem. Cytochem. 22, 401—413.

Eskeland, T. and Brandtzaeg, P. (1974). "Does J Chain Mediate the Combination of I9S IgM and Dimeric IgA with the Secretory Component rather than Being Necessary for Their Polymerization?" Immunochemistry 11, 161—163.

Fordtran, J.S., Rector, F.C., Locklear, T.W. and Ewton, M.F. (1967). "Water and Solute Movement in the Small Intestine of Patients with Sprue." J. Clin. Invest. 46, 287—298.

Fubara, E.S. and Freter, R. (1972). "Availability of Locally Synthesized and Systemic Antibodies in the Intestine." Inf. Immun. 6, 965—981.

Fubara, E.S. and Freter, R. (1973). "Protection Against Enteric Bacterial Infection by Secretory IgA Antibodies." J. Immunol. 111, 395—403.

Gaafar, S.M. (1971). "Pathology of Parasitic Diseases." Purdue University Press, La Fayette, Indiana.

Gallagher, N.D., Playoust, M.R. and Symons, L.E.A. (1971). "Mechanism of Fat Malabsorption in Rats Infected with *Nippostrongylus brasiliensis*." Gut 12, 1007—1010.

Genco, R.J., Evans, R.T. and Taubman, M.A. (1974). In: "The Immunoglobulin A System." (J. Mestecky and A.R. Lawton III, eds.) pp. 327—336. Plenum Press, New York and London.

Gibbons, R.J. (1974). In: "The Immunoglobulin A System." (J. Mestecky and A.R. Lawton III, eds.) pp. 315—325. Plenum Press, New York and London.

Green, H.H., Steinmetz, P.R. and Frazier, H.S. (1970). "Evidence for Proton Transport by Turtle Bladder in Presence of Ambient Bicarbonate." Am. J. Physiol. 218, 845—850.

Gros, G. and Moll, W. (1974). "Facilitated Diffusion of CO_2 Across Albumin Solutions." J. Gen. Physiol. 64, 356—371.

Guy-Grand, D., Griscelli, C. and Vassalli, P. (1974). "The Gut-Associated Lymphoid System: Nature and Properties of Large Dividing Cells." Eur. J. Immunol. 4, 435—443.

Hamilton, J.D., Dawson, A.M. and Webb, J.P.W. (1968). "Observation Upon Small Gut "Mucosal" pO_2 and pCO_2 in Anesthetized Dogs." Gastroenterology 55, 52—60.

Harpur, R.P. (1974a). "Pathways of Anaerobic Energy Metabolism in Helminths." Proc. Can. Soc. Zool., (M.D.B. Burt ed.).

Harpur, R.P. (1974b). "Haemolymph Gases and Buffers in *Ascaris lumbricoides*." Comp. Biochem. Physiol. 48A, 133—143.

Heremans, J.F. (1974). In: "The Immunoglobulin A System." (J. Mestecky and A.R. Lawton III, eds.) pp. 3—11. Plenum Press, New York and London.

Heyneman, D. (1963). "Host-Parasite Resistance Patterns — Some Implications from Experimental Studies with Helminths." Ann. N.Y. Acad. Sci. 113, 114—129.

Hill, I.R. and Porter, P. (1974). "Studies of Bactericidal Activity to *Escherichia coli* of Porcine Serum and Colostral Immunoglobulins and the Role of Lysozyme with Secretory IgA." Immunology 26, 1239—1250.

Hochachka, P.W., Fields, J. and Mustafa, T. (1973). "Animal Life Without Oxygen: Basic Biochemical Mechanisms." Amer. Zool. 13, 543—555.

Hochachka, P.W. and Mustafa, T. (1972). "Invertebrate Facultative Anaerobiosis." Science 178, 1056—1060.

Hopkins, C.A. and Stallard, H.E. (1974). "Immunity to Intestinal Tapeworms: the Rejection of *Hymenolepis citelli* by Mice." Parasitology 69, 63—76.

Hsung, J.C. and Haug, A. (1975). "Intracellular pH of *Thermoplasma acidophila*." Biochim. Biophys. Acta 389, 477—482.

Hubscher, T. and Eisen, A.H. (1973). In: "Mechanisms in Allergy: Reagin-Mediated Hypersensitivity." (L. Goodfriend, A.H. Sehon and R.P. Orange, eds.) pp. 413—437. Marcel Dekker, New York.

Ismail, M.M. and Tanner, C.E. (1972). "*Trichinella spiralis*: Peripheral Blood, Intestinal, and Bone-Marrow Eosinophilia in Rats and Its Relationship to the Inoculating Dose of Larvae, Antibody Response and Parasitism." Expl. Parasit. 31, 262—272.

Kasbekar, D.K. (1966). "Effect of Carbon Dioxide-Bicarbonate Mixtures on Rat Liver Mitochondrial Oxidative Phosphorylation." Biochim. Biophys. Acta 128, 205—208.

Kleinerman, E.S., Snyderman, R. and Daniels, C.A. (1974). "Depression of Human Monocyte Chemotaxis by *Herpes simplex* and Influenza Viruses." J. Immunol. 113, 1562—1567.

Komuniecki, R. and Roberts, L.S. (1975). "Developmental Physiology of Cestodes. XIV. Roughage and Carbohydrate Content of Host Diet for Optimal Growth and Development of *Hymenolepis diminuta*." J. Parasit. 61, 427—433.

Köhler, P. and Hanselmann, K. (1974). "Anaerobic and Aerobic Energy Metabolism in the Larvae (Tetrathyridia) of *Mesocestoides corti*." Expl. Parasit. 36, 178—188.

Larsh, J.E., Jr. and Race, G.J. (1975). "Allergic Inflammation as a Hypothesis for the Expulsion of Worms from Tissues: A Review." Expl. Parasit. 37, 251—266.

Larsh, J.E., Jr. and Weatherly, N.F. (1974). "Cell-Mediated Immunity in Certain Parasitic Infections." Curr. Top. Microbiol. Immunol. 67, 113—137.

Larsch, J.E., Jr. and Weatherly, N.F. (1975). "Cell-Mediated Immunity Against Certain Parasitic Worms." Adv. Parasit. 13, 183—222.

Larsh, J.E., Jr., Ottolenghi, A. and Weatherly, N.F. (1974). "*Trichinella spiralis*: Phospholipase in Challenged Mice and Rats." Expl. Parasit. 36, 299—306.

Levin, D.M., Rosenstreich, D.L., Wahl, S.M. and Reynolds, H.Y. (1974). "Peyer's Patch Lymphocytes: Demonstration of the Integrity of Afferent and Efferent T-Cell Functions in the Guinea-Pig and Rat." J. Immunol. 113, 1935—1941.

Liburd, E.M., Armstrong, W.D. and Mahrt, J.L. (1973). "Immunity to the Protozoan Parasite *Eimeria nieschulzi* in Inbred CD-F Rats." Cell. Immunol. 7, 444—452.

Lifson, N., Hakim, A.A. and Lender, E.J. (1972). "Effects of Cholera Toxin on Intestinal Permeability and Transport Interactions." Am. J. Physiol. 223, 1479—1487.

Lindh, E. (1975). "Increased Resistance of Immunoglobulin A Dimers to Proteolytic Degradation after Binding of Secretory Component." J. Immunol. 114, 284—286.

Livingstone, D.R. and Bayne, B.L. (1974). "Pyruvate Kinase from the Mantle Tissue of *Mytilus edulis* L." Comp. Biochem. Physiol. 48B, 481—497.

Mestecky, J. and Lawton, A.R. III. (1974). "The Immunoglobulin A System." Plenum Press, New York and London.

Mettrick, D.F. and Cannon, C.E. (1970). "Changes in the Chemical Composition of *Hymenolepis diminuta* (Cestoda: Cyclophyllidea) During Prepatent Development in the Rat Intestine." Parasitology 61, 229—243.

Mettrick, D.F. and Podesta, R.B. (1974). "Ecological and Physiological Aspects of Helminth-Host Interactions in the Mammalian Gastrointestinal Canal." Adv. Parasit. 12, 183—278.

Morag, A., Beutner, K.R., Morag, B. and Ogra, P.L. (1974). "Development and Characteristics of In Vitro Correlates of Cellular Immunity to Rubella Virus in the Systemic and Mucosal sites in Guinea Pigs." J. Immunol. 113, 1703—1709.

Mounib, M.S. and Eisan, J.S. (1972). "Fixation of Carbon Dioxide by the Testes of Rabbit and Fish." Comp. Biochem. Physiol. 43B, 393—401.

Murray, M. (1972). In: "Immunity to Animal Parasites." (E.J.L. Soulsby, ed.) pp. 155—190. Academic Press, New York and London.

Ogilvie, B.M. and Jones, V.E. (1973). "Immunity in the Parasitic Relationship between Helminths and Hosts." Progr. Allergy 17, 93—144.

Ogilvie, B.M. and Love, R.J. (1974). "Co-operation between Antibodies and Cells in Immunity to a Nematode Parasite." Transplant. Rev. 19, 147—168.

Ottolenghi, A. (1973). "High Phospholipase Content of Intestines of Mice Infected with *Hymenolepis nana*." Lipids 8, 426—428.

Parrott, D.M.V. and Ferguson, A. (1974). "Selective Migration of Lymphocytes within the Mouse Small Intestine." Immunology 26, 571—588.

Pierce, N.F. and Reynolds, H.Y. (1974). "Immunity to Experimental Cholera. I. Protective Effect of Humoral IgG Antitoxin Demonstrated by Passive Immunization." J. Immunol. 113, 1017—1023.

Plaut, A.G., Genco, R.J. and Tomasi, T.B., Jr. (1974). "Isolation of an Enzyme from *Streptococcus sangius* which Specifically Cleaves IgA." J. Immunol. 113, 289—291.

Podesta, R.B. and Mettrick, D.F. (1974a). "Pathophysiology of Cestode Infections: Effect of *Hymenolepis diminuta* on Oxygen Tensions, pH and Gastrointestinal Function." Int. J. Parasit. 4, 277—292.

Podesta, R.B. and Mettrick, D.F. (1974b). "Components of Glucose Transport in the Host-Parasite System, *Hymenolepis diminuta* (Cestoda) and the Rat Intestine." Can. J. Physiol. Pharmacol. 52, 183—197.

Podesta, R.B. and Mettrick, D.F. (1975). "*Hymenolepis diminuta*: Acidification and Bicarbonate Absorption in the Rat Intestine." Expl. Parasit. 37, 1—14.

Podesta, R.B. and Mettrick, D.F. (1976a). "The Inter-Relationships between the In Situ Fluxes of Water, Electrolytes and Glucose by *Hymenolepis diminuta.*" Int. J. Parasit. 6, 163—172.

Podesta, R.B. and Mettrick, D.F. (1976b). "Pathophysiology and Compensatory Mechanisms in a Compatible Host-Parasite System." Can. J. Zool. 54, 794—703.

Porter, R. and Knight, J. (1974). "Parasites in the Immunized Host: Mechanisms of Survival." Associated Scientific Publishers, Amsterdam, London and New York.

Rahaman, R. and Meisner, H. (1973). "Respiratory Studies with Mitochondria from the Rat Tapeworm *Hymenolepis diminuta.*" Int. J. Biochem. 4, 153—162.

Rehm, W.S., Sanders, S.S., Shoemaker, R.L., O'Callaghan, J., Tarvin J.T. and Friday, E.A. (1973). "Proton Conductance of Cell Membranes." J. Theor. Biol. 39, 131—153.

Roitt, I. (1974). "Essential Immunology." Blackwell, Oxford.

Rothwell, T.L.W. and Dineen, J.K. (1972). "Cellular Reactions in Guinea-Pigs Following Primary and Challenge Infection with *Trichostrongylus colubriformis* with Special Reference to the Roles Played by Eosinophils and Basophils in Rejection of the Parasite." Immunology 22, 733—745.

Rothwell, T.L.W., Prichard, R.K. and Love, R.J. (1974). "Studies on the Role of Histamine and 5-Hydroxy-tryptamine in Immunity against the Nematode *Trichostrongylus colubriformis*. I. In Vivo and In Vitro Effects of the Amines." Int. Arch. Allergy 46, 1—13.

Rubinstein, P., DeCary, F. and Streun, E.W. (1974). "Quantitative studies on Tumor Enhancement in Mice. I. Enhancement of Sarcoma I Induced by IgM, IgG_1 and IgG_2." J. Exp. Med. 140, 591—596.

Saz, H.J. (1971). "Facultative Anaeribiosis in the Invertebrates: Pathways and Control Systems." Am. Zool. 11, 125—135.

Saz, H.J. (1972). In: "Comparative Biochemistry of Parasites." (H.Van den Bossche, ed.) pp. 33—48. Academic Press, London and New York.

Schultz, D.R. and Miller, K.D. (1974). "Elastase of *Pseudomonas aeruginosa*: Inactivation of Complement Components and Complement-Derived Chemotactic and Phagocytic Factors." Inf. Immun. 10, 128—135.

Scofield, A.M. (1974). "Intestinal Absorption of D-Glucose and D-Galactose in Rats Infected with *Nematospiroides dubius*." Comp. Biochem. Physiol. 47A, 219—231.

Selby, G.R. and Wakelin, D. (1973). "Transfer of Immunity Against *Trichuris muris* in the Mouse by Serum and Cells." Int. J. Parasit. 3, 717—722.

Soulsby, E.J.L. (1972). "Immunity to Animal Parasites." Academic Press, New York and London.

Spencer, J.C., Waldman, R.H. and Johnson, J.E. III. (1974). "Local and Systemic Cell-Mediated Immunity after Immunization of Guinea Pigs with Live or Killed *M. tuberculosus* by Various Routes." J. Immunol. 112, 1322—1328.

Symons, L.E.A. (1969). "Pathology of Gastrointestinal Helminthiasis." Int. Rev. Trop. Med. 3, 49—100.

Tomasi, T.B. (1971). In: "The Secretory Immunologic System." (D.H. Dayton, Jr., P.A. Small, Jr., R.M. Chanock, H.E. Kaufman and T.B. Tomasi Jr., eds.) pp. 3—10. United States Government Printing Office, Washington, D.C.

Tomasi, T.B. and Grey, H.M. (1972). "Structure and Function of Immunoglobulin A." Progr. Allergy 16, 81—213.

Veldkamp, J., Van Der Gaag, R. and Willers, J.M.N. (1973). "The Role of Peyer's Patch Cells in Antibody Formation." Immunology 25, 761—771.

Waksman, B.H. (1973). "The Homing Pattern of Thymus-Derived Lymphocytes in Calf and Neonatal Mouse Peyer's Patches." J. Immunol. 111, 878—884.

Waldram, R. (1975). Mechanisms of Lipid Loss from the Small Intestinal Mucosa. Gut 16, 118—124.

Walker, W.A. and Isselbacher, K.J. (1974). "Uptake and Transport of Macromolecules by the Intestine. Possible Role in Clinical Disorders." Gastroenterology 67, 531—550.

Webster, L.A. and Wilson, R.A. (1970). "The Chemical Composition of Protonephridial Canal Fluid from the Cestode *Hymenolepis diminuta.*" Comp. Biochem. Physiol. 35, 201—209.

Weicker, J. and Underdown, B.J. (1975). "A Study of the Association of Human Secretory Component with IgA and IgM Proteins." J. Immunol. 114, 1337—1344.

Westergaard, H. and Dietschy, J.M. (1974). "Delineation of the Dimensions and Permeability Characteristics of the Two Major Diffusion Barriers to Passive Mucosal Uptake in the Rabbit Intestine." J. Clin. Invest. 54, 718—732.

Wilson, F.A. and Dietschy, J.M. (1974). "The Intestinal Unstirred Layer: Its Surface Area and Effect on Active Transport Kinetics." Biochim. Biophys. Acta 363, 112—126.

Zinneman, H.H. and Kaplan, A.P. (1972). "The Association of Giardiasis with Reduced Intestinal Secretory Immunoglobulin A." Am. J. Dig. Dis. 17, 793—797.

Ecological Aspects of Parasitology
Editor: C.R. Kennedy
© *North-Holland Publishing Company, Amsterdam, 1976*

CHAPTER 16

PARASITES OF THE EYE AND BRAIN

G. RICHARD O'CONNOR

Francis I. Proctor Foundation for Research in Ophthalmology and the Department of Ophthalmology, University of California, San Francisco (U.S.A.)

Contents

16.1. Introduction

The fact that both the brain and the eye provide a highly suitable environment for the growth and development of a number of animal parasites is not, most likely, a matter of mere happenstance. The vertebrate eye must be looked upon as an embryonic outgrowth of the diencephalon, with a few later additions of ectodermal and mesodermal components. The retina, in particular, possesses certain striking resemblances to cerebral tissue as regards vasculature and biochemical constituents. The fact that both tissues are the principal target of certain parasites is probably not, therefore, purely coincidental.

Because of limitations of space, only those parasites which affect *both* the brain and the eye will be discussed in this chapter. A distinction will also be made between those parasites which affect the inner eye exclusively, as opposed to those which affect the epibulbar or orbital tissues, for these later are clearly unrelated to those which affect the brain

16.2. The Protozoa

16.2.1. The Amoebae

Although Harris and Birch (1960) described a choroiditis in an 18-year-old Caucasian male as attributable to infection with *Entamoeba histolytica*, there is no histologic proof that the ocular condition was caused by this parasite. The ocular inflammation accompanied amoebic colitis, for which there was adequate parasitological evidence, and the eye lesion improved after the successful eradication of the colonic infection. Braley and Hamilton (1957) and King et al. (1964) described central choroidosis with cystic and serous lesions of the retina as manifestations of amoebic infection. Although there were no cultural or histological data on support their conclusions, they also found that the ocular disease improved after their patients were given courses of Diodoquin®, carbarsone, and chloroquine. In none of their cases was there evidence of retinal penetration, and in all cases the overlying vitreous remained clear. Knowing what we do about the invasive quality of *Entamoeba histolytica* it seems strange that the organism, if present, would have been halted by the lamina vitrea (Bruch's membrane) of the retina. Host tissues appear to be damaged only by direct contact with amoebae. Lysosomal enzymes are released, probably by the activation of contact-sensitive surface organelles. Relative anoxia and the acid conditions of cell necrosis favor the penetration of amoebae into the tissues. Conditions of low oxygen tension and low pH (6.0—6.5) are encountered in the intestinal lesions (Knight, 1975), but not in the choroid, retina, and vitreous. Indeed, Davson and Luck (1956) state that the pH of the vitreous is about 7.2, making it somewhat less than normally attractive to advancing trophozoites. These may not be the critical reasons, however, for the failure of amoebae to penetrate the retina and vitreous. According to Hogan et al. (1971), Bruch's membrane, the all-important anatomic barrier between the choroid and retina, is composed of alternating laminae of specialized collagen and elastic tissue. *Entamoeba* may not have the specialized enzyme equipment necessary for the digestion of this layer.

Because of the location of the fundus lesions in and around the posterior pole of the eye, the parasites are through to gain access by way of the posterior ciliary vessels. Although no amoebae have been seen in these posterior lesions, Krümmel is said by Duke-Elder (1966a) to have found organisms with the morphologic characteristics of *Entamoeba* in the iris and ciliary body. The pictures which Duke-Elder shows of these organisms are highly convincing, although data relating to the isolation of the parasite are lacking.

If the case for amoebic invasion of the eye is rather weak, a much stronger case can be made for invasion of the brain. More than 100 cases of amoebic brain abscess have been described, all of them secondary to amoebic abscesses of the liver, and virtually all of them fatal. Clearly, the parasite has the ability to digest cerebral tissue, restricted only by the inflammatory reaction which occurs at the edges of the lesion. It seems likely that the eye and brain would be seeded by the same metastatic, blood-borne processes; yet hepatic abscesses are not regularly described in the articles dealing with the ocular lesions. The apparent contradiction may be resolved by the concept that the posterior ocular lesions represent hypersensitivity phenomena. Animal models of the disease must, however, be firmly established before this concept can be accepted.

The facultative parasites, *Naegleria* and *Acanthamoeba*, are well known for their ability to cause fatal meningoencephalitis. Duma et al. (1971) suggest that *Naegleria gruberi* gains access to the cerebral meninges through invasion of the nasal mucosa and penetration of the cribiform plate. A history of swimming or diving in lakes and streams infested by these free-living parasites has been obtained universally in the cases described. It seems clear that these parasites have acquired the enzyme equipment to digest mammalian tissues.

Schlaegel and Culbertson (1972) have described experimental infections of the rabbit uvea and optic nerve with *Hartmanella*. The highly destructive lesions leave no doubt that these amoebae can damage the inner ocular structures. In the absence of cerebral lesions it is doubtful, however, that they would play much of a role in naturally occurring uveitis in man. The organisms should, perhaps, be looked for in a systematic way in pathologic specimens of eyes that have been labelled "idiopathic uveitis".

Recently Nagington et al. (1974) and Visvesvara et al. (1975) have described corneal lesions due to *Acanthamoeba polyphaga*. It seems clear that these organisms do have the ability to cause indolent ulcers of the cornea, particularly when a corneal abrasion has been treated with local medications containing corticosteroids. It is more than likely that the facultative amoebas can gain a foothold on the cornea only when the epithelial barrier has been broken and only under conditions of local immunosuppression.

16.2.2. Trypanosomes

Trypanosomiasis is not a very common cause of ocular morbidity, although at least three cases of iridocyclitis associated with African sleeping sickness have been described by Bargy (1928). Ikede (1974) has also shown that *Trypanosoma brucei* can cause severely destructive lesions of the ovine uvea when the organism is injected intravenously into sheep. The lesions showed perivascular infiltrations of lymphocytes and plasma cells along with a proliferative endarteritis. Trypanosomes were seen in the stroma of the iris and ciliary body in positions analogous to their location in brain substance. The breakdown of blood vessel barriers is signaled by the perivascular location of the parasites and by the hemorrhagic quality of the lesions.

16.2.3. The Sporozoa

Unquestionably, the most important sporozoan parasite infecting both the brain and the eye is *Toxoplasma gondii*. Although this organism was originally described by Nicolle and Manceaux (1908) in the first decade of this century, it received no attention as an ocular pathogen until Jankû (1923) found organisms morphologically identical to *Toxoplasma* in tissue sections of an eye that had been removed from an infant with congenital toxoplasmosis. The child was thought to have a "coloboma" of the macula, i.e. a congenital defect of the eye caused by failure of fetal fissure to close during embryonic life. Jankû surprised the ophthalmic community by finding an infectious etiology for this condition. His work was later confirmed by Wolf, Cowen, and Paige (1939) and Jacobs et al. (1954) who isolated living *Toxoplasma* parasites from the eyes and brains of human subjects, transmitted the infection to experimental animals, and thus essentially fulfilled Koch's postulates.

Although toxoplasmosis was once considered an ophthalmic rarity, it is now thought of as a common cause of retinochoroiditis, being responsible for 30—50% of all granulomatous inflammations of the posterior segment of the eyes (Jacobs et al., 1956) (Woods, 1960).

Typically the lesion presents itself as a focal necrotizing retinitis in or near the macula. The intense inflammatory reaction that accompanies the lesion is responsible for the throwing off of large clouds of inflammatory cells into the overlying vitreous. While the retinal lesions may be solitary, they are usually seen in clusters, among which lesions of various ages can be discerned, the older ones appearing as atrophic or hyperpigmented spots in the fundus. Although peripheral lesions have been observed by Chesterton and Perkins (1967) and by Hogan et al. (1964) the great bulk of the lesions are in the region of the macula and optic nerve. This had led to the hypothesis that the parasite enters the eye through the posterior ciliary arteries. A significant number of lesions have affected the optic nerve itself or an area of the retina that is immediately adjacent to the optic nerve head. These papillary and juxtapapillary lesions have led Berengo and Frezzotti (1962) to suppose that the parasite passes from the brain into the eye through the optic nerve or its meningeal sheaths. In considering this concept we must remember that the eye is an embryonic outpocketing of the brain and that the potential space between the sensory retina and the retinal pigment epithelium is really a continuation of the primitive ventricular system of the brain. Perkins (1973) and many other authors believe that all ocular toxoplasmosis is a manifestation of congenital toxoplasmosis. The brain and the eye are believed to be infected simultaneously at some time during fetal life, most probably during the first trimester of pregnancy. Although such lesions often appear to be already healed at the time of birth, late relapses of inflammation are thought to occur in adolescent or adult life as a result of the breakdown of *Toxoplasma* cysts. These resistant forms of the organism, so characteristic of chronic infection, contain hundreds or thousands of dormant parasites within a refractile cyst wall that is composed of both host and parasite components (Scholtyseck, 1973). The cysts attract little or no inflammatory reaction (Fig. 1), but when they rupture, intense inflammation ensues, either as a result of renewed multiplication of the parasite or as a result of hypersensitivity reactions.

Frenkel (1961) is largely responsible for a theory that links recurrent ocular inflammations with hypersensitivity phenomena. He believes that the organisms released from *Toxoplasma* cysts in the later stages of the disease do not invade new, previously uninvolved cells. Local immune factors in the tissues of a host that previously experienced active disease are said to prevent this from happening. Dr. Robert Nozik and I (Nozik and O'Connor, 1970) (O'Connor, 1970) have been largely unable to substantiate this hypothesis. In an animal model of ocular toxoplasmosis (Nozik and O'Connor, 1968) we have been unable to produce focal necrotizing retinitis by the injection of either soluble *Toxoplasma* antigens or heat killed organisms into areas adjacent to previously active lesions. For this reason we have preferred to think that all focal necrotizing retinitis of whatever age or duration is the result of the multiplicative activity of the organism itself.

While inflammation of the anterior segment of the uvea occurs quite frequently in association with or in the wake of activity in the posterior (retinal) lesions, *Toxoplasma* organisms have never been found in the iris or ciliary body of man. This is in contradistinction to the situation observed in many domestic animals where Piper et al. (1970) have demonstrated the organism in the anterior uvea. Some fundamental difference in the vascular supply of animal eyes or some other important environmental factors must account for the differences in the distribution of organisms in the various ocular tissues of man and lower animals. Huldt (1966) has demonstrated that infective organisms (trophozoites) are disseminated throughout the body in the protective milieu of blood leukocytes, principally lymphocytes. Encased

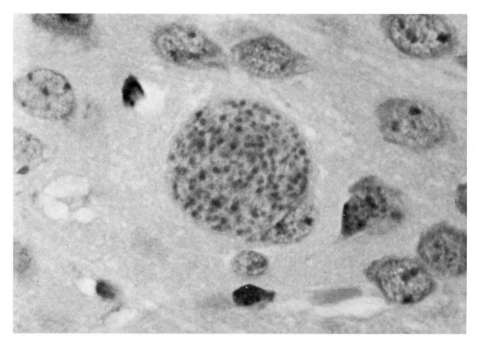

Fig. 1. Cyst of *Toxoplasma gondii* in mouse brain. Note absence of inflammatory reaction. H and E, 1000 ×.

within these cells, where they are protected against the activities of antibodies and complement, the parasites eventually reach the terminal capillaries of many tissues. When the leukocyte becomes temporally halted in one of these capillary beds, the trophozoites may emerge from the white cell, erode their way through the capillary wall, and attack whatever parenchymal cells are available. From this point onward the interplay of many environmental factors determines the extent to which the tissue is damaged. These include the virulence of the organism, the efficiency of the host's immunologic defense system, and the vascular supply of the target tissue.

It is a known fact that some strains of *Toxoplasma* (e.g. the RH strain) have a tendency to rapid asexual multiplication in the tissues with little spontaneous tendency to cyst formation. In contrast to these virulent strains, others, such as the Beverley strain, multiply slowly in mammalian tissues and form cysts early in the course of the disease. Animals experimentally infected with the latter tend to survive indefinitely, acting as reservoirs of infection, whereas inoculation of a mouse with even one RH strain organism will usually result in death.

The competence of the immunological defense system to protect the host against the ravages of *Toxoplasma* is probably the single most important factor. Since antibodies cannot penetrate host cells under ordinary circumstances, and since *Toxoplasma* is an obligate intracellular parasite, antibodies seem to serve relatively little function in the protection of the tissues. At most they are thought to rid the bloodstream of free-swimming parasites that have not yet sought the protection of cells. Cell-mediated immunity, on the other hand, seems to be all-important in the defense of the host. Frenkel (1967) emphasized the role of the sensitized lymphocyte in *Toxoplasma* infections, and Remington et al. (1972) showed that "activated" macrophages are the most important single element in the eradication of *Toxoplasma.* Although it was long suspected that some kind of lymphocyte-macrophage interaction was necessary for the "activation" of macrophages Borges and Johnson (1975) have conclusively demonstrated that lymphokines secreted by sensitized T-lymphocytes activate macrophages. The multiplication of organisms taken up by these macrophages is arrested and the parasites ultimately disintegrate. Macrophages from non-immune animals, on the other hand, will be destroyed by multiplying Toxoplasmas in the normal course of events. Jones (1974) has shown that living Toxoplasmas, like living tubercle bacilli and living Chlamydial organisms, come to lie within parasitophorous vacuoles within the cytoplasm of the macrophage. These vacuoles fail to fuse with lysosomal bodies within the macrophage's cytoplasm unless the organism has been killed or coated with antibody. It has been suggested that the living organism may secrete a substance (possibly cyclic AMP) which inhibits the fusion of the parasitophorous vacuole with the lysosome. The dead or antibody-coated organism fails to exert this influence; lysosomal enzymes are then emptied into the parasitophorous vacuole, and the parasite is digested.

Individuals deprived of T-cell immunity are the special prey of *Toxoplasma.* Patients suffering from Hodgkin's disease or other lymphoproliferative disorders such as chronic lymphatic leukemia may succumb to *Toxoplasma* infections despite the presence of large amounts of circulating antibodies in their bloodstreams. The same is true of patients who have received total-body irradiation for the treatment of malignant neoplasms or of patients who have received extensive immunosuppressive therapy as a deterrant to the rejection of kidney transplants. It has been suggested that intrauterine infections, particularly those that occur during the first trimester of pregnancy, result in highly destructive lesions of the eye and brain for the very reason that effective T-cell immunity has not yet been developed.

The story is not yet complete, however. There may still be a role for antibody even in the interstitial spaces. Nozik and O'Connor (1968) found large numbers of plasma cells in the subsiding retinal lesions that they had induced in rabbit eyes with the Beverley strain of *Toxoplasma.* Although the Beverley strain has a normal tendency to form cysts rather early in the course of the disease, the plasma cells were thought to play some role in the shift from trophozoite to cyst. Since the only known function of plasma cells is to secrete antibody, separate experiments were performed with serum antibodies, both with and without complement, in tissue cultures infected with the RH strain of *Toxoplasma*. The trophozoites of the RH strain, which normally destroy an entire tissue culture monolayer within a few days, were largely converted to cysts when fresh supplies of antibody and complement were added to the tissue culture medium every other day. Antibody alone or complement alone failed to mediate this change. Shimada, O'Connor, and Yoneda (1974) concluded that antibody, supplied to the interstitial spaces, may play an important role in the healing of toxoplasmic lesions simply by converting trophozoites to the relatively harmless cysts. Whether this is done through circulating antibodies or through plasma cells that have infiltrated the lesions may be of no ultimate consequence.

If antibodies play some protective role in the tissues, is the permeability of the blood vessels which supply those tissues important? Hogan and Feeney (1963) have shown that the permeability of the blood vessels supplying the brain and retina is markedly different from that of the vessels supplying the choroid and other structures of the eye. The vessels of the retina and brain are impermeable to even such small molecules as fluorescein and would certainly not permit the egress of significant amounts of Immunoglobulin G into the tissue spaces. The anatomic basis of this low permeability is found in the so-called "tight junctions" that can be observed between the endothelial cells of the retinal and cerebral capillaries. This is probably the basis of the well known "blood-brain barrier" that inhibits the cross-over of antibodies from the blood stream to the cerebrospinal fluid and vice versa. The relative lack of antibodies in the anterior third of the retina and in the cerebral grey matter may account for the persistence of *Toxoplasma* organisms in these tissues long after *Toxoplasma* has been eradicated from every other tissue of the body, including the choroid. It is a matter of record that the majority of *Toxoplasma* parasites located in histologic sections of the eye have been found in the anterior third of the retina, i.e. the portion nourished by the central retinal artery and its branches. The posterior two-thirds of the retina, which is nourished by the relatively permeable choroidal vessels, shows relatively few organisms. Thus the distribution and quality of the ocular blood vessels may play an important role in the fate and location of the parasite.

Asymptomatic infection with *Toxoplasma gondii* is widespread among the populations of most cities in the temperate and tropical zones of the world. Judging from antibody surveys, some 30—50% of the people of the San Francisco Bay area are infected with *Toxoplasma*, but very few of these have symptoms. Among adults who are known to have acquired the febrile, lymphadenopathic form of the disease at some particular time, less than 1% are reported to have developed ocular lesions. This and other bits of evidence led Perkins (1973) to the conclusion that nearly all ocular toxoplasmosis represents an early or late relapse of congenital disease. Whether this conclusion is correct remains to be seen. The work of Desmonts and Couvreur (1974) shows that infants may, indeed, be born with the infection (as indicated by the isolation of parasites from the cord blood), and yet manifest none of the classical signs of the disease. Since a certain number of these children later developed chorioretinal lesions, we know that *Toxoplasma* infection may be a latent disease. The presence of small numbers of cysts in the retina could never be detected by ophthalmoscopic examination initially, yet the breakdown of these cysts might bring on florid inflammatory disease later in life. The cause of cyst breakdown and the nature of the host's response to the organisms released from the cysts are crucial issues. As Frenkel (1971) puts it, the eye and the brain seem to be immunologically compromised tissues; for one reason or another, they are poorly defended against *Toxoplasma*. The nature of their blood supply, their isolation from other organs, their lack of lymphatic drainage, and their special vulnerability during early fetal life, when T-cell protection is not available, may be important factors in determining their special susceptibility to infection.

16.3. The Helminths

With few exceptions, parasitism of the eye and brain by helminths appears to reflect an errant wandering of larvae into these tissues. In such cases the eye and brain are not thought to offer any special attractants to the parasite. It is, rather, a matter of fortuitous wandering and, perhaps, of trapping of these parasites in the affected tissues. In the case of the eye, structural damage may be suffered as the result of movements of the parasite or as the result of an expanding mass (e.g. the bladder of a cysticercus). In some instances the larva is well tolerated by the eye or brain so long as the parasite remains living, but at the instant of death inflammatory reactions of a highly destructive nature may ensue.

16.3.1. Nematode parasites

16.3.1.1. Trichinella spiralis
Trichinosis is a common, though generally paucisymptomatic, infection affecting large segments of the population of North America and Central Europe. Perpetuated through the habit of eating undercooked pork or wild game, the disease continues to cause eye symptoms and, occasionally, fatal encephalitis. Fortunately the major ocular involvement is epibulbar, the extraocular muscles being extensively parasitized by larvae which undergo encystment in striated muscle fibers. This process is accompanied by considerable edema both of the eye muscles and of the periorbital tissues.

In their search for striated muscle fibers the larvae of this worm have been known to penetrate many other tissues including the optic nerve and retinal vessels. Edwards (1954) has described retinal hemorrhages and papilloedema as manifestations of *Trichinella* infection. Although such complications must be rare, their occurrence indicates that trichinosis may occasionally be vision threatening. The major effect on the eyes is, of course, the inflammation that occurs in the extraocular muscles. Although the symptoms of pain on motion of the eyes and photophobia generally last no longer than 10 days, the larvae may remain viable in the tissues for up to 10 years.

16.3.1.2. Toxocara canis and cati

Manifest disease of the eye and brain from infection with *Toxocara* is mainly limited to children ranging from 4 to 16 years of age. Ingestion of the eggs of *Toxocara* can generally be traced to the eating of "mud pies" or to close domestic association with young puppies or kittens that have not been de-wormed. Hand to mouth transfer of the ova is thought to be an important epidemiologic factor. The eggs hatch in the intestine and invade the mucosa, whence they are disseminated via the portal circulation to the liver and ultimately to the right ventricle and lungs. Unable to penetrate the alveoli, as the normal ascarid cycle would dictate, the larvae continue to wander through the body, emerging from the blood vessels of the eye and brain among other organs.

Although many of the lesions may be asymptomatic, the ocular manifestations of visceral larva migrans take three major forms: (1) an elevated granuloma in the macular region; (2) a fulminant endophthalmitis; and (3) an eosinophilic abscess in the peripheral retina or ciliary body. The latter two forms are associated with extensive cellular reaction in the vitreous, possibly associated with the death of the larva.

Byers and Kimura (1974) reported a case of intraocular nematode infection (believed to be *Toxocara)* in an adolescent Caucasian male. The highly motile larva was seen to emerge from the optic nerve head and to proceed across the vitreous cavity. It produced no symptoms other than "floating spots" while in this state. However, the vitreous was apparently unable to provide sufficient nutrition for the larva, and upon the death of the organism a violent cellular reaction occurred in the vitreous. The patient was treated for this with systemic corticosteroids and experienced a satisfactory return of vision.

Wilder (1950) and Ashton (1960) are responsible for finding *Toxocara* larvae in the so-called pseudogliomas of childhood. Such lesions, originally thought to represent degenerative conditions in the central retinas of young children, caused no reaction in the overlying vitreous. They were occasionally associated with bizarre pigmentary disturbances, traction on the retina, and hemorrhage. Several eyes enucleated because of the apparent growth of a "glioma" in the retina yielded granulomas that surrounded partially degenerated worms. Hogan, Kimura, and Spencer (1965) demonstrated

Toxocara in an eosinophilic abscess of the peripheral retina in a six year old girl who gave no history of contact with kittens or puppies. This patient had no other symptoms of visceral larva migrans, and her blood count showed no eosinophilia. The eye was not inflamed or painful, and it manifested no signs other than those associated with chronic inflammation of the ciliary body. O'Connor (1972) showed that migration of a *Toxocara* larva could cause retinal detachment, and Phillips and Mackenzie (1973) showed that the parasite could cause inflammation of the optic nerve head as well.

16.3.1.3. Ascaris lumbricoides

Ascaris infection is only a minor cause of ocular and cerebral morbidity. Since man is the natural host of this ascarid, the ingested eggs, after hatching in man's intestine, yield larvae that will eventually leave the alveoli, ascend the trachea, and be swallowed. Errant larvae that had remained in the pulmonary vessels and were returned to the left ventricle have the possibility of reaching the eye and the brain. The first recorded case of ocular ascariasis was described by Calhoun (1937). It was associated with a severe iridocyclitis, dislocation of the lens, and secondary glaucoma. Parsons' (1952) case showed a clearly discernible *Ascaris* larva in the sub-retinal space that remained viable for a period of 3 years. In the course of its wanderings the larva caused the disintegration of the macula and pigmentary changes. It would seem that this larva did not have the enzymic equipment or the mechanical force to penetrate the retina. It remained confined to the natural cleavage plane that exists between the sensory retina and the retinal pigment epithelium.

16.3.1.4. Onchocerca volvulus

Onchocerciasis is a highly important cause of blindness in equatorial Africa and to a lesser extend in Central and South America. Corneal opacity, associated with the migration of microfilariae into the stroma of the cornea, is the principal cause of blindness. The microfilariae are also found in the anterior chamber and iris, where they cause an intense inflammatory reaction, especially upon death of the organisms. Until recently the chorioretinal atrophy and optic nerve atrophy so commonly observed in the posterior segment of the eye were thought to represent degenerative effects. Neumann and Gunders (1973) recently showed that the posterior lesions of oncherciasis are associated with the passage of microfilariae through the perivascular spaces of the short posterior ciliary vessels. They observed plasma cell and lymphocyte infiltrations in the perivascular and perineural spaces, always in association with degenerating microfilariae. Intact microfilariae, seen in the sclera and retrobulbar tissues, were completely free of inflammatory cells.

Invasion of the ocular structures by the microfilariae of *Onchocerca* is clearly related to the presence of one or more subcutaneous nodules (containing adult worms) on the head or neck. The microfilariae are primarily parasites of the dermis and reach the ectodermal coats of the eye by an extension of skin parasitism. Once they have gained access to the inner eye, the parasites swim freely in the aqueous humor and in the relatively loose tissue of the iris.

Optic nerve atrophy, erroneously considered by some to be a manifestation of central nervous system involvement, is probably a direct effect of Wallerian degeneration, since extensive numbers of retinal ganglion cells are destroyed by the chorioretinal inflammation.

16.3.1.5. Other filarial worms of the eye

These will be mentioned only briefly since they do not generally affect the brain and are of relatively little importance as causes of ocular morbidity.

Loa loa, the so-called "eye worm", is primarily a cutaneous pathogen. Its blood borne microfilariae may occasionally invade the uvea and optic nerve as well as many other organs of the body. The parasite is best known for its occasional presentation as an adult under the bulbar conjunctiva of the eye. Here it may be anaesthetized with cocaine drops, grasped with forceps, and removed, following incision of the conjunctiva. Patients are aware of the movements of the worm under the conjunctiva, but the worm causes no significant pain or visual loss.

The filariae responsible for elephantiasis, *Wuchereria* and *Brugia*, may also present themselves as young adults in the eye. Normally parasites of the lymphatics, they are occasionally carried off into the blood stream and deposited in various organs. Rose (1965) has photographed a young adult of *Brugia malayi* in the anterior chamber of the eye. The worm swam vigorously in the aqueous humor for a number of days, causing a moderate iritis. It was eventually removed through a small incision at the limbus of the cornea.

16.3.2. Cestodes

16.3.2.1. Cysticercoids

Migration of the larval forms of *Taenia* (principally *Taenia solium*) throughout the body, following the hatching of the eggs in the upper gastrointestinal tract, results in grave disease of the eye and brain. Cysticercosis is a cause of considerable morbidity in India, Mexico, and South America where drinking water and food contaminated with the onchospheres of *Taenia* are ingested.

In their classical treatise on the subject, Dixon and Lipscomb (1961) analysized 450 cases of epilepsy attributable to cysticercosis. Most of their patients consisted of military veterans who had returned from service in British India. Central nervous system symptoms were generally noted within 4 years of infection, but the range of the latent period extended from

9 months to 30 years. Eight of their 450 patients had ocular involvement, the organism being identified with certainty in 4 cases. The larvae are believed to enter the eye through the choroidal circulation, whence they migrate forward to a sub-retinal location. With increasing fluid accumulation in the bladder of the cysticercus, the retina may become detached (Fig. 2). The craning movements of the head and neck of the larva (Fig. 3) have resulted in penetration of the retina in some cases, the cysticercus gaining free access to the vitreous cavity at this time. The larva may later cross through the pupillary opening and appear in the anterior chamber.

In its subretinal location the parasite is somewhat immobilized and operations to remove the parasite can be performed without permanent damage to the eye. Once the cysticercus gains access to the vitreous, the prognosis is much worse, for the free motion of the parasite within the vitreous gel makes it much more difficult to remove. Should the parasite die within the vitreous and release its toxic cyst fluid into the immediate environment, severe inflammation will ensue, and degeneration of the entire eye is almost inevitable. Belfort and his colleagues in Brazil have treated subretinal cysticerci with xenon arc photocoagulation in an attempt to kill

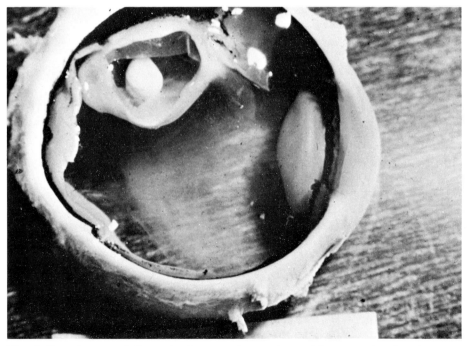

Fig. 2. Detachment of the retina by a cysticercus located between the choroid and the retina. (Kindness of Dr. Carlos Siverio, Lima, Peru).

340

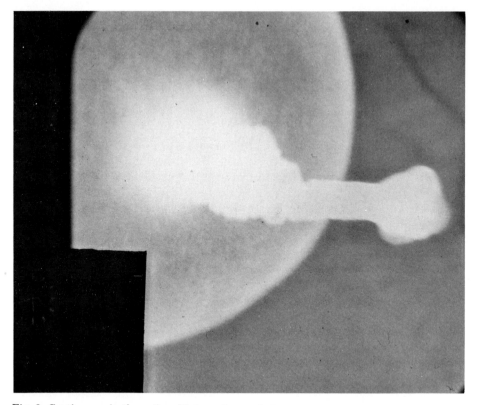

Fig. 3. Cysticercus in the retina. Note scolex and elongated neck.

the organism and provide a firm scar (Fig. 4). They have generally taken the precaution of pre-treating such patients with systemic corticosteroids in order to minimize the inflammatory reaction.

The epibulbar structures, including the conjunctiva, extraocular muscles, and lids may also be extensively parasitized. In the case of the eye muscles extensive hypertrophy may be produced.

Cysticercosis of the brain may cause a wide range of symptoms ranging from epileptic fits to partial paralysis. Endocrine abnormalities and paranoid psychoses have also been noted. The symptoms are manifestations of mechanical and irritative effects of the development of the cysticerci. Within a few months after the ingestion of the ova of *Taenia solium* the cysticerci attain palpable size in subcutaneous locations. In silent areas of the brain parasites of this size would cause no symptoms. After an unpredictable interval the larvae die and degenerative changes take place. The factors which determine the death of the larvae are not known, but Dixon and Lipscomb (1961) state that they tend to affect a large number of larvae at one time. During the degenerative process the fluid content, and therefore the size and

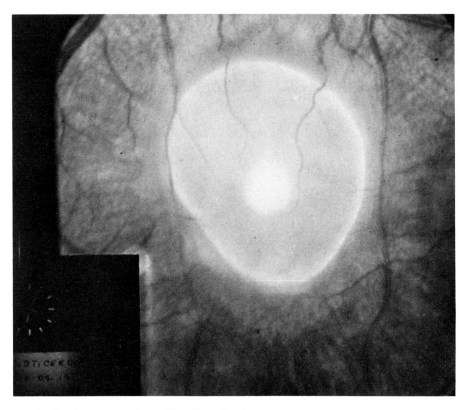

Fig. 4. Cysticercus of the retina following treatment with xenon arc photocoagulation. (Kindness of Dr. Rubens Belfort, Jr., Saõ Paulo, Brazil).

tension, of the cyst increases. Afterwards the cyst may remain enlarged, undergo partial shrinkage, or return to its original size and softness. The latter events account for the fact that some central nervous system lesions seem to resolve spontaneously, while other lesions progress relentlessly, causing blockage of the ventricles, hydrocephalus, and death (Fig. 5). The number of lesions, of course, depends strictly on the number of eggs ingested.

16.3.2.2. Echinococcus spp.

When eggs of the canine cestode, *Echinococcus* sp., are ingested, the hexacanth larvae penetrate the upper intestinal wall and reach the liver principally via the portal circulation. Some will be arrested and develop in the liver, and a few may pass on directly to heart, lungs, and eventually the brain. Arana-Iñiguez (1973) states that the brain is affected in only 2% of all cases of hydatid disease. This is in marked contrast to the situation observed in cysticercosis where the brain is a highly important target of the disease. While selective tissue tropism has been suggested as the explanation for the differences observed, the factors responsible for it have not been elucidated. It may be that the thick outer structure of the hydatid cyst does not transmit the nutrients available from cerebral tissue as efficiently to the cyst of *Echinococcus* as it does to the cyst of *T. solium*. Other factors must be involved, possibly related to the trapping of the organism. Although *Echinococcus* affects bone very often, the cranium is affected in only 2% of the cases while the vertebrae are affected in 50%.

The hydatid cyst of the nervous system, as well as cranial and vertebral hydatidosis, is an example of primary hydatidosis. Secondary hydatidosis is that form of the disease which results from the rupture of a primary hydatid cyst in an organic surrounding. In the rare cases where several hydatid brain cysts have been encountered, the heart has generally been involved. Rupture of a primary hydatid cyst into the left ventricle gives rise to widespread dissemination of the scolices ("hydatid sand") and the establishment of multiple lesions in the brain.

In the brain, increased intracranial pressure, associated with obstruction of

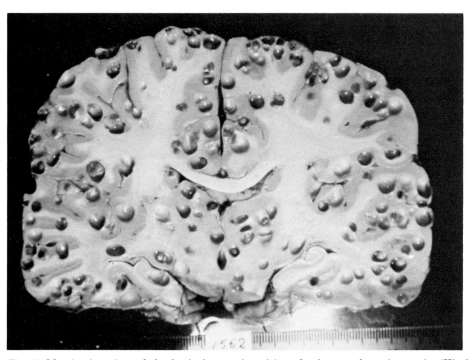

Fig. 5. Massive invasion of the brain by cysticerci in a fatal case of cysticercosis. (Kindness of Dr. Francisco Contreras, Lima, Peru).

the ventricular system and papilloedema, is the usual manifestation of the disease. Young patients, usually children in the age group of 4 to 12, who manifest intractible headache, dizziness, and papilloedema are highly suspect candidates for hydatidosis in endemic areas. Because the cerebral tissue is soft and highly compressible, hydatid cysts of the brain may grow to tremendous size exceeding 600 grams in weight. Their surgical removal is frought with danger, for inadvertant rupture of the cyst will give rise to multiple new lesions in the area.

The eye is only rarely affected by internal parasitism with *Echinococcus*. When it occurs, single cysts are seen most often in the space between the retina and choroid. The choroidal circulation is believed to be the route of access to the inner eye. A large, elevated mound may be seen in the macular area, initially resembling the lesion caused by *Toxocara*. It rapidly expands in volume, however, detaching the retina and occasionally rupturing into the vitreous. When the latter happens, multiple grape-like clusters of small cysts develop in the vitreous. These may extend forward to the lens, causing cataract, and may give rise to large increases in intraocular pressure. Scherz et al. (1973) have described *Echinococcus* in the anterior chamber.

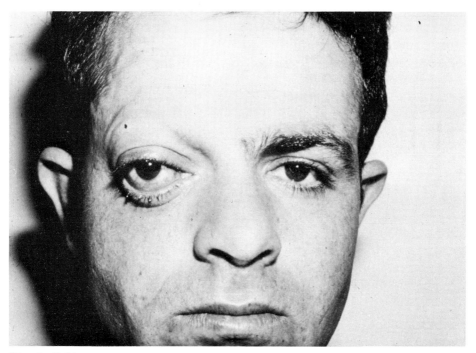

Fig. 6. *Echinococcus* cyst of the orbit producing marked proptosis of the right eye. (Kindness of Dr. Sahag Baghdassarian, Beirut, Lebanon).

The eye is, of course, much more confining than the brain in terms of expanding lesions. Intractible pain may force the enucleation of the eye, as reported by Arora et al. (1964). More often, as in the cases of Litricin (1953), Rapaport et al. (1957) and Aberastain (1963), the eyes have been enucleated because of suspected malignant neoplasm.

Involvement of the orbit is quite common in echinococcosis (Fig. 6). Baghdassarian and Zakharia (1971) described three cases in Lebanon, each presenting with unilateral proptosis of the eye. Laboratory tests for hydatid disease (Casoni test and Weinberg complement fixation test) were negative in all three patients. The authors pointed out the importance of suspecting hydatid disease in endemic areas, for the surgical approach to such lesions is very different when the nature of the pathology is suspected in advance. The usual approach to orbital tumors causing proptosis might result in the rupture of the cyst and the establishment of multiple lesions in the orbit. The eventual loss of an otherwise normal eye could be prevented by fore-thought. Injection of hydatid cysts with absolute ethanol prior to their surgical excision apparently prevents spread of the lesion.

16.3.3. Trematodes

16.3.3.1. Schistosoma sp.

Cerebral and ocular complications of schistosomiasis have been reported but are generally poorly documented. Haddock (1973) attributes the cerebral and spinal cord lesions principally to *Schistosoma haematobium*, and most of the reported ocular lesions have been caused by this organism. Eggs of this fluke, upon release from the parent female in the finer radicals of the portal circulation, may fail to reach their normal outlet in the urinary tract. Transported to the eye and brain in the blood stream, the eggs have been known to lodge in the conjunctiva where a granulomatous inflammation has been caused by reaction to constituents in the shell of the egg. Hollwich et al. (1972) described a case of recurrent severe iridocyclitis as attributable to *Schistosoma haematobium* infection. The eggs were recovered from the urine, but not from the eye of their patient. Treatment of the systemic infection brought about resolution of the eye disease. It is difficult for me to make a causal connection between the two.

Duke-Elder (1966b) quotes Cecchetto as describing a case of embolism of the central retinal artery caused by an egg of *Schistosoma mansoni*. From the point of view of size, this egg would more appropriately cause the occlusion of a branch (arteriole) of the central retinal artery.

Acknowlegement

This work was supported in part by Grant EY-01597 from the National Institutes of Health and in part by an unrestricted grant from Research to Prevent Blindness, Inc., New York City.

16.4. REFERENCES

Aberastain, T.G. (1963) "Hidatidosis endocular". Arch. Oftalmol. B. Air. 38, 142—146.

Arana-Iñiguez, R. (1973) In: "Tropical Neurology" (J.D. Spillane, ed.) pp. 408—415, Oxford University Press, London.

Arora, M.M., Dhanda, R.P., Bhagwat, A.G. and Kalevar, V.K. (1964) "Intraocular hydatid cyst". Brit. J. Ophthalmol. 48, 507—509.

Ashton, N. (1960) "Larval granulomatosis of the retina due to *Toxocara*". Brit. J. Ophthalmol. 44, 129—148.

Baghdassarian, S.A. and Zakharia, H. (1971) "Report of three cases of hydatid cyst of the orbit". Am. J. Ophthalmol. 71, 1081—1084.

Bargy, M. (1928) "Trois cas d'iridocyclite à trypanosomes". Clin. Ophthalmol. 32, 271—278.

Berengo, A. and Frezzotti, R. (1962) "Active neuro-ophthalmic toxoplasmosis. A clinical study on nineteen patients". Adv. Ophthalmol. 12, 265—343.

Borges, J.S. and Johnson, W.D. (1975) "Inhibition of multiplication of *Toxoplasma gondii* by human monocytes exposed to T-lymphocyte products". J. Exp. Med. 141, 483—496.

Braley, A.E. and Hamilton, H.E. (1957) "Serous choroiditis associated with amebiasis". Arch. Ophthalmol. 58, 1—14.

Byers, B. and Kimura, S.J. (1974) "Uveitis after death of a larva in the vitreous cavity". Am. J. Ophthalmol. 77, 63—66.

Calhoun, F.P. (1937) "Intraocular invasion by the larva of the *Ascaris*". Arch. Ophthalmol. 18, 963—970.

Chesterton, J.R. and Perkins, E.S. (1967) "Ocular toxoplasmosis among Negro immigrants in London". Brit. J. Ophthalmol. 51, 617—621.

Davson, H. and Luck, C.P. (1956) "A comparative study of the total carbon dioxide in the ocular fluids, cerebrospinal fluid, and plasma of man and mammalian species". J. Physiol. 132, 454—464.

Desmonts, G. and Couvreur, J. (1974) "Congenital toxoplasmosis: A prospective study of 378 pregnancies". N. Engl. J. Med. 290, 1110—1116.

Dixon, H.B.F. and Lipscomb, F.M. (1961) "Cysticercosis: An analysis and follow-up of 450 cases". Medical Research Council Special Report Series: No. 299. Her Majesty's Stationery Office, London.

Duke-Elder, S. (1966a) "System of Ophthalmology" Vol. IX, p. 441, Mosby, St. Louis.

Duke-Elder, S. (1966b) "System of Ophthalmology". Vol. IX, p. 473, Mosby, St. Louis.

Duma, R.J., Rosenblum, W.I., McGehee, R.F., Jones, M.M. and Nelson, E.C. (1971) "Primary amoebic meningoencephalitis caused by Naegleria". Ann. Int. Med. 74, 923—931.

Edwards, J.D. (1954) "The ocular manifestations of trichinosis". Trans. Ophthalmol. Soc. U.K. 74, 495—497.

Frenkel, J.K. (1961) "Pathogenesis of toxoplasmosis with a consideration of cyst rupture in *Besnoitia* infection". Surv. Ophthalmol. 6, 799—825.

Frenkel, J.K. (1967) "Adaptive immunity to intracellular infection". J. Immunol. 98, 1309—1319.

Frenkel, J.K. (1971) "Toxoplasmosis: Mechanisms of infection, laboratory diagnosis, and management". Curr. Top. Pathol. 54, 28—75.

Haddock, D.R.W. (1973) In: "Tropical Neurology", (J.D. Spilane, ed.) p. 155, Oxford University Press, London.

Harris, D. and Birch, C.L. (1960) "Bilateral uveitis associated with gastrointestinal *Entamoeba histolytica* infection: Case report". Am. J. Ophthalmol. 50, 496—500.

Hogan, M.J. and Feeney, L. (1963) "The ultrastructure of the retinal vessels. II. The small vessels". J. Ultrastruct. Res. 9, 29—46.

Hogan, M.J., Kimura, S.J. and O'Connor, G.R. (1964) "Ocular toxoplasmosis". Arch. Ophthalmol. 72, 592—600.

Hogan, M.J., Kimura, S.J. and Spencer, W.H. (1965) "Visceral larva migrans". J.A.M.A. 194, 1345—1347.

Hogan, M.J., Alvarado, J. and Weddell, J. (1971) "Histology of the human eye: An atlas and textbook". pp. 328—333, Saunders, Philadelphia.

Hollwich, F., Dieckhues, B., Junemann, G. and Aretz, H. (1972) "Bilharziose des Auges". Klin. Monatsbl. Augenheilk. 161, 430—434.

Huldt, G. (1966) "Experimental toxoplasmosis: Studies of the multiplication and spread of *Toxoplasma* in experimentally infected rabbits". Acta Path Microbiol. Scand. 67, 401—423.

Ikede, B.O. (1974) "Ocular lesions in sheep infected with *Trypanosoma brucei*". J. Comp. Pathol. 84, 203—213.

Jacobs, L., Fair, J.R. and Bickerton, J.H. (1954) "Adult ocular toxoplasmosis". Arch. Ophthalmol 52, 63—71.

Jacobs, L., Naquin, H., Hoover, R. and Woods, A.C. (1956) "A comparison of the toxoplasma skin tests, the Sabin-Feldman dye tests, and the complement fixation tests for toxoplasmosis in various forms of uveitis". Bull. Johns Hopkins Hosp. 99, 1—15.

Jankû, J. (1923) "Pathogenesa a Pathologická Anatomie T. Zv. Vrozeneho Kolobomu Žlute Śkurny Oku Normál ně Velikém a Mikrophtalmickěm s Nǎlezem Parasitu v. Sitnici". Cǎs. Lék. Čes. 62, 1021—1027.

Jones, T.C. (1974) "Macrophages and intracellular parasitism". J. Reticuloendothel. Soc. 15, 439—450.

King, R.E., Praeger, D.L. and Hallett, J.W. (1964) "Amebic choroidosis". Arch. Ophthalmol. 72, 16—22.

Knight, R. (1975) In: "Textbook of Medicine". (P.B. Beeson and W. McDermott, eds.) p. 298, Saunders, Philadelphia.

Litricin, O. (1953) "*Echinococcus* cyst of the eyeball". Arch. Ophthalmol. 50, 506—509.

Nagington, J., Watson, P.G., Playfair, T.J., McGill, J. and Jones, B.R. (1974) "Amoebic infection of the eye". Lancet 2, 1537—1540.

Neumann, E. and Gunders, A.E. (1973) "Pathogenesis of the posterior segment lesions of ocular onchocerciasis". Am. J. Ophthalmol. 75, 82—89.

Nicolle, C. and Manceaux, L. (1908) "Sur une infection à corps de Leishman (ou organismes voisins du gondi)". Compt. Rend. Acad. Sci. (Paris) 147, 763—766.

Nozik, R.A. and O'Connor, G.R. (1968) "Experimental toxoplasmic retinochoroiditis". Arch. Ophthalmol. 79, 485—489.

Nozik, R.A. and O'Connor, G.R. (1970) "Studies on ocular toxoplasmosis in the rabbit. I. The effect of antigenic stimulation". Arch. Ophthalmol. 83, 724—728.

O'Connor, G.R. (1970) "The influence of hypersensitivity on the pathogenesis of ocular toxoplasmosis". Trans. Am. Ophthalmol. Soc. 68, 501—547.

O'Connor, P.R. (1972) "Visceral larva migrans of the eye". Arch. Ophthalmol. 88, 526—529.

Parsons, H.E. (1952) "Nematode chorioretinitis. Report of a case with photographs of a viable worm". Arch. Ophthalmol. 47, 799—800.

Perkins, E.S. (1973) "Ocular toxoplasmosis". Brit. J. Ophthalmol. 57, 1—17.

Phillips, C.I. and Mackenzie, A.D. (1973) "Toxocaral larval papillitis". Brit. Med. J. 1, 154—155.

Piper, R.C., Cole, C.R. and Shadduck, J.A. (1970) "Natural and experimental toxoplasmosis in animals". Am. J. Ophthalmol. 69, 662—668.

Rapaport, M., Mieres, A., Cicolini, J. (1957) "Hidatidosis endocular". Rev. Assoc. Med. Argent. 71, 189—191.

Remington, J.S., Krahenbuhl, J.L. and Mendenhall, J.W. (1972) "A role for activated macrophages in resistance to infection with *Toxoplasma*". Infect. Immun. 6, 829—834.

Rose, L. (1965) "Filarial worm in anterior chamber of eye in man". Arch. Ophthalmol. 75, 13—15.

Scherz, W., Meyer-Schwickerath, G., Piekarski, G. and Waubke, T.N. (1973) *"Echinococcus* in der Vorderkammer". Klin. Monatsbl. Augenheilk. 163, 66—70.

Schlaegel, T.F. and Culbertson, C.G. (1972) "Experimental *Hartmanella* optic neuritis and uveitis". Ann. Ophthalmol. 4, 103—115.

Scholtyseck, E. (1973) In: "The Coccidia". (D.M. Hammond and P.L. Long, eds.), p. 81, University Park Press, Baltimore.

Shimada, K., O'Connor, G.R. and Yoneda, C. (1974) "Cyst formation by *Toxoplasma gondii* (RH strain) in vitro". Arch. Ophthalmol. 92, 496—500.

Visvesvara, G.S., Jones, D.B. and Robinson, N.M. (1975) "Isolation, identification, and biological characterization of *Acanthamoeba polyphaga* from a human eye". Am. J. Trop. Med. Hyg. 24, 784—790.

Wilder, H.C. (1950) "Nematode endophthalmitis". Trans. Am. Acad. Ophthalmol. Otolaryngol. 55, 99—109.

Wolf, A., Cowen, D. and Paige, B.H. (1939) "Human toxoplasmosis: occurrence in infants as an encephalomyelitis: verification by transmission to animals". Science 89, 226—227.

Woods, A.C. (1960) "Modern concepts of the etiology of uveitis". Am. J. Ophthalmol. 50, 1170—1187.

Ecological Aspects of Parasitology
Editor: C.R. Kennedy
© *North-Holland Publishing Company, Amsterdam, 1976*

CHAPTER 17

BLOOD FLUIDS — HELMINTHS

S.R.SMITHERS and M.J. WORMS

National Institute for Medical Research, Mill Hill, London NW7 1AA (Great Britain)

Contents

In this chapter blood fluid parasites will be limited to those parasites which spend all or a major part of their existence within the fluid contained in the specialized vascular systems of the vertebrate body — the blood circulatory system and the lymphatic system.

17.1. The habitat

The immediate environment of a parasite is similar throughout the vascular habitat; it is a cavity lined with serous endothelium which contains a fluid having for the most part a uni-directional flow. There are however physical and chemical differences within the habitat in different regions of the body. The blood circulatory system is organized into two major series of vessels, the arterial and venous systems, and within each system there are subcirculations which serve major organs or regions of the body. The two series of vessels differ functionally and structurally both from each other and in different regions of each series. The lymphatic system is an extensive network of vessels which contain lymph fluid and cells which may be aggregated into masses of lymphoid tissue. The complexity of this system varies in different classes of vertebrates. Structurally the lymphatic vessels resemble those of the venous series of the blood vascular system. The system functions in parallel with the venous system for the return of fluid and cells from the extravascular compartment to the intravascular.

The fluid within the vascular systems is not in direct continuity with the interstitial fluid and cells. The flow characteristics of the fluid are determined by the anatomical position of the vessel and by the structure of the vessel itself. Flow in arteries decreases in velocity and pressure with distance from the heart and is transformed from pulsatile to almost uniform flow in the capillary bed. There is a small increase in the velocity of flow in the venous system from the capillary bed to the larger veins but pressure remains low. Flow in the lymphatics is relatively sluggish and is brought about principally by body movement.

The endothelium is, in general, a barrier to large molecules, but ions and small molecules cross this barrier readily and there is an active exchange between intra and extravascular compartments. There is a net loss of fluid from the arterial vascular system into the tissues and interstitial fluid may be considered as a filtrate of blood plasma. The major portion of this fluid is returned to the intravascular compartment via the lymphatic system. The lymphatic system may also take up large protein molecules, cells and complexes, or particles which enter the tissue by any route. This results in considerable differences in protein content of fluid in different compartments of the vascular habitat. The chemical composition of the fluid varies also and is in a state of continuous modification during its passage through the vessels under the influence of the metabolic state of the organ or region traversed [see Altman and Dittmer (1974) and Baker, J.R. this volume]. Further modification may also be caused by the presence of the parasite itself.

17.2. The parasites

The intravascular habitat is exploited by many parasitic helminths for at least part of their life cycle. In many species the period spent within this habitat is limited to the use of the vascular system as a migration pathway within the host. This use may occur more than once at different periods in the life cycle, as in *Angiostrongylus cantonensis* for example, which migrates in the blood both in the third larval and adult stages (Mackerras and Sandars, 1955). In most species, however, vascular migration appears to be confined to a single movement from the point of entry and may be of less than 24 hours duration. This relatively rapid passage occurs in the early invasive phase of many helminths of medical and economic importance, for example, *Necator, Ascaris, Trichinella* and *Crenosoma* among the nematodes, *Taenia, Multiceps, Echinococcus* among the cestodes, and *Diplostomulum* among the trematodes.

During this brief passage, the parasites adjust to rapid changes in the environment and their sensory mechanisms function with a precision which enables them to leave the system at a selected point. Surprisingly this phase in the life cycle has received relatively little study although it may prove to be of great importance in parasite ecology. There is increasing evidence that it is in this early period of host-parasite contact, either immediately prior to or during the intravascular phase, that the acquired immune responses of the host are effective.

In other parasites, passage through the vascular system may be prolonged until a partial development has been completed. *Strongylus vulgaris* in the horse completes its 3—4 month larval development within the mesenteric arteries before moving into the lumen of the intestine in which the adult stage is passed (Duncan and Pirie, 1972).

A third group of helminths spend almost their entire existence in the vertebrate host within the blood and lymphatic systems. It is with these species that this chapter is mainly concerned, in particular with the two groups which have exploited this habitat most successfully — the filarial worms and the blood flukes. These two groups exhibit major differences in the manner in which the problems of life in the vascular habitat have been overcome.

17.2.1. The filarial worms (Superfamily Filariodea Weinland, 1858)

Grouped in this superfamily is an assemblage of parasitic nematodes probably of polyphyletic origin [see Anderson, 1957 for discussion]. Some members are parasitic in nonvascular sites from which the eggs gain direct access to the external environment. In these species larval development occurs in non-haematophagous terrestrial hosts. Other members in the families *Setariidae* (Yorke and Maplestone, 1926) Anderson, 1958 and *Onchocercidae* (Leiper, 1911) Anderson, 1958 inhabit sites which have no direct access to the external environment. The problem of transmission in these parasites has been overcome by the development of viviparity, the movement of larvae to peripheral sites and the use of arthropods as intermediate hosts.

The adult worms (Fig. 1) are typically elongate slender worms which range in length from less than 1 cm to more than 40 cm according to species. The tapered extremities are devoid of lips, pseudolabia and ornament, but bear papillae which are believed to have a sensory function. The body surface is smooth except for fine ridges in the outer layer of the 'chitinous' cuticle (Rogers et al., 1974). The nervous and muscular systems are well developed and the worms are capable of vigorous movement. The alimentary tract is relatively simple in form. There is a small oral opening which leads, via a buccal cavity, to a straight oesophagus. The oesophagus contains both muscular and glandular elements. The intestine is long and relatively narrow and opens via the rectum at the subterminal anus or cloaca. In members of some genera the anus is absent and the posterior portion of the gut is atrophied.

The worms are viviparous and the uteri are packed with embryos which become progressively advanced as they pass anterior from the ovaries. In the terminal portions of the uteri the larvae, or microfilariae, are long, slender and apparently freed from the eggshell except for a specialized portion which may be retained as an investing sheath.

Recent studies have revealed the microfilaria to be a much modified first stage nematode larva adapted to an existence in the vascular habitat and for passage through the narrow mouthparts of haematophagous arthropods. The basic nematode body structure is present though in a partially developed state, the reproductive system is undifferentiated and the alimentary canal is

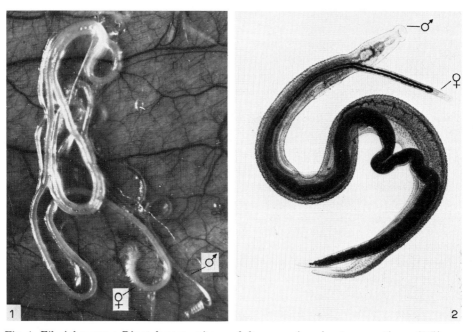

Fig. 1. Filarial worms: *Dipetalonema viteae* adult worms in subcutaneous tissue (× 5).

Fig. 2. *Schistosoma mansoni* adult worms in copula (× 19.5).

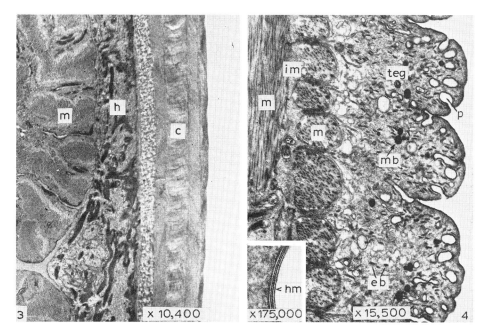

Fig. 3. T/S showing cuticle, hypodermis and somatic muscles of *Dipetalonema viteae* (×10,400). (Photograph by courtesy of Dr. D.J. McLaren).

Fig. 4. L/S showing tegument and muscle layers of *Schistosoma mansoni* (×15,000). Inset — showing heptalaminate nature of outer tegumental membrane (×175,000). (Photograph by courtesy of Dr. D.J. Hockley).

Abbreviations: c, cuticle; e.b, elongate bodies; h, hypodermis; h.m, heptalaminate membrane; i.m, interstitial material; m, muscle; m.b, membraneous bodies; n, nerves; p, pits; teg, tegument.

much reduced and probably non-functional (Laurence and Simpson, 1974; McLaren, 1972; Simpson and Laurence, 1972; Johnston and Stehbens, 1973).

The adult worms of members of the *Onchocercidae* parasitize a wide variety of sites in the vertebrate host although each species is confined to one or a small number of sites. A number of species of several genera inhabit the blood vascular systems in the adult stage. Members of the genera *Brugia* and *Wuchereria* parasitize the lymphatic system both within the vessels and in the lymph nodes. Although few species parasitize the vascular system in the adult stage, the microfilariae of the majority of members of the *Onchocercidae*, irrespective of the location of the adult worms, inhabit the blood, lymphatics or skin.

The prolonged existence of active progeny alongside the parent population for relatively long periods is an unusual phenomenon in the metazoa

(Johnston and Stehbens, 1973). The relationships between the two populations are obscure. Studies on immunity and response to drug action indicate that in many ways the two populations are distinct. The life cycle is continued when the microfilariae are ingested by a blood-sucking host, within which the larvae penetrate the gut wall and move to the preferred site of development characteristic for a particular species (see reviews of Hawking and Worms, 1961; Lavoipierre, 1958). The larvae increase in size and undergo two moults to the third stage larva which is infective to the vertebrate host.

17.2.2. The blood flukes (Order Schistosomatida) (Skryabin and Shults, 1937)

The blood flukes comprise an order of the Trematoda which are adapted for the most part to the venous system of vertebrates. No other trematode group presents such unusual variety and diversity. The schistosomes (suborder Schistosomatata) are parasites of mammals and birds, the sanguinicolids, (family Sanguinicolidae) are parasites of turtles and the spirorchids (family Spirorchidae) parasitize fish. The basic life cycle has an alternation of generations, with the sexual generation of adult worms in the vertebrate host and an asexual multiplicative stage generally in a molluscan host. (Certain sanguinicolids use marine polychaete annelids as their intermediate host.) Eggs laid by mature worms ultimately hatch to give rise to miracidia. The miracidia penetrate the mollusc and cercariae develop in sporocysts or rediae. The cercariae emerge from the intermediate host and are usually found in the surface layers of the aquatic environment where they remain viable for only a few hours; once contact has been made, they directly penetrate the final host. The adult flukes become located in vessels within an organ system through which access to the external environment may be achieved.

All known sanguinicolids and spirorchids are hermaphrodite whereas the schistosomes are unisexual dimorphic flukes with the male worm generally larger and more muscular than the long slender female (Fig. 2). Lateral flaps of the male fold together, possibly locked by minute spines in some species, to form a gynaecophoric canal within which the female is held.

The tegument of these parasites is a cytoplasmic syncitium bounded by a plasma membrane which may be covered in spines or tubercles (Fig. 4). Suckers may be absent, but when present include a poorly developed oral sucker which is used to anchor the worms against the blood flow. The mouth leads into a blind digestive tract which may divide into two caecae. The excretory-water balance system consists of flame cells, collecting tubules and an excretory bladder with a terminal pore.

The schistosomes of birds and mammals appear to be relatively conservative in their choice of final habitat. With the exception of *Dendritobilharzia pulverulenta* which occurs in the arterial system of anseriform birds, the schistosomes are restricted to the venous system, commonly the mesenteric veins; *S. haematobium* however is found in the veins of the vesical plexus of man and *S. nasalis* in those of the nasal fossa of cattle. The spirorchids are found in the blood vascular system including both mesenteric and visceral blood vessels and in the heart. They occur also in the lymphatic vessels and extra-vascular sites in the brain, spinal cord and lungs. Sanguinicolids are generally parasitic in the gill vessels of fish but have been reported from the coelom. The location of some sanguinicolids appears to vary both with season and the stage of maturity of the worm. Sexually mature *Sanguinicola inermis* are localized in the large blood vessels of the gills of the carp during the summer but are concentrated mainly in the bulbus arteriosus in the winter and spring. Immature *Cardicola klamathensis* occur throughout the circulatory system of *Salmo clarkii* but the mature worms occur in the efferent renal vein (Smith, 1972).

17.3. Problems of the blood fluid habitat

Many of the problems of parasitism of the intravascular habitat are common to all endoparasites of vertebrates. It is essential for the parasite to establish and maintain itself in the host, meet with another member of the same species, mate, reproduce and ensure transmission of its progeny to a new host. These goals should be achieved with as little insult to the host as possible.

The problems particularly associated with parasitism of the vascular system are (1) those of entrance and exit from a habitat which has no direct opening to the external environment, (2) the maintenance of the parasite in a flowing medium and (3) the avoidance of damage to the vascular system, especially in relation to blood coagulation and mechanical impedence of flow.

Similarly, the advantages of the intravascular habitat, e.g. constancy of environment, protection, facilitated respiratory and excretory exchange, are, in the main, those common to all parasite habitats. One advantage which appears to apply more especially to the intravascular habitat however is the availability of nutrients such as glucose and amino acids which are present in a form suitable for direct absorption.

Our knowledge of the biology of helminths of blood fluids has been derived mainly from studies on parasites of medical and economic importance, or those which can be studied in the laboratory, notably the human schistosomes *Schistosoma mansoni*, *S. haematobium* and *S. japonicum* and filariae of the genera *Brugia*, *Dirofilaria*, *Dipetalonema* and *Litomosoides*. Much of the information available is a synthesis of scattered observations on a relatively small number of species and in some instances may not reflect the general situation.

17.3.1. Entrance to vertebrate host and migration to final habitat

The infective stages of both schistosomes and filarial worms enter the vertebrate host through the skin. Cercariae of the blood flukes effect direct penetration into their host unaided by any outside source. The sanguinicolids and spirorchids most probably succeed in penetrating only the softer mucus membranes. *Sanguinicola inermis* cercariae for example attach themselves to the mucus membrane of the buccal cavity and the gills of carp and penetrate to the blood vessels; spirorchids generally penetrate the conjuctiva, oral, nasal and anal membranes (Smith, 1972). The cercariae of schistosomes usually enter the host through fully keratinized skin. Apparently there is no host selection by the cercariae; the human schistosome *S. mansoni* for example will penetrate the epidermis of the pigeon (Continho-Abath and Jampolsky, 1957) and many bird schistosomes penetrate mammalian skin, an activity which can give rise to cercarial dermatitis in man.

The biology of penetration has been mainly studied in *S. mansoni*. Cercariae penetrate the skin of the definitive host with the assistance of lytic substances from the penetration glands (Stirewalt and Fregeau, 1966). Certain factors in skin lipids are responsible for triggering the penetration behaviour; in *S. mansoni* and *S. haematobium* this factor is believed to be unsaturated and saturated fatty acids (Shiff et al., 1972), whereas in the bird schistosome *Austrobilharzia terrigalensis*, cholesterol has this stimulatory effect (Clegg, 1969). Penetration is rapid, the dead keratinized layers of the stratum corneum are penetrated within a few minutes and the cercariae turn horizontally, to tunnel just above the cells of the Malpighian layer. Ten minutes after first contact with the skin, the young schistosomes begin to penetrate this living layer and enter the dermis (Clegg and Smithers, 1968). During this process the fluke rapidly becomes adapted to survival in the vertebrate; the cercarial tail is cast, the contents of the penetration glands are shed and the glycocalyx covering the cercarial body is lost. Now the surface of the organism will no longer react with antibodies to form the cercarienhüllenreaktion (Stirewalt, 1963). Once within the dermis the surface membrane of the parasite changes from that of a conventional trilaminate structure, to that of a multilaminate membrane typical of the adult worm (Hockley and McLaren, 1973) (Fig. 4). As a result of these changes, the fluke is no longer able to survive in fresh water but only in a physiological medium and at this stage the parasite is called a schistosomulum.

The schistosomula of *S. mansoni* spend 2—3 days in the dermis before migrating to the lungs; *Trichobilharzia ocellata*, a parasite of ducks, leaves the skin within 24 h (Bournes et al., 1973). The manner of entry of schistosomes into the venous system is not well understood; it has been reported to occur directly into the veins, or indirectly via lymphatics or by both routes. *T. ocellata* appear to enter exclusively through small veins. Once in the venous system the schistosomula are carried to the heart and thence to the lungs where their size allows them to migrate through the small capillaries. *T. ocellata* leave the vascular system to occupy air spaces, but later re-enter the veins in bronchial tissue. The significance of this migration, which is not reported for other schistosomes, is unexplained (Bournes et al., 1973).

There is much controversy on the route taken by schistosomes when they migrate from the lungs to the portal system. A vascular migration has been reported for *S. mansoni* and *S. japonicum* but there is also evidence of an extravascular route through the mediastinum, diaphragm and peritoneal cavity (Wilks, 1967).

The principal route of migration from the portal vessels of the liver to the small mesenteric veins of the intestine is a direct one, against blood flow. This final migration takes place once the worms have matured and paired and the large muscular male schistosomes of mammals, with their well-developed ventral sucker, aid the more delicate female in this operation. It is suggested that frail schistosomes such as *T. ocellata* become coiled during this migration and that this posture enables the worms to press against the inner walls of the vessels and make headway against the blood flow (Bournes et al., 1973).

In contrast to the schistosome cercariae, the infective larvae of filariae are unable to penetrate the fully keratinized epidermis unaided although *B. pahangi* is capable of penetration of the conjunctival epithelium and buccal mucosa (Ah et al., 1974; Gwadz and Chernin, 1972). The typical mode of entry through the skin is by active movement of the larvae from the site of deposition on the skin surface through the break made in the skin by the arthropod whilst feeding. Among mosquito borne species this movement is rapid and may be accomplished in less than 2 minutes. The larvae are protected from dessication by the deposition, apparently induced by the parasite, of a small volume of fluid by the insect at the time of feeding (Ewert, 1967; McGreevy et al., 1974). There is little information on the reactivity of the nematode cuticle on entering upon this parasitic mode of life, although observations on *Necator americanus* and *Ascaris suum* indicate that the parasite surface is altered in some way shortly after entry to the vertebrate host (see Lumsden, 1975).

Within the tissues of the vertebrate larvae move to a preferred site in which further development takes place, during which the parasite undergoes two moults. The site may vary with the species of parasite and the host. The speed with which this translocation is accomplished largely depends on the anatomical relationship between the site of entry and the final location of the parasite. The species which inhabit the lymphatic system probably move solely within it and arrive at the final development site soon after entry to the host. Several species are known to use the intravascular pathway to move to non-vascular sites and translocation may be very rapid. Larvae of *B. pahangi* in the jird may gain access to the lymphatic systems and be carried to the lungs within 5 min of entry to the tissues and penetrate to the pleural cavity within 15 min (Ah et al., 1974). *Litomosoides carinii* travels whilst still in the third larval stage via the lymphatic pathway in the cotton rat and can be demonstrated in the pleural cavity in which development to the adult stage is completed after 2—3 days (Wenk, 1967). *Dirofilaria immitis*, however, apparently does not use an intravascular pathway in the dog and movement from the site of entry in the skin to the final site in the heart occupies more than 60 days during which time the larvae complete development to the adolescent stage, principally in the subcutaneous and muscle tissues (Orihel, 1961). The manner in which the large worm (18 mm in length) gains access to the heart is not known.

The rapidity of movement of filariae (adult *Loa loa* may move one inch in 2 min according to Manson-Bahr, 1951), coupled with the observation that the filariae do not possess well developed glandular structures other than those of the oesophagus, would indicate that movement is primarily by mechanical means.

On the problems of site location and mate finding we have very little information either specifically for parasites of the vascular system or for helminths in general (see review of Ulmer, 1971). It is known that site recognition is well developed and filariae may occur only in very circumscribed locations, e.g. the tarsal bursa. It is known also that infection with only 2—3 infective stages may result in a patent infection of filariae or schistosomes. Whether this is achieved by the influence of the worms upon each other, e.g. by pheromones, or by the chance meeting of worms at the same location is not known.

17.3.2. Life in the vascular environment

The physical characteristics of the vascular environment must influence the morphology of the parasite. This habitat, a tube with unidirectional flow, is akin to a river and it is not surprising to find both in filariae and blood flukes a tendency to elongate shape, cylindrical cross section and a reduction of surface projections, i.e. a streamlining, which serves to reduce friction and turbulence.

Filarial worms lack organs of attachment or cuticular spines and it is not known how they maintain themselves in situ within the vascular system. Some species burrow into the lining of the vessel but in most species, it is probable that the worms maintain themselves by coiling and bracing against the walls of the vessels. Schistosomes possess both spines and suckers. The ventral sucker of the male worm is larger and more muscular than that of the female and serves to maintain the position of the paired worms within the blood vessels against the blood flow (Smyth, 1962). The tegumental spines make indentations into, but do not damage, the endothelium and probably assist in anchoring the parasite (Smith and Lichtenberg, 1974). All movement of paired worms is effected by the male. In vitro, male worms move by an elongation of the portion of the body anterior to the ventral sucker. The oral sucker is then attached firmly to the substrate and the ventral sucker brought into close proximity to it. There is no reason to doubt that this 'looping' movement also occurs in vivo.

Some spirorchids possess both oral and ventral suckers, others an oral sucker only. According to Holliman et al. (1971) *Spirorchis parvus* which lacks a ventral sucker, moves through the blood vessels or free in the tissues of its host by a peristaltic wave of contractions which displaces one or several bulb-like enlargements of the anterior end towards the hind body. Dorso ventral undulating swimming movements of the lateral body margins may also assist.

Adult sanguinicolids lack suckers and their ability to move and attach themselves in selected habitats appears to be related to the possession of variously shaped and arranged tegumental spines (Smith, 1972). The short stout bodied *Aporocotyle macfarlani* lies in the host's blood vessels with its anterior extremity directed into the blood flow; the worm drifts back with the flow until it becomes lodged by the tegumental spines in a narrow vessel. The distribution of the parasite in the body is apparently determined by the architecture of the ventral aorta and its branches and by the fact that flow rates are less in some regions than in others.

Psettarium sebastodorum is a long narrow fluke in which the posterior lobe of the body is asymmetrical and bears combs of spines along the lateral margin. This species inhabits the auricle or ventricle of the heart in which it is found threaded into the intertrabecular spaces and held in place by loops of the body aided by the tegumental spines (Holmes, 1971).

These methods of localization within the vascular systems do not appear to injure the endothelium. Cells as well as fluids are able to pass between the parasite and the endothelium and flow does not appear to be mechanically interrupted. Where flow is impeded substantially as in long term infections of *W. bancrofti* in man, pathological changes of elephantiasis are found. The failure of the parasites to activate the clotting mechanisms of the host is a mystery. The tegument of schistosomes contains large amounts of charged groups which under normal conditions are potent activators for the intrinsic pathway of blood coagulation. Tsang and Damian (1975) provide evidence that adult *S. mansoni* possess an inhibitor of the intrinsic pathway of blood coagulation. Their findings suggest that this inhibition may act specifically on the enzymatic conversion step of factor XI to factor XIa.

17.3.3. Nutrition and metabolism

Current understanding of the nutrition and metabolism of the filariae and blood flukes is meagre. The little information which has accrued is derived mainly from experimental studies on the human schistosomes and the filarial worms *Litomosoides carinii* and *Dirofilaria* spp.

The living surface of the worm and the blood constituents of the host constitute the all important host—parasite contact. A parasite surface surrounded by a medium rich in nutrients might be expected to have an absorptive function and such appears to be the case in the blood flukes. Well developed absorptive surfaces often bear microvilli which serve to increase the surface area. No adult schistosome bears surface microvilli, however, and it is probable that this particular morphological feature is unsuitable to parasites which live closely applied to the walls of the small capillaries. Nevertheless, the surface area of schistosomes may be considerably increased due to infoldings or ridges. In *S. mansoni* for example, there are extensive invaginations of the plasma membrane resulting in the formation of cavernous channels which extend deeply into the tegumental matrix, and the surface of adult male *S. japonicum* is secondarily amplified by convolutions (Lumsden, 1975). Such features would allow at least some flow of blood through vessels 'blocked' by flukes and at the same time effect a considerable increase in surface area.

Schistosomes are heavily dependent on glucose for their principal source of energy; adult worms use 15—20% of their body weight of glucose per hour. Rogers and Bueding (1975) used an apparatus which isolated the mouth of *S. mansoni* from the rest of the body and were able to demonstrate that the tegument is the primary site of glucose absorption in this parasite. Worms whose glycogen stores were depleted exhibited an increase in the rate of glucose uptake via the tegument but there was no demonstrable glucose uptake through the mouth. When in copula it is possible that the male worm provides glucose to the female.

Asch and Read (1975) have provided similar evidence for the absorption of the amino acids glycine and proline by *S. mansoni*. Their experiments indicated that the transtegumental absorption is the predominant, if not exclusive, mode of transport of these amino acids and probably other low molecular weight compounds; it is highly unlikely that absorption is occurring via the caecum.

The majority of the glucose absorbed by the schistosome is degraded to lactic acid via the Emden-Myerhof pathway (Senft, 1965). It is possible however that oxidative phosphorylation is important in egg production. When *S. mansoni* are cultured in the presence or absence of oxygen, the utilization of glucose and the production of lactic acid does not alter, yet under anaerobic conditions, virtually no eggs are produced, compared with some egg production observed in the presence of oxygen (Schiller et al., 1975). It would be of interest to know whether the role of oxygen assumes a greater importance in blood flukes living in the gills of fish for example.

The gut is another important route for nutrients. Red blood cells are ingested by schistosomes and haemoglobin is broken down to haematin which becomes visible as a black pigment within the gut (Rogers, 1940). *S. mansoni* possesses a protease, with optimal activity at pH 3.9, which is specific for haemoglobin. The breakdown of globin by this enzyme yields mainly small peptides rather than free amino acids and it has been suggested that these products might be absorbed and used directly for eggs formation (Timms and Bueding, 1959). Skin tests on human patients with this purified enzyme have shown that it induces specific antibody formation, presumably because the gut contents of the worm are void into the bloodstream (Senft and Maddison, 1975).

In unmated female S. mansoni, which do not mature fully, there is little pigment in the intestine, which suggests that a major role in the supply of nutrients for reproduction comes via the gut. However, the relative importance of the gut and the tegument in the overall nutrition of schistosomes remains unknown.

The cuticle of relatively few species of filarial worms has been examined but all appear to possess both in adult and larval, including microfilarial, stages the basic nematode structure of three layers divided into a number of sub-layers. The nematode cuticle may be considered as a modified cell component derived from the underlying syncytial hypodermis. Despite a somewhat rigid structure necessary for its role in body movement it remains metabolically active and is capable of growth and differentiation, may bear electro-negative charges and be antigenic. The major role of the cuticle appears to be protective and except for water and certain small ions is impermeable. The principal route of entry for nutritive materials appears therefore to be via the alimentary canal. The alimentary canal of microfilariae is much modified, however, and probably non-functional. Johnston and Stehbens (1973) suggest that as the microfilariae do not undergo development in the vertebrate host, their nutritional demands are probably limited to those of energy only and it is possible that the small molecules required for this may pass across the cuticle. Evidence for the passage of glucose across the cuticle has been obtained for Ascaris and it appears likely that in larval nematodes such as Toxocara canis, which lack a patent digestive tract yet live for long periods in vertebrate tissues, transfer of nutrients across the cuticle occurs (Lumsden, 1975).

The alimentary tract of larval and adult filariae is well developed. The intestine has a villous lining and the oesophagus has contractile elements which would enable material to be actively ingested. Whether the worms ingest cells, fluid or both is not clear. Vincent et al. (1975) state that the lumen of the alimentary tract of Brugia malayi is without any detectable content. In the larger Dirofilaria immitis, however, erythrocytes have been observed in the oesophagus (McLaren, 1971). Little is known of the process of feeding in filarial worms and it would be of interest to study those species in which part of the alimentary tract is lacking.

Several studies have been made of the metabolism of filarial worms (Bueding, 1949; Hutchison and McNeill, 1970; Wang and Saz, 1974). L. carinii is able to utilize glucose both aerobically and anaerobically but oxidative metabolism is essential for survival. Under conditions of anaerobiasis, glycogen stores in the muscles and hypodermis of the worm are utilized. Other species *(D. viteae, Brugia pahangi)* are homolactate fermenters and do not require oxygen for energy metabolism. Enzymes of the glycolytic and tricarboxylic acid cycles have been identified but it appears that the tricarboxylic acid cycle is not utilized.

17.3.4. Exit from the vertebrate host and transmission

The female blood flukes lay their eggs still within the confines of the closed vascular system. For the continuation of the life cycle, these eggs or the larvae they contain (the miracidia) must reach the external environment. The larvae of spirorchids and schistosomes leave the host still within the eggshell which is relatively thick and hardened by quinone tanning. The eggs are shed from the host with the faeces or urine depending on the location of the adult worms in relation to the intestine or bladder. To obtain access to these systems, the eggs must traverse the endothelium and wall of the blood vessel, cross any intermediate tissue and penetrate into the lumen of the intestine or bladder. Many schistosome eggs possess a spine, the position of which depends on the species. This spine may help the egg to resist blood flow when it is released into the blood vessel and may also assist penetration of the vessel wall. Eggs of *S. mansoni* have been shown to release at least one proteolytic enzyme which has true collagenase activity (Smith, 1974). Presumably the function of this enzyme is to facilitate the passage of the egg through the tissues. During this passage the embryos develop and most of the eggs which are excreted contain mature miracidia which hatch in fresh water. In *S. haematobium*, there is evidence that eggs are passed in the urine more frequently at mid-day that at other times, i.e. when the host is more likely to be near water (Farooq and Mallah, 1966).

Eggs are distributed by the blood and are most abundant in the organs which receive blood from the vessels in which flukes are lodged. Sanguinicolid, spirorchid and schistosome eggs are thus found in a variety of habitats from which there is little possibility of reaching the external environment. In the case of human schistosomes it is these trapped eggs which are the main cause of the serious pathology of these infections; the liver and intestines are the main organs affected by *S. mansoni* and *S. japonicum*, whereas it is the bladder and urinogenital tract which suffer damage in *S. haematobium* infection. This type of pathology is seldom seen in birds infected with *T. ocellata* in which the eggs are laid in very small veins of the intestinal mucosa within a few cells of the lumen, from which the possibility of carriage to other areas is minimal (Bourns et al., 1973).

The eggs of sanguinicolid flukes are commonly present in the gill filaments of their fish hosts but so far as is known they are not shed but hatch within the filaments. They are therefore not subjected to the action of digestive enzymes and this feature may relate to the absence of hardening by quinone tanning in the elastic thin-shelled eggs. The miracidia, at least in the case of *Sanguinicola davisi* and *S. inermis*, leave the host by breaking out through the gill tissue (Smith, 1972).

Ultimately the exit of microfilariae from the body of the vertebrate host is achieved through the agency of a blood feeding arthropod which removes them with the blood meal. The achievement of this removal poses some problems for the parasite. First, that of producing microfilariae at a density high enough to ensure their presence in the relatively small volume of blood taken up by the vector. Secondly, as most blood feeding arthropods have definite feeding periods and preferences for feeding on certain areas, the microfilariae must be present at the correct time and site to ensure their uptake.

Microfilaria production may persist for long periods and as the microfilariae may be long-lived and are produced in large numbers (75,000 per worm per day in *W. bancrofti* (Hairston and Jachowski, 1968)), they are able to populate the host to very high levels. Although the host appears to be able to tolerate very large numbers of microfilariae, the presence of large numbers in the blood may adversely affect transmission by causing the death of the arthropod vectors. The microfilarial population is not distributed uniformly, however, within the vascular system: only a small proportion appears in the peripheral circulation; the majority appear to be within the 'deep' capillaries, especially those of the lungs (Pacheco, 1974).

The level of peripheral microfilaraemia in chronic infections generally remains relatively stable for considerable periods. Whether this stability is controlled by the host or the parasite remains unknown. In some species the number of microfilariae in the peripheral blood fluctuates regularly and follows a 24-h cycle. This phenomenon is called periodicity and has been developed to different degrees in different species of filariae. Some species, e.g. *W. bancrofti*, exhibit complete periodicity in which microfilariae are almost totally absent from the peripheral blood during the daylight hours but appear there each night. Other species, e.g. *D. immitis, B. malayi*, exhibit incomplete periodicity in which microfilariae are present in the peripheral blood throughout the 24-h period but are present in greater numbers at one time of day than at others. This period of maximum concentration may occur at night or by day according to species. In many species periodicity does not appear to be present and in such aperiodic species the number of microfilariae may fluctuate irregularly throughout the 24-h period.

Periodicity appears to be caused by the active migration of microfilariae between the peripheral blood and a reservoir in the microcirculation of the lungs within which the microfilariae are able to accumulate. The mechanisms by which these periodicities are achieved have received much attention (see Hawking, 1975; Worms, 1972). It appears that the microfilariae respond to cues, particularly oxygen tension and body temperature, from the host. Different species respond to different cues or to the same cues in different ways to produce periodicities of different amplitude and phase. The period when maximum numbers of microfilariae are present in the peripheral blood of the host is usually coincident with the period of feeding activity of the arthropod host. Thus it would appear that two factors operate to maximise the chances of successful transmission; a control of the numbers of microfilariae in the peripheral circulation and a behavioural adaptation of the microfilariae which ensures that those in the peripheral blood are present at the optimum period for uptake by the vector.

17.3.5. Evading the immune response

Many protozoan and helminth parasites generate chronic and long-lasting infections. Following a relatively long-prepatent period, there often exists an extended phase of patency thus providing a greater opportunity for the successful transmission of the progeny and the continuing survival of the species. The vertebrate immune response functions to detect and destroy genetically foreign organisms. Persistence in the vertebrate host for more than a week or two implies that the parasite has evolved a mechanism which circumvents host immunity. From the majority of studies there is clear indication that the host responds to the presence of the parasite by cell mediated activities and antibody production, yet it appears that protective immunity is often effective only against the young stages of a new infection. In this way the mature parasite continues to survive and reproduce, whilst the early death of the host from superinfection is avoided.

The problem of circumventing host immunity is one that all parasites have to overcome; the manner in which it is solved varies according to the morphology and physiology of the parasite and the habitat. The helminths of blood fluids are of particular interest in this respect as they are continually bathed in the undiluted components of the immune response; antibodies, complement, and immunologically competent cells.

Studies in recent years on experimental animal models have greatly improved our understanding of the schistosome host/parasite immune relationship (WHO, 1974). About 3 months after infection with S. mansoni, the mouse will destroy 70—80% of a cercarial challenge infection. Despite the development of this acquired resistance, the mature worms from the initial infection remain apparently unharmed and continue to produce eggs (Sher et al., 1974). Immunity in the presence of an active infection has been called concomitant immunity (Smithers and Terry, 1969) and this situation is now known to exist in several schistosome/host combinations; it may also occur in man (Smithers and Terry, 1976).

The main stimulus to the development of acquired immunity in schistosomes is derived from the living adult worm and protective immunity is directed against the young schistosomula of the challenge, usually within a few days after penetration. The exact mechanism which allows the adult worm to remain unharmed by the immunity it has itself engendered is not completely understood, but there are several indications.

The schistosome tegument is bounded by a living plasma membrane. Under special conditions it is susceptible to antibody damage (Smithers, Terry and Hockley, 1969) and when host immunity damages the worm, the lesion occurs at the surface (Hockley and Smithers, 1970). It follows that under normal conditions the adult schistosome tegument is protected from the immune response.

The plasma membrane of *S. mansoni* and other schistosomes so far examined is multilaminate. [It is interesting that the plasma membrane of *Fasciola hepatica*, a fluke which does not reside in blood fluids, has the conventional trilaminate structure] (McLaren, personal communication). Evidence from isotope studies has shown that the surface membrane turns over and particulate membrane antigens are released into culture (Kusel et al., 1975). It has also been shown that the outer layer of the multilaminate plasma membrane turns over faster than the basal layer (Wilson, 1975). It is very probable that the specialized morphological and physiological properties of the schistosome tegument are an adaptation to life in the bloodstream and assist in counteracting antibody attack on the surface. In vitro experiments support this idea; when schistosomes are exposed to an antibody directed against surface components, there is intense activity of the surface membrane. It appears that fragments of the surface are being cast off and that the damaged membrane.is being rapidly removed and replaced (Perez and Terry, 1973). This phenomenon may represent an important defence mechanism against the host's immune attack.

Other surface phenomena may be involved. It is now known that molecules, believed to be of host origin, are incorporated into the surface of the schistosome. Schistosomes cultured in human blood, for example, will acquire the A, B or H blood group antigens corresponding to the blood type of the culture media. These blood group substances are believed to be acquired as glycolipids (Clegg, 1974). There is no evidence of an acquisition of host molecules by the related liver fluke *F. hepatica* (Hughes and Harness, 1973) and this phenomenon among schistosomes may be another adaptation to life in the blood. Schistosomula recovered from mice after 4 days possess mouse antigens on their surface and they fail to absorb anti-schistosome antibody; on the other hand, young schistosomula which have recently transformed from cercariae do not possess mouse antigens and their surfaces readily absorb antibody (McLaren et al., 1975). This indirect evidence suggests that the acquisition of host antigens onto the surface of schistosomes prevents attack by antibody. Host antigens may therefore effect some form of immunological blocking or disguise, but the molecular basis for host antigen acquisition and its exact role in circumventing host immunity remains one of the intriguing problems of modern parasitology.

Filarial worms, whether adults or microfilariae, are particularly noted for their longevity but their mechanism of evading host immunity remains un-known. The cuticle of the filarial worm is very different from the tegument of the blood fluke and mechanisms of evasion which operate in schistosomes are unlikely to resemble those which are effective in the filariae. A low grade immunity is developed in response to infection with several filariae and in many instances it is the actively migrating 3rd stage larvae which both pro-voke and are most affected by this immunity (Ogilvie and Worms, 1976). Nematodes develop through clearly defined stages, delineated by sudden moulting, and clearly because of this there is the possibility of stage specific immunity, i.e. one particular phase in the life cycle which is highly suscep-tible to the immune response. It is conceivable that concomitant immunity in filarial infections operates through stage specific immunity directed at one of the early moults. However, more research is needed to explain how the non-susceptible stages escape the host's response.

17.4. Conclusion

The problems associated with entry into and life within the vascular habi-tat have been successfully overcome both by the filariae and the blood flukes, including the ability to survive in the presence of immunity. Trans-mission is effected, as in many parasites, by excessive production of larvae or eggs; but a large proportion of blood helminths at this stage fail to gain exit from the closed habitat and it is the retention of the larvae or eggs which cause host pathology. As transmission is adversely affected by serious dis-ease, it would appear that this problem has been less successfully solved. Concomitant immunity however, helps to delay the onset of pathological changes in these chronic infections by avoiding superinfection and unless ini-tial exposures are unusually heavy, severe disease is delayed and a period of successful transmission is achieved.

17.5. REFERENCES

Ah, H.S., Klei, T.R., McCall, J.W. and Thompson, P.E. (1974). "*Brugia pahangi* infections in Mongolian jirds and dogs following the ocular inoculation of infective larvae". J. Parasit. 60 (4), 643—648.

Altman, P.L. and Dittmer, D.S. (1974) (Eds.). "Federation of American Societies for Experimental Biology, Bethesda, Maryland." Biology Data Book, Vol. III, 2nd Edn.

Anderson, R.C. (1957). "The life cycles of Dipetalonematid nematodes (Filarioidea: Dipetalonematidae). The problem of their evolution." J. Helminth., 31 (4) 203—224.

Asch, H.L. and Read, C.P. (1975). "Transtegumental absorption of amino acids by male Schistosoma mansoni". J. Parasit. 61, 378—379.

Bourns, T.K.R., Ellis, J.C. and Rau, M.E. (1973). "Migration and development of *Tricho-*

bilharzia ocellata (Trematoda:Schistosomatidae) in its duck hosts". Can. J. Zool. 51, 1021—1030.

Bueding, E. (1949). "Studies on the metabolism of the filarial worm *Litomosoides carinii*". J. exp. Med. 89, 107—130.

Clegg, J.A. (1969). "Skin penetration by cercariae of the bird schistosome *Austrobilharzia terrigalensis:* the stimulatory effect of cholesterol". Parasitology 59, 973—989.

Clegg, J.A. (1974). "Host antigens and the immune response in schistosomiasis". In: "Parasites in the immunized host: mechanisms of survival". Ciba Foundation Symposium 25 (new series) pp. 161—183.

Clegg, J.A. and Smithers, S.R. (1968). "Death of schistosome cercariae during penetration of the skin. II. Penetration of mammalian skin by *Schistosoma mansoni*". Parasitology 58, 111—128.

Coutinho-Abath, J.O. and Jampolsky, R. (1957). "Comportamento das cercarias de *Schistosoma mansoni* na infectacao experimental de animais refractarios. I. Histopatologia das reacoes cutaneas observadas no pombo domestico (*Columba livia domestica*)". Anais Soc. Biol. Pernamb. 15, 93—125.

Duncan, J.L. and Pirie, H.M. (1972). "The life cycle of *Strongylus vulgaris* in the horse". Res. Vet. Sci. 13, 374—379.

Ewert, A. (1967). "Studies on the transfer of infective *Brugia pahangi* larvae from vector mosquitos to the mammalian host". Trans. R. Soc. trop. Med. Hyg., 61, 110—113.

Farooq, M. and Mallah, M.B. (1966). "The behavioural pattern of social and religious water-contact activities in the Egypt-49 Bilharziasis project area". Bull. Wld. Hlth. Org. 35, 377—387.

Gwadz, R.W. and Chernin, E. (1972). "Oral transmission of *Brugia pahangi* to jirds. (*Meriones unguiculatus*)". Nature 239, 524—525.

Hairston, N.G. and Jachowski, L.A. (1968). "Analysis of the *Wuchereria bancrofti* population in the people of American Samoa". Bull. Wld. Hlth. Org. 38, 29—59.

Hawking, F. (1975). "Circadian and other rhythms of parasites". Adv. Parasit., 13, 123—182.

Hawking, F. and Worms, M. (1961). "Transmission of filarioid nematodes". Ann. Rev. Entom. 6, 413—432.

Hockley, D.J. and Smithers, S.R. (1970). "Damage to adult *Schistosoma mansoni* after transfer to a hyperimmune host". Parasitology 61, 95—100.

Hockley, D.J. and McLaren, D.J. (1973). "*Schistosoma mansoni* changes in the outer membrane of the tegument during development from cercaria to adult worm". Int. J. Parasit. 3, 13—25.

Holliman, R.B., Fisher, J.E. and Parker, J.C. (1971). "Studies on *Spirorchis parvus* (Stunkard 1923) and its pathological effects on *Chrysemys picta picta*". J. Parasit. 57, 71—77.

Holmes, J.C. (1971). "Habitat segregation in sanguinicolid blood flukes (Digenea) of scorpaenid rock fishes (Perciformes) on the Pacific coast of North America". J. Fish Res. Bd. Can. 28, 903—909.

Hughes, D.L. and Harness, E. (1973). "The experimental transfer of immature *Fasciola hepatica* from donor mice and hamsters to rats immunized against the donors". Res. Vet. Sci. 14, 220—222.

Hutchinson, W.F. and McNeill, K.M. (1970). "Glycosis in the adult dog heartworm *Dirofilaria immitis*". Comp. Biochem. Physiol. 35, 721—727.

Johnston, M.R.L. and Stehbens, W.E. (1973). "Ultrastructural studies on the microfilaria of *Cardianema* sp. Alicata, 1933 (Nematoda;Onchocercidae)". Int. J. Parasit. 3, 243—250.

Kusel, J.R., Sher, A., Perez, H., Clegg, J.A. and Smithers, S.R. (1975). "Use of radioactive isotopes in the study of specific schistosome membrane antigens". In "Nuclear Techniques in Helminthology Research" IV, pp. 127—143. Int. Atomic Energy Agency, Vienna.

Laurence, B.R. and Simpson, M.G. (1974). "The ultrastructure of the microfilaria of *Brugia.* Nematoda: Filarioidea". Int. J. Parasit. 4, 523—536.

Lavoipierre, M.M.J. (1958) "Studies on the host parasite relationships of filarial nematodes and their arthropod hosts II. The arthropod as a host to the nematode: a brief appraisal of our present knowledge based on a study of the more important literature from 1878 to 1957". Ann. trop. Med. Parasit. 52, 326—345.

Lumsden, R.H. (1975). "Surface ultrastructure and cytochemistry of parasitic helminths". Exp. Parasit. 37, 267—339.

McGreevy, P.B., Theis, J.H., Lavoipierre, M.M.J. and Clark, J. (1974). "Studies on filariasis III. *Dirofilaria immitis:* emergence of infective larvae from the mouthparts of *Aedes aegypti".* J. Helminth., 48, 221—228.

Mackerras, M.J. and Sandars, D.F. (1955). "The life history of the rat lungworm, *Angiostrongylus cantonensis* (Chen) (Nematoda: Metastrongylidae)". Aust. J. Zool. 3, 1—21.

McLaren, D.J. (1971). "Ultrastructural studies on filarial worms". Ph. D. Thesis Brunel University.

McLaren, D.J. (1972). "Ultrastructural studies on microfilariae (Nematoda: Filarioidea)". Parasitology 65, 317—332.

McLaren, D.J., Clegg, J.A. and Smithers, S.R. (1975). "Acquisition of host antigens by young *Schistosoma mansoni* in mice: correlation with failure to bind antibody in vitro". Parasitology 70, 67—75.

Manson-Bahr, P. (1951). "Mansons Tropical Diseases". 13th Edn., Cassell, London.

Ogilvie, B.M. and Worms, M.J. (1976). "Immunity to nematode parasites of man with special reference to *Ascaris* hookworms and filariae". In: "Immunology of Parasitic Infections" (Eds. S. Cohen and E.H. Sadun) Blackwell, Oxford, England, pp. 380—407.

Orihel, T.C. (1961). "Morphology of the larval stages of *Dirofilaria immitis* in the dog". J. Parasit. 47, 251—262.

Pacheco, G. (1974). "Relationship between the number of circulating microfilariae and the total population of microfilariae in a host". J. Parasit. 60, 814—818.

Perez, H. and Terry, R.J. (1973). "The killing of adult *Schistosoma mansoni* in vitro in the presence of antisera to host antigenic determinants and peritoneal cells". Int. J. Parasit. 3, 499—503.

Rogers, R., Denham, D. and Nelson, G.S. (1974). "Studies with *Brugia pahangi* 5. Structure of cuticle". J. Helminth. 48, 113—117.

Rogers, S.H. and Bueding, E. (1975). "Anatomical localization of glucose uptake by *Schistosoma mansoni* adults". Int. J. Parasit. 5, 369—371.

Rogers, W.P. (1940). "Haemological studies on the gut contents of certain nematode and trematode parasites". J. Helminth. 18, 53.

Schiller, E.L., Bueding, E., Turner, V.M. and Fisher, J. (1975). "Aerobic and anaerobic carbohydrate metabolism and egg production of *Schistosoma mansoni* in vitro". J. Parasit. 61, 385—389.

Senft, A.W. (1965). "Recent developments in the understanding of amino acid and protein metabolism by *Schistosoma mansoni,* in vitro". Ann. trop. Med. Para. 59, 164—168.

Senft, A.W. and Maddison, S.E. (1975). "Hypersensitivity to parasite proteolytic enzyme in schistosomiasis". Am. J. trop. Med. Hyg. 24, 83—89.

Sher, F.A., Mackenzie, P.E. and Smithers, S.R. (1974). "Decreased recovery of invading parasites from the lungs as a parameter of acquired immunity to schistosomiasis in the laboratory mouse". J. Infect. Dis. 130, 626—633.

Shiff, C.J., Cmelik, S.H.W., Ley, E.H. and Kriel, R.L. (1972). "The influence of human skin lipid on the cercarial penetration responses of *Schistosoma haematobium* and *Schistosoma mansoni".* J. Parasit. 58, 476—480.

Simpson, M.G. and Laurence, B.R. (1972). "Histochemical studies on microfilariae". Parasitology 64, 61—88.

Smith, J.H. and Lichtenberg, F. von, (1974). "Observations on the ultrastructure of the tegument of *Schistosoma mansoni* in mesenteric veins". Am. J. trop. Med. Hyg. 23, 71—77.

Smith, J.W. (1972). "The blood flukes (Digenea: Sanguinicolidae and Spirorchidae) of cold-blooded vertebrates and some comparison with the schistosomes". Helm. Abs. Series A 41, 161—204.

Smith, M.A. (1974). "Radioassays for the proteolytic enzyme secreted by living eggs of *Schistosoma mansoni*". Int. J. Parasit. 4, 681—683.

Smithers, S.R. and Terry, R.J. (1969). "Immunity in Schistosomiasis". Ann. N.Y. Acad. Sci. 160, 826—840.

Smithers, S.R. and Terry, R.J. (1976). "The Immunology of Schistosomiasis". In "Advances in Parasitology" (Ben Dawes, ed.) 14, 399—422.

Smithers, S.R., Terry, R.J. and Hockley, D.J. (1969). "Host antigens in schistosomiasis. Proc. R. Soc. B. 171, 483—494.

Smyth, J.D. (1962). "Introduction to animal parasitology". E.U.P. London.

Stirewalt, M.A. (1963). "Cercaria vs schistosomula *(Schistosoma mansoni):* absence of the pericercarial envelope in vivo and the early physiological and histological metamorphosis of the parasite". Expl. Parasit. 13, 395—406.

Stirewalt, M.A. and Fregeau, W.A. (1966). "An invasive enzyme system present in cercariae but absent in schistosomules of *Schistosoma mansoni*". Expl. Parasit. 19, 206—215.

Timms, A.R. and Bueding, E. (1959). "Studies of a proteolytic enzyme from *Schistosoma mansoni*". Br. J. Pharmacol. Chemother. 14, 68—73.

Tsang, V.C.W. and Damian, R.T. (1975). "Demonstration and mode of action of an inhibitor of the intrinsic pathway of blood coagulation from adult *Schistosoma mansoni* [Abstract]". 10th Joint Conf. on Parasit. Dis. US-Japan Coop. Med. Sci. Prog. pp. 147—148.

Ulmer, M.J. (1971). "Site finding behaviour in helminths in intermediate and definitive hosts". In Fallis, A.M. (ed.) 'Ecology and Physiology of Parasites'. Adam Hilger, London.

Vincent, A.L., Ash, L.R. and Frommes, S.P. (1975). "The ultrastructure of adult *Brugia malayi* (Brug, 1927) (Nematoda: Filarioidea)". J. Parasit. 61, 499—512.

Wang, E.J. and Saz, H.J. (1974). "Comparative biochemical studies on *Litomosoides carinii, Dipetalonema viteae* and *Brugia pahangi* adults". J. Parasit. 60, 316—321.

Wenk, P. (1967). "Der Invasionsweg der metazyklischen Larven von *Litomosoides carinii* Chandler (1931) (Filariidae)". Z. f. Parasitenkunde 28, 240—263.

W.H.O. (1974). "Immunology of Schistosomiasis". Bull. WHO 51, 553—595.

Wilks, N.E. (1967). "Lung to liver migration of schistosomes in the laboratory mouse". Am. J. trop. Med. Hyg. 16, 599—605.

Wilson, R.S. (1975). "Synthesis of macromolecules by the epithelial surfaces of *Schistosoma mansoni*, with particular reference to the fate of secretory vesicles in the worm tegument". Proc. Int. Conf. on Schistosomiasis, Cairo 1975 (in press).

Worms, M.J. (1972). "Circadian and seasonal rhythms in blood parasites". Zool. J. Linn. Soc. 51 (Suppl. 1) 53—67.

Ecological Aspects of Parasitology
Editor: C.R. Kennedy
© North-Holland Publishing Company, Amsterdam, 1976

CHAPTER 18

BLOOD FLUIDS — PROTOZOANS

J.R. BAKER

M.R.C. Biochemical Parasitology Unit, Molteno Institute, Downing Street, Cambridge,
CB2 3EE (Great Britain)

Contents

Introduction

Members of only two genera of protozoa have become adapted to
dwelling in blood plasma during a substantial portion of their life cycle,
Trypanosoma and *Cryptobia* (or *Trypanoplasma*); the former are widespread common parasites of vertebrates of all classes, and include two
notable pathogens of man — *Trypanosoma brucei* sspp. in tropical Africa
and *T. cruzi* in South and Central America. *Cryptobia* appears less successful as an intravascular parasite of vertebrates, being found there only in
fishes and possibly a salamander, though many species occur in the gut and

other organs of vertebrates and invertebrates. As *Cryptobia* has been far less studied than *Trypanosoma*, most of the discussion in this chapter will relate primarily to the latter. Both genera belong to the order Kinetoplastida, distinguished by possessing an unusually large mass of aggregated intra-mitochondrial deoxyribonucleic acid (DNA) termed the kinetoplast, and are usually placed in distinct suborders — the Bodonina and Trypanosomatina, respectively. Further information is given by Hoare (1972) on *Trypanosoma*, Kudo (1966), Grassé (1952) and Wenyon (1926) on *Cryptobia* and Diem and Lentner (1970) on blood plasma. This chapter considers how the parasites have benefited from, or overcome, the advantages and dis-advantages of sanguicolous life.

18.1. Advantages

18.1.1. Nutrition

Certain solutes, particularly small inorganic ions, may traverse the pellicle by osmosis, but it is now known that trypanosomes feed mainly by pino-cytosis. This apparently occurs only in the flagellar pocket or "reservoir", either through the limiting membrane (pellicle), as in haematozoic forms of *T. brucei* (Brown et al., 1965; Steiger, 1973), *T. congolense* (Vickerman, 1969a; 1971), *T. vivax* (Vickerman, 1969b) and perhaps all salivarian try-panosomes (Rudzinska and Vickerman, 1968), or via a cytopharynx opening through a cytostome in the wall of the flagellar pocket or, sometimes in intracellular *T. cruzi*, the cell surface near the pocket (Milder and Deane, 1969). Cytostomal-cytopharyngeal complexes have been reported from forms cultivated in vitro of the following species: *T. mega* (Steinert and Novikoff, 1960), *T. conorrhini* (Brooker and Preston, 1967; Milder and Deane, 1969), *T. raiae* (Preston, 1969) and *T. cruzi* (Milder and Deane, 1969). Wéry and de Groodt-Lasseel (1966) published an electronmicro-graph of cultivated *T. cruzi* which Preston (1969) thought indicated the existence of a cytopharynx, although Brack (1968) had not seen one in *T. cruzi* in vitro. Amongst haematozoic trypomastigotes, the only report has been of a "suggestion of a short cytostome" in *T. "avium"* (Baker and Bird, 1968), a species now reclassified as *T. corvi* (Baker, in press) — a suggestion which Preston (1969) thought "may be premature". Brooker (1971) thought that a micrograph of haematozoic *T. lewisi* published by Anderson and Ellis (1965) strongly indicated the presence of a cytopharynx. Milder and Deane (1969) reported a cytostome and cytopharynx in intra-cellular forms (probably amastigotes) of *T. cruzi* in tissue culture, as well as in epimastigotes grown in vitro. A thorough study of the morphology of the complex was made by Preston (1969) in *T. raiae*. Pinocytotic uptake of material in the flagellar pocket of *T. theileri* and *T. melophagium* in vitro has been postulated by Herbert (1965) but not conclusively demonstrated; no evidence was presented for the existence of a cytostome or cytopharynx. Thus, as Brooker (1971) speculated, it may be that all stercorarian * trypanosomes of mammals and non-mammals have a cytostome in some or all of their developmental stages, whereas this organelle is absent from the salivarian species. The dearth of records from haematozoic stercorarians may be due to the fact that low parasitaemias, generally characteristic of these species, render difficult the examination of much (or any!) material by electron microscopy. Pinocytosis in trypanosomes has been discussed by Jadin (1971). In *T. brucei* Langreth and Balber (1975) demonstrated ingestion of ferritin (as an indicator) from the flagellar pocket of haematozoic (slender and stumpy) and culture forms. The process was most active in stumpy forms and least so in culture forms, in which typical "spiny-coated-vesicles" (pinosomes) were not seen, the ferritin entering "small smooth cisternae" instead. The process in all forms was much reduced at low temperature and so was assumed to require energy. Langreth and Balber (1975) reported secretion of acid phosphatase into the flagellar pocket of haematozoic *T. brucei* (but not into that of culture forms), possibly via acanthosomes (small vesicles which may be involved in transport and exocytosis of lysosomal enzymes), which presumably indicates the occurrence of extracellular digestion (as proposed by Jadin, 1971); following Milder and Deane (1969), they suggested that the pocket also served as a storage organelle for plasma proteins.

After ingestion, proteins are presumably digested by lysosomal enzymes; pinocytotic vacuoles (pinosomes) apparently fuse with tubular or spherical vesicles containing hydrolases (primary lysosomes) to form secondary lysosomes in the usual way (see Langreth and Balber, 1975). There is evidence of extracellular digestion by acid phosphatase secreted into the flagellar pocket (Jadin, 1971; Langreth and Balber, 1975) in haematozoic but not cultured forms of *T. brucei*. Preston (1969) found no acid phosphatase within the flagellar pocket, cytopharynx or pinocytotic vesicles of *T. raiae* cultivated in vitro, though its presence has been demonstrated there in other trypanosomatid genera in vitro — e.g. *Crithidia fasciculata* (Brooker, 1971) — as reviewed by Jadin (1971). Acanthosomes have been reported from haematozoic trypomastigotes of *T. brucei* (Langreth and Balber, 1975), epimastigotes of *T. brucei* and *T. vivax* in tsetse flies (Steiger, 1973; Vickerman, 1973), and metacyclic trypomastigotes of *T. brucei* in tsetse flies (Steiger, 1973); none was seen by Langreth and Balber (1975) in *T. brucei* trypomastigotes cultivated in vitro. Perhaps extracellular (or "intra-pocket") digestion is fairly common in the genus in vivo but not in

---

* Hoare (1964) introduced the term stercoraria as the name of a taxonomic group of species of *Trypanosoma* parasitizing mammals only. In common with some other workers, I am here using it adjectivally without its original taxonomic restriction.

vitro. Amongst the compounds acquired by pinocytosis of plasma is presumably haemoglobin; this is a compound of the protein globin with four molecules of the ferrous iron porphyrin haem, an unstable substance rapidly oxidized to its hydroxide haematin (containing trivalent ferric iron). Haematin, or the related chloride haemin, is an essential nutrient for probably all species of *Trypanosoma* and the other members of the Trypanosomatidae (Taylor and Baker, 1968). How it is digested by the parasites is not known. In *Plasmodium* species some or all of the ingested haem is unused and, in the form of haematin, probably combined with degraded protein or polypeptide, remains as an end product of haemoglobin digestion (haemozoin or "malarial pigment"; Deegan and Maegraith, 1956a, b). However, all or most trypanosomatids must extend its breakdown beyond this stage since no insoluble iron-containing residue has been demonstrated except possibly in *T. cyclops* Weinman, 1972 of Malaysian monkeys, which produces in vitro a brownish-black pigment derived from haemoglobin; however, it is not known to contain iron and, unlike malarial pigment, is not birefringent (Weinman, 1972). If the pigment represents unused haem from ingested haemoglobin, this might suggest that *T. cyclops* in vitro does not require haemin; it was maintained serially in cultures said to be without haemoglobin though, since the medium contained 5% foetal calf serum, there must be doubt as to its absolute absence (Lwoff, 1951).

The carbohydrate metabolism of trypanosomes will be discussed in section 18.1.3. Their lipid metabolism is less well understood; it has been reviewed by Von Brand (1973) and, for trypanosomes of Anura, by Bardsley and Harmsen (1973) and will not be considered further here.

18.1.2. Excretion

Little is known of this. Soluble excretory products of small molecular size (e.g. carbon dioxide, organic acids and salts) presumably diffuse outwards through the trypanosomes' pellicle (see 18.1.3). A contractile vacuole has been reported by Clark (1959) from *Herpetomonas*, *Crithidia*, *Leishmania* and seven species of *Trypanosoma* including *T. cruzi* in vitro, and by Brooker (1971) from *Crithidia fasciculata* but, as Langreth and Balber (1975) point out, physiological evidence for its existence has been obtained only in *C. fasciculata* (Cosgrove and Kessel, 1958). Such an organelle would no doubt assist in the excretion of soluble metabolic waste products. Insoluble matter — "membranous whorls" and other debris — may be excreted into the flagellar pocket by exocytosis (Langreth and Balber, 1975) or "cellular defaecation" (Brooker, 1971), the reverse of pinocytosis. The "streamers" of pellicle plus surface coat, or plasmanemes, produced by haematozoic *T. brucei*, may be a means of getting rid of superfluous surface coat, perhaps as a part of the process of antigenic change (Vickerman, 1971) (see section 18.2.3).

18.1.3. Respiration

The respiratory metabolism of the subspecies of *T. brucei* is the most thoroughly studied. Although, as far as is known (not very far), all stages of most species of *Trypanosoma* have an active intramitochondrial respiratory system involving a conventional Kreb's cycle and cytochrome chain at least in outline, some haematozoic trypomastigotes of *T. brucei* (the "long slender" multiplicative forms) do not; they "display an active aerobic glycolysis [and] break down glucose to pyruvic acid, but no further ...." (Vickerman, 1971; q.v. for further references). This squanders both glucose and oxygen, and indeed terminal hypoglycaemia is the "primary cause of death in acute murine trypanosomiasis" due to *T. brucei* (Herbert et al., 1975) — though not in infections of man or other large mammals; the blood glucose level fell to less than 25 mg/100 ml, one-seventh of the normal value. However, blood plasma is richly supplied with oxygen and glucose, and it is presumably only because it inhabits such an environment that *T. brucei* can be so profligate. In its vector, and in culture in vitro, where presumably both glucose and oxygen are less readily available, it reverts to the mitochondrial system and breaks down glucose fully to carbon dioxide and water (Bowman et al., 1972). The other salivarian species, *T. vivax* and *T. congolense*, apparently still rely to some extent on mitochondrial respiration when in the vertebrate host's blood, though less so than they do in the invertebrate vector (Vickerman, 1971). Evidence of a cyanide-insensitive, extra-mitochondrial respiratory pathway has also been obtained in *T. mega* of Anura (Ray and Cross, 1972) and *T. corvi* of Aves (= *T. "avium"*) (Shinondo, 1975) in vitro and one may occur in other (perhaps all) species. The extent to which this component contributes to the whole in different species or stages of the life cycle is controlled by the kinetoplast deoxyribonucleic acid (Simpson, 1972); apparently only in long slender haematozoic trypomastigotes of *T. brucei* (and the closely related *T. evansi*, *T. equinum* and *T. equiperdum*), is it the only functional pathway. Even though they may utilize the less-wasteful Kreb's cycle-cytochrome oxidase system, the population densities of some stercorarian trypanosomes in the blood may depend (at least in part) on the level of their hosts' glycaemia. Rhythmic changes in glycaemia, possibly controlled by secretion of catecholamines and other substances, may be responsible for diurnal and annual fluctuations of parasitaemia in anuran amphibians (Bardsley and Harmsen, 1973) and — perhaps — other species of *Trypanosoma* — e.g. *T. corvi*, which seems to have a "spring relapse" of parasitaemia (Baker, 1956b).

Haematozoic *T. brucei*, if maintained for even relatively short periods in vitro, rapidly "poisons" itself by the accumulation of its own excreted pyruvic acid in the medium (Lumsden et al., 1965). The rapid removal of acidic excretory products, the buffering effect of the surrounding plasma and the presumed removal of the excreted pyruvic acid from the blood by its involvement in the host's own metabolic processes, would all tend to mitigate this and thus furnish another example of the advantages (to *T. brucei* et alia in particular) of an intravascular habitat. The absence of these processes from the invertebrate host's gut may be another reason why *T. brucei* reverts to mitochondrial respiration in that environment.

18.1.4. Transport

Although many haematozoic parasites which live in their vertebrate hosts elsewhere than in the circulating plasma use the latter as a vehicle to carry them to and from their alternative niches, or their invertebrate vectors, only *Trypanosoma* and *Cryptobia* have species which are truly haematozoic extracellular parasites (i.e., spend all or a substantial part of their life cycle in the plasma). However, *Cryptobia* may be found, as well as in blood, in "loose connective tissue" including subcutaneous reticular tissue (Putz, 1972). The salivarian trypanosome *T. brucei* lives in lymph and tissue fluids of its vertebrate hosts to a substantial extent (Ssenyonga and Adam, 1975) and also invades the cerebrospinal fluid of man (Ormerod, 1970) and wild or domestic mammals (McCulloch, 1967 and Losos and Ikede, 1972, respectively). *T. equiperdum*, causative agent of the venereally-transmitted disease "dourine" of equines, is almost restricted to oedematous patches in the skin of the host and only rarely enters the blood (Hoare, 1972).

Except for *T. equiperdum*, all species of *Trypanosoma* (and *Cryptobia* of vertebrates) use the blood as a route of entry to their vector, a haematophagous invertebrate (insect, leech or — rarely — mite for *Trypanosoma* and leech for *Cryptobia*). After re-entry into the vertebrate host, they may either directly invade (or be injected by the vector into) the blood, or they may enter it after an initial period of multiplication in tissue fluid local to the point of entry — as in the "chancre" of early *T. brucei* sspp. infection in man. After entering the blood, those species which dwell almost exclusively therein remain there — though not necessarily evenly distributed throughout the vascular system; many species, including *T. lewisi* of rats, other members of the subgenus *Herpetosoma* (Molyneux, 1970), *T. (Megatrypanum) hoarei* of shrews (Hoare, 1972) and probably many trypanosomes of frogs (Bardsley and Harmsen, 1973), reproduce more-or-less sequestered in the small blood vessels of internal organs; other species of *Herpetosoma* divide only in the fluid within lymphoid tissue and not at all in the blood (Molyneux, 1970). *T. brucei* sspp. reproduce preferentially though not exclusively in lymph (Ssenyonga and Adam, 1975). Trypanosomes of the subgenus *Schizotrypanum* (including *T. (S.) cruzi*, which causes Chagas's disease in south and central America) reproduce in their vertebrate hosts only within tissue cells; after 6—6½ days (in vitro at 38°C: Dvorak and Poore, 1974), small trypomastigotes are liberated into the blood and are either transported to other tissue cells or — growing into the "adult" haematozoic form — await ingestion by a suitable vector (Reduviidae). Other species, such as *T. congolense*, are apparently virtually solely haematozoic (Losos and Ikede, 1972; Ssenyonga and Adam, 1975).

The blood-inhabiting stages of *Cryptobia* and *Trypanosoma* have a well-developed undulating membrane — a "fin-like" expansion of the surface pellicle produced by the undulation of the recurrent flagellum, which is attached to the pellicle at intervals by connections resembling hemidesmosomes (Vickerman, 1969b). Perhaps this membrane facilitates locomotion in a viscous medium such as plasma.

18.2. Disadvantages

18.2.1. Access

One of the problems facing an intravascular parasite is access to, or exit from, its habitat. Within the body, the blood is not rigidly separated from cells and fluids of other tissues. The lymphatics connect with the veinous system via the thoracic duct on the left and the lymphatic duct on the right and the subclavian veins (Hewer, 1949). Blood and lymph capillaries are often closely associated with one another and are lined in their smallest extremities by only a single layer of endothelial cells, though the blood capillaries are also surrounded by a thin connective tissue sheath derived from mesenchymal pericytes. Hewer (1949, p. 142) states that "amoeboid cells pass through the [capillary] wall by actively pushing between the cells, or even occasionally by passing through them"; doubtless trypanosomes and trypanoplasms — both highly motile organisms — travel thus between the blood and tissue spaces. All trypanosomes and trypanoplasms (except *T. equiperdum*, p. 376) rely on a vector to transport them to their vertebrate host. Transfer from the former to the latter may be effected by the release of the parasites from the vector into the buccal cavity of the vertebrate when the latter eats the vector, or at least crushes it in its mouth, as probably occurs with *T. corvi* of birds (= *T. "avium"*; Baker, 1956a), some trypanosomes of rodents including *T. lewisi* (Hoare, 1972), *T. grayi* of crocodiles (Hoare, 1929, 1972) and possibly others; the released trypanosomes are thought to penetrate actively the mucous membrane lining the buccal cavity or pharynx. Alternatively, trypanosomes released from the vector when it defaecates upon its host's skin shortly after feeding may gain entry through the puncture caused by the vector's proboscis; this is thought to be a common route of entry of *T. cruzi* (Hoare, 1972). Perhaps the most sophisticated method (though not necessarily evolutionarily recent) is the "anterior station" development in the mouthparts or salivary glands of the vector resulting in injection of the parasites directly into the host's blood by the vector when it feeds, adopted by salivarian trypanosomes of mammals, *T.*

rangeli (Hoare, 1972), and trypanosomes and *Cryptobia* of fish, which are transmitted by leeches — Annelida, class Hirudinea (Wenyon, 1926; Becker and Katz, 1965). Leeches feed either by rasping through the epidermis by means of chitinous toothed jaws and sucking blood from the resulting haemorrhage, or by inserting a protrusible proboscis into the host's dermis; an anticoagulant is secreted in the leech's saliva, which ensures a copious blood flow. The insect vectors of salivarian trypanosomes of mammals (*Glossina* for species of the subgenera *Trypanozoon*, *Nannomonas* and *Duttonella*, reduviid Hemiptera for *T. rangeli;* see Hoare, 1972) feed by insertion of a proboscis which, in the case of *Glossina* at least, erodes the dermal capillary bed leading to a localized haemorrhage; an anticoagulant is injected with the insect's saliva, taking with it metatrypomastigotes if these are present in the insect's salivary glands or hypopharynx, and blood is sucked from the ensuing pool. Thus the infective forms are deposited into the dermal tissue rather than directly into a capillary, which may account in part for the subsequent initial development of *T. brucei* in a more-or-less localized chancre. However, this must also reflect the tissue tropism of this species referred to above (p. 376), since species of the subgenera *Nanno-monas* and *Duttonella* are presumably deposited similarly and there is no record of their undergoing restricted local initial development.

18.2.2. Exit

Haematozoic trypanosomes and trypanoplasms depend entirely on their vectors as a means of exit from their vertebrate hosts. When the vectors feed (section 18.2.1), parasites are inevitably ingested with the blood meal, whether the latter is acquired directly from a capillary or by "pool feeding". Thus exit poses no problem for the parasites, apart from the need to maintain an adequate parasitaemia to ensure infection of the vectors when they feed. Some of the ways in which this is done are discussed in the next two sections (18.2.3 and 18.2.4).

18.2.3. Humoral antibodies

Specific antibodies are secreted by the host's B-lymphocytes into the blood plasma in response to stimulation by parasite antigen (Humphrey and White, 1970; Holborow, 1973; Roitt, 1974) and may attack the parasite in various ways, including complement-mediated lysis (lysins), agglutination (agglutinins) and facilitating phagocytosis (opsonins). An unusual antibody is directed against certain species of *Trypanosoma* belonging to the subgenus *Herpetosoma* — "ablastin", which interferes with parasite division in some fashion not understood but has no other harmful effect (D'Alesandro, 1970, 1975).

In order for a parasite strain or species to survive, it must persist in its host for as long as possible at a level adequate to ensure a good chance of onward transmission but low enough to avoid undue risk of killing the host. Thus the production of antibodies by the host is not entirely disadvantageous to the parasite, provided that the latter can evolve a compromise preventing its rapid total elimination. Thus an antibody like "ablastin" is ideal from the parasite's viewpoint — by preventing multiplication it reduces parasitaemia to a level which does not seriously endanger the host * and, probably, does not furnish too great an antigenic stimulus to the latter; because of this, and because "ablastin" is not lethal, the "adult" or differentiated (non-reproductive) trypanosomes may persist for long periods in the blood of their vertebrate hosts, ready and waiting to continue their life cycle after ingestion by the appropriate vector. It has been suggested that "ablastic" antibodies may occur in Amphibia (Lom, 1969), and they may be more widespread than is generally known; many stercorarian trypanosomes of mammals and non-mammals have life cycles suggestive of their occurrence, with a brief initial period of multiplication (if any) in the vertebrate host followed by a more-or-less extended period of circulation in the vertebrate's blood of apparently differentiated (i.e., non-multiplicative) trypomastigotes. However, at least with *T. lewisi*, a trypanosomicidal antibody is eventually produced, usually in a matter of weeks, and the entire parasite population within that host is wiped out. With such a biocoenose, in which infected vertebrates are in close and virtually continual contact with a highly susceptible vector population of fleas (Siphonaptera), this does not matter; ample opportunity will have been afforded for onward transmission by the time the death of the parasite population occurs. But in other situations where vertebrate-vector contact is transitory, and where the vector population may be less susceptible to infection, the "ablastin-mechanism" may not be adequate. This applies perhaps especially to the trypanosomes transmitted by *Glossina*, where the vector susceptibility is so low (especially with *T. brucei*), and contact between vertebrate and vector so sporadic, that parasitaemias persisting for months or even years may be necessary to ensure parasite survival and are probably the rule for *T. b. gambiense*, *T. b. rhodesiense* and *T. b. brucei* in their natural hosts (man, wild ungulates and wild ungulates respectively). The association of *T. b. rhodesiense* and man is probably of relatively recent evolutionary origin and not adequate to ensure the survival of the species, the intervention of a wild ungulate host being necessary (Baker, 1974). Thus these parasites have needed to evolve a mechanism of surviving the host's trypanosomicidal antibody response. Three ways are known by which intravascular parasites have succeeded in doing this — antigenic disguise, antigenic variation and intracellular sequestration (or combinations of some of these). The first

* There is, in fact, some evidence that *T. lewisi*, an "ablastin-inducer" par excellence, is beneficial to its host the rat (Lincicome, 1971).

method, not shown conclusively to occur amongst protozoa but thoroughly elucidated in the case of *Schistosoma* spp. (Clegg, 1974), may be used, in combination with antigenic variation, by *T. vivax;* host protein has been identified in the extra-pellicular cell coat of this species in the vertebrate blood (Vickerman, 1974), but is more likely to the host antibody.

Antigenic variation is practised by a variety of intravascular parasites — the spirochaete *Borrelia recurrentis* (Wilson and Miles, 1964), *Plasmodium* (Brown, 1974) and salivarian trypanosomes — *T. brucei* sspp., *T. vivax* and *T. congolense* (reviewed by Vickerman, 1974). To summarise briefly this well-known story, the haematozoic trypomastigotes of all three salivarian subgenera of *Trypanosoma (Trypanozoon, Nannomonas* and *Duttonella)* possess a glycoprotein coat about 12—15 nm thick overlying the unit membrane at their cell surface. This coat presents to the host antigens in response to which agglutinating and probably opsonizing (and other?) antibodies — initially amongst the IgM molecules — are produced. In a way not fully understood, but apparently related to this production of specific antibody, in a proportion of the trypanosome population a change occurs in the molecular configuration of the antigenic determinants in this coat so that the IgM antibody molecules no longer react with them. Thus is that particular parasite population saved from destruction, and its numbers increase until, some 2 to 4 days later, the host has elaborated a new IgM antibody specific to the changed antigen; and so on, two dozen or more variants of one strain of *T. brucei* having been identified (see Gray, 1970). It has been suggested that the "plasmanemes" or "filopodia" which *T. brucei* develops at least under certain conditions in vitro represent coat (and hence antigen) being discarded (Vickerman, 1971). There is as yet no good evidence that this process of variation occurs in other *Trypanosoma* subgenera; in forms like *T. lewisi* in which sterile immunity finally develops in the vertebrate host, it almost certainly does not. The situation regarding *T. cruzi* is less clear cut. Specific anti-*T. cruzi* IgM is detectable in the serum of infected persons only during the first nine months or so after infection (Camargo and Amato Neto, 1974), which suggests that variation is not continually taking place; but Dzbenski (1974) found that antisera collected on different occasions from mice infected with *T. cruzi* only occasionally precipitated soluble antigen ("exoantigen") prepared from the blood of mice infected with the homologous strain, which suggests that antigenic variation might occur (and is, so far as I know, the first piece of evidence to do so). The repeated recrudescences of low-level parasitaemia characteristic of chronic *T. cruzi* infection suggest some kind of antigenic change or disguise of the extracellular trypomastigotes.

There is no available evidence concerning exoantigens or antigenic variation of trypanoplasms — nor of those species of *Trypanosoma* which parasitise vertebrate hosts other than mammals, apart from the development of immunity against homologous reinfection in some experimentally infected birds (Baker, 1956b; Molyneux and Gordon, 1975), which suggests absence of variation; further study is clearly needed.

18.2.4. Phagocytes

Phagocytosis of haematozoic trypomastigotes of mammals has been shown to occur in vitro or in vivo with *T. brucei* (Goodwin, 1970; Takayanagi et al., 1974a, b), *T. cruzi* (Scorza and Scorza, 1972; Kierszenbaum et al., 1974) and *T. lewisi* (Lange and Lysenko, 1960; Patton, 1972); Takayanagi et al. (1974a, b) clearly showed that it depended on the presence of agglutinating antisera, though whether of the agglutinin itself or a concurrently occurring opsonin is uncertain. The workers using *T. lewisi* also showed the need for, or the enhancing effect of, specific antisera, which lead to the postulated presence therein of an opsonin. The involvement of antiserum in the phagocytic attack on trypanosomes suggests that antigenic variation (section 18.2.3) may also serve to protect those species which practise it against ingestion by macrophages — at least to some extent. *T. cruzi* somehow evades, not phagocytosis itself, but subsequent digestion; for, at least under certain conditions and in certain macrophages, it can not only survive but also multiply, eventually destroying the destroyer — the macrophage (Scorza and Scorza, 1972; Milder et al., 1973). Perhaps associated with this, *T. cruzi* penetrates actively into macrophages in vitro in addition to being passively ingested by them (Dvorak and Schmunis, 1972; Milder et al., 1973; Alexander, 1975). Similar behaviour is shown by other species of *Schizotrypanum* (Liston, 1975; Baker and Selden, 1975) and other protozoan parasites including *Leishmania* (Mauel et al., 1974) and *Toxoplasma* (Hirsch et al., 1974).

18.3. Other influences

There is not much information about the influence of other plasma constituents (e.g. hormones) on trypanosomes, and none at all relating to their effect (if any) on trypanoplasms. It has been suggested that periodic fluctuations of parasitaemia of trypanosomes in Anura may be partially controlled by hormonal secretion (Bardsley and Harmsen, 1973), probably acting indirectly by controlling the level of other plasma components (e.g. glucose). Hormones such as corticosteroids can indirectly influence the trypanosome population by affecting the host's immune responses (see Petana, 1964). Ashcroft (1959) suggested that cortisone might additionally have "a direct toxic action ... on the trypanosomes" but this has not been substantiated. The presence in the plasma of non-specific cross-reacting antibodies to other antigens may also influence adversely a population of trypanosomes; some species of the spirochaete *Borrelia* confer at least partial protection against *T. brucei* sspp. in rats (Trautmann, 1907; Galliard et al., 1957, 1958, 1959a; Lapierre et al., 1958) and — to a lesser extent — *T. cruzi* (Galliard et al., 1959b, 1963).

18.4. Conclusions

There is no doubt that trypanosomes (and presumably the sadly neglected *Cryptobia* species) have adapted successfully to life in the vertebrate's blood. Most live more-or-less in harmony with their hosts, neither killing them nor being killed by them in undue haste. They are geographically widespread; the only continent from which I am unaware of their having been reported is Antarctica — and I should be surprised if they do not occur there in fish and, possibly, birds. They parasitize a wide range of hosts of all the major vertebrate classes. They live bathed constantly in a nutrient soup. And most people are unaware of their existence. What more could any parasite desire?

18.5. REFERENCES

Alexander, J. (1975). "Effect of the antiphagocytic agent cytochalasin B on macrophage invasion by *Leishmania mexicana* promastigotes and *Trypanosoma cruzi* epimastigotes". J. Protozool. 22, 237—240.

Anderson, W.A. and Ellis, R.A. (1965). "Ultrastructure of *Trypanosoma lewisi:* flagellum, microtubules and the kinetoplast". J. Protozool. 12, 483—499.

Ashcroft, M.T. (1959). "The effect of cortisone on *Trypanosoma rhodesiense* infections of albino rats". J. infect. Dis. 104, 130—137.

Baker, J.R. (1956a). "Studies on *Trypanosoma avium* Danilewsky 1885. II. Transmission by *Ornithomyia avicularia* L.". Parasitology 46, 321—334.

Baker, J.R. (1956b). "Studies on *Trypanosoma avium* Danilewsky 1885. III. Life cycle in vertebrate and invertebrate hosts". Parasitology 46, 335—352.

Baker, J.R. (1974). In: "Trypanosomiasis and Leishmaniasis with special reference to Chagas' disease". CIBA Foundation Symp. 20, 28—50.

Baker, J.R. (in press). In: "The biology of the Kinetoplastida". (W.H.R. Lumsden, ed.). Academic Press, London and New York.

Baker, J.R. and Bird, R.G. (1968). "*Trypanosoma avium:* fine structure of all developmental stages". J. Protozool. 15, 298—308.

Baker, J.R. and Selden, L.F. (1975). "*Trypanosoma (Schizotrypanum) dionisii* in macrophage cultures: 2. Temperature sensitivity of intracellular parasites". Parasitology 71, xviii.

Bardsley, J.E. and Harmsen, R. (1973). "The trypanosomes of Anura". Adv. Parasit. 11, 1—73.

Becker, C.D. and Katz, M. (1965). "Transmission of the hemoflagellate, *Cryptobia salmositica* Katz, 1951, by a rhynchobdellid vector". J. Parasit. 51, 95—99.

Bowman, I.B.R., Srivastava, H.K. and Flynn, I.W. (1972). In: "Comparative Biochemistry of Parasites". (H. van den Bossche, ed.) pp. 329—342. Academic Press, New York and London.

Brack, C. (1968). "Elektronenmikroskopische Untersuchungen zum Lebenszyklus von *Trypanosoma cruzi*". Acta trop. 25, 289—356.

Brooker, B.E. (1971). "The fine structure of *Crithidia fasciculata* with special reference to the organelles involved in the ingestion and digestion of protein". Z. Zellforsch. mikrosk. Anat. 116, 532—563.

Brooker, B. and Preston, T.M. (1967). "The cytostome in trypanosomes and allied flagellates". J. Protozool. 14, Suppl., 41—42.

Brown, K.N. (1974). In: "Parasites in the Immunized Host: Mechanisms of Survival". CIBA Foundation Symp. 25, 35—51.

Brown, K.N., Armstrong, J.A. and Valentine, R.C. (1965). "The ingestion of protein molecules by blood forms of *Trypanosoma rhodesiense*". Expl Cell Res. 39, 129—135.

Camargo, M.E. and Amato Neto, V. (1974). "Anti-*Trypanosoma cruzi* IgM antibodies as serological evidence of recent infection". Revta Inst. Med. trop. S. Paulo 16, 200—202.

Clark, T.B. (1959). "Comparative morphology of four genera of Trypanosomatidae". J. Protozool. 6, 227—232.

Clegg, J.A. (1974). In: "Parasites in the Immunized Host: Mechanisms of Survival". CIBA Foundation Symp. 25, 161—183.

Cosgrove, W.B. and Kessel, R.G. (1958). "The activity of the contractile vacuole of *Crithidia fasciculata*". J. Protozool. 5, 296—298.

D'Alesandro, P.A. (1970). In: "Immunity to Parasitic Animals". (G.J. Jackson, R. Herman and I. Singer, eds.) Vol. 2, pp. 691—738. North-Holland, Amsterdam.

D'Alesandro, P.A. (1975). Ablastin: the phenomenon". Expl Parasit. 38, 303—308.

Deegan, T. and Maegraith, B.G. (1956a). "Studies on the nature of malarial pigment (haemozoin). I. — The pigment of the simian species, *Plasmodium knowlesi* and *P. cynomolgi*". Ann. trop. Med. Parasit. 50, 194—211.

Deegan, T. and Maegraith, B.G. (1956b). "Studies on the nature of malarial pigment (haemozoin). II. — The pigment of the human species, *Plasmodium falciparum* and *P. malariae*". Ann. trop. Med. Parasit. 50, 212—222.

Diem, K. and Lentner, C., eds. (1970). "Documenta Geigy Scientific Tables". Edn. 7. Geigy, Basle.

Dvorak, J.A. and Poore, C.M. (1974). "*Trypanosoma cruzi*: interaction with vertebrate cells in vitro. IV. Environmental temperature effects". Expl Parasit. 36, 150—157.

Dvorak, J.A. and Schmunis, G.A. (1972). "*Trypanosoma cruzi*: interaction with mouse peritoneal macrophages". Expl Parasit. 32, 289—300.

Dzbenski, T.H. (1974). "Exoantigens of *Trypanosoma cruzi* in vivo". Tropenmed. Parasit. 25, 485—491.

Galliard, H., Lapierre, J. and Rousset, J.-J. (1957). "Essai de classification des *Borrelia* par leur pouvoir protecteur contre *Trypanosoma brucei* chez la souris". Bull. Soc. Path. exot. 50, 663—666.

Galliard, H., Lapierre, J. and Roussett, J.-J. (1958). "Comportement spécifique de différentes espèces de *Borrelia* au cours de l'infection mixte avec *Trypanosoma brucei*. Son utilisation comme test d'identification des spirochètes récurrents". Ann. Parasit. hum. comp. 33, 177—208.

Galliard, H., Lapierre, J. and Rousset, J.-J. (1959a). "Complément à la note sur le test d'identification des spirochètes récurrents". Bull. Soc. Path. exot. 52, 269—271.

Galliard, H., Lapierre, J. and Roussett, J.-J. (1959b). "Atténuation de l'infection à *Trypanosoma cruzi* chez la souris blanche par différentes souches de Borrelia". Bull. Soc. Path. exot. 52, 272—276.

Galliard, H., Lapierre, J. and Coste, M. (1963). "Contribution à l'étude d'une souche pathogène de *Trypanosoma cruzi* (souche Tulahuen, Chili). III. Effets des spirochètes récurrents (*Borrelia duttoni* et *B. persica*) sur l'évolution de l'infection à *Trypanosoma cruzi*". Ann. Parasit. hum. comp. 38, 1—9.

384

Goodwin, L.G. (1970). "The pathology of African trypanosomiasis". Trans. R. Soc. trop. Med. Hyg. 64, 797—812.

Grassé, P.-P., ed. (1952). "Traité de Zoologie", 1 (1). Masson, Paris.

Gray, A.R. (1970). In: "The African Trypanosomiases". (H.W. Mulligan, ed.) pp. 113—116. Allen and Unwin, London.

Herbert, I.V. (1965). "Cytoplasmic inclusions and organelles of in vitro cultured *Trypanosoma theileri* and *Trypanosoma melophagium* and some speculations on their function". Expl Parasit. 17, 24—40.

Herbert, W.J., Mucklow, M.G. and Lennox, B. (1975). "The cause of death in acute murine trypanosomiasis". Trans. R. Soc. trop. Med. Hyg. 69, 4. [Laboratory demonstration]

Hewer, E.E. (1949). "Textbook of Histology for Medical Students". Heineman, London.

Hirsch, J.G., Jones, T.C. and Len, L. (1974). In: "Parasites in the Immunized Host: Mechanisms of Survival". CIBA Foundation Symp. 25, 205—220.

Hoare, C.A. (1929). "Studies on *Trypanosoma grayi*. 2. Experimental transmission to the crocodile". Trans. R. Soc. trop. Med. Hyg. 23, 39—56.

Hoare, C.A. (1964). "Morphological and taxonomic studies on mammalian trypanosomes. X. Revision of the systematics". J. Protozool. 11, 200—207.

Hoare, C.A. (1972). "The trypanosomes of mammals". Blackwell, Oxford and Edinburgh.

Holborow, E.J. (1973). "An ABC of Modern Immunology". Edn. 2. The Lancet, London.

Humphrey, J.H. and White, R.G. (1970). "Immunology for Students of Medicine". Edn. 3. Blackwell, Oxford.

Jadin, J.-M. (1974). "Cytologie et cytophysiologie des Trypanosomidae". Acta zool. pathol. Antwerp. 53, 3—168.

Kierszenbaum, F., Knecht, E., Budzko, D.B. and Pizzimenti, M.C. (1974). "Phagocytosis: a defense mechanism against infection with *Trypanosoma cruzi*". J. Immunol. 112, 1839—1844.

Kudo, R.R. (1966). "Protozoology". Edn. 5. Thomas, Springfield.

Lange, D.E. and Lysenko, M.G. (1960). "In vitro phagocytosis of *Trypanosoma lewisi* by rat exudative cells". Expl Parasit. 10, 39—42.

Langreth, S.G. and Balber, A.E. (1975). "Protein uptake and digestion in bloodstream and culture forms of *Trypanosoma brucei*". J. Protozool. 22, 40—53.

Lapierre, J., Larivière, M. and Rousset, J.J. (1958). "Protection de la souris contre une souche virulente de *Trypanosoma gambiense* par certaines espèces de *Borrelia*". Bull. Soc. Path. exot. 51, 173—176.

Lincicome, D.R. (1971). In: "Aspects of the Biology of Symbiosis". (T.C. Cheng, ed.) pp. 139—227. University Park Press, Baltimore and Butterworths, London.

Liston, A.J. (1975). "*Trypanosoma (Schizotrypanum) dionisii* in macrophage cultures: 1. Preliminary observations on entry into cells". Parasitology 71, xviii.

Lom, J. (1969). In: "Immunity to Parasitic Animals". (G.J. Jackson, R. Herman and I. Singer, eds) Vol. 1, pp. 249—265. North-Holland, Amsterdam.

Losos, G.J. and Ikede, B.O. (1972). "Review of pathology of diseases in domestic and laboratory animals caused by *Trypanosoma congolense, T. vivax, T. brucei, T. rhodesiense* and *T. gambiense*". Vet. Path. 9, Suppl., 1—71.

Lumsden, W.H.R., Cunningham, M.P., Webber, W.A.F., van Hoeve, K., Knight, R.H. and Simmons, V. (1965). "Some effects of hydrogen ion concentration on trypanosome numbers and infectivity". Expl Parasit. 16, 8—17.

Lwoff, M. (1951). In: "Biochemistry and Physiology of Protozoa". (A. Lwoff, ed.) Vol. 1, pp. 129—176. Academic Press, New York.

Mauel, J., Behin, R., Biroum-Noerjasin and Doyle, J.J. (1974). In: "Parasites in the Immunized Host: Mechanisms of Survival". CIBA Foundation Symp. 25, 225—238.

McCulloch, B. (1967). "Trypanosomes of the brucei subgroup as a probable cause of disease in wild zebra (*Equus burchelli*)". Ann. trop. Med. Parasit. 61, 261—264.

Milder, R. and Deane, M.P. (1969). "The cytostome of *Trypanosoma cruzi* and *T. conorhini*". J. Protozool. 16, 730—737.

Milder, R.V., Kloetzel, J. and Deane, M.P. (1973). "Observation on the interaction of peritoneal macrophages with *Trypanosoma cruzi*. I. — Initial phase of the relationship with blood stream and culture forms in vitro". Revta Inst. Med. trop. S. Paulo 15, 386—392.

Molyneux, D.H. (1970). "Developmental patterns in trypanosomes of the subgenus *Herpetosoma*". Ann. Soc. belge Méd. trop. 50, 229—238.

Molyneux, D. and Gordon, E. (1975). "Studies on immunity with three species of avian trypanosomes." Parasitology 70, 181—187.

Ormerod, W.E. (1970). In: "The African Trypanosomiases". (H.W. Mulligan, ed.) pp. 587—601. Allen and Unwin, London.

Patton, C.L. (1972). "*Trypanosoma lewisi:* influence of sera and peritoneal exudate cells". Expl Parasit. 31, 370—377.

Petana, W.B. (1964). "Effects of cortisone upon the course of infection of *Trypanosoma gambiense*, *T. rhodesiense*, *T. brucei* and *T. congolense* in albino rats". Ann. trop. Med. Parasit. 58, 192—198.

Preston, T.M. (1969). "The form and function of the cytostome — cytopharynx of the culture forms of the elasmobranch haemoflagellate *Trypanosoma raiae* Laveran and Mesnil". J. Protozool. 16, 320—333.

Putz, R.E. (1972). "Biological studies on the hemoflagellates *Cryptobia cataractae* and *Cryptobia salmositica*". Technical Papers of the Bureau of Sport Fisheries and Wildlife, no. 63, pp. 3—25. U.S. Dept of the Interior, Washington, D.C.

Ray, S.K. and Cross, G.A.M. (1972). "Branched electron transport chain in *Trypanosoma mega*". Nature New Biol. 237, 174—175.

Roitt, I.M. (1974). "Essential Immunology". Edn. 2. Blackwell, Oxford.

Rudzinska, M.A. and Vickerman, K. (1968). In: "Infectious blood diseases of man and animals". (D. Weinman and M. Ristic, eds.) pp. 217—306. Academic Press, New York, London.

Scorza, C. and Scorza, J.V. (1972). "The role of inflammatory macrophages in experimental acute chagasic myocarditis". J. reticuloendothelial Soc. 11, 604—616.

Shinondo, C.J. (1975). "The development in vitro of haematozoic trypomastigotes of *Trypanosoma avium*". Thesis accepted for M.Sc. degree, University of Cambridge, England.

Simpson, L. (1972). "The kinetoplast of the hemoflagellates". Int. Rev. Cytol. 32, 139—207.

Ssenyonga, G.S.Z. and Adam, K.M.G. (1975). "The number and morphology of trypanosomes in the blood and lymph of rats infected with *Trypanosoma brucei* and *T. congolense*". Parasitology 70, 255—261.

Steiger, R. (1973). "On the ultrastructure of *Trypanosoma (Trypanozoon) brucei* in the course of its life cycle and some related aspects". Acta trop. 30, 64—168.

Steinert, M. and Novikoff, A.B. (1960). "The existence of a cytostome and the occurrence of pinocytosis in the trypanosome, *Trypanosoma mega*". J. biophys. biochem. Cytol. 8, 563—569.

Takayanagi, T., Nakatake, Y. and Enriquez, G.L. (1974a). "*Trypanosoma gambiense:* phagocytosis in vitro". Expl Parasit. 36, 106—113.

Takayanagi, T., Nakatake, Y. and Enriquez, G.L. (1974b). "Attachment and ingestion of *Trypanosoma gambiense* to the rat macrophage by specific antiserum". J. Parasit. 60, 336—339.

Taylor, A.E.R. and Baker, J.R. (1968). "The cultivation of parasites in vitro". Blackwell, Oxford and Edinburgh.

Trautmann, R. (1907). "Etude expérimentale sur l'association du spirille de la tick-fever et de divers trypanosomes". Ann. Inst. Pasteur 21, 808—824.

Vickerman, K. (1969a). "The fine structure of *Trypanosoma congolense* in its bloodstream phase". J. Protozool. 16, 54—69.

Vickerman, K. (1969b). "On the surface coat and flagellar adhesion in trypanosomes". J. Cell Sci. 5, 163—194.

Vickerman, K. (1971). In: "Ecology and physiology of parasites". (A.M. Fallis, ed.) pp. 58—89. University of Toronto Press, Toronto.

Vickerman, K. (1973). "The mode of attachment of *Trypanosoma vivax* in the proboscis of the tsetse fly *Glossina fuscipes:* an ultrastructural study of the epimastigote stage of the trypanosome". J. Protozool. 20, 394—404.

Vickerman, K. (1974). In: "Parasites in the Immunized Host: Mechanisms of Survival". CIBA Foundation Symp. 25, 53—80.

Von Brand, T. (1973). "Biochemistry of Parasites". Edn. 2. Academic Press, New York and London.

Weinman, D. (1972). "*Trypanosoma cyclops* n.sp.: a pigmented trypanosome from the Malaysian primates *Macaca nemestrina* and *M. ira*". Trans. R. Soc. trop. Med. Hyg. 66, 628—636.

Wenyon, C.M. (1926). "Protozoology". Baillière, Tindall and Cox, London. (Reprinted 1965 by Baillière, Tindall and Cassel, London.)

Wéry, M. and de Groodt-Lasseel, M. (1966). "Ultrastructure de *Trypanosoma cruzi* en culture sur milieu semi-synthétique". Ann. Soc. belge Méd. trop. 46, 337—348.

Wilson, G.S. and Miles, A.A. (1964). "Topley and Wilson's Principles of Bacteriology and Immunity". Edn. 5. Vol. 1, pp. 1104—1133. Edward Arnold, London.

Ecological Aspects of Parasitology
Editor: C.R. Kennedy
© *North-Holland Publishing Company, Amsterdam, 1976*

CHAPTER 19

BLOOD CELLS

F.E.G. COX

Department of Zoology, King's College, Strand, London WC2R 2LS (Great Britain)

Contents

19.1. Introduction

The parasitic way of life is a dangerous one for although the parasite is surrounded by everything it needs it frequently stimulates an immune response which turns its environment into an extremely hostile one. This problem is partially overcome when a parasite becomes intracellular because hosts do not normally mount immune responses against their own cells. The blood cells of vertebrates provide discrete and generally safe habitats for parasites but the number that have actually taken advantage of the facilities offered are relatively few. The small size of the cells excludes metazoan parasites and all the intracellular protozoa, except one genus, belong to the sub-phylum Apicomplexa. This group contains the malaria parasites of man, and practically everything that is known about intracellular blood parasites has been derived from studies on a few forms of malaria in experimental animals. The need to discover as much as possible about the metabolism of malaria parasites in order to develop effective drugs has overshadowed the study of the ecology of blood parasites as a whole, and there is considerable disparity between the amount of information available on, for example, *Plasmodium berghei* in mice on one hand and economically unimportant blood parasites on the other. So great is this difference that although the total metabolism of *P. berghei* is now almost fully understood it is not even possible to classify some of the less important, but much more prevalent, parasites, even those in the blood of mammals.

It is intended that this chapter should not only summarise what is known about the ecological of parasites of blood cells but also identify the areas where little is known. Before this can be done it is necessary to look at the background of these parasites and the nature of the blood cells themselves.

19.2. Protozoa that inhabit blood cells

The genera of protozoa that live in blood cells are listed in Table 1. All but *Endotrypanum* are members of the Apicomplexa. *Endotrypanum schaudinni* and *E. monterogeii* are trypanosomes that occupy the red blood cells of sloths in South and Central America. Apart from the monograph by Shaw (1969) these trypanosomes have been little studied and practically nothing is known of the relationships that exist between them and their host cells. They will, therefore, not be considered further in this chapter.

All the remaining parasites in the blood can be divided into two major groups, the Eucoccidiida, comprising the three sub-orders Adeleina, Eimeriina and Haemosporina, of which only in the latter do all species occur in the blood, and the Piroplasmea. The life-cycle of all the eucoccidians is basically the same and is summarised in Fig. 1. The stages that occur in the blood, however, differ from group to group and the ecology of these stages is strongly influenced by their role in the life-cycles. In the Eimeriina, sporozoites and merozoites occur in blood cells. These forms, which are probably fully differentiated when they enter the blood cell, have no need to derive anything from it and merely use the cell as a means of transport to reach the peripheral circulation. In all other species, gametocytes occur in the blood. These develop into gametes and therefore have a number of metabolic requirements which have to be met from the host cell. The greatest demand on the host cell is made by those species which multiply within it. These include the Garniidae and Plasmodiidae and some of the Haemogregarinidae.

The remaining group of blood parasites belong to the class Piroplasmea. Although traditionally separated from the Coccidia, there is a growing tendency to classify them close to Haemosporina because of similarities at the ultrastructural level. The life-cycles, however, are quite different and

TABLE 1

The genera of protozoa that occur in blood cells. Some of these genera are not universally accepted and are indicated in brackets under the genus that is generally accepted. *Haemohormidum* was formerly called *Babesiosoma*. A, Amphibians; F, Fish; R, Reptiles; M, Mammals.

Class	Order or Suborder	Family	Genus	Host range
Mastigophora	Kinetoplastida Eucoccidiida	Trypanoso-matidae	*Endotrypanum*	sloths
Sporozoea	Adeleina	Haemogre-garinidae	*Hepatozoon*	R, B, M
(Subclass Coccidia)			*Haemogre-garina*	F, A, R
			Karyolysus	lizards
	Eimeriina	Lankesterel-lidae	*Lankesterella*	A, B
			Schellackia	A, lizards
			Lainsonia	lizards
		Eimeriidae	*Isospora*	B
	Haemosporina	Haemoproteidae	*Haemoproteus* (*Parahaemo-proteus*)	B, R, M B
			Haemocysti-dium	lizards
			Simondia	chelonians
			Hepatocystis	M
			Nycteria	bats
			Polychromo-philus	
		Leucocyto-zoidae	*Leucocytozoon* (*Akiba*) (*Saurocytozoon*)	B B lizards
		Plasmodiidae	*Plasmodium*	R, B, M
Piroplasmea	Piroplasmida	Dactylosomidae	*Haemohormi-dium*	F, A
			Dactylosoma	F, A, R
			Sauroplasma	R
			Anthemosoma	rodents
		Babesiidae	*Babesia* (*Entopolypoides*)	R, B, M monkeys
			Echinozoon	rock hyrax
		Theileriidae	*Theileria*	ungulates
			Cytauxzoon	ungulates
			Haematoxenus	ungulates

there is no sexual reproduction. The infective stages, which are not sporozoites, are injected into the vertebrate host. In the Theileriidae, most Dactylosomidae and a few of the Babesiidae, schizogony occurs in fixed cells and

390

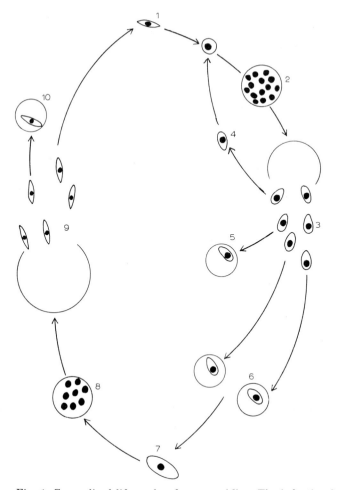

Fig. 1. Generalised life-cycle of a eucoccidian. The infection begins when a sporozoite (1) enters a host cell. Schizogony occurs within the cell (2) and merozoites are produced (3). This schizogonic cycle may be repeated (4) and in the Plasmodiidae schizogony occurs in the blood. In some Eimeriina, merozoites may enter blood cells (5). After one or more schizogonic cycles some merozoites enter cells in which they differentiate into gametocytes (6). The gametocytes of all the blood parasitic Adeleina and Haemosporina occur in the blood. The gametocytes become gametes which fuse to produce a zygote (7). A further cycle of asexual division occurs (8) resulting in the production of sporozoites (9) which completes the cycle. The sporozoites of most blood parasitic Eimeriina occur in blood cells (10). The stages infective to the invertebrate vectors are the gametocytes in the Adeleina and Haemosporina and the Sporozoites in the Eimeriina.

the products, which are equivalent to merozoites and will be regarded as such here, enter blood cells and multiply by binary fission producing two (*Babesia, Theileria, Sauroplasma*), four (*Haemohormidium, Dactylosoma*) or sixteen (*Anthemosoma*) merozoites. These asexual stages are taken up by

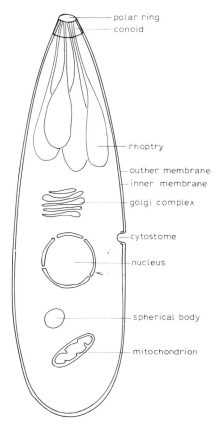

polar ring
conoid

rhoptry

outher membrane
inner membrane

golgi complex

cytostome

nucleus

spherical body

mitochondrion

Fig. 2. Diagram to show the main features of a generalised motile stage of a member of the sub-phylum Apicomplexa.

blood sucking invertebrates in which a series of asexual schizogonic phases occur culminating in the production of stages infective to the vertebrate in the salivary glands. The relationships between the Piroplasmea or piroplasms and the Coccidia are difficult to assess at the present time because of the relatively recent discoveries of parasites such as *Anthemosoma*, *Garnia* and *Fallisia* which do not fit easily into the recognised categories. Landau (1974) has produced an evolutionary tree of blood parasites quite different from that of Baker (1965) and at present it is probably best to consider the various groups of parasites separately rather than trying to seek evolutionary similarities which may not really exist.

The basic structure of the motile form that actually enters the blood cell is similar in all the Apicomplexa (Aikawa and Sterling, 1974; Corliss, 1974). It is an elongate structure surrounded by one or two membranes (Fig. 2). The anterior end is cone shaped and contains a complex of micronemes and electron dense rhoptries· which may be important in entry into the blood cell. On the side of the organism there is a pore, the cytostome or micropore, which plays a role in nutrition. The similarities that exist at the ultrastructural level provide a base to which the changes that occur during the parasite's metabolism may be referred. An understanding of the ecology of blood parasites, therefore, begins with changes at the ultrastructural and associated biochemical levels.

19.3. The nature and dynamics of blood cells

The blood cells of vertebrates are broadly divided into groups, the red cells or erythrocytes and the white cells or leucocytes. In mammals there are three series of white cells, lymphocytes, monocytes and polymorphonuclear granulocytes (Hardisty and Weatherall, 1974). In reptiles and birds there is a fourth series, the thrombocytes, which are equivalent to the platelets in mammals (Lucas and Jamroz, 1961; Saint-Girons, 1970). Erythrocytes are the cells which most commonly harbour parasites, followed by monocytes and, least frequently, lymphocytes. Some of the Garniidae occur in thrombocytes and one species at least occurs in neutrophils belonging to the polymorphonuclear granulocyte series (Lainson et al., 1975). There is also a report that malaria parasites occur in platelets (Fajardo and Tallent, 1974).

All blood cells originate in the bone marrow. Lymphocytes begin as stem cells, differentiate into lymphoblasts and mature into lymphocytes which exist in a number of distinct populations in the peripheral blood and in lymphoid tissues (Elves, 1972). There is considerable flux between the fixed and wandering parts of the lymphocyte pool. In the monocyte series, promonocytes become monocytes which mature into macrophages characteristically found in fixed tissues, such as the liver and lungs, and in the peritoneal cavity. It is generally held that the monocytes in the blood represent a transient stage in the migration between the bone marrow and a particular tissue (Whitelaw and Batho, 1975). There is some evidence to suggest that macrophages from the tissues may return to the circulation. It is extremely difficult, using the standard techniques of light microscopy, to distinguish between small lymphocytes and monocytes, and this is particularly true in birds and reptiles.

Erythrocytes also originate in the bone marrow. In all vertebrates the earliest identifiable stage is the erythroblast which is a nucleated cell capable of carrying out almost any metabolic function. In mammals, the late erythroblast extrudes its nucleus and becomes a reticulocyte. The reticulocyte retains a number of metabolic functions and its mitochondria, but as it differentiates into a mature erythrocyte the mitochondria are lost, the metabolic functions diminish and haemoglobin is synthesised until it forms the bulk of the cell (Harris and Kellermeyer, 1970). In vertebrates other than mammals the nuclei and much of the metabolic potential are retained.

The cells in the blood, therefore, represent a heterogeneous collection of leucocytes that are passing through the circulation and erythrocytes that will spend all their lives there. It is difficult to estimate how long lymphocytes remain in the circulation because so many distinct populations with different life spans occur there. Most are fairly short lived and produced in bursts but others are produced daily and live for months (Elves, 1972; Greaves et al., 1973). Monocytes probably only circulate for a short time, between 36 and 104 hours in mammals (Whitelaw and Batho, 1975). Erythrocytes are the permanent inhabitants of the blood and their longevity varies from species to species. In man the life span is 100—120 days, in mice about 40 days, in chickens 28 days but in poikilotherms it is much longer, 600—800 days in turtles and over 1,000 days in toads (Berlin, 1964; Berlin and Berk, 1975). In a healthy animal the numbers and the proportions of the various kinds of cells in the blood remain relatively constant. This is because loss is balanced by the recruitment of new cells from the bone marrow. No parasite is able to invade a blood cell until it has left the bone marrow and, in the case of erythrocytes, the parasite perishes when the cell in which it lives is destroyed.

The life span of red cells in many species has been carefully estimated and the dynamics of red cell survival have been discussed by Berlin (1964), Bergner (1965), Wagner (1972) and Berlin and Berk (1975). In a stable system the number of cells recruited (birth) equals the number lost (death) and there is no complication due to any other kind of gain (immigration) or loss (emigration). The recruitment is of reticulocytes and in most mammals 1—2% of the total red cells are newly arrived reticulocytes which mature to erythrocytes within 48 hours thus keeping the proportion of identifiable young cells to less than 5% of the total population. Any increased destruction of erythrocytes decreases their mean survival time (Wagner, 1972). If this destruction is minimal the bone marrow increases its output of reticulocytes while the total number of cells remains constant. This means that the number of red cells in the blood remains constant while the proportion of reticulocytes increases. The haemopoietic system may stabilise at this level or, if the amount of destruction increases, the imput of new cells may fail to keep up with the loss, resulting in anaemia. When this happens an increasing proportion of the circulating red cells will be reticulocytes.

Red cell destruction can be induced in a number of ways using chemical substances such as phenylhydrazine but parasites that feed on or live in these cells can also be instrumental in their destruction. Hookworms, for example, cause considerable blood loss and anaemia. The parasites that live within red cells often have distinct preferences for reticulocytes or mature cells and thus the availability of young and mature cells has a marked effect on the progress of any infection while the parasites themselves affect the availability of suitable cells.

Monocytes and lymphocytes are transient cells in the circulating blood and fluctuations in their numbers are less important to intracellular parasites than fluctuations in red cells. Protozoan parasites are not known to affect the dynamics of the populations of the white cells in which they live.

19.4. Physico-chemical conditions

The physico-chemical conditions that exist within any cell are related to the normal maintenance of that cell and to its specific functions. Blood cells can supply parasites with the amino acids they need for protein synthesis, simple sugars for respiration and a large number of macromolecule precursors. There are also supplies of substances from the plasma that may or may not be used by the host cells themselves (Bessis, 1973; Hardisty and Weatherall, 1974).

Monocytes are the most actively metabolic cells that harbour parasites (Axline, 1970). They are phagocytic, and use large amounts of glucose for respiratory oxidative metabolism. They also have considerable hydrolase activity which is associated with their role in taking up and destroying organic debris and foreign organisms (Nichols and Bainton, 1975). It is not known how parasites survive the action of lysosomes inside monocytes although this problem has been investigated using organisms such as yeasts (Hart and Young, 1975). Lymphocytes also have considerable capacities for oxidative metabolism and glucose utilization but they are not phagocytic and therefore do not possess enzymes for the intracellular digestion of other organisms (Elves, 1972). Antibody producing lymphocytes (B lymphocytes) are extremely active cells geared to producing the protein for antibodies. T lymphocytes, which are involved in cell mediated immunity, have less cytoplasm and are less active metabolically than B cells (Greaves et al., 1973).

Erythrocytes are the least active of the blood cells (comprehensive reviews in Harris and Kellermeyer, 1970; Bessis, 1973; Surgenor, 1974, 1975). The normoblasts in the bone marrow are very active synthesising DNA and RNA and manufacturing the precursors of haemoglobin, 80% of which is formed during this stage. Haemoglobin eventually constitutes about a third of the mature erythrocyte. Some of the functions of the normoblast persist during the reticulocyte stage, in which mitochondria are still present, but most are lost in the mature mammalian erythrocyte although retained in fishes, amphibians, birds, and reptiles. Glucose metabolism involves the Embden-Meyerhof and pentose phosphate pathways. The main functions of the erythrocyte are the carrying of oxygen from the lungs or gills to the tissues and the carrying of carbon dioxide in the reverse direction.

19.5. Entry into blood cells

The sporozoan parasites in the blood resemble one another at the ultra-structural level particularly with respect to the apical complex (Fig. 2). It is thought that this complex is important in effecting entry into the host cell. The complex consists of polar rings, rhoptries and micronemes which disappear when the parasite has entered its cell. At one time it was thought that the parasites bored through the cell membrane but it is now clear that, in the Haemosporina at least, the parasite is engulfed by the cell membrane and in effect phagocytosed (Ladda et al., 1969). Bannister et al. (1975), using the scanning electron microscope, have found that the merozoite of *Plasmodium* knowlesi first attaches to an appropriate red cell, then comes to lie in a shallow depression and is finally drawn into the cell. This is probably what happens in all species of malaria parasites and in their close relatives. It is likely that substances, as yet unidentified and possibly associated with the rhoptries, are produced at the apical complex which facilitate passive entry into a red cell by altering the properties of its membrane. In the case of *Babesia* spp. the situation is apparently quite different and it is likely that the parasite bores directly into the cell. No trace of a host cell membrane, representing the remains of a phagotrophic vacuole, surrounding the parasite has been identified (Aikawa and Sterling, 1974). The parasites that occur in monocytes are probably taken up actively as part of the normal phagocytic activities of these cells. Parasites are very selective in the choice of the cells they invade. Butcher et al., (1973) suggest that merozoites of malaria parasites recognise receptors in the membrane of the red cell and attach to these by means of the apical complex. This binding is very specific and would account for the fact that parasites are able to select, bind to and enter cells appropriate for their further development. Miller et al. (1973a) reject the possibility that these receptors may be sialoglycoproteins and suggest a protein or lipoprotein complex. As well as surface receptors there is a difference in charge between the parasite and its host cell and this probably assists entry by allowing the two to come together and remain in close contact (Miller et al., 1973b).

19.6. Growth and metabolism

Sporozoan parasites may enter blood cells and remain there as fully differentiated sporozoites or merozoites awaiting passive uptake by a blood sucking invertebrate, as in the Eimeriina. The same is probably true of the Hepatozoidae. Such parasites require little from their host cells and practically nothing is known about their metabolism. In the Haemosporina, young merozoites enter blood cells and grow and differentiate into gametocytes. This is an active process in which the parasites build up materials and energy for the subsequent development of the gametes. At one time it was thought that the gametocytes circulated for long periods, thus requiring a continued source of energy, but it is now known that in the mammalian malaria parasites this period is short, less than twelve hours (Hawking, 1975; Hawking et al., 1968). The gametocytes of avian malaria parasites circulate for about eighteen hours (Hawking et al., 1972) so it is unlikely that any of the gametocytes of the Haemosporina circulate for any length of time. The most active phase in the metabolism of these gametocytes must be during their early development and feeding, and is similar to that in the trophozoites of the malaria parasites (Aikawa and Sterling, 1974). A major difference between the metabolism of gametocytes and trophozoites of the parasites of mammals is that gametocytes possess cristate mitochondria and Krebs cycle enzymes while trophozoites do not (Howells and Maxwell, 1973a). At the present time there is insufficient information available to allow one to make any general conclusions about the nutritional requirements and metabolism of gametocytes.

Parasites that multiply within their host cells need materials for protein synthesis, nucleic acid synthesis and respiration. Members of the Garniidae undergo schizogony in circulating monocytes, lymphocytes, thromocytes and erythrocytes (Lainson et al., 1971, 1974) but nothing is known of their metabolism. Practically everything that is known about the metabolism of blood parasites comes from experiments on the malaria parasites and this subject has been reviewed a number of times (McKee, 1951; Moulder, 1962; Garnham, 1966; Peters, 1969, 1970, 1974; Fletcher and Maegraith, 1972; Howells et al., 1972; Oelshlegel and Brewer, 1975). These reviews are best regarded as interim reports on a subject which is still only partly understood.

After it has entered the host cell the malaria parasite lies in a parasitophorus vacuole surrounded by host cell membrane. Such vacuoles are characteristic of many intracellular parasites (Aikawa and Sterling, 1974). The distinctive features of the merozoite, such as the polar ring, rhoptries, micronemes and the inner of the three membranes, disappear and it becomes a trophozoite or feeding stage. The cytostome becomes functional, a food vacuole forms and haemoglobin is taken into the parasite. At this stage the food vacuole is limited by a single membrane so one membrane, presumably that originating from the host cell, is lost. Further small vacuoles form at the periphery of the original one and these become pinched off and pass into the cytoplasm of the parasite. Lysosomes become associated with the food vacuoles and digestion occurs in the presence of acid phosphatases; the food vacuoles are therefore phagolysosomes. In some species, pinocytosis vesicles also occur around the periphery of the parasite and haemoglobin digestion occurs within these simultaneously with digestion in the food vacuoles. The patterns of feeding and digestion in malaria parasites are shown diagrammatically in Fig. 3 but there is considerable variation in detail from species to species (Aikawa, 1971; Aikawa and Sterling, 1974).

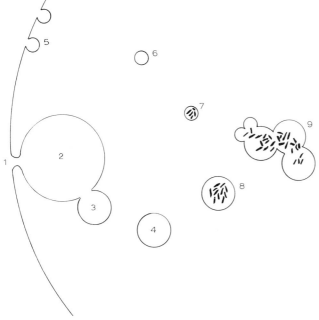

Fig. 3. Diagrammatic representation of feeding mechanisms in the Plasmodiidae. Cell contents are taken in through the cytosome (1) and into a major food vacuole (2). Other subsidiary vacuoles form (3) and break free (4). Pinocytosis vacuoles also form (5) and circulate (6). Digestion occurs within all of these vacuoles and eventually pigment granules are formed (7), (8). The vacuoles containing pigment coalesce (9) and are incorporated into a single residual body when the host cell ruptures.

The parasite digests the protein portion of the haemoglobin at low pH using an acid phosphatase that is specific to the haemoglobin of the host cell (Levy and Chou, 1973; Levy et al., 1974; Chan and Lee, 1974). Haemoglobin that is not utilised becomes the characteristic malaria pigment sometimes called haemozoin. Pigment has proved difficult to analyse as even individual granules differ from one another (Morselt et al., 1973) but in very pure samples it is a homogeneous crystalline substance consisting of haemin bound in a regular manner to a protein matrix (Moore and Boothroyd, 1974). At one time it was thought that pigment was merely an unused by-product of haemoglobin breakdown but it now seems that it is actively synthesized by the parasite (Homewood et al., 1972; Moore and Boothroyd, 1974).

When digestion is completed the food vacuoles coalesce to form a residual body containing a single mass of pigment (Aikawa and Sterling, 1974). Unused amino acids pass into the free amino acid pool of the plasma where they are available for use by other generations of the parasite.

The free amino acid pools of the plasma and serum are important sources of protein precursors for the parasite. Normally amino acids pass into the red cell under the control of the cell itself but infected cells become "leaky" to amino acids, most of which diffuse into the cell in large amounts and also into the parasite itself without any expenditure of energy on its part (Sherman and Tanigoshi, 1972, 1974a). The parasite causes this change, which is to its advantage, by depleting the host cell ATP and altering the ion and amino acid transport systems (Sherman and Tanigoshi, 1974a).

All malaria parasites depend on plasma glucose as a source of energy and glucose consumption by infected cells is some ten times greater than by uninfected cells. This is because malaria parasites are able to render their host cells permeable to glucose, thus ensuring a plentiful supply (Homewood and Neame, 1974; Sherman and Tanigoshi, 1974b). Trophozoites of the mammalian malaria parasites do not possess cristate mitochondria (avian parasites do and their glucose metabolism is quite different) and all the available evidence suggests that they have no Krebs cycle enzymes (Howells and Maxwell, 1973a). Most of the glucose is metabolised directly to lactate and the parasites contain the enzymes necessary for this (Howells and Maxwell, 1973a, 1973b). In parasites free from their cells 20% of the glucose may pass to the pentose phosphate pathway (Shakespeare and Trigg, 1973). The inability to carry out oxidative phosphorylation probably leaves the parasite short of ATP, which is depleted while it is dividing. Some ATP may be redirected from the host cell. Malaria parasites produce pyruvate kinase which increases the amounts of ATP in the cell at the expense of DPG (2,3-diphosphoglycerate). ATP is required by the parasite but DPG is not, being used for red cell oxygen transport (Oelschlegel et al., 1975). The parasite, therefore, manipulates the metabolism of the host cell to its advantage. It also depends on the cell for the provision of glucose-6-phosphate dehydrogenase (G-6-PD) and probably other enzymes (Fletcher and Maegraith, 1972).

The trophozoite of the malaria parasite spends about half its life in the cell metabolising various host cell materials and synthesising the nucleic acids it needs for subsequent division (Vickerman and Cox, 1967; Siddiqui and Schnell, 1972). Division consists of a series of rapid binary fissions which typically produce merozoites. The whole cycle, in the most studied species, takes 24 hours. There is considerable variation from sprecies to species although the time taken for the whole cycle is usually a multiple of 24 hours and the number of merozoites produced a multiple of eight. Malaria parasites make their own RNA and synthesise the bulk of their DNA prior to division although there is some evidence that they also make DNA after division (Conklin et al., 1973). Some malaria parasites are able to make their own pyrimidines (cytosine and thymine) de novo but have to rely on the host for purines (adenine and guanine) (Neame et al., 1974). They are, however, able to take advantage of the metabolism of the red cell because they can use purine metabolites such as hypoxanthine which would normally be excreted by mammalian cells (Madhu et al., 1975).

This brief review of the metabolism of malaria parasites is based largely on studies on experimental malarias in rodents and monkeys. Even in these extensively studied species there is no clear overall pattern of metabolism, but what is clear is that the malaria parasites do not simply sit in red cells surrounded by everything they need which they take in with little expenditure of energy. These parasites have to work for their living and have to manipulate the host cell to their own needs. If one regards the red cell as a host within a host, there can be few parasites that have lives so totally integrated with those of their hosts as do the malaria parasites.

So little is known about the metabolism of blood parasites other than malaria parasites that no coherent pattern emerges. Food vacuoles and pigment are formed by members of the Haemoproteidae so it is likely that the gametocytes of these parasites digest haemoglobin in the same way as the malaria parasites. The metabolism of the avian members of the Haemoproteidae probably resembles that of avian malarias but nothing is known about the mammalian forms. The Leucocytozoidae and Piroplasmea possess cytostomes and form what appear to be food vacuoles but pigment has never been found in the former and only in isolated cases in the latter. Whether this indicates a more efficient utilization of haemoglobin or a completely different kind of metabolism is not at all clear. A tubule connecting the parasite with the host plasma has been seen in *Babesia equi* (Frerichs and Holbrook, 1974) but its significance is not understood and it is quite likely that piroplasms take in the food material they require all over their surfaces. Carbohydrate metabolism in the piroplasms is similar to that in the malaria parasites (Rickard, 1969). As for the other blood parasites, virtually nothing is known about the ways in which they feed and digest their food.

It is, of course, important to remember that the stages of blood parasites that actually occur in the blood represent only part, and often only a small part, of the whole life-cycle. These parasites are continually preparing for phases of division and for transfer from one host to another. Overall patterns of change in the ultrastructure (Aikawa and Sterling, 1974) and metabolism (Howells and Maxwell, 1973a) of intracellular protozoa have been described and these changes provide much more insight into the ways of life of parasites that live in blood than detailed studies of isolated stages ever can.

19.7. Host responses

It is assumed, with no really good reasons, that parasites within blood cells are isolated from the immune response of the host. The only recorded effects of antibodies on blood parasites are that malaria parasites are agglutinated in the presence of immune serum (Brown et al., 1968). There is no proven role for cell mediated responses in any blood parasitic infection nor is there any indication of the mechanisms of the immune response in any parasites other than malaria parasites and piroplasms. The account of immunity to blood parasites that best accords with all the known facts in this. Parasites are not affected by the immune response while they are in their host cells and they are subject to antibody attack only while they are free in the serum on their way to another cell. This means that it is the merozoites that are the susceptible stages and these represent the weak link in the life-cycle. The merozoite is therefore exposed to a hostile environment and must gain access to a new cell as quickly as possible. This is where the rapid location of the host cell by means of the apical complex and subsequent entry into the cell are important.

There are suggestions that infected cells are destroyed by the immune response and that this destruction may also extend to uninfected cells. The experiments to prove this and to eliminate non-immunological lytic factors (Fletcher and Maegraith, 1972) have not yet been conclusive. At present it seems reasonable to conclude that it is the merozoite that is at risk while it is on its way from cell to cell and this corresponds to a normal ecological situation in which free living animals are in particular danger when they are moving from one safe part of their habitat to another. It is characteristic of blood parasites that infections persist for a considerable time. This indicates that these parasites are able to outwit the host's immune response and there is considerable evidence that, in *Plasmodium knowlesi* in monkeys, this is done by the parasites undergoing antigenic variation (Brown and Hills, 1974). The evidence for antigenic variation in other malaria parasites is less convincing and in other blood parasites virtually non existent. Nevertheless, antigenic variation is the most plausible explanation of long term chronic infections with blood parasites.

19.8. Population dynamics and control

The blood system of a vertebrate is, from an ecological point of view, a balanced system in which the populations of blood cells remain relatively constant, the loss from deaths being balanced by the gain from births. Within the red cell population, there are young, mature and senescent cells in fixed proportions and thus the whole system is extremely simple to study. Parasites that live and multiply in red cells also have fairly simple population dynamics. They enter a cell, multiply a given number of times and then leave the cell to reinvade new ones. Malaria parasites markedly affect their host cells which rapidly become distorted and have a reduced survival time (Kreier et al., 1972). There is some evidence that parasites are removed from red cells (pitted) in the spleen (Schnitzer et al., 1972) and uninfected cells show signs of this pitting (Balcerzak et al., 1972). It is not known how decreased survival time of infected cells and the removal of parasites from cells affects the dynamics of the infection.

The non-fatal parasites of mice are particularly convenient models of blood infections (Cox, 1975). Malaria parasites in mice multiply eight-fold every 24 hours. Starting from a single parasite, the infection is capable of increasing unchecked until almost half the mouse's red blood cells are invaded, representing a total parasite load of about 8.5×10^9. It is only at this point that the immune response comes into play and the infection is brought under control. An earlier check on the rate of increase in parasite numbers occurs if the parasite has a particular predilection for either reticulocytes or mature cells. Most strains of *Plasmodium berghei* prefer reticulocytes and normally as the infection progresses the parasite destroys more and more cells bringing about the release of new reticulocytes into the circulation. The parasite is therefore instrumental in procuring a supply of its favourite cells but does not exert any fine control on erythropoesis by retarding the production of erythropoietin, the hormone responsible for stimulating erythrocyte production (Rencricca et al., 1974). If the supply of new reticulocytes is diminished the parasite fails to realise its full potential (Cox, 1974). In the case of *Plasmodium vinckei* and *P. chabaudi* the preferred cells are mature and increased reticulocytosis tends to check the infection.

The rodent piroplasm, *Babesia rodhaini*, divides into two every eight hours thus achieving the same multiplication rate as the malaria parasites. This parasite prefers mature cells and induced reticulocytosis is inimical to its development and prevents it from reaching its frequently fatal conclusion (McHardy, 1972). The intrinsic rate of multiplication of the parasite itself also has a marked effect on the infection produced. *Babesia microti* multiplies at a rate of 2.8 every 24 hours. This relatively slow rate of increase means that it takes 22 days for a single parasite to produce an infection in which half the host's blood cells are infected and the immune response has ample time to bring the infection under control.

Populations of parasites in red cells are controlled by the basic reproductive rate of the parasite, the immune response and the availability of suitable cells. All of these can be measured but the problem of antigenic variation contributes an unknown parameter. Theoretically, an infection should either progress to a fatal conclusion or should be eliminated by the immune response. In fact, chronic latent infections are the rule and these are often characterised by irregular recrudescences. Brown and Hills (1974) have presented evidence to suggest that *Plasmodium knowlesi* in monkeys is able to manipulate the immune response and they postulate two antibodies, a "killing" one and a "switch" one. The killing one has been evolved by the host to combat the parasite while the switch one, although produced by the host, is used by the parasite to trigger the production of a new antigenic variant that can survive in the immune host. This is another example of a way in which a parasite can use the host for its own ends.

Populations of parasites in the blood are affected, although not regulated, by coincident or prior infections with other organisms. These other organisms exert a variety of effects. They may alter the balance of reticulocytes to mature cells and thus indirectly influence the infection, they may possess antigens in common with the parasites and thus bring about an enhanced immune response or they may depress the immune response causing exacerbation of the parasite infection (Cox, 1975).

19.9. Transmission

The main problem which has to be solved by a blood parasite is to ensure transmission to a new host. In all cases transmission is effected by a blood sucking vector, usually an arthropod in the case of terrestrial forms and a leech in aquatic ones. In the Eimerina, *Lankesterella*, *Schellackia* and *Lainsonia*, sporozoites enter blood cells and these are simply taken up by the vector, in which they undergo no change, and introduced into a new host when the vector is eaten. In the piroplasms, asexual merozoites are taken up by the vector, usually a tick in the case of terrestrial hosts, in which development to the infective stages takes place. Transmission occurs when the vector bites a new host and the infective stages are injected from the salivary glands. In all other sporozoa in the blood the stages infective to the vector are the gametocytes. Hawking et al. (1968, 1972) have shown that the gametocytes of malaria parasites are infective to mosquitoes for only a few hours. The gametocytes mature periodically according to the circadian rhythm of the host so that they are most abundant when the mosquito is biting. The periodicity of the parasites is inherently 24 hours and the actual timing of gametocyte production is triggered by the host's daily rhythms. *Plasmodium gallinaceum* in fowl has a 36 hour cycle and the infectivity of its gametocytes extends over a period of 18 hours.

In the Plasmodiidae, the production of gametocytes is associated with schizogony in the blood but in the other Haemosporina this phase does not occur in the life-cycle and gametocyte are produced solely from exo-erythrocytic schizonts. Nevertheless, periodicity has been demonstrated in *Leucocytozoon* (Roller and Desser, 1973) so it is quite possible that this is a general characteristic of all Haemosporina.

As well as a daily periodicity, some blood parasites also show marked seasonal periodicities. These have been discussed by Landau (1973). In the malaria parasites the infection is essentially a permanent one with gametocytes being produced as long as schizogony occurs. Seasonal transmission is ensured not by the seasonal production of gametocytes but by the biting of the mosquitoes that transmit the infection. Other seasonal transmissions are brought about by periodic bursts of gametocyte production as in *Leuco-*

cytozoon, which is transmitted by Simuliidae, *Akiba* by *Culicoides*, *Parahaemoproteus* by *Culicoides* and probably species of *Hepatocystis* and *Nycteria* in which the life-cycle is not completely known. In *Haemoproteus* and *Polychromophilis* there is a permanent association between the host and the vectors, hippoboscid flies in the case of the former and nycteribid flies in the latter. Presumably new hosts become infected while fledglings in their nests.

The transmission of piroplasms of mammals is complicated by the fact that the vectors are one-host, two-host or three-host ticks. Seasonal transmission is brought about by the activities of the ticks while the parasites continuously circulate in the blood in small numbers. In the Babesiidae, transovarian transmission is the general rule, while this does not occur in the Theileriidae. Transmission of the piroplasms is therefore comparable with the malaria parasites in that it is largely determined by the activities of the vector rather than by the parasites themselves.

9.10. Envoi

The parasites that live within blood cells survive there not because they have any particular adaptations to life in this habitat but because of their overall versatility. They are all opportunists. They have complex life-cycles and have to live in a variety of types of cell in both vertebrate and invertebrate hosts and in extreme habitats where they have to survive digestive enzymes in the gut or antibodies in the serum. Not only do they have to survive in each kind of habitat, but also they have to manipulate each to their own ends and thereby enter relationships more complex than in most other parasite situations. Blood parasites are, nevertheless, among the most successful of parasites but few people have marvelled at their success; most have simply sought ways to destroy them because of the dangers some of them present to man and his domesticated animals. What we know about the ecology of blood parasites is largely derived from studies on malaria, yet less economically important species might well provide useful information both on the basic ecology of blood parasites and on the reasons for the success of those species that disturb man. One need not look too far for suitable parasites to study. Lainson and his colleagues (1971, 1974, 1975) and Telford (1975) have uncovered a wealth of parasitological material in the lizards of South America and the Caribbean and if these should receive a fraction of the attention lavished on the malaria parasites of rodents it should be possible to carry on from the position as it was in 1903 when Minchin wrote "... it is amongst the Haemosporidia of cold-blooded Vertebrata that researches are most needed".

19.11. REFERENCES

Aikawa, M. (1971). *Plasmodium:* "The fine structure of malarial parasites". Expl. Parasit. 30, 284–320.

Aikawa, M. and Sterling, C.R. (1974). "Intracellular Parasitic Protozoa". Academic Press, London.

Axline, S.G. (1970). "Functional biochemistry of the macrophage". Seminars in Hematology, 7, 142–160.

Baker, J.R. (1965). In: "Evolution of Parasites". (A.E.R. Taylor, ed.) pp. 1–27. Blackwell, Oxford.

Balcerzak, S.P., Arnold, J.D. and Martin, D.C. (1972). "Anatomy of red cell damage by *Plasmodium falciparum* in man". Blood, 40, 98–104.

Bannister, L.H., Butcher, G., Dennis, E.D. and Mitchell, G.H. (1975). "Invasion of erythrocytes by *Plasmodium knowlesi* merozoites: an electron microscope study". J. Protozool., 22, 48A (abstract).

Bergner, P.E. (1965). "On stationary and non-stationary red cell survival curves". J. theor. Biol., 9, 366–388.

Berlin, N.I. (1964). In "The Red Blood Cell". (C. Bishop and D.M. Surgenor, eds.) pp. 423–450. Academic Press, New York and London.

Berlin, N.I. and Berk, P.D. (1975). In "The Red Blood Cell" 2nd Edn., Vol. 2 (D.M. Surgenor ed.) pp. 958–1020. Academic Press, New York.

Bessis, M. (1973). "Living Blood Cells and their Ultrastructure" (Translated from "Cellules du Sang, Normale et Pathologique". Masson, Paris by R.I. Weed). Springer Verlag, Berlin.

Brown, I.N., Brown, K.N. and Hills, L.A. (1968). "Immunity to malaria: the antibody response to antigenic variation by *Plasmodium knowlesi.*" Immunology. 14, 127–138.

Brown, K.N. and Hills, L.A. (1974). "Antigenic variation and immunity to *Plasmodium knowlesi:* antibodies which induce antigenic variation and antibodies which destroy parasites." Trans. R. Soc. trop. Med. Hyg. 68, 139–142.

Butcher, G.H., Mitchell, G.H. and Cohen, S. (1973). "Mechanism of host specificity in malarial infections". Nature, Lond. 244, 40–42.

Chan, V.L. and Lee, P.Y. (1974). "Host-cell specific proteolytic enzymes in *Plasmodium berghei* infected erythrocytes." S.E. Asian J. trop. Med. Publ. Hlth. 5, 447–449.

Conklin, K.A., Chou, S.C., Siddiqui, W.A. and Schnell, J.V. (1973). "DNA and RNA syntheses by intraerythrocytic stages of *Plasmodium knowlesi.*" J. Protozool. 20, 683–688.

Corliss, J.O. (1974). "Classification and phylogeny of the Protista" In: "Actualités Protozoologiques: Résumé des discussions des tables rondes du 4e Congress International de Protozoologie." (P. de Puytorac and J. Grain, eds.) pp. 251–264. Université de Clermont.

Cox, F.E.G. (1974). "A comparative account of the effects of betamethasome on mice infected with *Plasmodium vinckei chabaudi* and *Plasmodium berghei yoelii.*" Parasitology, 68, 19–26.

Cox, F.E.G. (1975). "Factors affecting infections of mammals with intraerythrocytic protozoa." Symp. Soc. exp. Biol. 29, 429–451.

Elves, M.W. (1972). "The lymphocytes". 2nd Edn. Lloyd—Luke Medical Books, London.

Fajardo, L.F. and Tallent, C. (1974). "Malarial parasites within human platelets". J. Am. Med. Ass. 229, 1205–1207.

Fletcher, A. and Maegraith, B. (1972). "The metabolism of the malaria parasite and its host." Adv. Parasit. 10, 31–48.

Frerichs, W.M. and Holbrook, A.A. (1974). "Feeding mechanisms of *Babesia equi.*" J. Protozool. 21, 707–709.

Garnham, P.C.C. (1966). "Malaria Parasites and other Haemosporidia." Blackwell, Oxford.

Greaves, M.F., Owen, J.J.T. and Raff, M.C. (1973). "T and B Lymphocytes." Excerpta Medica, Amsterdam.

Hardisty, R.M. and Weatherall, D.J. eds (1974). "Blood and its disorders." J.P. Lippincott, Philadelphia.

Harris, J.W. and Kellermeyer, R.W. (1970). "The Red Cell." Harvard University Press, Cambridge, Mass.

Hart, P. D'A. and Young, M.R. (1975). "Interference with normal phagosome-lysosome function in macrophages, using ingested yeast cells and suramin." Nature, Lond. 256, 47—49.

Hawking, F. (1975). "Circadian and other rhythms of parasites." Adv. Parasit. 13, 123—182.

Hawking, F., Gammage, K. and Worms, M.J. (1972). "The asexual and sexual circadian rhythms of *Plasmodium vinckei chabaudi*, of *P. berghei* and of *P. gallinaceum*." Parasitology, 65, 189—201.

Hawking, F., Worms, M.J. and Gammage, K. (1968). "24- and 48-hour cycles of malaria parasites in the blood; their purpose, production and control." Trans. R. Soc. trop. med. Hyg. 62, 731—760.

Homewood, C.A., Jewsbury, J.M. and Chance, M.L. (1972). "The pigment formed during haemoglobin digestion by malarial and schistosomal parasites." Comp. Biochem. Physiol. 43B, 517—523.

Homewood, C.A. and Neame, K.D. (1974). "Malaria and the permeability of the host erythrocyte." Nature, Lond. 252, 718—719.

Howells, R.E. and Maxwell, L. (1973a). "Further studies on the mitochondrial changes during the life cycle of *Plasmodium berghei*: electrophoretic studies on isocitrate dehydrogenases." Ann. trop. Med. Parasit., 67, 279—283.

Howells, R.E. and Maxwell, L. (1973b). "Citric acid cycle activity and chloroquine resistance in rodent malaria parasites: the role of the reticulocyte." Ann. trop. Med. Parasit. 67, 285—300.

Howells, R.E., Peters, W. and Homewood, C.A. (1972). In: "Comparative Biochemistry of Parasites." (H. van den Bossche, ed.) pp. 235—258. Academic Press, New York and London.

Kreier, J.P., Mohan, R., Seed, T. and Pfister, R.M. (1972). "Studies of the morphology and survival characteristics of erythrocytes from mice and rats with *Plasmodium berghei* infections." Z. Tropenmed. Parasit. 23, 245—255.

Ladda, R.L., Aikawa, M. and Sprinz, H. (1969). "Penetration of erythrocytes by merozoites of mammalian and avian malarial parasites." J. Parasit. 55, 633—644.

Lainson, R., Landau, I. and Shaw, J.J. (1971). "On a new family of non-pigmented parasites in the blood of reptiles: Garniidae fam. nov. (Coccidiida: Haemosporidiidea). Some species of the new genus *Garnia*." Int. J. Parasit. 1, 241—250.

Lainson, R., Landau, I. and Shaw, J.J. (1974). "Further parasites of the family Garniidae (Coccidiida Haemosporidiidea) in Brazilian lizards. *Fallisia effusa* gen. nov., sp. nov. and *Fallisia modesta* gen. nov., sp. nov." Parasitology 68, 117—125.

Lainson, R., Landau, I. and Shaw, J.J. (1975). "Some parasites of the Brazilian lizards *Plica umbra* and *Uranoscodon superciliosa* (Iguanidae)." Parasitology, 70, 119—141.

Landau, I. (1973). "Diversité des méchanismes assurant la pérennité de l'infection chez les sporozoaïres coccidiomorphes." Mem. Mus. natn. Hist. nat. Paris 77, 1—62.

Landau, I. (1974). "Hypothèses sur le phylogénie des coccidiomorphes de vertébrés." Z. Parasitenk. 45, 63—75.

Levy, M.R. and Chou, S.C. (1973). "Activity and some properties of an acid proteinase from normal and *Plasmodium berghei* infected red cell." J. Parasit. 59, 1064—1070.

Levy, M.R., Siddiqui, W.A. and Chou, S.C. (1974). "Acid protease activity in *Plasmodium falciparum* and *P. knowlesi* and ghosts of their respective host red cells." Nature, Lond. 247, 546—549.

Lucas, A.M. and Jamroz, C. (1961). "Atlas of Avian Hematology." United States Department of Agriculture, Agriculture Monograph 25.

McHardy, N. (1972). "Protective effect of haemolytic serum on mice infected with *Babesia rodhaini*." Ann. trop. Med. Parasit. 66, 1—5.

McKee, R.W. (1951). In: "Biochemistry and Physiology of Protozoa" Vol. 1. (A. Lwoff ed.) pp. 251—322. Academic Press, New York.

Madhu, S., Manandhar, P. and van Dyke, K. (1975). "Detailed purine salvage metabolism in and outside the free malarial parasite." Expl. Parasit. 37, 138—146.

Miller, L.H., Dvorak, J.A., Shiroishi, T. and Durocher, J.R. (1973a). "Influence of erythrocyte membrane components on malaria merozoite invasion." J. exp. Med. 6, 1597—1601.

Miller, L.H., Powers, K.G., Finerty, J. and Vanderburg, J.P. (1973b) "Difference in surface charge between host cells and malaria parasites." J. Parasit. 59, 925—927.

Minchin, E.A. (1903). In: "A treatise on Zoology." Part 1. 2nd Fasc. (E.R. Lankester ed.) p. 240. Adam and Charles Black, London.

Moore, G.A. and Boothroyd, B. (1974). "Direct resolution of the lattice planes of malarial pigment." Ann. trop. Med. Parasit. 68, 489.

Morselt, A.F.W., Glastra, A. and James, J. (1973). "Microspectrophotometric analysis of malarial pigment." Expl. Parasit. 33, 17—22.

Moulder, J.W. (1962). "The Biochemistry of Intracellular Parasites." University of Chicago Press, Chicago.

Neame, K.D., Brownbill, P.A. and Homewood, C.A. (1974). "The uptake and incorporation of nucleosides into normal erythrocytes and erythrocytes containing *Plasmodium berghei*." Parasitology, 69, 329—335.

Nichols, B.A. and Bainton, D.F. (1975). In: "Mononuclear phagocytes in Immunity, Infection and Pathology." (R. van Furth ed.) pp. 83—92. Blackwell, Oxford.

Oelshlegel, F.J. and Brewer, G.J. (1975). In: "The Red Blood Cell." 2nd Edn., Vol. 2 (D.M. Surgenor ed.) pp. 1263—1305. Academic Press, New York.

Oelshlegel, F.J., Sander, B.J. and Brewer, G.J. (1975). "Pyruvate kinase in malaria host-parasite interaction." Nature, Lond., 255, 345—347.

Peters, W. (1969). "Recent advances in the physiology and biochemistry of plasmodia." Trop. Dis. Bull. 66, 1—29.

Peters, W. (1970). "Chemotherapy and Drug Resistance in Malaria." Academic Press, London and New York.

Peters, W. (1974). "Recent advances in antimalarial chemotherapy and drug resistance." Adv. Parasit. 12, 69—114.

Rencricca, N.J., Stout, J.P. and Coleman, R.M. (1974). "Erythropoietin production in virulent malaria." Infection and Immunity, 10, 831—833.

Rickard, M.D. (1969). "Carbohydrate metabolism in *Babesia rodhaini*: Difference in the metabolism of normal and infected rat erythrocytes." Expl. Parasit., 25, 16—31.

Roller, N.F. and Desser, S.S. (1973). "Diurnal periodicity in peripheral parasitaemias in ducklings (*Anas boschas*) infected with *Leucocytozoon simondi* Mathis and Leger." Can. J. Zool. 51, 1—9.

Saint-Girons, M.C. (1970). In: "Biology of the Reptilia, Vol. 3." (C. Gans and T.S. Parsons, eds.) pp. 73—91. Academic Press, New York.

Schnitzer, B., Sodeman, T., Mead, M.L. and Contacos, P.G. (1972). "Pitting function of the spleen in malaria: ultrastructural observations." Science 177, 175—177.

Shakespeare, P.G. and Trigg, P.I. (1973). "Glucose catabolism by the simian malaria parasite *Plasmodium knowlesi*." Nature, Lond., 241, 538—540.

Shaw, J.J. (1969). "The Haemoflagellates of Sloths." (London School of Hygiene and Tropical Medicine Memoir No 13). H.K. Lewis, London.

Sherman, I.W. and Tanigoshi, L. (1972). "Incorporation of ^{14}C-amino acids by malaria (*Plasmodium lophurae*). V. Influence of antimalarials on the transport and incorporation of amino acids." Proc. helminth. Soc. Wash. 39, 250—260.

Sherman, I.W. and Tanigoshi, L. (1974a). "Incorporation of ^{14}C-amino acids by malarial plasmodia (*Plasmodium lophurae*). VI. Changes in the kinetic constants of amino acid transport during infection." Expl. Parasit. 35, 369—373.

Sherman, I. and Tanigoshi, L. (1974b). "Glucose transport in the malarial (*Plasmodium lophurae*) infected erythrocyte." J. Protozool. 21, 603—607.

Siddiqui, W.A. and Schnell, J.V. (1972). "In-vitro and in-vivo studies with *Plasmodium falciparum* and *Plasmodium knowlesi*." Proc. helminth. Soc. Wash. 39, 204—210.

Surgenor, D.M. ed. (1974—1975). "The Red Blood Cell". (2nd Edn., Vol. 1, 1974; Vol. 2, 1975, pp 1372). Academic Press, New York.

Telford, S.R. (1975). "Saurian malaria in the Caribbean: *Plasmodium azurophilum* sp. nov., a malarial parasite with schizogony and gametogony in both red and white blood cells." Int. J. Parasit. 5, 383—394.

Vickerman, K. and Cox, F.E.G. (1967). "Merozoite formation in the erythrocytic stages of the malaria parasite *Plasmodium vinckei*." Trans. R. Soc. trop. Med. Hyg. 61, 303—312.

Wagner, W.M. (1972). "The erythrocyte: survival-modeling concepts." Expl. Parasit. 31, 39—52.

Whitelaw, D.M. and Batho, H.F. (1975). Kinetics of monocytes. In: "Mononuclear Phagocytes in Immunity, Infection and Pathology." (R. van Furth, ed.) pp. 175—187. Blackwell, Oxford.

Ecological Aspects of Parasitology
Editor: C.R. Kennedy
© *North-Holland Publishing Company, Amsterdam, 1976*

CHAPTER 20

INTRACELLULAR-COCCIDIA

P.L. LONG

Houghton Poultry Research Station, Houghton, Huntingdon, Cambs (Great Britain)

Contents

20.1. Introduction

Members of the subclass coccidia belong to the phylum Protozoa and the subphylum Sporozoa. With few exceptions, the organisms are intracellular parasites of the intestinal epithelium. Coccidia have features common to members of the suborder Haemosporina to which *Plasmodium* belong; they differ however from *Plasmodium* in that they have only one host in the life cycle. Multiplicative phases of the life cycle (schizogony) and the reproductive phase (gametogony) occur in the same host, culminating in the production of the zygotes (oocysts). A diagrammatic representation of the life cycle of a sporozoan is given in Chapter 19. The life cycle of *Eimeria* differs to the extent that sporogony occurs outside the host and not in an intermediate host.

There are about 25 genera in the family Eimeriidae but the most widely studied members of this family belong to four genera, *Eimeria*, *Isospora*, *Tyzzeria* and *Wenyonella*, many species of which are parasites of medical and veterinary importance. The major characteristics of members of the *Eimeria* are: (a) the structure of the oocysts, which always contain four sporocysts within which are two sporozoites; (b) marked host specificity, there being very few exceptions to the general rule that species from one animal will not develop in closely related hosts; (c) marked species specificity, in which host resistance acquired to one species does not protect against infection with another and (d) marked predilection for development at specific sites within the host.

The most extensively studied species of *Eimeria* occur in domestic animals, in which many species cause disease (Coccidiosis). Intensive methods used for the production of farm and laboratory animals favour the reproduction of *Eimeria* and the usual well balanced host/parasite relationship, which commonly occurs in wild animals, breaks down. Low level infections occur in most animals in the wild because the intake of infective (sporulated) oocysts is small but husbandry methods used in intensive animal production allow large numbers of sporulated oocysts to be available for infection or re-infection. The situation in poultry production, where birds are kept in large numbers on litter floors, is sufficiently acute to make continuous medication of the feed a routine procedure. This method of control is not free from problems and major research efforts on coccidiosis have been devoted to chemotherapeutic control. Less work has been done on the parasite and on host/parasite relations; some aspects of this research have been reviewed (Horton-Smith and Long, 1963; Long and Horton-Smith, 1968).

Numerous attempts have been made to infect foreign hosts with species of *Eimeria* (see review by Levine, 1961). There have been conflicting reports but most of the recent work has confirmed the strong host specificity of the *Eimeria*. Pellerdy (1969) thought biochemical mechanisms were responsible for the lack of development of species of *Eimeria* in foreign hosts. However, McLoughlin (1969) obtained limited development of *E. meleagrimitis* (from the turkey) in chickens treated with the corticosteroid dexamethasone indicating that immune mechanisms might be involved in the establishment of foreign species of *Eimeria*.

20.2. Limitations imposed on coccidia by the habitat

20.2.1. Factors affecting site of development

The site specificity of *Eimeria* within the host is so well established that only a few species of the genus *Eimeria* are known to develop in sites other than the intestine. *Eimeria truncata* and *E. stiedae*, develop in the kidney of the goose and the liver of the rabbit respectively, whilst *E. neitzi* develops in the uterus of the Impala. These parasites are not known to develop in other sites. Studies on site specificity of species of *Eimeria* parasitising the chicken were made by Horton-Smith and Long (1965; 1966) and these indicated that for most species parasitising the chick the site of parasitism was close to the point in the alimentary tract where sporozoites emerged from the oocysts. This suggested that selection is determined largely by the interplay of host and parasite factors which bring about excystation of the invasive stage (sporozoites). With *E. tenella* the sporozoites emerge from the oocysts in the small intestine, yet they initiate infection of the caeca less than 1.5 h after the introduction of the oocysts by crop intubation. If the sporozoites are introduced intravenously they initiate infection of the normal intestinal site (Sharma and Reid, 1962; Davies and Joyner, 1963; Long and Rose, 1965). In another study, Leathem (1969) produced only minimal infection with *E. tenella* in the small intestine of ceacectomised chickens which suggests that sporozoites are normally attracted to the caeca by chemotaxis.

In chickens treated with the corticosteroid dexamethasone schizogony of *E. tenella* occurred in the liver bile ducts and, in one case, in the mid-intestine (Long, 1970a). No stages beyond the 2nd generation schizonts were found and the parasites were surrounded by host defence cells. Rose (1970) and Long and Rose (1970), who used another corticosteroid, betamethasone, showed that treatment with this compound greatly diminished acquired immunity to *Eimeria acervulina* var. *mivati* and that extended schizogany occurred in the treated birds. Chick embryos have poorly developed immune mechanisms and it is possible that successful infection of the embryo CAM by several species of *Eimeria* (Long, 1965) may be due to this fact.

When embryos were inoculated intravenously with large numbers of cleaned sporozoites of *E. tenella*, schizogony occurred in the bile duct epithelium of the liver; no infection occurred at other sites. It is thought that they were removed from the circulation by the liver and consequently developed there; the liver of chickens appears to be highly efficient in the removal of foreign particles from the circulation (Dobson, 1957).

Horton-Smith and Long (1965, 1966) examined six intestinal species from the fowl and showed that with three species the whole endogenous cycle could occur in the caeca if sporozoites were inoculated directly into this site. We have noted that those species of *Eimeria* which developed the whole of

their endogenous cycle in the caeca were those in which development occured to some extent in the caeca as a result of the usual (oral) route of infection.

Eimeria praecox, a parasite of the duodenum, does not develop in the
large intestine and sporozoites hatched in vitro and introduced via the cloaca
or inoculated into the caeca produced infection in the usual (duodenum) site
(Long, 1967). Recent work with *E. praecox* in this laboratory confirms that
sporozoites of this species inoculated directly into the caeca fail to develop
in this site. However, they initiate infection of the upper small intestine with
little or no delay in the normal prepatent time. Our experiments suggest that
the sporozoites migrate vertically from the large intestine to the duodenum.
Assuming a delay of up to four hours, this would mean that the sporozoites
travel a distance of approximately 16 cms in this time. This would mean that
a sporozoite is capable of moving a distance approximately 2,500 times its
own length in one hour.

Thus it seems clear that special factors are involved in the habitat which
attract the invasive stages of coccidia. These factors are decisive in determining the precise part of the intestine in which the parasite develops; changes in
the "normal" site only occurring when the host is greatly stressed or is highly incompetent immunologically.

20.2.2. *Factors affecting choice of host cell, and development of the para-sites*

With a few exceptions, parasites of the genus *Eimeria* develop in epithelial
cells and most species develop in epithelial cells of the intestinal villi. With
two of these exceptions (*E. stiedae* in the liver of rabbits and *E. truncata* in
the kidney of the goose) the parasites develop in epithelial cells of the bile
ducts and kidney tubules respectively. Of the nine species of *Eimeria* which
are known to be parasitic in the chicken, seven are found only in epithelial
cells of the intestinal villi. The first-generation schizonts of *E. tenella* and *E.
necatrix* develop in the crypts of Lieberkühn but the merozoites of these
schizonts invade connective tissue cells between the glands and give rise to
the large second-generation schizonts which are so characteristic of these
species. The merozoites of this generation are thought to invade epithelial
cells of the crypts and the superficial villi where gamounts are formed. The
question arises as to whether these merozoites migrate away from the epithelium because of the unfavourable environment or whether they are carried
away by host defence cells (macrophages). The grossly enlarged nuclei of the
host cells harbouring second-generation schizonts of *E. tenella* led Gresham
and Cruickshank (1959) to conclude that development of this stage occurred
in macrophages. It is possible that some of these stages do develop in macrophages but it is equally possible that gross structural changes are brought
about by the growth of these large schizonts within the cells and that this

gives them the appearance of macrophages. This view was expressed by Long (1970d) and Fernando et al. (1974) as a result of studies on *E. tenella* and *E. necatrix*. Bergman (1970) studied the ultrastructure of cells from chicken caecal tissues infected with second-generation schizonts of *E. tenella* and concluded that growth of the parasites brought about changes in the host cells giving them the appearance of macrophages.

Long (1971) demonstrated that schizogony and gametogony of *E. tenella* occurred in the liver of chick embryos inoculated intravenously with sporozoites and that gametogony occurred more freely in embryos which had been treated with dexamethasone. These schizonts and gametocytes were found in epithelial cells of the bile duct but many gametocytes were also found within cells which appeared to be hepatocytes. Because of the difficulty in identifying tissue cells and macrophages this work has been extended to the ultrastructural level as the infected cells can be characterised by their organelles. We have shown that second-generation merozoites of *E. tenella* will invade and develop into gametocytes in hepatocytes and development of asexual stages occurs in fibrocytes of the portal tract and in the endothelial cells which line the sinusoids. Merozoites which were occasionally seen within macrophages were always in varying stages of destruction. Schizonts developed in all of these different cells, except macrophages, but gametocytes were detected only in hepatocytes (Lee and Long, 1972). Recently Long and Rose (1975) have made a number of attempts to grow *E. tenella* in in vitro cultured macrophages. First-generation schizonts were grown in only two experiments and only very few schizonts were found.

Eimeria tenella may develop in a variety of host cells and it is interesting to question why they are normally restricted to specific cells in the caeca of the fowl. As already mentioned, the sporozoites of *E. tenella* are capable of invading a wide variety of cells in vitro but development of schizonts, although occurring in fibrocytes, occurs most freely in epithelial cells. Studies in vitro show that some sporozoites may remain within host cells without developing. These were not destroyed because, when introduced into a more favourable site after remaining in the cells for 4 days, they produced infection (Long, 1974b).

In in vitro studies, gametocytes have been shown to develop only in 'islands' of epithelial-like cells and have not been found in fibrocytes (Doran, 1969; Strout and Ouellette, 1969; Long, 1969).

Our observations that second generation schizonts develop in cells lining the sinusoids of the liver of chick embryos is consistent with the occurrence of this stage in mesodermal cells of the lamina propria of intestinal tissue. Similarly, the observations that gametocytes of this species develop in hepatocytes is consistent with the development of this stage in endodermal cells. This work suggests that unfavourable conditions for invasive stages are present in most situations. However, once sporozoites or merozoites become intracellular they have excellent opportunities for survival and replication.

The mechanisms involved in cell invasion are not understood but host cell factors probably stimulate sporozoites to invade. For example with *Toxoplasma* the zoites may be phagocytosed or they may induce the host cell to phagocytose them. This process appears to occur with macrophages HeLa cells and mouse fibroblasts (Jones et al., 1972). The trophozoite is then separated from the cell cytoplasm by its own membrane and the cell membrane. *Eimeria* sporozoites invade epithelial cells actively by making a small hole in the cell membrane and squeezing through. A parasitophorous vacuole is then formed when growth of the parasite occurs. When macrophages are involved our studies show that phagocytosis, or some form of induced phagocytosis, occurs.

These parasites are not destroyed by macrophages, at least not unless the macrophages are sensitised and antibodies and/or immunologically competent cells are also present. Jones and Hirsh (1972) showed that *Toxoplasma* trophozoites are able to block the movement of lysosomal agents into the phagocytic vacuole. When dead *Toxoplasma* were phagocytosed lysosomal enzymes were able to enter the vacuoles and digest the parasites. *Eimeria* sporozoites may have similar properties because the parasites can survive for several days within macrophages and some development may occur (Long and Rose, 1975).

Coccidian parasites produce a good deal of nucleic acid during their development; very much more than that produced by host cells. They also need to elaborate large amounts of polysaccharides which they produce de novo from glucose acquired from the host. The storage polysaccharide in the macrogametocytes and oocysts is in the form of amylopectin (Ryley et al., 1969).

Most of this storage polysaccharide in the sporozoites is used up during the excystation process and penetration of host cells. The oxygen uptake of sporozoites is stimulated by fructose or glucose.

Freeman (1970) found a significant reduction in glycogen stores of voluntary muscle during *E. tenella* infection and discussed the possibility that toxins might be produced by the parasites. Toxins, if they are involved, are more likely to affect the host itself rather than individual cells harbouring the parasites. At the ultrastructural level we have not observed gross changes in host cells until near to the time of the maturation of the parasite. Thus it is necessary for the parasite to avoid interference with cell function at least until it is mature or has acquired energy stores. However, these aspects of the host parasite relationships have not been actively studied and deserve full investigation.

20.3. Factors affecting development of *Eimeria* in the intestine

20.3.1. *Physiological and morphological changes in the gut*

The host has complete control upon the ability of *Eimeria* sporozoites to invade intestinal cells. Successful excystation of sporozoites depends firstly upon suitable conditions in the stomach of mammals (or gizzard of birds) in bringing about changes in the oocyst membrane so that later treatment with trypsin (or chymotrypsin) and bile stimulates sporozoites to be freed from the oocyst and sporocyst.

The speed of these reactions determines where, in the intestine, the sporozoites will be able to penetrate. Thus if aged oocysts are ingested fewer sporozoites will emerge and these will take longer to be freed from oocysts (Long, 1970b; Doran and Vetterling, 1969). In very young chickens, where digestive enzyme activity is low, sporozoites will not be available in the anterior regions of the intestine and sporozoites of duodenal species of *Eimeria* are able to develop further down the intestine.

The caeca appear to have a minor role in digestion of food but the mechanisms by which they fill and discharge have been widely studied. It has been found that the caeca fill at fairly regular intervals and that valve-like structures at the entrances to the tubes act as filters so that only fluids with small particles can gain entrance. The evacuation of caecal contents occurs as a result of contractions which originate at the base of the caecum and pass along the body of this organ. These contractions occur at fairly regular intervals and depend upon the degree of caecal distension. Coccidia inhabiting the caeca appear to have overcome the effects of caecal evacuation mainly because the invasive stages (sporozoites and merozoites) are within cells or they stay close to the mucosa. It is only when present in the caecal lumen that they are lost in the caecal evacuations.

Recently, Witlock et al. (1975) have shown with the aid of a scanning electron microscope that four major types of surface conformation exist in the caeca and that *E. tenella* produces only slight lesions in the neck region, severe lesions in the dilated portions and only moderate lesions in the distal (blind) portion. Development of *E. tenella* was best in areas where low ridges and protruding collarlike structures occur. It is in these areas that epithelial cell turnover is likely to be lowest.

Another factor affecting the movement of coccidial stages within the large intestine is the fact that material from the cloaca, including urine, can be drawn into the caeca as a result of antiperistaltic movements originating at the cloaca. These movements have the effect of transferring the merozoites from the large intestine to the caeca and enable the life cycle to be completed in another site, and with *E. necatrix*, schizogony occurs in the small intestine and gametogony only in the caeca.

20.3.2. Morphological changes of the intestine due to parasitism

Replication of *Eimeria* in the intestine causes gross changes in the morphology of the tissues. Shortening of villus height leading to flat mucosa or villous atrophy is a common effect of parasitism well described by Pout (1967), and common to a number of intestinal maladies including malabsorption syndromes.

The effects of coccidial infection upon the mucosal architecture of the upper intestine may vary from slight shortening of the villi to clubbed shaped villi. In severe infections of the lower intestine, the mucosa may be extremely flat. Pout pointed out that villus atrophy results in a reduction of the surface area for absorption and a reduction in the number of functional epithelial cells as well as a reduction in alkaline phosphatase activity in the villi. In addition, during the recovery phase of infection the lamina and tunica propria become infiltrated by lymphoid cells (Long, 1973b). This results in gross thickening of the mucosa with a consequent reduction in epithelial tissue.

These anatomical changes result in reduced areas for absorption of nutrients and a slowing of peristalsis. Intracellular stages of *Eimeria* may not be affected by these changes initially but as fewer epithelial cells are available in the shortened villi there may be a restrictive influence later on in the infections and merozoites may not readily find new cells to invade. A fourth complication is that, as peristalsis is reduced, bacteria from the lower gut (particularly bacteria of the genus *Clostridium*) migrate anteriorly and grow actively because these areas contain exudates rich in serum proteins and blood. This may lead to a condition known as necrotic enteritis where large areas of villus tissue becomes necrotic and are sloughed away leaving only a few coccidial stages within these lesions.

20.3.3. Epithelial cell regeneration

Pout (1967) suggested that the life cycles of species of *Eimeria* are synchronised with the rate of epithelial turnover of their respective hosts, the rate of epithelial cell turnover and endogenous cycles of *Eimeria* being shorter for smaller as compared with larger animals (Leblond and Walker, 1956). A study of the life cycle of a number of *Eimeria* affecting chickens confirms this view in that first generation schizogony occurs in the crypts of Lieberkühn and successive generations of smaller schizonts occur along the sides of the villi in the next 1—3 days with gametocytes and oocysts being produced in cells at the tips of the villi. It therefore appears that the endogenous stages of *Eimeria* have adapted to develop at rates which fit in with epithelial cell turnover and that after initial invasion by sporozoites of epithelial cells of the crypts of Lieberkühn the developing stages find themselves moving to more superficial areas in the epithelial cell 'shunt' towards the tips of the villi. As each schizont generation matures the merozoites invade new cells adjacent to the old host cells so that by the 4—6th day (the time needed for most species of *Eimeria* affecting chickens to complete their endogenous cycle) gametocytes and oocysts are produced at the tips of the villi, the oocysts are then released into the intestinal lumen. In the chicken, epithelial cells comprising the whole crypt of Lieberkühn are replaced in about 2 days (Imondi and Bird, 1966), and the endogenous cycles of chicken coccidia, comprising several asexual generations with gametogony, are accordingly of short duration (4—6 days). Despite the rapid turnover rate of epithelial cells, there is still a considerable loss of endogenous stages because many developing stages reach the tips of villi too early and are shed into the intestinal lumen.

Schizonts of some species (*E. acervulina*, *E. praecox*) take only 6—12 hours to reach maturity but gut epithelial cells must surely divide every few hours. It would seem to be disadvantageous, from the parasite's standpoint, to develop within dividing cells and observation of the intracellular stages does not suggest this to be a common phenomenon. However, it may be that *Eimeria* parasites are capable of slowing down epithelial cell turnover to suit their own needs. Long and Millard (1968) showed that sporozoites of *E. acervulina* and *E. tenella* within cells of the crypts of Lieberkühn of chickens treated with an anticoccidial drug could survive for at least 60 days. The coccidia may inhibit the division of the progenitor cells which they invade or affect the rate of division of cells in the crypt as a whole. Speer et al. (1971) observed sporozoites leaving host cells shortly after invading them so a less likely explanation is that sporozoites, inhibited by anticoccidial drugs, could still leave cells and re-invade new ones.

20.4. Effect of parasite dose

It is now well established that the reproduction of *Eimeria* within the host depends upon the number of oocysts ingested, adverse effects on the host judged by body weight changes, clinical signs of disease and mortality is increased by administering more oocysts. Krassner (1963) coined the term "crowding factor" to describe a phenomenon in which fewer oocysts were produced as the initial dose was increased, conditions in the intestine having an adverse effect on the ability of the parasites to complete their cycle. This could be caused by a reduction in the number of host cells available for parasitisation. However, this may not be the sole reason for the reduction of development of *Eimeria* in heavy infections. We have observed that in the heavy infections, many of the intracellular sporozoites fail to grow and these could therefore be lost to the intestinal lumen, if they are unable to escape, as the epithelial cells move to the tips of the villi during the "shunt". The "crowding factor" probably operates well in infections with the duodenal parasite *E. praecox*. The oocyst production of this species in young chickens

TABLE 1

Oocyst production in infections with *E.praecox* in chickens of different ages

Dose of oocysts	Age of chickens (weeks)	Total oocysts produced per bird (1×10^6)	Reproductive potential *
100	3	14.35	143,000
250	3	47.62	190,000
1×10^5	1.5	22.93	229
1×10^6	1.5	23.50	23.5
2×10^6	1.5	15.15	7.57
1×10^4	3	3.13	313
1×10^5	3	17.44	174
1×10^6	3	15.27	15
1×10^5	6	279.43	2,794
1×10^6	6	199.59	199
2×10^6	6	332.56	166

* Number of oocysts produced per oocyst fed

given a range of doses of oocysts is given in Table 1 (Long, 1967). These results show that:

(i) reduction in development is not very evident when small doses (100–250 oocysts) were given.

(ii) the so-called "crowding factor" was seen in 1.5—6-week-old birds when doses of 10^4—2×10^6 were given and

(iii) reproduction was always better in older birds, probably due to more living space (suitable host cells) being available. If the "crowding factor" was brought about by better host immunity factors, one might expect this to be better expressed in older (6-week-old) birds than in 1.5—3-week-old birds, but the reverse was true. In addition, *E. praecox* completes its life cycle in less than four days and it would be surprising if acquired immunity was involved to any great extent at this time. This parasite seems to have adapted very well in the chicken host to achieve a high reproduction rate from a small dose in a short time, leaving the host as the strong immunity is developed. Not all species of *Eimeria* have achieved this ideal, and Lotze and Leek (1970) demonstrated that as the dose of oocysts of *E. intricata* to sheep was increased, the late life cycle stages (gametocytes) were inhibited. Most of the experimental work on parasite life cycles has been carried out using single doses of oocysts, but recently Joyner and Norton (1973, 1976) have shown that if extremely small doses (1—20 oocysts) are given daily, they induce a much greater immune response which may have a longer duration. They argue that 'trickle' infections may be more analogous to the field situation. However, if this were so, one would expect the majority of chickens under field conditions to be effectively resistant. It is probably more likely that chickens living in an infective environment are partially resistant and that light infections continue.

Recent studies in our laboratory (Long and Rowell, 1975; Long et al., 1975) show that in broiler chickens, oocyst numbers in the litter rise to a peak at 4—5 weeks of age and then rapidly decline at 8 weeks, but do not disappear from the litter.

Under "natural" conditions chickens discharge oocysts regularly and the host/parasite relationship seems best in 'barn yard' chickens. These may be constantly picking up oocysts but only achieve a partial immune state because of other factors, poorer nutrition or lowered immunological competance or interrelationship of other disease producing agents.

It has been shown that infection of chickens with Marek's disease virus, the cause of an infectious lymphomatous disease, lowers the resistance of chickens to *Eimeria* infection (Biggs et al., 1968). The recent extensive use of an effective vaccine for this disease in the field has profoundly reduced the incidence of severe outbreaks of intestinal coccidiosis in growing chickens. It thus seems clear that the host normally places some restriction upon the parasite which is reflected by a less than optimum replication of the parasite, and that these restrictions may be relaxed by the intervention of other disease producing agents.

However, this does not mean that *Eimeria* are incapable of causing severe pathogenic effects without the intervention of other disease agents because Clark et al. (1962) showed that the gross pathology of *E. tenella* infection was similar in gnotobiotic and conventionally reared chickens. In addition, Long (1965, 1970c) has shown that proliferation of *E. tenella* and mortality associated with severe haemorrhage occur in the bacteria-free environment of the developing chicken embryo. Of course, the proliferation of *Eimeria* in intestinal tissues provokes cell and protein exudates in which bacteria such as *Clostridium* are able to grow, causing necrotic enteritis. Similarly, Stephens and Barnett (1964) found that *Salmonella typhimurium* produced greater effect on the host when introduced into chickens infected simultaneously with *E. necatrix*. Williams (1973) showed that the reproductive potential of *E. tenella* was significantly reduced in concurrent infections with *E. acervulina*. However, the reproductive potential of *E. acervulina* was not reduced when it developed with four other species. The mechanisms involved are not understood especially since these two parasites do not share the same site for development.

20.5. Host responses

20.5.1. Changes in the vascular permeability at the site of invasion and development

Preston-Mafham and Sykes (1967) showed that *E. acervulina* coccidiosis affects the uptake of nutrients and fluids and detected the loss of serum pro-

teins into the gut lumen. This loss of serum proteins detectable by Ponta-
mine sky blue dye test, started with a slight reaction about the 48th h after
inoculation with oocysts, and became more severe 72—144 h as epithelial
cell destruction increased. Long (1968) showed that the vascular leak occurred as early as 3.5 h and continued for a few hours in response to sporozoite invasion. These changes occurred only with heavy infections and the
vascular leak occurring after 3.5—7 h could only be produced when 1×10^7
oocysts were given. Thus with moderate inocula, such as would occur in natural
conditions, vascular leakage and loss of viable parasites is minimal. Rose and
Long (1969) showed that vascular leak was more rapid and severe in immune
chickens and that the reaction could be "blocked" by treating the birds with
corticosteroids.

The development of second generation schizonts of *E. tenella* and *E.
necatrix* is accompanied by gross tissue disruption, cellular reactions and haemorrhage. Large numbers of both developing and invasive stages are lost at
this time. The effect of this is to restrict the gametocytes and oocyst formation by the loss of these stages in the faeces and by the reduction in suitable
epithelial tissue for the development of those that remain. Many of the parasites which remain are able to withstand host defence cells of various types
including granulocytes and macrophages. Macrophages may engulf invasive
stages but they probably do not destroy them.

20.5.2. *Effect of antibodies*

It is difficult to be precise about the extent to which antibodies of various
classes affect replication of coccidia in the intestine. During a primary infection the effects may be minimal because antibodies produced in response to
early developmental forms may not profoundly affect later stages. However,
immune responses may influence the length of the patent period i.e. the
length of time oocysts are produced. Long and Rose (1970) showed that
treatment with the corticosteroid betamethasone caused the extended
schizogony of *E. acervulina* var. *mivati* which allowed oocysts to be shed for
up to 50 days. During immunisation antibodies of different classes can be
detected in the serum, but Long and Rose (1970) found that serum antibodies did not protect against infections initiated by the usual (oral) route.
However, they found that serum antibodies were highly effective in blocking
infection initiated by inoculating sporozoites into the peripheral blood. It
would therefore seem reasonable to suggest that antibodies may be involved
in restricting the development of *Eimeria* to their usual intestinal site. Long
(1970a) found that in chickens treated with the corticosteroid dexamethasone, schizogony of *E. tenella* occurred in the liver and examination of serum samples from these birds were negative for antibody. However, it seems
reasonable to assume that antibodies play a minor role in inhibiting replication of *Eimeria* during a primary infection.

20.5.3. Host cell responses

Penetration of the host cell occurs with the mimimum loss of host cell constituents but if sporozoites leave cells then the loss of material is greater. A
parasitophorous vacuole is formed between the parasite and the host cell
cytoplasm and this contains particulate matter. Intake of nutrients is
thought to occur by ingestion of substances through micropores when the
formation of food vacuoles has been observed (Hammond, 1973). Responses
appear to be relatively slight during sporozoite invasion but within 24 hours
numerous granulocytes (polymorphonuclear leucocytes) infiltrate the invasion site. These cells, and later on plasma cells, infiltrate the lamina propria
during the next few days and by the 5th—6th days there is considerable infiltration of the tunica propria and lamina propria and muscularis mucosa by
lymphoid cells; some foci develop within these sites. During this time the
parasites may complete several sexual generations within epithelial cells.
It is doubtful if the cellular responses have any effect upon parasite replication whilst the parasites remain intracellular. Merozoites, released from schizonts, may be destroyed before they invade new cells, but most of them
probably invade adjacent host cells. However, when development of *Eimeria*
occurs beneath the epithelium, e.g. first generation schizonts of *E. nina-
kohlyakimovae*, granulocytes invade the schizonts and eventually destroy
them (Wacha et al., 1971). Changes in the host cells may occur and cells infected with *E. necatrix* and *E. tenella* have enlarged nuclei with hypertrophy
of the cytoplasm (Long, 1973b).

Macrophages have an interesting role in the coccidian life cycle; Tyzzer et
al. (1932) were the first to suggest that sporozoites were ingested by macrophages and transported to the crypts of Lieberkühn where they were destroyed. I believe that macrophages are often involved in transport of sporozoites, but that most sporozoites migrate to the crypts directly. Having
reached the crypts I believe that sporozoites develop within epithelial cells
(sporozoites freely invade and leave cells) and that development of *Eimeria*
within macrophages probably does not occur. It is of interest that *Eimeria*
appear to be unaffected by this rather important host defence cell. The giant
nuclei present in host cells parasitised by large schizonts arise as a result of
the host-parasite interaction and the nuclei contain increased amounts of nucleic acid (Long, 1970 and Fernando et al., 1974); I consider these cells to
be modified epithelial cells and not macrophages.

20.6. Immunogenicity and immunological variation in coccidia

Rose (1973) has reviewed existing knowledge on the antigens of *Eimeria*
and it appears that antigens detected by agglutination or lysis of sporozoites,
fluorescent labelling, serum neutralisation and agar-gel precipitation, cross

react with immune serum from animals given other species of *Eimeria*. There is also slender evidence that *Eimeria* have some antigens similar to host tissues and this may assist them to evade the immune response.

It is difficult to obtain information on the relative immunogenicity of the different species of *Eimeria* occurring in animals. However, a single exposure to infection may confer a substantial immunity. Such immunogenic species include *E. maxima* and *E. praecox* from the chicken, *E. scabra* (pig,) *E. ahsata* (sheep) and *E. stiedai* (rabbit). Examples of species with poor immunogenicity include *E. necatrix* and *E. tenella* (chickens), *E. alabamensis* (cow), *E. intestinalis* (rabbit), *E. callospermophili* (ground squirrel), *E. utahensis* (kangeroo rat) and *E. anseris* (goose).

It is of interest that one of the highly immunogenic species affecting chickens (*E. praecox*) is the least pathogenic, whereas the most pathogenic species occurring in the chicken, *E. necatrix* and *E. tenella* are the least immunogenic.

With highly immunogenic species the magnitude of the dose is clearly not very important. For example, complete immunity to *E. maxima* was achieved with a dose of between 50—5,000 oocysts (Long, 1959; Biggs et al., 1968; Rose, 1974). However, with *E. tenella*, a poorly immunogenic species, 3—5 doses of between 500—100,000 were necessary to achieve complete immunity (Horton-Smith and Long, 1963). Chickens which survived an initial dose of 60,000 oocysts were highly resistant for 2—3 weeks (Horton-Smith et al., 1961). Recently, Joyner and Norton (1973) have made the interesting observation that five oocysts of *E. tenella* given daily to chickens for 28 days resulted in a solid immunity.

An attenuated strain of *E. tenella* has been developed by serial embryo passage which has very low pathogenicity but high immunogenicity. During the period of "embryo adaptation" it was noted that the large subepithelial 2nd generation schizonts were replaced by small schizonts restricted to the epithelium. This strain cross-protected against a chicken maintained strain of *E. tenella* (Long, 1972).

It is of interest that two highly immunogenic species of *Eimeria*, (*E. maxima* and *E. praecox*) are (a) restricted to development within epithelial cells, (b) show marked predilection for development within specific sites, (c) have a short patent period and (d) do not develop in chicken embryos or cell cultures. In contrast, *E. tenella* and *E. necatrix* are poorly immunogenic and are characterised by (a) high pathogenicity, (b) development of second generation schizonts beneath the epithelium, (c) have a long patent period and (d) development in both chicken embryos and cell cultures.

A fundamental characteristic of species of *Eimeria* is that immunisation by one species does not protect against challenge with another species and indeed, this characteristic has been used to establish new species of *Eimeria*.

Heterogenicity of immunogenicity within a species has been demonstrated by Joyner (1969) and by Long (1973a) who showed that birds made solidly immune to one strain of *E. acervulina* were not completely protected against infection with another strain of the same species. This immunological variation within strains of *E. acervulina* may be accompanied by morphological differences in the oocyst.

Two strains of *E. maxima* from Great Britain did not completely cross-protect, and a strain of *E. maxima* isolated from the Malaysian Jungle fowl differed from these two strains (Long, 1974a). Conversely, a strain of *E. praecox* from the Ceylon Jungle fowl (Long et al., 1974) was immunologically similar to a strain isolated in Britain in 1966. Similarly, strains of *E. tenella* isolated from various places appear to cross-protect.

Immunological variation may be an important reason why populations completely resistant to coccidia are not produced under natural conditions. Although at present we do not know the extent of immunological variation under natural conditions, it seems probable that coccidia, especially highly immunogenic ones, must modify antigens to escape rejection by host factors.

20.7. Concluding remarks

A particularly interesting aspect in the biology of the host-parasite relations of *Eimeria* is the site specificity shown by the different species. Is this apparent selection determined by the parasite, or is it brought about mainly by host mechanisms acting upon the parasite? That is, does the parasite choose the most suitable from a range of possible sites, or is it forced, by the host, to occupy one site only, all others being unsuitable for invasion and development?

Research seems to show that many intestinal species of *Eimeria* develop in sites where sporozoites emerge from the oocysts whilst other species must migrate, or be carried, to other sites before they can develop. Thus it seems clear that some species are, to some degree, flexible in their site for development in the intestine but other species showed a marked degree of site specificity.

The results so far discussed indicate that *Eimeria* parasites are capable of developing in a wide range of host cells in vitro. Carefully conducted in vivo experiments suggest that at least with some species the site for development is determined by host factors. The *Eimeria* are highly immunogenic, a single infection with most species inducing substantial immunity to reinfection. Thus it is interesting to speculate that *Eimeria* were formerly parasites of other tissues (perhaps more vascular ones) and that host defences forced them to occupy less vascular regions as they induced greater immune responses.

When immune mechanisms are diminished by treatment with immunodepressants, (betamethasone, dexamethasone, cyclophosphamide) host and site specificity and also reproduction of the parasites can be affected (Mc-Loughlin, 1969; Long, 1970; Long and Rose, 1970; Long et al., 1974 and Long and Fernando, 1974). This work indicates that the strong site and host specificity of these parasites is most probably determined by the host and not the parasite.

Other coccidia such as *Toxoplasma* are much less immunogenic and this parasite develops in a wide range of cells and sites in a number of hosts. Within specific sites, species of *Eimeria* develop in epithelial cells and there are only a few exceptions to this general rule. Gametocytes and oocysts of *Eimeria* are always found in epithelial cells and although *Toxoplasma* develops in a wide range of cells gametocytes appear to develop only in epithelial cells of the cat intestine.

Having found a suitable site (and suitable cells within this site), the sporozoites must penetrate the cell; sporozoites may enter and leave cells of various types before they settle in a suitable cell. *Eimeria* parasites appear not to be destroyed by host cells as they are capable of leaving and entering other cells. The parasites appear to be able to arrest the division of the host cell and this may lead to hypertrophy of the host cell nucleus, and such cells appear to have enhanced nucleic acid synthesis. This is probably necessary in order to sustain the nucleic acid requirements of *Eimeria* during the rapid nuclear divisions leading up to the formation of mature schizonts and microgametes. Most merozoites appear to escape from schizonts and manage to avoid being destroyed by host defence mechanisms by rapidly invading new epithelial cells.

The degree of replication of coccidia within their hosts depends upon many factors. The dose of oocysts is obviously important because coccidia have a fixed number of schizont generations preceding gametogony. Exceedingly high doses lead to crowding effects which severely limit the reproduction potential of the parasite. Trickle infections (very small regular doses of oocysts) perhaps analogous to infections acquired under natural conditions, stimulate immune responses best of all. As coccidia tend to be highly immunogenic their success as animal parasites perhaps depends upon their ability to undergo immunological variation. Immunity developed to one species does not confer protection against another and within a given species immunological variation between strains is probably very extensive.

The life cycle of coccidia is well suited for surviving adverse conditions. When immune responses develop in the host, oocysts are formed which leave the host.

The oocysts discharged in the faeces are fairly resistant to adverse conditions and some can survive for up to two years; a single oocyst is sufficient to initiate infection.

In spite of rigid preferences for development within specific hosts and sites and ability to evoke substantial host responses, the *Eimeria* are highly successful animal parasites.

20.8. REFERENCES

Bergmann, V. (1970). "Elektronenmikroskopische Untersuchungen zur pathogenese der Blinddarmkokzidiose der Huhnerkuken." Arch. Exp. Vet. Med. 24, 1170—1184.

Biggs, P.M., Long, P.L., Kenzy, S.G. and Rootes, D.G. (1968). "Relationship between Marek's disease and coccidiosis. II. The effect of Marek's disease on the susceptibility of chickens to coccidial infection." Vet. Rec. 83, 284—289.

Clark, D.T., Smith, C.K. and Dardas, R.B. (1962). "Pathological and Immunological changes in gnotobiotic chickens due to *Eimeria tenella.*" Poult. Sci. 41, 1635—1636.

Davies, S.F.M. and Joyner, L.P. (1963). "Infection of the fowl by the parenteral inoculation of oocysts of *Eimeria.*" Nature, Lond. 194, 996—997.

Dobson, E.L. (1957). In: "Physiopathology of the reticuloendothelial system." pp. 80—114. Blackwell, Oxford.

Doran, D.J. (1969). "*Eimeria tenella:* from sporozoites to oocysts in cell culture." Proc. Helminth. Soc. Wash. 37, 84—92.

Doran, D.J. and Vetterling, J.M. (1969). "Influence of storage period on excystation and development in cell cultures of sporozoites of *Eimeria meleagrimitis,* Tyzzer, 1929." Proc. Helminth. Soc. Wash. 36, 33—35.

Fernando, M.A., Pasternak, J., Barrell, R. and Stockdale, P.H.G. (1974). "Induction of host nuclear DNA synthesis in coccidia-infected chicken intestinal cells." Int. J. Parasit. 4, 267—276.

Freeman, B.M. (1970). "Carbohydrate stores in chickens infected with *Eimeria tenella.*" Parasitology, 61, 245—251.

Gresham, G.A. and Cruikshank, J.G. (1959). "Protein synthesis in macrophages containing *Eimeria tenella.*" Nature, Lond. 184, 1153.

Hammond, D.M. (1973). In: "The Coccidia" (D.M. Hammond and P.L. Long, eds.) pp. 45—79. University Park Press, Baltimore.

Horton-Smith, C. and Long, P.L. (1963). In: "Advances in Parasitology," Vol. 1. (Ben Dawes, ed.) pp. 67—107. Acad. Press, London and New York.

Horton-Smith, C. and Long, P.L. (1965). "The development of *Eimeria necatrix,* Johnson, 1930, and *Eimeria brunetti,* Levine 1942, in the caeca of the domestic fowl (*Gallus domesticus*)." Parasitology, 55, 401—405.

Horton-Smith, C. and Long, P.L. (1966). "The fate of the sporozoites of *Eimeria acervulina, Eimeria maxima* and *Eimeria mivati* in the caeca of the fowl." Parasitology, 56, 569—574.

Horton-Smith, C., Beattie, J. and Long, P.L. (1961). "Resistance to *Eimeria tenella* and its transference from one caecum to the other in individual fowls." Immunology, 4, 111—121.

Imondi, A.R. and Bird, F.H. (1966). "The turnover of intestinal epithelium in the chick." Poult. Sci. 45, 142—147.

Jones, T.C. and Hirsch, J.G. (1972). "The interaction between *Toxoplasma gondii* and mammalian cells. II. The absence of lysosomal fusion with phagocytic vacuoles containing living parasites." J. exp. Med. 136, 1173—1194.

Jones, T.C., Yeh, S. and Hirsch, J.G. (1972). "The interaction between *Toxoplasma gondii* and mammalian cells. I. Mechanism of entry and intracellular fate of the parasite." J. exp. Med. 136, 1157—1172.

Joyner, L.P. (1969). "Immunological variation between two strains of *Eimeria acervulina.*" Parasitology, 59, 725—732.

Joyner, L.P. and Norton, C.C. (1973). "The immunity arising from continuous low level infection with *E. tenella.*" Parasitology, 67, 333—340.

Joyner, L.P. and Norton, C.C. (1976). "The immunity arising from continuous low level infection with *Eimeria maxima* and *E. acervulina.*" Parasitology, 72, 115—125.

426

Krassner, S.M. (1963). "Factors in host susceptibility and oocyst infectivity in *Eimeria acervulina* infections." J. Protozool. 10, 327—332.

Leathem, W.D. (1969). "Tissue and organ specificity of *Eimeria tenella* (Raillet and Lucet, 1891, Fantham, 1909) in cecectomised chickens." J. Protozool. 16, 223—226.

Leblond, C.P. and Walker, B.E. (1956). "Renewal of cell populations." Physiol. Rev. 36, 255—276.

Lee, D.L. and Long, P.L. (1972). "An electron microscopal study of *Eimeria tenella* grown in the liver of the chick embryo." Int. J. Parasit. 2, 55—58.

Levine, N.D. (1961). "Protozoan parasites of domestic animals and of man." Burgess, Minn. USA.

Long, P.L. (1959). "A study of *Eimeria maxima* Tyzzer, 1929, a coccidium of the fowl (*Gallus gallus*)." Ann. trop. Med. Parasit. 53, 325—333.

Long, P.L. (1965). "Development of *Eimeria tenella* in avian embryos." Nature, Lond. 208, 509—510.

Long, P.L. (1967). "Studies on *Eimeria praecox*, Johnson 1930, in the chicken." Parasitology, 57, 351—361.

Long, P.L. (1968). "The pathogenic effects of *Eimeria praecox* and *E. acervulina* in the chicken." Parasitology, 58, 691—700.

Long, P.L. (1969). "Observations on the growth of *Eimeria tenella* in cultured cells from the parasitised chorioallantoic membranes of developing chicken embryo." Parasitology, 59, 757—765.

Long, P.L. (1970a). "Development (schizogony) of *Eimeria tenella* in the liver of chickens treated with corticosteroid." Nature, Lond. 225, 290—291.

Long, P.L. (1970b). "Studies on the viability of sporozoites of *Eimeria tenella.*" Z. Parasitkde. 35, 1—6.

Long, P.L. (1970c). "Some factors affecting the severity of infections with *Eimeria tenella* in chicken embryos." Parasitology, 60, 435—447.

Long, P.L. (1970d). "In vitro culture of *Eimeria tenella.*" J. Parasit. 56, (4) Sect. 2, 214—215.

Long, P.L. (1971). "Schizogony and gametogony of *Eimeria tenella* in the liver of chick embryos." J. Protozool, 18, 17—20.

Long, P.L. (1972). "*Eimeria tenella:* reproduction, pathogenicity and immunogenicity of a strain maintained in chick embryos by serial passage." J. comp. Path. Ther. 82, 429—437.

Long, P.L. (1973a). "Studies on the relationship between *Eimeria acervulina* and *E. mivati.*" Parasitology, 67, 143—155.

Long, P.L. (1973b). In: "The Coccidia." (D.M. Hammond and P.L. Long, eds.) pp. 253—294. University Park Press, Baltimore.

Long, P.L. (1974a). "Experimental infection of chickens with two species of *Eimeria* isolated from the Malaysian Jungle fowl." Parasitology, 69, 337—347.

Long, P.L. (1974b). "The growth of *Eimeria* in cultured cells and in chicken embryos:" A review. Proc. Symp. "Coccidia and Related Organisms." University of Guelph. pp. 57—84.

Long, P.L. and Fernando, M.A. (1974). "Infection of chickens with *Eimeria* from jungle fowl." Parasitology 69, (Suppl.) xiii—xix.

Long, P.L., Fernando, M.A. and Remmler, O. (1974). "Experimental infections of the domestic fowl with a variant of *Eimeria praecox* from the Ceylon Jungle fowl." Parasitology, 69, 1—9.

Long, P.L. and Horton-Smith, C. (1968). In: "Advances in Parasitology." Vol. 6. (Ben Dawes, ed.) pp. 313—325. Acad. Press, London and New York.

Long, P.L. and Millard, B.J. (1968). "*Eimeria:* effect of meticlorpindol and methyl benzoquate on endogenous stages in the chicken." Expl. Parasit. 23, 331—338.

Long, P.L. and Rose, M.E. (1965). "Active and passive immunisation of chickens against intravenously induced infections of *Eimeria tenella.*" Expl. Parasit. 16, 1—7.

Long, P.L. and Rose, M.E. (1970). "Extended schizogony of *Eimeria mivati* in beta-methasone-treated chickens." Parasitology, 60, 147—155.

Long, P.L. and Rose, M.E. (1975). "Growth of *Eimeria tenella* in vitro in macrophages from chicken peritoneal exudates." 48, 291—294.

Long, P.L. and Rowell, J.G. (1975). "Sampling broiler house litter for coccidial oocysts." Br. Poult. Sci. 16, 583—592.

Long, P.L., Tompkins, R.V. and Millard, B.J. (1975). "Coccidiosis in broilers: evaluation of infection by the examination of broiler house litter for oocytes." Avian Path. 4, 287—294.

Lotze, J.C. and Leek, R.G. (1970). "Failure of development of the sexual phase of *Eimeria intricata* in heavily inoculated sheep." J. Protozool. 17, 414—417.

McLoughlin, D.K. (1969). "The influence of dexamethasone on attempts to transmit *Eimeria meleagrimitis* to chickens and *E. tenella* to turkeys." J. Protozool. 16, 145—148.

Pellerdy, L.P. (1969). "Attempts to alter the host specificity of Eimeriae by parenteral infection experiments." Acta. Vet. (Brno). 38, 43—46.

Pout, D.D. (1967). "Villous atrophy and coccidiosis." Nature, Lond. 213, 306—307.

Preston-Mafham, R.A. and Sykes, A.H. (1967). "Changes in permeability of the mucosa during intestinal coccidiosis infections in the fowl." Experientia, 23, 972—973.

Rose, M.E. (1970). "Immunity to coccidiosis: effect of betamethasone treatment of fowls on *Eimeria mivati* infections." Parasitology, 60, 137—146.

Rose, M.E. (1973). In: "The Coccidia". (D.M. Hammond and P.L. Long, eds.) pp. 295—341. University Park Press, Baltimore.

Rose, M.E. (1974). "The early development of immunity to *Eimeria maxima* in comparison with that to *Eimeria tenella.*" Parasitology, 68, 35—45.

Rose, M.E. and Long, P.L. (1969). "Immunity to coccidiosis: gut permeability changes in response to sporozoite invasion." Experientia, 25, 183—184.

Ryley, J.F., Bentley, M., Manners, D.J. and Stark, J.R. (1969). "Amylopectin, the storage polysaccharide of the coccidia *Eimeria brunetti* and *E. tenella.* J. Parasit. 55, 839—845.

Sharma, N.N. and Reid, W.M. (1962). "Successful infection of chickens after parental inoculation of oocysts of *Eimeria* spp." J. Protozool. 48 (Suppl.): 33.

Speer, C.A., Davis, L.R. and Hammond, D.M. (1971). "Cinemicrographic observations on the development of *Eimeria larimerensis* in cultured bovine cells." J. Protozool. 18, (Suppl.) 11.

Stephens, J.F. and Barnett, B.D. (1964). "Concurrent *Salmonella typhimurium* and *Eimeria necatrix* infections in chicks." Poult. Sci. 45, 353—356.

Strout, R.G. and Quellette, C.A. (1969). "Gametogony of *Eimeria tenella* (coccidia) in cell cultures." Science, N.Y. 163, 695—696.

Tyzzer, E.E., Theiler, H. and Jones, E.E. (1932). "Coccidiosis in gallinaceous birds. II. A comparative study of species of *Eimeria* of the chickens." Am. J. Hyg. 15, 319—393.

Wacha, R.S., Hammond, D.M. and Miner, M.L. (1971). "The development of the endogenous stages of *Eimeria ninakoklyakimovae* (Yakimoff and Rastegaieff, 1930) in domestic sheep." Proc. Helminth. Soc. Wash. 38, 167—180.

Williams, R.B. (1973). "The effect of *Eimeria acervulina* on the reproductive potentials of four other species of chicken coccidia during concurrent infections." Br. vet. J. 129, 29—31.

Witlock, D.R., Lushbaugh, W.B., Danforth, H.D. and Ruff, M.D. (1975). "Scanning electron microscopy of the cecal mucosa in *E. tenella*-infected and uninfected chickens." Avian Dis. 19, 293—303.

Part III

HOST—PARASITE POPULATION INTER-RELATIONSHIPS

Ecological Aspects of Parasitology
Editor: C.R. Kennedy
© North-Holland Publishing Company, Amsterdam, 1976

CHAPTER 21

DYNAMIC ASPECTS OF PARASITE POPULATION ECOLOGY

Roy M. ANDERSON

*Department of Zoology, University of London, King's College, London WC2R 2LS
(Great Britain)*

Contents

21.1. Introduction

The population biology of single and multispecies animal communities has, in recent years, received considerable attention from both experimental and theoretical ecologists (e.g. Krebs, 1972; May, 1973; Maynard Smith, 1974).

Our gradually improving understanding of the processes which control the dynamics of animal populations and lead to their often observed stability and resilience has in certain areas been enhanced by theoretical studies of a mathematical nature (Hassell and May, 1973; Ayala et al., 1973; May, 1974b; Beddington et al., 1975). Progress in ecology often seems to occur when ideas and concepts are formalised into sets of parameters and relationships that can be built into a model which provides the means of investigating the dynamical behaviour of a specific system. Today, a justification of the use of mathematics in ecology is hardly necessary since it is becoming generally accepted by many biologists that genuine scientific advances in a diffuse area of investigation, such as ecology, will be slow and uncertain if conducted entirely in intuitive and verbal terms. In order to achieve quantitative insights into the dynamics of animal populations a more formal mathematical framework is ideally required. This type of approach should, however, proceed hand in hand with experimental and field studies. Mathematical models ideally provide a firm basis, not only for predictive purposes and generating discussion, but also for examining precisely the consequences of various biological assumptions and, most importantly, for suggesting areas in which experimental investigation is required.

A comment by Maynard Smith (1974) that "... ecology will not come of age until it has a sound theoretical basis and ... we have a long way to go before this happy state of affairs is reached" appears particularly apposite when considering the ecology of helminth and protozoan parasites. To the casual observer, it often appears surprising that so little is known concerning the dynamics of host-parasite interactions, in comparison to other areas of ecology such as the population biology of insect species (e.g. Varley et al., 1973). On more careful examination, this lacuna in our knowledge appears less surprising if we consider the complexities of many helminth and protozoan parasite life cycles, which often involve several distinct host and parasite populations. This inherent biological complexity has lead to the field of parasite ecology receiving more than its fair share of polemics, no doubt created by the paucity of quantitative information available in the parasitological literature. The lack of experimental and theoretical information is undoubtedly due to the difficulties in measurement which arise from the intimacy of the relationship between host and parasite. For instance an observational task, such as the counting of animal numbers, which is simple in the case of many freeliving populations, generates many practical difficulties when considering endoparasitic organisms.

In spite of these problems progress is beginning to be made in the understanding of the processes which regulate the dynamics of host parasite interactions, particularly in the case of medically important diseases. Theoretical studies of host-parasite systems have been primarily concerned with the dynamics of malaria (Ross, 1910; Macdonald, 1957), schistosomiasis (Macdonald, 1965; Hairston, 1965; Goffman and Warren, 1970; Cohen, 1973; Nasell and Hirsch, 1972, 1973) and helminth parasites of domestic animals (Tallis and Leyton, 1966, 1969; Leyton, 1968). More general mathematical and numerical studies on the dynamics of parasite populations have been carried out by Kostitzin (1934) and (1939), Crofton (1971b) and Anderson (1974a). The main difficulty at present, however, in assessing the biological relevance of these studies is directly related to the lack of detailed long term investigations of the dynamics of parasite populations. Furthermore, very few quantitative studies have been made of the rate parameters which control the flow of organisms through a parasite life cycle.

The aim of this chapter is to examine the biology and structure of parasite life cycles in terms of the dynamics of the populations involved. The procedure used to examine population growth is to adopt such concepts as birth, death and immigration rates and to treat these as operating continually so that given the rates, one can write down a differential equation whose solution specifies the population size at any instant in time. In certain cases stochastic models are formulated in which population events such as births or deaths are considered as chance processes which occur not with certainty but with a given probability. Particular attention is paid to the type and functional form of the population rate parameters which control the dynamics of the parasite life cycles. In addition, the processes which aid in the regulation and enhance the stability of host-parasite interaction are discussed in general terms.

In a chapter of this size it is obviously not possible to give a comprehensive account of the many biological mechanisms and processes which influence the dynamics of host and parasite populations. A few topics have therefore been singled out for fuller discussion while many equally important areas such as immunological mechanisms, time delays and density independent factors receive only brief mention.

The dynamics of host-parasitoid systems are not treated in this chapter and the interested reader is referred to the work of Hassell and Varley (1969), Hassell (1971), Rogers (1972), Hassell and Rogers (1972) and Hassell and May (1973).

21.2. Population biology of parasite life cycles

The majority of protozoan and helminth parasite life cycles are of a complex dynamical nature. They often involve more than one species of host and many distinct parasite populations occupying various ecological niches.

Parasite life cycles can in general be classified into two categories; direct cycles involving a single species of host, and indirect cycles utilising two or more host species. The host within or on which the parasite becomes adult and reproduces sexually is usually termed the definitive, or final host, while hosts in which either asexual or no reproduction occurs are termed intermediate hosts.

To aid in the clarification of the number and type of host and distinct parasite populations involved in a specific life cycle it is often desirable to express the dynamics of the system in the form of a flow chart. This type of qualitative representation not only defines the rate parameters controlling the flow of parasites through the system but also helps to delineate areas in

434

which the biological mechanisms involved are poorly understood. In addition, where sufficient biological information is available, flow charts can aid in the formulation of mathematical models to describe the population dynamics of a specific host-parasite interaction.

The nematode *Trichuris trichiura* has a direct life cycle, which can be presented in the form of a simple flow diagram (Fig. 1). The parasite becomes adult in the caecum of man and when gravid lays eggs which pass to the external habitat with the faeces of the host. The embryonated eggs are infective to man and thus the cycle is completed by their ingestion. Although this life cycle is of a very simple form, it involves three distinct populations; the host, the adult parasites within the gut and the eggs in the external habitat. In addition it can be seen from Fig. 1, that six rate parameters control the flow of parasites through the cycle, excluding the possibility of immigration and emigration occurring within the human population. Indirect cycles are correspondingly more complex due to the involvement of more host species. The structure of one such cycle, that of the cestode *Caryophyllaeus laticeps*, is illustrated in Fig. 2. The parasite is adult in the gut of cyprinid fish and when mature lays eggs which pass to the external aquatic environment in the faeces of the host. These eggs, when ingested by various species of tubificid oligochaete undergo further development and the cycle is completed by the consumption of the intermediate host by the fish. The life cycle of this species involves two host and three distinct parasite populations, the dynamics of which are controlled by a total of ten rate parameters. The framework of the cycle also contains two host-parasite interactions and a predator-prey interaction between final and intermediate hosts. More complex life cycles such as those of many digeneans and cestodes involve even more populations and rate parameters.

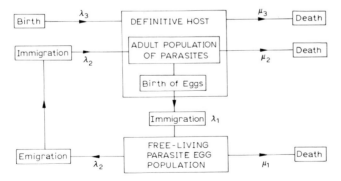

Fig. 1. Diagrammatic representation of a simple direct life cycle (e.g. *Trichuris trichiura*). The population parameters determining the flow of parasites through the system, indicated in the flow chart, are defined as rates per parasite per unit of time.

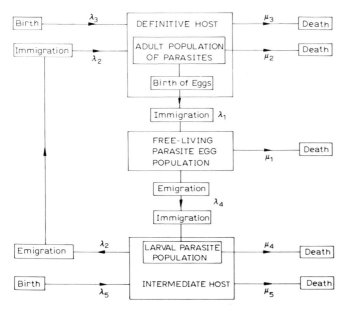

Fig. 2. Diagrammatic representation of an indirect life cycle (e.g. *Caryophyllaeus lati-ceps*). The population parameters determining the flow of parasites through the system, indicated in the flow chart, are defined as rates per parasite per unit of time.

When faced with such complexity it is apparent that a formal mathematical framework will greatly aid in the investigation of the rich dynamical properties which these populations interactions possess. It is possible to formulate systems of differential equation to describe the population change within direct and indirect life cycles (Anderson, 1974a); however, these are of necessity highly non linear if they are to bear any resemblance to reality, and thus difficult to investigate theoretically. It appears more realistic, at present, due to the paucity of quantitive information available in the literature concerning the many rate parameters involved in parasite life cycles, to consider the various populations, depicted as compartments in the flow charts, as separate entities.

This type of approach should help to clarify which population parameters require experimental determination and hence aid in the mathematical description of the dynamics of particular populations. However, it is important to note that the behaviour of one compartment or population is determined to a large extent by the behaviour of the preceeding compartment in the cycle (see Figs. 1 and 2). The dynamics of a complete life cycle can thus only be fully understood by regarding all the interconnecting compartments as a single unit possessing its own unique dynamical behaviour which is determined by the constituent parts of the system.

21.3. Population parameters

The flow of parasites through a life cycle is controlled by the birth, death, immigration and emigration rate parameters which determine the growth characteristics of each parasite and host population. These rates of change are generally termed population parameters and they may be of a complex functional form dependent on various physical and biological variables.

In the case of parasitic species which do not multiply within the definitive host, the numerical size of an adult parasite population within a single host is simply determined by the rate at which parasites enter, the immigration or infection rate, and the rate at which parasites die, the death rate. The dynamics of this part of a life cycle can thus be described as an immigration death process. It appears reasonable in most host parasite interactions to assume that the immigration rate is independent of the number of parasites already within the host. Density dependent factors within the parasite's microenvironment, or host generated immune responses, may influence the rate of establishment of invading stages or the death rate of adult forms, but are unlikely to affect the number of parasites gaining entry to the host. This assumption appears particularly valid in the case of helminth or protozoan parasites which gain entry to the final host by ingestion of the intermediate host. It is important to note, however, that this premise may be violated if the behaviour of the host is altered by previous or current parasite burdens.

For simplicity, if it is assumed that the immigration rate is constant and independent of population size, and that the death rate is also constant and simply proportional to the number of parasites within the host, then the following differential equation describes the rate of change of N_t, the number of adult parasites at time t:

$$\frac{dN_t}{dt} = \lambda - \mu N_t \tag{1}$$

where λ is the immigration rate per host per unit of time, and μ is the death rate per parasite per unit of time. Given that there are N_0 parasites within the host at time $t = 0$, the solution of this equation is as follows;

$$N_t = \frac{\lambda}{\mu} [1 - \exp(-\mu t)] + N_0 \exp(-\mu t). \tag{2}$$

The biological model, represented by this equation, predicts growth to an equilibrium population size N^*, equal to λ/μ. The time taken to reach the steady state situation is dependent on the values of N_0, λ and μ.

The behaviour of this parasite model contrasts markedly with the behaviour of a free-living population controlled by constant birth and death rates. The immigration-death process predicts regulated growth to an equilibrium, while the birth-death model predicts exponential growth or decline with no equilibrium unless the birth and death rates are equal.

When considering small numbers of organisms (when N* is small) chance fluctuations in the time of arrival of an immigrant or death of an adult parasite often give rise to an observed population process which appears to differ from the deterministic prediction. There exists only a certain probability that a population change will occur in a given time interval and thus a stochastic formulation allowing for chance events provides additional information about the dynamics of the adult parasite population. The analagous stochastic model for equation 2 predicts a Poisson distribution, at equilibrium, for the number of parasites present in a single host at time t (Cox and Miller, 1967). Thus where $P_N(t)$ is the probability of observing N parasites at time t, the individual terms of the distribution are given by;

$$P_N(t) = \left[\frac{\lambda}{\mu}(1 - \exp(-\mu t))\right]^N \left[\exp -\left[\frac{\lambda}{\mu}(1 - \exp(-\mu t))\right]\right] \cdot \frac{1}{N!} \tag{3}$$

The equilibrium mean and variance of this distribution are,

$$E\{N*\} = \text{Var } \{N*\} = \frac{\lambda}{\mu}. \tag{4}$$

Populations of free-living stages of parasitic species, such as protozoan spores, nematode larvae or digenean miracidia, are also controlled by immigration, emigration and death rates, birth processes rarely occurring in the free-living habitat. The preceeding comments about the dynamics of immigration-death processes thus apply to these populations.

Different population growth characteristics are exhibited by parasitic species which multiply within the definitive host, such as many protozoans or those which undergo asexual multiplication within an intermediate host such as the digenean flukes. The dynamics of this type of system can be broadly described by the following differential equation, where α is a constant birth rate per parasite per unit of time;

$$\frac{dN_t}{dt} = \lambda + (\alpha - \mu)N_t \tag{5}$$

The equilibrium population size of this model, $\hat{N}$, is given by

$$\hat{N} = \frac{\lambda}{-(\alpha - \mu)} \tag{6}$$

from which it can be seen that regulated growth to a steady state only occurs if μ the death rate is greater than the birth rate α. The model thus exhibits either exponential growth when $\alpha > \mu$ or damped growth to an equilibrium when $\alpha < \mu$. It is interesting to note that the stochastic analogue of the deterministic birth, death, immigration model (equation 5) predicts a negative binomial distribution for $P_N(t)$, the probability of observing N parasites in a host at time t, given that the starting population density is zero at time t = 0 (Bartlett, 1960).

Exponential growth of parasites is unlikely to occur for long in the real world, since it will either result in the death of the host or constraints will operate to reduce the rate of growth as a result of intraspecific competition or due to a host generated immunological response.

Density dependent constraints may act either on the birth, death or immigration parameters. However, before considering such mechanisms, a brief discussion of the basic types of parasite population rates appears relevant.

21.3.1. Death rates

The death rate per parasite per unit of time for both the models described above was assumed to be constant, the total death rate of the population within a host being proportional to the parasite burden.

A pure death process incorporating this assumption can be described by the following model, where N_t is the number of parasites surviving at time t;

$$\frac{dN_t}{dt} = -\mu N_t. \tag{7}$$

Given that there were N_0 individuals present at time t = 0, the solution of this model is;

$$N_t = N_0 e^{-\mu t} \tag{8}$$

an equation which predicts simple exponential decline.

The analogous stochastic model of a pure death process predicts a positive binomial probability distribution for the number of parasites surviving at time t (Bailey, 1964). If $P_N(t)$ is the probability of observing N parasites at time t, the individual terms of the distribution are given by;

$$P_N(t) = \binom{N_0}{N} \exp(-N\mu t) \left[1 - \exp(-\mu t)\right]^{N_0 - N} \tag{9}$$

where the mean and variance of the distribution are;

$$E\{N_t\} = N_0 e^{-\mu t} \tag{10}$$

and

$$\text{Var } \{N_t\} = N_0 e^{-\mu t} [1 - e^{-\mu t}]. \tag{11}$$

In general few populations either parasitic or free living exhibit simple exponential loss. Ignoring, for the moment, the role of density dependent processes, the constant death rate assumption will be invalid due to the dependence of this rate on other biological or physical variables.

Age dependent death rates appear to be common in populations of parasites, particularly in the case of free living infective stages, such as miracidia and cercariae which possess finite energy reserves.

The survival curves of the cercaria and adult of the ectoparasitic digenean

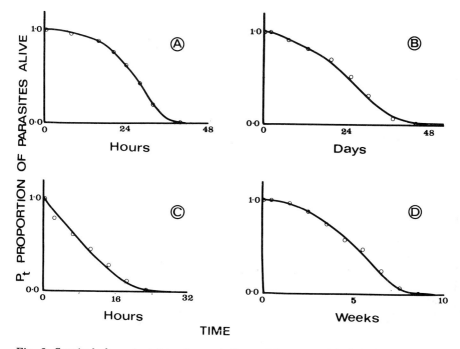

Fig. 3. Survival characteristics of populations of four parasitic helminths. *Graph A*, Cercaria of *Transversotrema patialense* (25°C) (data from Anderson and Whitfield, 1975). *Graph B*, L_3 larvae of *Bunostomum trignocephalum* (25°C) (data from Narain, 1965). *Graph C*, Miracidia of *Schistosomatium douthitti* (23°C) (data from Oliver and Short, 1956). *Graph D*, Adult *T. patialense* parasitic on *Brachydanio rerio* (25°C) (data from Anderson et al., 1976). Open circles, observed points. Solid line, survival curve predicted by age-dependent death model (equation 12).

Transversotrema patialense (Anderson and Whitfield, 1975, and Anderson et al., 1976), the miracidia of *Schistosomatium douthitti* (Oliver and Short, 1956) and the first stage larva of *Bunostomum trignocephalum* (Narain, 1965), clearly exhibit age dependent survival (Fig. 3). The functional forms of the death rates are illustrated in Fig. 4 from which it can be seen that the rates increase exponentially with time. The appropriate model for an age dependent death process has been described by Anderson and Whitfield (1975) and is of the form;

$$\frac{dN_t}{dt} = -\mu(t)N_t \tag{12}$$

where $\mu(t)$ is the age dependent death function. Various mechanisms cause these patterns, and care must be exercised in their interpretation. In the case of free living stages of parasitic species such as cercariae, the increasing death rate is most probably due to the progressive utilisation of a finite energy

440

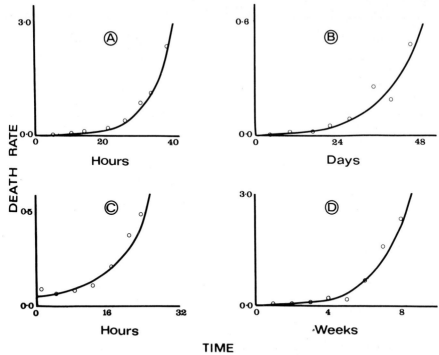

Fig. 4. Instantaneous age dependent death rates for populations of *Graph A*. Cercaria of *T. patialense*, death rate/4 hours. (25°C) (data from Anderson and Whitfield, 1975). *Graph B* L_3 larvae of *Bunostomum trignocephalum*, death rate/day. (25°C) (data from Narain, 1965). *Graph C* Miracidia of *Schistosomatium douthitti*. death rate/hour (23°C) (data from Oliver and Short, 1956). *Graph D* Adult *T. patialense*, death rate/week (25°C) (data from Anderson et al., 1976). Open circles; observed rates. Solid lines; best fit exponential model of the form $\mu(t) = a \exp(bt)$.

reserve. Adult parasitic populations within or on the host, however, may be influenced by a time dependent immune response generated by the host. Alternatively age dependence in survival may simply be due to senescence.

The basic immigration death model described in equation 1, can be modified to incorporate age dependency in the death rate in the following manner. If N_t is now a vector containing s elements, s being the maximum age class attained by the parasite, then

$$N_t = \begin{pmatrix} n_1(t) \\ n_2(t) \\ \vdots \\ n_i(t) \\ \vdots \\ n_s(t) \end{pmatrix}$$

where $n_i(t)$ is the number of parasites of age i at time t. In matrix form the appropriate model for an age structured immigration death process, formulated in discrete time, is

$$N_{t+1} = AN_t + b \tag{13}$$

In this equation the vector b is a constant immigration vector of the form;

$$b = \begin{pmatrix} \lambda \\ 0 \\ 0 \\ \vdots \\ 0 \\ 0 \end{pmatrix}$$

since on entering a host at rate λ per unit of time, immigrants become parasites of age zero. A is a transition matrix where;

$$A = \begin{pmatrix} 0 & 0 & 0-0 & 0 \\ P_0 & & & \\ & P_1 & & \\ & & P_2 & \\ & & \cdots & \\ & & & P_{s-1}\ 0 \end{pmatrix}$$

and P_i is the proportion of parasites of age i time units which survive to age i + 1.

The solution of equation (13) is

$$N_t = A^t N_0 + (I - A)^{-1} (I - A^t)b \tag{14}$$

where N_0 is the age structure of the parasite population at time zero, and I is the unit identity matrix. As the time parameter t becomes large, the stable age distribution of the parasite population within the host is given by;

$$N_t = (I - A)^{-1}b \tag{15}$$

and the equilibrium population size $\hat{N}$ is $\Sigma_{i=0}^s \hat{n}_i$.

Anderson et al. (1976) carried out a series of experiments to obtain estimates of the parameters contained in equation (14) for populations of *Transversotrema patialense* parasitic on the surface of a cyprinid fish *Brachydanio rerio*. The experimental fish were exposed to a constant number of cercariae per week and changes in the adult parasite populations were monitored over a four month period. Separate experiments were carried out to determine the age dependent survival rates of the adult parasites on the fish host (Fig. 3). The results of these population experiments are illustrated in Fig. 5 from which it can be seen that the observed mean number of parasites per host is closely mimiced by the predictions of the theoretical model defined in equation (14). At equilibrium, which is achieved after the maximum life span of the adult parasite has·elapsed, a stable age distribution exists.

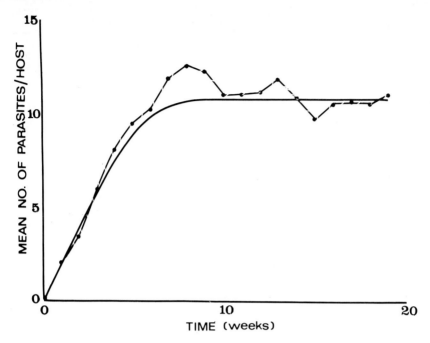

Fig. 5. Experimental investigation of a parasite population controlled by immigration and death processes. Stippled line; changes in the mean number of adult *T. patialense* per host over a 20 week period, each fish having been exposed to a constant number of cercariae per week (average immigration rate/fish/week. = 2.0). Solid line; predicted mean number of parasites per host from equation 14. (data from Anderson et al., 1976).

The death rate of the whole population thus remains constant, for although each age class has its own specific death rate, the proportions in which the age classes are present remains the same through time. The shape of this stable age distribution is identical to the shape of the survival curve for the adult parasites (Fig. 3). It is interesting to note that if each age class of adult parasites has an age specific egg production rate, then the total population at equilibrium will produce a constant number of eggs per unit of time.

The survival rates of parasites of poikilothermic hosts and of free living stages of parasitic species are invariably influenced by environmental variables such as temperature. For example Rose (1956) demonstrated that the longevity of the free living larvae of *Dictyocaulus viviparus* was markedly affected by both temperature and humidity. The survival of adult helminths within their definitive hosts is also affected by climatic factors in the host's environment. In some cases temperature dependent death rates will markedly influence the population dynamics of adult parasites, leading to cyclic oscillations in population size due to seasonal fluctuation in the temperature of the environment. Assuming for simplicity that the immigration rate is constant, then the appropriate model for such systems is;

$$\frac{dN_t}{dt} = \lambda - \mu(t)N_t \tag{16}$$

where $\mu(t)$ is a periodic function of time due to temperature dependence (Anderson, 1974b, 1976).

Other physical parameters of the environment may influence survival. Stromberg and Crites (1974), for example, demonstrated that the survival of the first stage larvae of *Camallanus oxycephalus* was highly dependent on the salinity of the aquatic habitat in addition to temperature. Farley (1962) reported that both temperature and pH were determinants of the survival rates of the miracidia of *Schistosomatium douthitti*.

The survival characteristics of many parasite populations may be both age and temperature dependent where the death rate is a function of two variables. Thus the rate of change in population size N_t would be of the form

$$\frac{dN_t}{dt} = -\mu(t,T)N_t. \tag{17}$$

If the function $\mu(t,T)$ is known then the partial differentiation of equation (17) gives rise to a surface in the three dimensional space created by the variables population size, temperature and time. In reality the death rate may be a function of more than two variables and thus careful experimental analysis is required to determine the precise form of this population parameter.

21.3.2. Immigration or infection rates

Populations of both free living stages of parasitic organisms and adult or larval parasites within a host are subject to immigration processes. Immigration into a larval or egg population in aquatic or terrestrial habitats is caused by the release of organisms from either the definitive or intermediate host. In contrast, the growth of parasite populations within a host results from passive or active infection, the rate of which is usually determined by the behavioural characteristics of the host or parasite or both. It is impossible in a chapter of this nature to describe the numerous functional forms of such population parameters; however the following examples demonstrate their diversity.

The simplest type of immigration process occurs when the rate of input into an adult population is directly proportional to the density of infective larvae or eggs in the habitat of the host. For example, the number of cercariae of *Transversotrema patialense* which attach to the fish host and become adult forms is approximately directly proportional to the density of the larvae to which the host is exposed. If we assume for simplicity that the death rate is constant then the rate of change in N_t, the number of adult parasites on a single host at time t is given by;

$$\frac{dN_t}{dt} = \lambda C_t - \mu N_t \tag{18}$$

where C_t is the number of cercariae in the habitat at time t. To solve this equation the dynamics of the cercarial population must be understood. This particular host parasite interaction, however, is of a much more complex nature. The immigration rate λ, for example, varies according to the size of the fish, the larger hosts presenting bigger targets. The dependency of parasite population parameters on host characteristics such as size or age is undoubtedly common (Anderson, 1974c).

Where infection of the host is achieved by means of a free living stage, the age structure of the larval population is often of importance. For example the rate of ingestion of the first stage larvae of *Camallanus oxycephalus* by the copeopod intermediate host is dependent on the activity and thus age structure of the larval population (Stromberg and Crites, 1974). Similarly the infectivity of the cercariae of *Transversotrema patialense* (Anderson and Whitfield, 1975) and *Cercaria floridensis* (Miller and McCoy, 1930) to their respective fish hosts is highly dependent on the age of the larvae.

Many helminth life cycles include predator prey interactions between definitive and intermediate hosts, such as that of the fish tapeworm *Caryophyllaeus laticeps*. In this specific example the feeding activity of the host will be related to the density of the intermediate host population in relation to alternative food resources, the age and thus the size of the host and in addition the season of the year (Anderson, 1974b).

The immigration of eggs, spores or larvae into free living populations of parasitic species will depend on many factors related to the size of the preceeding parasite population within the intermediate or final host. For example, the rate of emergence of cercariae from the molluscan intermediate host may not only be a function of the size of the redial or sporocyst populations but also of the size (Raisyte, 1968) and nutritional status of the snail (Sinderman et al., 1957; Coles, 1973), season of the year (James, 1968), temperature of the aquatic habitat (Dutt and Srivastava, 1962), pH of the water (Bauman et al., 1948) and even the time of day where daily emergence rhythms occur (Campbell, 1973a). The growth of egg populations of *Hymenolepis nana* in the external habitat is determined to a large extent by the dynamics of the adult parasite population within the final host. Egg production by adult parasites can be dependent on the age of the worm, the nutritional status of the host and the density of parasites within the gut (Ghazal and Avery, 1974).

It can be seen from the preceeding examples that it is difficult to draw general conclusions concerning the functional form of immigration parameters. Furthermore little quantitive information is available in the literature concerning these processes. It is obvious however that they constitute a very important group of parameters within a parasite life cycle since they determine the transmission rates from one parasite population to another.

21.3.3. Birth rates

Parasite birth processes can adopt a variety of forms and may either contribute directly to the population of parent organisms or provide immigrants to the next population within the life cycle.

It is interesting to note that the stochastic model of a pure birth process, where the likelihood of a birth occurring is subject to chance mechanisms, predicts a negative binomial distribution of organisms present at any point in time (Bailey, 1964). Thus from a theoretical standpoint, considerable variation in parasite population size may exist within a population of hosts each receiving the same inoculum of a rapidly multiplying protozoan parasite. Such a situation may arise during the early stages of parasite population growth, when few or no deaths occur.

Protozoan parasites such as species belonging to the genus *Plasmodium*, which multiply within their host, possess enormous reproductive potentials and often exhibit exponential growth during the early stages of infection in the vertebrate. *Plasmodium falciparum* for example is thought to produce in excess of 30,000 merozoites by mitotic divisions from one schizont during the exoerythrotic stage of its life cycle, within a period of approximately 5 days (Baker, 1968). Similar potential is also shown by helminth parasites such as digenean species within their molluscan hosts. Meyerhof and Rothschild (1940) observed cercarial production by *Cryptocotyle lingua* in a single *Littorina littorea* over a five year period and reported an average production rate of 830 cercariae per day.

Helminth parasites within their definitive host usually exhibit immense egg production potential. Two strains of *Schistosoma mansoni* in hamsters produced an average of between 100—120 eggs per female worm per day (Wright and Bennett, 1967). Amongst the nematode parasites *Haemonchus contortus* a parasite of sheep, is often quoted as being one of the most prolific egg-layers; it has been estimated that a single mature female has a peak production rate of 10,000 eggs per day (Crofton, 1966). Even larger rates have been reported for certain acanthocephalan parasites, for example *Macracanthorhynchus hirudinaceus*, a parasite of pigs, produces eggs at a maximum rate of 260,000 per female worm per day, egg production continuing for up to 12 months (Kates, 1944). Many ectoparasitic arthropods also have high reproductive potentials. The human louse *Pediculus humanus* produces an average 9 eggs per day for a period of approximately 34 days. (Buxton, 1939).

Although these examples illustrate the potential of parasitic species to reproduce at a rapid rate, which is always in excess of the host's birth rate, of more importance to the dynamics and stability of parasite life cycles is

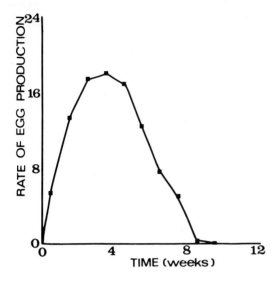

Fig. 6. The relationship between egg production per fluke per week by adult *T. patia-lense*, and the age of the parasites. (C. Mills, Zoology Dept. King's College London, unpublished data).

the functional dependence of such rates.

Age dependent birth rates are common, particularly among helminth parasites. *Gyrodactylus alexanderi*, for example, which reproduces viviparously on the surface of its fish host, on average gives birth to its first offspring after 1.6 days, while the second birth takes 6.9 days to occur (Lester and Adams, 1974).

Production of eggs by adult *Transversotrema patialense* on the fish host is similarly dependent on the age structure of the parasite population. Fig. 6 clearly illustrates the rise in the birth rate as the parasite ages followed by a gradual decline in egg production (C. Mills, Zoology Dept., King's College, London, unpublished data). It is interesting to note that egg production in *T. patialense* begins almost immediately the parasite has established itself on the host. This is not in general the case; the majority of helminths undergo considerable development within the final host before egg production commences. The duration of this development phase has important consequences on the dynamics of specific life cycles, since it introduces a time delay into the system. These delays may in certain cases have a destabilising effect on the dynamics of the host parasite interaction. Such effects, however, depend on the length of the lag (Maynard Smith, 1974; May et al., 1974).

The causal mechanisms of age dependent birth rates are unclear, although early build up in egg production is often associated with the growth of the parasite. The decline phase may be due either to natural senescence or to the effects of a time dependent immunological response produced by the host.

The birth rates of helminth and protozoan parasites, within poikilothermic hosts, are often influenced by density independent factors. Such responses are naturally associated with developmental periods, the time delays themselves being determined by climatic factors such as temperature. These effects are particularly noticeable in the reproduction and development of larval digeneans within their molluscan intermediate hosts (Wright, 1971; Nice and Wilson, 1974; Kendall, 1964). Other factors such as nutritional status of the host will also influence the birth rates of certain parasites. The production of the cercariae of *Transversotrema patialense*, for example, within the molluscan host *Melanoides tuberculata* declines steadily in starved snails (P.J. Whitfield and R.M. Anderson, unpublished data).

In the case of dioecious parasites such as nematodes and acanthocephala, the sex ratios existing within a specific parasite population in a single host will determine the frequency of copulations and thus will influence egg production rates. Little information is available at present, however, to indicate how frequently adult helminths copulate or how this behaviour influences egg production or is influenced by parasite density.

Undoubtedly, the most important group of processes which influence parasite birth rates are either density dependent or immunological in nature. The regulatory role of these factors is discussed in the following section.

21.4. Regulation of population growth

In theory all parasite species are capable of unlimited exponential growth in their host populations, but in reality such patterns of growth do no persist for long. What sort of mechanisms regulate parasite population growth and lead to the observed persistence and stability of host-parasite interactions? Before discussing this question further a few definitions will help to clarify later comments.

The phrase regulated growth is used in this chapter to imply that populations grow or decay towards an equilibrium or steady state. This equilibrium may either be constant or oscillatory in time, the latter pattern usually being referred to as a limit cycle. When a population is subject to a small displacement from its equilibrium, if it returns to that point the steady state is regarded as stable and it is said to possess neighbourhood stability. Alternatively, if the population moves away from the equilibrium after perturbation, the steady state is regarded as unstable. In reality populations will never remain at the equilibrium level for long, but will fluctuate around this state due to demographic and environmental stochasticity. (For a more detailed discussion of this topic see May (1973)).

Regulatory mechanisms can be divided into two broad categories, generally termed density dependent and density independent processes. In the former set of mechanisms the rate parameters such as birth and death rates, which determine the character of population growth, vary in response to population density. The death rate may increase as density rises or the birth rate decreases. In practice both may vary. Such factors are recognised in ecological theory today as being of unique importance in stabilizing the growth of animal populations. Density independent factors are primarily physical in character and thus of a climatic nature. They undoubtedly have a profound influence on the population dynamics of both free-living larval or egg stages and on adult or larval parasites within poikilothermic hosts. If growth were to be controlled entirely by density independent factors, however, it is unlikely that a parasite population would persist for long on an evolutionary time scale. During periods of adverse conditions, due either to long or short term climatic change, the probability of local or global extinction would be high due to the lack of compensatory mechanisms in the parameters controlling the population dynamics of the parasite life cycle. The majority of parasite populations are likely to be influenced by both types of factors.

Host-parasite interactions are in general characterised by two additional types of regulatory processes, which do not act on free-living populations. The first concerns the lethal level concept first proposed by Crofton (1971a), in which a given burden of parasites results in the death of the host. Thus the removal of a single host results in the loss of a number of parasites from the system. The second and perhaps most important type of mechanism concerns the regulatory role of host generated immunological responses. Immune responses may result either in a decrease in the reproductive rate of parasites or an increase in their death rate or both.

The complex nature of many parasite life cycles has tended to lead to confusion in the parasitological literature concerning the need for regulatory mechanisms. It has been argued that the complexity of the cycle itself, resulting from the many host and parasite populations involved (Fig. 2), generates the stability of the system. The general principle, well known in other areas of ecology, that a stable equilibrium will not exist unless some form of density dependent regulatory mechanisms is operating on one or more of the population parameters is of direct relevance to the dynamics of host-parasite interactions. The complexity of the parasite's life cycle may or may not enhance the stability of the system. The assumption that complexity leads to stability in multispecies interactions is a controversial topic in ecological theory at present and for detailed treatment of this area the reader is referred to May (1973), Roberts (1974) and Gilpen (1975).

Two types of regulatory mechanism are discussed more fully in the following two sections.

21.4.1. Density dependent processes

The influence of density dependent factors in parasite life cycles can best be considered by examining a few simple models of population growth. The dynamics of a simple direct life cycle, of the type shown by the nematode *Trichuris trichiura* parasitic in the gut of humans provides a convenient biological model (Fig. 1).

Assuming for simplicity that the population rates are constant, then the following differential equations describe the basic features of such an interaction in the case of a single host. In this model N_t represents the number of adult parasites in the gut of the host and W_t the egg population in the free living environment, both at time t.

$$\frac{dW_t}{dt} = \lambda_1 N_t - \mu_1 W_t - \lambda_2 W_t \tag{19}$$

$$\frac{dN_t}{dt} = \lambda_2 W_t - \mu_2 N_t \tag{20}$$

where the population rates λ_1, μ_1, λ_2 and μ_2 are defined in the life cycle flow chart illustrated in Fig. 1. It is assumed in this model that the rate of infection of the host is directly proportional to the density of the infective stages (eggs) present in the free living environment.

The solution of these equations demonstrates that the above population model, which contains no density dependent constraints, exhibits either exponential growth or decay of the two parasite populations, depending on the parameter values. No equilibrium state is achieved and thus regulated growth does not occur.

Similar models describing indirect and thus more complex life cycles, and models which determine the dynamics of the interaction within a population of hosts, exhibit the same patterns of behaviour provided no density dependent constraint is incorporated. Furthermore, the adult and free living populations of the parasite will not attain an equilibrium state, even if the host population is regulated by a density dependent constraint of the logistic form.

Some form of regulatory process, either host induced or resulting from intra specific competition in the parasite microenvironment, must be applied to the rate parameters controlling the dynamics of the parasite populations.

This point can be simply illustrated by again considering the direct life cycle illustrated in Fig. 1. Four population parameters determine the dynamics of this simple interaction and one or more of these rates can be density dependent. For example the parameter λ_1, the rate of egg production by the adult parasites may vary according to population density. Many reports of this type of response exist in the parasitological literature. Several studies have demonstrated the existence of density dependent egg production in cestode parasites, particularly in species of the genus *Hymenolepis* (Ghazal and Avery, 1974; Bailey, 1972; Jones and Tan, 1971). Some of the best examples concern nematode and digenean parasites of domestic animals, for instance, the egg production per fluke by *Fasciola hepatica* in sheep decreases as the fluke burden increases (Boray, 1969). Similarly, Michel (1967) reported that heavy infections of the nematode *Ostertagia ostertagi* in calves resulted in a decrease in the egg output per worm when compared with light infections. Michel (1967) also noted that the faecal egg counts of groups of animals harbouring varying worm burdens was very similar and he suggested that the egg production per host per unit of time was limited by the environment of the worms within the host, which could only sustain a certain rate. This latter observation suggests the existence of a carrying capacity of the parasites microenvironment within the host. It is not clear as yet whether this is determined by the host as a result of an immune response, or whether it is due to interspecific competition for limited resources. It appears probable that both types of mechanism operate in host parasite interactions.

The consequences of such processes on the dynamics of the parasite populations can be crudely examined in the direct life cycle case by rewriting equations 19 and 20, incorporating λ_1 the egg production rate as a function of N_t. A simple form of $\lambda_1(N_t)$ is as follows

$$\lambda_1(N_t) = a - bN_t$$

where the constant a represents the maximum egg production rate and the constant b determines the severity of the density dependent effect. Equations 19 and 20 become

$$\frac{dW_t}{dt} = (a - bN_t)N_t - W_t(\mu_1 + \lambda_2) \tag{21}$$

$$\frac{dN_t}{dt} = \lambda_2 W_t - \mu_2 N_t \tag{22}$$

The above model predicts regulated growth to the non zero equilibrium states,

$$\hat{N} = \left[a - \frac{\mu_2}{\lambda_2}(\mu_1 + \lambda_2) \right] \frac{1}{b} \tag{23}$$

and

$$\hat{W} = \left[a - \frac{\mu_2}{\lambda_2}(\mu_1 + \lambda_2) \right] \frac{\lambda_2}{\mu_2 b} \tag{24}$$

The positive and realistic values of $\hat{N}$ and $\hat{W}$ are stable with respect to small perturbations.

The behaviour of this simple model can be better understood by examining the phase plane of N_t and W_t. At any time t, the state of the interaction is fully described by the sizes of the adult parasite and egg populations, N_t and W_t and thus for each state there exists a point in the (N,W) plane. If the lines for which $dW/dt = 0$ and $dN/dt = 0$ are plotted within the phase plane, the point or points of intersection demonstrate the existence of equilibrium states for the parasite populations. Such a plot is illustrated in Fig. 7, for a given set of parameter values, from which it can be seen that the lines intersect once in the positive side of the plane and thus an equilibrium state exists. Also in Fig. 7, a further plot is shown for a different set of parameter values and in this case no intersection of the two lines occurs. It is thus apparent that only certain combinations of parameter values lead to regulated population growth and stable equilibria. Equations 23 and 24 yield the constraint that a $\geqslant \mu_2/\lambda_2(\mu_1 + \lambda_2)$ for positive equilibria to exist. The biological interpretation of this statement is that the maximum egg production rate, a, must be sufficiently large to overcome the effects of removal of eggs by infection and death rates of both eggs and parasites, in order to maintain the interaction.

Other rate parameters in parasite life cycles may also be density dependent. For example, the death rate of either adult parasites in the definitive host or larval parasites in the intermediate host, is an obvious candidate. J.F.

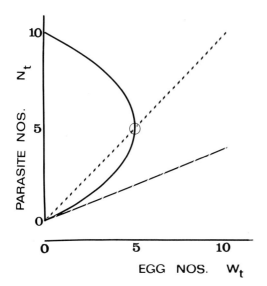

Fig. 7. Phase plane created by the time dependent variables, adult parasite population size N_t, and egg population size W_t. Solid line; trajectory predicted by the equation $(a-bN_t)N_t -W_t(\mu_1+\lambda_2) = 0$ where a = 1.0, b = 0.1, μ_1 = 0.1, λ_2 = 0.4. Finely stippled line; trajectory predicted by the equation $(\lambda_2 W_t-\mu_2 N_t) = 0$, where $\lambda_2 = \mu_2 = 0.4$. The equilibrium population sizes $\hat{N}$ and $\hat{P}$ are given by the point of intersection of the two lines i.e. $\hat{N} = \hat{P} = 5.0$. Stippled line; trajectory predicted by the equation $(\lambda_2 W_t-\mu_2 W_t) = 0$, where $\mu_2 = 1.0$, $\lambda_2 = 0.4$. Line does not intersect solid line in positive area of phase plane.

Michel (Central Veterinary Research Council, Wadebridge, unpublished data) administered infective larvae of *Ostertagia ostertagi* to four groups of calves, each group receiving a different dosage level varying from 8,330 up to 300,000 parasites. Samples of calves from each group were slaughtered at various periods after the initiation of the infection, and their worm burdens estimated. The results of these experiments and some date of Murray et al. (1970), are presented in Fig. 8, from which it can be seen that the death rate of the parasites increases in accord with the initial dosage level. The rate of change of N_t, the number of parasites present in the gut of a calf at time t, can be described by the following model

$$\frac{dN_t}{dt} = -\mu(N_t)N_t \qquad (25)$$

where $\mu(N_t)$ is the density dependent death rate. If this function is of linear form where $\mu(N_t) = a + bN_t$, a and b being constants, the solution of equation 25 adequately describes Michel's data (Fig. 8).

Similar responses have been observed in cestode infections. For example Befus (1975), found that single primary infections of *Hymenolepis diminuta* in laboratory mice survived longer than six worm infections.

A frequently reported density dependent phenomenon, the so-called crowding effect, in which the size of an adult parasite is determined by the density of parasites within a host (Read, 1951; Roberts, 1961; Dobson, 1965; Boray, 1969), may also have important consequences on the population dynamics of helminth infections. Small helminths, resulting from poor growth rates due to high densities of parasites, will almost certainly have a limited potential for egg production (Ghazal and Avery, 1974) but may in addition have lower probabilities of survival.

In certain instances population parameters appear to vary according to the density of parasites already present within a potential host. The majority of such cases appear to be caused by an immune response generated by the host, in response to already established parasites or previous experiences of infection. For example, Jenkins and Phillipson (1971) reported that low level infections of the rat nematode, *Nippostrongylus brasiliensis*, resulted in a decrease in the initial establishment of infective larvae, when compared with uninfected controls. In the case of this parasite the severity of the response appears to depend on the size of the infection already present in the host.

This type of process, however, may in certain cases be due to the failure of the invading parasitic stages to establish on or in the host as a direct result of lack of available space for attachment caused by the density of the existing population. Anderson (1974c) ascribed the observed underdispersed distribution of the gill monogenean *Diplozoon paradoxum* on the bream *Abramis brama* to such a mechanism.

If the infection process is essentially a chance phenomenon, resulting from

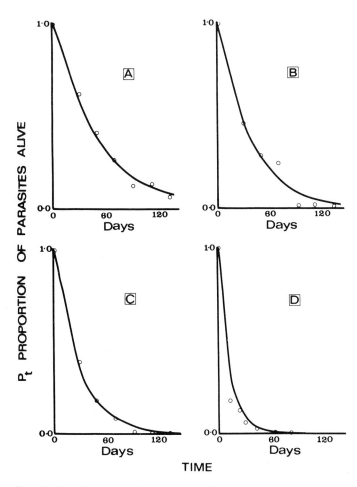

Fig. 8. Density dependent survival characteristics of the nematode *Ostertagia ostertagi*, parasitic in the gut of calves. The initial dosage of larvae given to each experimental animal, and thus the initial population density N_o is as follows. *Graph A* 8330 larvae. *Graph B* 25,000 larvae. *Graph C* 75,000 larvae. (J.F. Michel, Central Veterinary Research Council, Wadebridge; unpublished data). *Graph D* 3000,000 larvae (data from Murray et al., 1970) Open circles; observed points. Solid lines; survival curves predicted by density dependent death model (equation 25).

random contacts between host and parasite, a Poisson distribution of parasites per host is likely to be generated provided each host is equally susceptible to invasion and that the chance of a parasite establishing is independent of the parasite burden of the host.

When the immigration or infection rate is dependent on the number of parasites i already established, where $\lambda_i = \lambda_o + \beta i$, $x_i(t)$ is the number of hosts

containing i parasites and W_t the number of infective stages, then the following differential equations describe the dynamics of the infection process.

$$\frac{dW_t}{dt} = -W_t \sum_{i=0}^{s} \lambda_i x_i(t) \tag{26}$$

$$\frac{dx_0(t)}{dt} = -\lambda_0 x_0(t) W_t \tag{27}$$

$$\frac{dx_i(t)}{dt} = (\lambda_{i-1} x_{i-1}(t) - \lambda_i x_i(t)) W_t \tag{28}$$

$$\frac{dx_s(t)}{dt} = \lambda_{s-1} x_{s-1}(t) W_t \tag{29}$$

The parameter s represents a theoretical limit to the number of parasites
a host can harbour and can adopt any discrete value. The constant β in the
λ_i function can either be positive or negative. When positive, high densities
of parasites within the host will enhance the chances of further infections
occurring. In this case the solution of the above equations demonstrates that
an overdispersed distribution of parasite counts is generated. Conversely
when β is negative, a more realistic parasitological situation, the presence of
adult parasites within the host decreases the chances of invading stages establishing by lowering the immigration rate. This type of mechanism results in
the generation of underdispersion of parasite counts per host. Although
these theoretical results are of interest, particularly in the case of distribu-
tions generated by laboratory infections, it is important to note that in natural populations of parasites, the many processes which result in overdispersion, such as heterogeneity in infection rates, will tend to mask the distribution effects of density dependent infection.

In many reports of parasite population growth it is difficult to determine
which of the population rates is responding to the influence of density. Mc-
Connachie (1968) demonstrated that 'in vitro' cultivation of an axenic strain
of *Entamoeba invadens* in a Trypticase-liver broth resulted in regulated
population growth to an equilibrium density. It is not clear whether the
death rate is responding to density as a result of accumulation of toxic
wastes in the culture medium or whether the rate of division of the parasites
is restricted by limited food resources at high population density. The precise nature of the response, however, is unimportant since the observed regulated growth implies that the natural intrinsic rate of increase is density
dependent in this species in the 'in vitro' environment.

Density dependent survival or immigration can be encorporated into the
direct life cycle model (equations 19 and 20) in a similar manner to that employed for the egg production rate. The consequences of such a modification
is regulated growth to a stable equilibrium for certain combinations of parameter values. It thus becomes apparent that any form of density dependent

constraint, whether applied to one or more population parameters, leads to regulated population growth for given values of the birth, death and immigration rates. Furthermore, this principle applies to models describing both a single or an entire population of hosts.

When dealing with models of both host and parasite populations some form of density dependent constraint is required to regulate the host population. This may either be due to the effects of interspecific competition acting on the host's birth and death rate or due to the operation of some form of lethal level concept, where specific burdens of parasites result in the death of the host. The parasite may in addition have sublethal effects, such as modification of the host reproduction potential (Pan, 1965). One further theoretical prediction is of interest. In the absence of constraints on host population growth, where either exponential growth or decay occurs, provided that a density dependent mechanism is operating on one or more of the parameters controlling the parasite populations, the mean parasite burden per host approaches an equilibrium state, independently of the size of the host population.

As mentioned previously, it is difficult at present to decide whether the observed complexity of many parasite life cycles enhances or decreases their stability. It is true to say, however, that complexity resulting from numerous host and distinct parasite populations, leads to the presence of large numbers of rate parameters in a given system (Fig. 2). In complex life cycles these rates create many opportunities for density dependent responses to occur. The number of density dependent constraints on population growth required within a life cycle may be directly related to the number of birth processes which occur. However, it is tempting to draw the speculative conclusion that the more parameters there are of this form, in a given life cycle, the more stable the host parasite interaction.

21.4.2. Lethal level concept

The lethal level concept was first proposed by Crofton (1971a) who considered that the majority of parasite species are capable of killing their hosts if present in large enough numbers. Crofton (1971a and 1971b) argued that the death of a single host in the tail of an overdispersed distribution of parasite counts within the host population resulted in the removal of a large number of parasites from the system. He considered that this type of mechanism played an important regulatory role in the population dynamics of host parasite interactions. Crofton thought that an overdispersed distribution of parasites per host was essential for such a process to lead to regulation in the interaction. Theoretically, however, it can be demonstrated that this assumption is not necessary, the form of the distribution being relatively unimportant.

One of the major difficulties in the practical interpretation of Crofton's lethal level is whether such a level exists with a reasonable degree of constancy even within a small stratum of a host population. The precise density of parasites which result in the death of a host will vary not only with temporal and spatial factors but also within strata of the host population due to genotypic differences. In addition death may not be a direct consequence of the influence of the parasite, but resulting from secondary infections which are capable of establishment due to the presence of the parasite.

These practical difficulties in envisaging the precise meaning, and in measurement of a lethal level are in general unimportant to the theoretical consequences of the concept. An expected mean lethal level must be assumed for a specific population in one locality at a specific point in time.

The theoretical effectiveness of Crofton's concept as a regulatory influence in host parasite interactions can be crudely examined by considering the following model.

If $x_i(t)$ is the number of hosts with i parasites at time t, then the host may harbour i = 0, 1, 2 ..... s, parasites, where (s + 1) is the lethal level required to kill the host. For simplicity in the model s is regarded as constant, in time, and for each individual within the population of hosts. The $x_i(t)$'s at any one point in time form the frequency distribution of parasites within the host population. This form of notation was first used by Kositzin (1934) to aid in the description of the dynamics of host parasite interactions and was further developed by Anderson (1974a).

In the case of a simple direct life cycle (Fig. 1) where $\Sigma_{i=0}^s x_i(t) = H_t$, the host population size at time t and $\Sigma_{i=0}^s ix_i(t) = P_t$, the parasite population size also at time t, then the following differential equations describe the essential dynamics of the interaction.

$$\frac{dw}{dt} = \lambda_1 \sum ix_i - \mu_1 w - \lambda_2 w \sum x_i \tag{30}$$

$$\frac{dx_0}{dt} = \lambda_3 \sum x_i - \mu_3 x_0 - \lambda_2 wx_0 + \mu_2 x_i \tag{31}$$

$$\vdots$$

$$\frac{dx_i}{dt} = \lambda_2 wx_{i-1} - \mu_3 x_i - \lambda_2 wx_i + \mu_2(i + 1)x_{i+1} - \mu_2 ix_i \tag{32}$$

$$\vdots$$

$$\frac{dx_s}{dt} = \lambda_2 wx_{s-1} - \mu_3 x_s - \lambda_2 wx_s - \mu_2 sx_s \tag{33}$$

In this model w_t represents the size of the egg population and λ_1 .... λ_3 and μ_1 .... μ_3 are defined in Fig. 1. All these rates are regarded as constants. The major assumptions incorporated in these equations are (a) the infection rate of a host is directly proportional to the density of the egg population and is unaffected by the number of parasites already present, (b) no density dependent constraints are operating on any of the host or parasite rate parameters, (c) the lethal level mechanism operates at a parasite density of (s + 1).

The behaviour of the solutions of these equations, for specific parameter values is illustrated in Fig. 9, from which it can be seen that the host, parasite and egg populations exhibit regulated growth to equilibrium levels. It is thus apparent that the lethal level mechanism can, in the absence of any density dependent constraints on host or parasite birth or death rates, regulate the interaction between host and parasite populations.

This interesting result is however subject to tight constraints on the parameter values of the model. In general the stability of the system is greatly enhanced in terms of parameter space by the addition of density dependent constraints, particularly on the host population, such that in the absence of the parasite regulated growth to a stable equilibrium occurs. It is interesting to note that in the crude computer models which were used to test the relevance of the lethal level mechanism, density dependent host population growth was incorporated by Crofton (1971b).

One further point of interest concerns the distribution of parasites within the host population at equilibrium. In the simple model outlined above, the rate parameters, in particular λ_2 the host infection rate, determine the form of the distribution. Whether it is over or underdispersed is immaterial to the regulatory influence of the lethal level concept. The form of the distribution of parasite numbers per host does, however, determine the magnitude of the regulatory mechanism which is, in addition, influenced, as is the distribution itself, by the numerical value of (s + 1) the lethal level. In reality if (s + 1) is

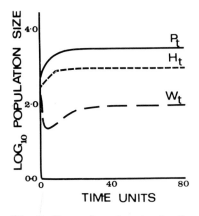

Fig. 9. Dynamics of a simple direct life cycle, host-parasite interaction as predicted by equation 30—33. This model incorporates the lethal level concept, where (s + 1) = 6 parasites/host, but no density dependent constraints. N_t = adult population size, H_t = host population size and W_t = egg population size, all at time t.

very large, and the distribution highly overdispersed, then the regulation of the parasite population is likely to be achieved by other means, such as immunological responses. Furthermore, the parasite will contribute little to the mortality rate of the host population and thus other constraints such as density dependent host population growth will stabilise the dynamics of the host-parasite interaction.

21.5. Concluding remarks

The precise structure of a helminth or protozoan life cycle, with respect to the number and type of both host and parasite populations involved, is likely to determine the general dynamical properties of a host parasite system.

In many cases a number of population regulatory mechanisms, of the density dependent type, operate within a single life cycle. Larval populations of digenean parasites within the molluscan intermediate host, for example, may be influenced by intra-specific competition within their finite microenvironment. In contrast the growth of adult fluke populations within a definitive host may be constrained by the host's immune response. The stabilising influences of such density dependent responses will act in general to suppress the destabilizing effect of both time delays and density independent factors.

At first sight it may appear that constraints on growth are required for each population within a parasite life cycle. This, however, is untrue since as demonstrated earlier, the growth of any one parasite population is dependent to some extent on the size of the preceeding population in the life cycle, the immigration and emigration parameters determining the degree of this interdependence (Figs. 1 and 2). It thus appears that only a single constraint is required to regulate the growth of all the populations.,

These general comments on the regulation of parasite population growth, however, do not fully take into account the interactive aspects of the host parasite relationship. It is important to remember that any such interaction must be considered at the level of both parasite and host populations since the parasites themselves will influence the parameters controlling the growth of the host population which in turn determines the absolute number of environments available for the parasite.

These interactive effects may be of the form envisaged by Crofton (1971) when he defined the lethal level concept, or of a more subtle nature, the parasite influencing the reproduction and mortality rates of the host population in a gradual density dependent manner. They must often play a crucial population regulatory role, as illustrated by the effects of the eradication of Malaria on the island of Mauritius on human population growth (Bruce-Chwatt and Bruce-Chwatt, 1974). The balance between such regulatory effects, whether resulting from parasite induced host mortalities, density dependent regulation in both host and parasite populations as a result of intra specific competition, or host generated immunological effects, will undoubtedly vary between different types of parasite life cycle.

It is difficult, at present, to assess the relative merits of the diverse range of population regulatory mechanisms exhibited within helminth and protozoan parasite life cycles. Furthermore it is not possible as yet to correlate the existence of a particular type of constraint within a specific life cycle to the dynamical behaviour of that cycle.

Hopefully, however, advances in the theoretical analysis of population interactions will provide the means of investigating the rich dynamical properties which such complex biological systems possess. Such work should help to answer important general questions concerning, for example, the comparative stability of direct and indirect parasite life cycle.

Mathematical models are bound to be too simplistic to exactly mimic the dynamics of observed host parasite systems. Their function, however, is not initially to aim at realism in detail, but to provide insights into the broad classes of behaviour exhibited by such interacting populations. It is essential, though, that mathematical studies of this kind, proceed hand in hand with experimental and field investigations in order to ensure close links with biological reality.

21.6. REFERENCES

Anderson, R.M. (1974a). "Mathematical models of host-helminth parasite interactions". Ecological stability (Ed. by M.B. Usher and M.H. Williamson) pp. 43—59. Chapman and Hall, London

Anderson, R.M. (1974b). "Population dynamics of the cestode *Caryophyllaeus laticeps* (Rallas, 1781) in the bream (*Abramis brama*)". J. Anim. Ecol. 43, 305—321.

Anderson, R.M. (1974c). "An analysis of the influence of host morphometric features on the population dynamics of *Diplozoon paradoxum* (Nordmann, 1832). J. Anim. Ecol. 43, 873—887.

Anderson, R.M. (1976). "Seasonal variation in the population dynamics of *Caryophyllaeus laticeps*". Parasitology. In press.

Anderson, R.M. and Whitfield, P.J. (1975). "Survival characteristics of the free-living cercarial population of the ectoparasitic digenean *Transversotrema patialensis*". (Soparkr, 1924) Parasitology, 70, 295—310.

Anderson, R.M., Whitfield, P.J. and Mills, C. (1976). "Experimental investigations of the population dynamics of the ectoparasitic digenean *Transversotrema patialense* on the fish host *Brachydanio rerio:* Consequences of a constant cercarial infection rate". In preparation.

Ayala, F.J., Gilpen, M.E. and Ehrenfeld, J.G. (1973). "Competition between species: Theoretical Models and Experimental Tests". Theoret. Pop. Biol. 4, 331—356.

Bailey, G.N.A. (1972) "Energetics of a host parasite system". Unpublished Ph.D. Thesis. University of Exeter.

Bailey, N.T.J. (1964). The elements of stochastic processes. John Wiley, London.

Baker, J.R. (1969). "Parasitic Protozoa". Hutchinson, London.

Bartlett, M.S. (1960). "Stochastic Population Models in Ecology and Epidemiology". Metheun, London.

Bauman, P.M., Bennett, H.J. and Ingallis, J.W. (1948). "The molluscan intermediate host and schistosomiasis japonica. Observations on the production and rate of emergence of cercariae of *Schistosoma japonica* from the molluscan intermediate host *Oncomelania quadrasi*". Am. J. Trop. Med. 28, 567—575.

Beddington, J.R., Free, C.A. and Lawton, J.H. (1975). "Dynamic complexity in predator prey models formed in difference equations". Nature, 255, 58—60.

Befus, A.D. (1975). "Secondary infections of *Hymenolepis diminuta* in mice: effects of varying worm burdens in primary and secondary infections". Parasitology, 71, 61—76.

Boray, J.C. (1969). "Experimental fascioliasis in Australia". Advances in Parasitology, 7, 96—210.

Bruce-Schwatt, L.J. and Bruce-Schwatt, J.M. (1974). "Malaria in Mauritius as dead as the dodo". Bull. N.Y. Acad. Med. 50, 1069—80.

Buxton, P.A. (1939). "The Louse". Edward Arnold, London.

Campbell, R.A. (1973a). "Studies on the biology of the life cycle of *Cotylurus flabelliformis* (Trematoda Strigeidae)". Trans. Am. micros. Soc. 92, 629—640.

Cohen, J.E. (1973). "Selective host mortality in a catalytic model applied to schistosomiasis". Amer. Natur., 107, 199—212.

Coles, C.C. (1973). "The effect of diet and crowding on the shedding *Schistosoma mansoni* cercariae by *Biomphalaria glabrata*". Ann. trop. Med. Parasit. 67, 419—423.

Cox, D.R. and Miller, H.D. (1967). "The Theory of Stochastic Processes". Methuen, London.

Crofton, H.D. (1966). Nematodes. Hutchinson, London.

Crofton, H.D. (1971a). "A quantitative approach to parasitism". Parasitology, 62, 179—193.

Crofton, H.D. (1971b). "A model of host-parasite relationships". Parasitology, 63, 343—364.

Dobson, C. (1965). "The relationships between worm population density, host survival and the growth of the third-stage larva of *Amplicaecium robertsi* Sprent and Mines, 1960, in the mouse". Parasitology, 55, 183—193.

Dubois, G. (1929). "Les cercaires de la region de Neuchatel". Bull. Soc. neuchatel. Sci. nat. 53, 3—177.

Dutt, S.C. and Srivastava, M.D. (1962). "Studies on the morphology and history of the mammalian blood fluke. *Orientobilharzia dattai* (Dutt and Srivastava) II. The molluscan phases of the life cycle and the intermediate host specificity". Indian J. vet. Sci. 32, 33—43.

Farley, J. (1962). "The effect of temperature and pH on the longevity of *Schistosomatium douthitti* miracidia". Can. J. Zool. 40, 615—620.

Ghazal, A.M. and Avery, R.A. (1974). "Population dynamics of *Hymenolepis nana* in mice: fecundity and the crowding effect". Parasitology, 69, 403—416.

Gilpin, M.E. (1975). "Stability of feasible predator-prey systems". Nature, 254, 137—139.

Goffman, W. and Warren, K.S. (1970). "An application of the Kermack-McKendrick theory to the epidemiology of schistosomiasis". Am. J. trop. Med. Hyg. 19, 278—283.

Hairston, N.G. (1965). "On the mathematical analysis of schistosome populations". Bull. Wld. Hlth. Org. 33, 45—62.

Hassell, M.P. (1971). "Mutal interference between searching insect parasites". J. Anim. Ecol. 40, 473—486.

Hassell, M.P. and May, R.M. (1973). "Stability in insect host parasite models". J. Anim. Ecol. 42, 693—726.

Hassell, M.P. and Rogers, D.J. (1972). "Insect parasite responses in the development of population models". J. Anim. Ecol. 41, 661—676.

Hassell, M.P. and Varley, G.C. (1969). "New inductive population model for insect parasites and its bearing on biological control". Nature, 223, 1133—1137.

James, B.L. (1968). "The occurrence of Parvatrema homoeotecnum James 1964 (Trematoda: Gymnophallidae) in a population of Littorina saxatilis tenebrosa (Mont.)" J. nat. Hist. 2, 21—37.

Jenkins, D.C. and Phillipson, R.F. (1971). "The kinetics of repeated low-level infections of Nippostrongylus brasiliensis in the laboratory rat". Parasitology, 62, 457—666.

Jones, A.W. and Tan, B.D. (1971). "Effect of crowding upon growth and fecundity in the mouse bile duct tapeworm, Hymenolepis microstoma". J. Parasit. 57, 88—93.

Kates, K.C.C. (1944). "Some observations on experimental infections of pigs with the thorn-headed worm, Macracanthorhynchus hirudinaceus". Am. J. vet. Res. 5, 166—172.

Kendall, S.B. (1964). "Some factors influencing the development and behaviour of nematodes in their molluscan hosts". (A.E.R. Taylor, ed.) pp. 51—73. Br. Soc. Parasit. Symp. 2, 51—73.

Kostitzin, V.A. (1934). "Symbiose, parasitisme et evolution". Hermann, Paris.

Kostitzin, V.A. (1939). "Mathematical Biology". George C. Harrap, London.

Krebs, C.J. (1972). "Ecology: The experimental analysis of distribution and abundance". Harper and Row, London.

Leyton, M.K. (1968). "Stochastic models in populations of helminthic parasites in the definitive host — II. Sexual mating functions". Maths. Biosci. 3, 413—419.

Lester, R.J.G. and Adams, J.R. (1974). "A simple model of a Gyrodactylus population". Int. J. Parasit, 4, 497—506.

McConnachie, E.W. (1968). "The growth of an axenic strain of Entamoeba invadens in different media". Parasitology, 58, 733—740.

MacDonald, G. (1957). "The Epidemiology and Control of Malaria". Oxford University Press, London.

MacDonald, G. (1965). "The dynamics of helminth infections with special reference to schistosomes". Trans. R. Soc. trop. Med. Hyg. 59, 489—506.

May, R.M. (1973). "Stability and Complexity in Model Ecosystems". Princeton University Press, Princeton.

May, R.M. (1974b). "Ecological systems in randomly fluctuating environments". "Progress in Theoretical Biology" (Ed. by R. Rosen and F. Snell). Academic Press, New York.

May, R.M., Conway, G.R., Hassell, M.P. and Southwood, T.R.E. (1974). "Time delays, density-dependent and single-species oscillations". J. Amin. Ecol. 43, 747—770.

Maynard Smith, J. (1974). "Models in Ecology". Cambridge University Press, London.

Meyerhof, E. and Rotheschild, M. (1940). "A prolific trematode". Nature, Lond. 146, 367.

Michel, J.F. (1967). "Regulations of egg output of populations of Ostertagia ostertagi". Nature, Lond. 215, 1001—1002.

Miller, H.M. and McCoy, O.R. (1930). "An experimental study of the behaviour of Cercaria floridensis in relation to its fish intermediate host". J. Parasit. 16, 185—197.

Murray, M., Jennings, F.W. and Armour, J. (1970). "Bovine Ostertagiasis: Structure, Function and Mode of Differentiation of the Bovine Gastric Mucosa and Kinetics of the Worm Loss". Res. Vet. Sci. 11, 417—427.

Narain, B. (1965). "Survival of the first-stage larvae and infective larvae of Bunostomum trignocephalum Rudolphi, 1808". Parasitology, 55, 551—558.

Nasell, I. and Hirsch, W.M. (1972). "A mathematical model of some helminthic infections". Com. Pure. Appl. Math. 25, 459—477.

Nasell, I. and Hirsch, W.M. (1973). "The transmission dynamics of schistosomiasis". Com. Pure. Appl. Math. 26, 359—453.

Nice, N.G. and Wilson, R.A. (1974). "A study of the effect of temperature on the growth of *Fasciola hepatica* in *Lymnaea truncatula*". Parasitology, 68, 47—56.

Oliver, J.H. and Short, R.B. (1956). "Longevity of miracidia of *Schistosomatium douthitti*". Expl. Parasit. 5, 238—249.

Pan, C.T. (1965). "Studies on the host-parasite relationship between *Schistosoma mansoni* and the snail *Australorbis glabratus*". Am. J. trop. Med. Hyg. 14, 931—976.

Raisyte, D. (1968). "On the biology of *Apatemon gracilis* (Rud, 1819), a trematode parasitizing in domestic and wild ducks". Acta. parasit. Lith. 7, 71—84.

Read, C.P. (1951). "The 'crowding effect' in tapeworm infections". J. Parasit. 37, 174—178.

Roberts, A. (1974). "The stability of a feasible random ecosystem". Nature, 251, 607—608.

Roberts, L.S. (1961). "The influence of population density on patterns and physiology of growth in *Hymenolepis diminuta* (Cestoda: Cyclophyllidea) in the definitive host". Expl. Parasit. 11, 332—371.

Rogers, D. (1972). "Random search and insect population models". J. Anim. Ecol. 41, 369—383.

Rose, J.H. (1956). "The Bionomics of the free-living larvae of *Dictyocaulus viviparus*". J. comp. Path. Ther. 66, 228—240.

Ross, R.C. (1910). "The Prevention of Malaria". John Murray, London.

Sindemann, C., Rosenfield, A. and Atrom, L. (1957). "The ecology of marine dermatitis-producing schistosomes. II. Effects of certain environmental factors on emergence of cercariae of *Asutralobilharzia variglandis*". J. Parasit. 43, 382.

Stromberg, P.C. and Crites, J.L. (1974). "Survival, activity and penetration of the first stage larvae of *Cammallanus oxycephalus* Ward and Magath, 1916". Int. J. Parasit, 4, 417—421.

Tallis, G.M. and Leyton, M. (1966). "A stochastic approach to the study of parasite populations". J. Theor. Biol. 13, 251—260.

Tallis, G.M. and Leyton, M. (1969). "Stochastic models of populations of helminthic parasites in the definitive host — I". Math. Biosci. 4, 39—48.

Varley, G.C., Gradwell, G.P. and Hassell, M.P. (1973). "Insect Population Ecology: an analytical approach". Blackwell, Oxford.

Wright, C.A. (1971). "Flukes and Snails". Allen and Unwin, London.

Wright, C.A. and Bennett, M.S. (1967). "Studies on *Schistosoma haematobium*. I: A strain from Durban, Natal, South Africa". Trans. R. Soc. trop. Med. Hyg. 61, 221—227.

SUBJECT INDEX